# Ground Deformation Patterns Detection by InSAR and GNSS Techniques

# Ground Deformation Patterns Detection by InSAR and GNSS Techniques

Editor

**Mimmo Palano**

MDPI • Basel • Beijing • Wuhan • Barcelona • Belgrade • Manchester • Tokyo • Cluj • Tianjin

*Editor*
Mimmo Palano
Istituto Nazionale di
Geofisica e Vulcanologia,
Osservatorio Etneo
Catania
Italy

*Editorial Office*
MDPI
St. Alban-Anlage 66
4052 Basel, Switzerland

This is a reprint of articles from the Special Issue published online in the open access journal *Remote Sensing* (ISSN 2072-4292) (available at: https://www.mdpi.com/journal/remotesensing/special_issues/Ground_Deformation_Patterns_Detection_by_InSAR_GNSS).

For citation purposes, cite each article independently as indicated on the article page online and as indicated below:

LastName, A.A.; LastName, B.B.; LastName, C.C. Article Title. *Journal Name* **Year**, *Volume Number*, Page Range.

**ISBN 978-3-0365-6886-7 (Hbk)**
**ISBN 978-3-0365-6887-4 (PDF)**

# Contents

# About the Editor

**Mimmo Palano**

Mimmo Palano (Ph.D.) Education: Master's Degree in Geology (cum laude), University of Catania, 2000; Ph.D. in Geological Sciences, University of Catania, 2004. Qualifications: National Scientific Qualification of the Italian Ministry of Education, University, and Research (MIUR) as Full Professor of Geophysics, 2020. Professional Appointments: Researcher at INGV since 2006. His research focuses on the physical processes associated with natural hazards in general, and seismotectonic and volcanic applications in particular, through the study of both geodetic and seismic signals.

*Editorial*

# Editorial for the Special Issue: "Ground Deformation Patterns Detection by InSAR and GNSS Techniques"

**Mimmo Palano**

Istituto Nazionale di Geofisica e Vulcanologia, Osservatorio Etneo—Sezione di Catania, 95125 Catania, Italy; mimmo.palano@ingv.it

**Citation:** Palano, M. Editorial for the Special Issue: "Ground Deformation Patterns Detection by InSAR and GNSS Techniques". *Remote Sens.* **2023**, *14*, 1104. https://doi.org/10.3390/rs14051104

Received: 20 January 2022
Accepted: 20 February 2022
Published: 24 February 2022

**Publisher's Note:** MDPI stays neutral with regard to jurisdictional claims in published maps and institutional affiliations.

In the last two decades, the rapid growth in continuous Global Navigation Satellite Systems (GNSS) networks and improvements in Interferometric Synthetic Aperture Radar (InSAR) imaging allowed the acquisition of continuous and spatially extensive datasets over large regions of Earth, significantly increasing the range of geoscience applications. In addition, the promising results obtained by the scientific community and the free availability of data, which permitted drastic cost reductions, have drawn increasing interest from the administrative managing office for the mapping and monitoring of ground deformation issues.

This Special Issue aims to provide a general overview of some geoscience applications of GNSS and InSAR techniques which are commonly used to study the surface deformation related to co- and post-seismic deformation, subsurface movements of magma beneath active volcanoes, soil deformation (e.g., natural/anthropic uplift or subsidence), monitoring of landslide, monitoring of industrial settlements, the motion of ice sheets, etc. The GNSS technique provides a set of 3D geodetic observations at a limited number of points on the ground surface. The continuous technological development in GNSS equipment currently allows collecting measurements at higher rates (up to 100 Hz), offering a wide range of new applications for solid and fluid Earth investigations. The InSAR technique provides a spatially dense set of geodetic observations of ground deformation in the viewing geometry of the satellite sensor, and with a temporal sampling limited to the satellite orbital revisit (up to 6 days with the Sentinel constellations). Any deformation of the ground surface can be measured by comparing two radar images of the same area, collected at different times from approximately the same position in space. InSAR processing advancements also allowed multi-temporal analyses, which sensibly improved the investigation of long-term deformation events.

GNSS and InSAR measurements can complement each other and are generally combined to infer the 3D surface deformation over a target region. A review of more than 190 studies dealing with InSAR and GNSS combined measurements has been proposed in Del Soldato et al. [1]. The ground deformation measurements coming from both techniques have been combined for different purposes [1], evidencing how their joint use has been readily employed by the scientific community as well as by stakeholders and environmental managers. In turn, the increasing range of applications started to push the development of new approaches aimed at fast and robust combinations of GNSS and InSAR measurements. In such a frame, Xiong et al. [2] proposed an iterative least squares approach for virtual observation (VOILS) based on the maximum a posteriori estimation criterion of Bayesian theorem while Parizzi et al. [3] developed an approach accounting for the spectral properties of the errors of InSAR and GNSS measurements, hence preserving all spatial frequencies of the deformation detected by the two techniques. Both methods have been tested and validated with both synthetic and real data. Achieved results highlighted that both methods led to significant improvement of the spatial accuracy of the combined deformation field, therefore allowing accurate detection of the ongoing deformations.

Several studies included in this Special Issue focused on the co-seismic deformation related to moderate to large earthquakes. De Novellis et al. [4] focused on the March 2021 Thessaly seismic sequence (Central Greece) highlighting the activation of unknown distinct blind fault segments in a sort of domino effect within the seismogenic crustal volume. Caporali et al. [5] analyzed the seismic sequence of November 2019 in Albania and inferred a NE-dipping reverse seismogenic fault located at a depth of $8 \pm 2$ km. Sakkas [6] focused on the 30 October 2020 $M_w6.9$ Samos Island (Aegean Sea) earthquake and suggested that the earthquake nucleated on a two-segments north-dipping listric fault characterized by a predominant dip-slip component and a minor lateral one. The complex deformation field associated with the April 2016 Kumamoto (Japan) seismic sequence was analyzed by He et al. [7] which modelled a four-segment fault geometry with right-lateral strike-slip kinematics coupled with a minor normal slip component. Valerio et al. [8] focused on the 7 November 2019 $M_w5.9$ earthquake hitting the East-Azerbaijan region and proposed a shallow NE-SW striking and SE-dipping fault as the seismogenic source. All these studies clearly proved that GNSS and InSAR data analysis and modelling are extremely useful tools in helping to constrain the causative fault of moderate to large earthquakes, especially in the case of blind and unknown faults, therefore providing useful information on the seismic hazard estimation of the investigated areas.

Active faults can be also affected by long-term creeping during the interseismic period. Geodetic observations are used to investigate co- and post-seismic deformations as well as transient deformations at least when these phenomena yield deformations high enough to be discriminated from long-term trends. However, there could be the possibility that the whole amount of observed long-term deformation could be partially or totally caused by inelastic processes instead of related to the building of elastic stress preparing the next earthquakes. Cambiotti et al. [9] focused on this topic by proposing a novel inverse method aimed at the discrimination of regional deformation and of long-term fault creep by inverting available GNSS measurements. Sparacino et al. [10] performed a seismic and geodetic moment-rates comparison for the western Mediterranean to identify that regions where the total deformation-rate budget is entirely released by crustal seismicity, and the ones where the excess deformation-rate can be released either in aseismic slip across active faults or through large future earthquakes. Achieved results by both studies proven that the geodetic measurements represent an essential part of the seismic-hazard analysis on highly deforming regions.

Other studies included in this Special Issue focused on the surface deformation related to the migration of fluids along the magmatic system of active volcanoes. Galvani et al. [11] analyzed twenty years of GNSS and levelling measurements collected on Ischia Island (Italy) and found a deflating source located at a depth of 4 km below the southern flank of Mt. Epomeo. Battaglia et al. [12] studied the subsidence of Dallol volcano (Erta Ale ridge of Afar, Ethiopia) and inferred a deflating source located beneath the volcano edifice at a depth ranging in the 0.5–1.5 km interval and characterized by a volume decrease between $-0.63$ and $-0.26 \times 10^6$ km$^3$/year. Boixart et al. [13] focused on the Sabancaya volcano (southern Perú), detecting an active deep source of deformation located between the Sabancaya and Hualca volcanoes with a volume change rate of $26 \times 10^6$–$46 \times 10^6$ m$^3$/yr. These studies evidenced that GNSS and InSAR techniques can detect and track with high detail the spatial and temporal evolution of the magmatic system during a volcanic crisis. Both techniques are essential tools for the continuous monitoring of active volcanoes as well as to understand magmatism, refine volcano models, and mitigate volcanic hazards.

Another topic addressed in this Special Issue is that of land subsidence which can occur for both natural and anthropic causes. Land subsidence represents a relevant issue that might affect highly developed urban and industrialized areas. Cando Jácome et al. [14] focused on the land subsidence due to the underground mining which is causing the collapse of many buildings in the urban area of Zaruma in Ecuador. The authors proposed a forecasting methodology for the continuous monitoring of the long-term soil subsidence in target areas, largely improving the traditional detection performed with total stations and geodetic marks. Mohamadi et al. [15] designed a PS-InSAR-based workflow on the detection of unusual vertical surface motions in urban areas in order to create temporal vulnerability maps for building collapse monitoring. Both studies highlight that the development of methodologies for the continuous monitoring of the land subsidence is strictly required to improve security standards aimed at the building collapse risk reduction in densely urbanized areas.

**Funding:** This research received no external funding.

**Acknowledgments:** The Guest Editor of this Special Issue would like to thank all authors who have contributed to this volume for sharing their scientific results and for their excellent collaboration. Special thanks are due to the community of distinguished reviewers for their valuable and insightful inputs. The *Remote Sensing* editorial team is gratefully acknowledged for its support during all phases related to the succesfully completion of this volume.

**Conflicts of Interest:** The author declares no conflict of interest.

# References

1. Del Soldato, M.; Confuorto, P.; Bianchini, S.; Sbarra, P.; Casagli, N. Review of Works Combining GNSS and InSAR in Europe. *Remote Sens.* **2021**, *13*, 1684. [CrossRef]
2. Xiong, L.; Xu, C.; Liu, Y.; Wen, Y.; Fang, J. 3D Displacement Field of Wenchuan Earthquake Based on Iterative Least Squares for Virtual Observation and GPS/InSAR Observations. *Remote Sens.* **2020**, *12*, 977. [CrossRef]
3. Parizzi, A.; Rodriguez Gonzalez, F.; Brcic, R. A Covariance-Based Approach to Merging InSAR and GNSS Displacement Rate Measurements. *Remote Sens.* **2020**, *12*, 300. [CrossRef]
4. De Novellis, V.; Reale, D.; Adinolfi, G.M.; Sansosti, E.; Convertito, V. Geodetic Model of the March 2021 Thessaly Seismic Sequence Inferred from Seismological and InSAR Data. *Remote Sens.* **2021**, *13*, 3410. [CrossRef]
5. Caporali, A.; Floris, M.; Chen, X.; Nurce, B.; Bertocco, M.; Zurutuza, J. The November 2019 Seismic Sequence in Albania: Geodetic Constraints and Fault Interaction. *Remote Sens.* **2020**, *12*, 846. [CrossRef]
6. Sakkas, V. Ground Deformation Modelling of the 2020 Mw6.9 Samos Earthquake (Greece) Based on InSAR and GNSS Data. *Remote Sens.* **2021**, *13*, 1665. [CrossRef]
7. He, Z.; Chen, T.; Wang, M.; Li, Y. Multi-Segment Rupture Model of the 2016 Kumamoto Earthquake Revealed by InSAR and GPS Data. *Remote Sens.* **2020**, *12*, 3721. [CrossRef]
8. Valerio, E.; Manzo, M.; Casu, F.; Convertito, V.; De Luca, C.; Manunta, M.; Monterroso, F.; Lanari, R.; De Novellis, V. Seismogenic Source Model of the 2019, Mw 5.9, East-Azerbaijan Earthquake (NW Iran) through the Inversion of Sentinel-1 DInSAR Measurements. *Remote Sens.* **2020**, *12*, 1346. [CrossRef]
9. Cambiotti, G.; Palano, M.; Orecchio, B.; Marotta, A.M.; Barzaghi, R.; Neri, G.; Sabadini, R. New Insights into Long-Term Aseismic Deformation and Regional Strain Rates from GNSS Data Inversion: The Case of the Pollino and Castrovillari Faults. *Remote Sens.* **2020**, *12*, 2921. [CrossRef]
10. Sparacino, F.; Palano, M.; Peláez, J.A.; Fernández, J. Geodetic Deformation versus Seismic Crustal Moment-Rates: Insights from the Ibero-Maghrebian Region. *Remote Sens.* **2020**, *12*, 952. [CrossRef]
11. Galvani, A.; Pezzo, G.; Sepe, V.; Ventura, G. Shrinking of Ischia Island (Italy) from Long-Term Geodetic Data: Implications for the Deflation Mechanisms of Resurgent Calderas and Their Relationships with Seismicity. *Remote Sens.* **2021**, *13*, 4648. [CrossRef]
12. Battaglia, M.; Pagli, C.; Meuti, S. The 2008–2010 Subsidence of Dallol Volcano on the Spreading Erta Ale Ridge: InSAR Observations and Source Models. *Remote Sens.* **2021**, *13*, 1991. [CrossRef]
13. Boixart, G.; Cruz, L.F.; Miranda Cruz, R.; Euillades, P.A.; Euillades, L.D.; Battaglia, M. Source Model for Sabancaya Volcano Constrained by DInSAR and GNSS Surface Deformation Observation. *Remote Sens.* **2020**, *12*, 1852. [CrossRef]
14. Cando Jácome, M.; Martinez-Graña, A.M.; Valdés, V. Detection of Terrain Deformations Using InSAR Techniques in Relation to Results on Terrain Subsidence (Ciudad de Zaruma, Ecuador). *Remote Sens.* **2020**, *12*, 1598. [CrossRef]
15. Mohamadi, B.; Balz, T.; Younes, A. Towards a PS-InSAR Based Prediction Model for Building Collapse: Spatiotemporal Patterns of Vertical Surface Motion in Collapsed Building Areas—Case Study of Alexandria, Egypt. *Remote Sens.* **2020**, *12*, 3307. [CrossRef]

*remote sensing*

*Review*

# Review of Works Combining GNSS and InSAR in Europe

Matteo Del Soldato *, Pierluigi Confuorto, Silvia Bianchini, Paolo Sbarra and Nicola Casagli

Earth Science Department of the University of Firenze, Via La Pira 4, 50121 Firenze, Italy; pierluigi.confuorto@unifi.it (P.C.); silvia.bianchini@unifi.it (S.B.); paolo.sbarra@stud.unifi.it (P.S.); nicola.casagli@unifi.it (N.C.)
* Correspondence: matteo.delsoldato@unifi.it; Tel.: +39-0552757551

**Abstract:** The Global Navigation Satellite System (GNSS) and Synthetic Aperture Radar Interferometry (InSAR) can be combined to achieve different goals, owing to their main principles. Both enable the collection of information about ground deformation due to the differences of two consequent acquisitions. Their variable applications, even if strictly related to ground deformation and water vapor determination, have encouraged the scientific community to combine GNSS and InSAR data and their derivable products. In this work, more than 190 scientific contributions were collected spanning the whole European continent. The spatial and temporal distribution of such studies, as well as the distinction in different fields of application, were analyzed. Research in Italy, as the most represented nation, with 47 scientific contributions, has been dedicated to the spatial and temporal distribution of its studied phenomena. The state-of-the-art of the various applications of these two combined techniques can improve the knowledge of the scientific community and help in the further development of new approaches or additional applications in different fields. The demonstrated usefulness and versability of the combination of GNSS and InSAR remote sensing techniques for different purposes, as well as the availability of free data, EUREF and GMS (Ground Motion Service), and the possibility of overcoming some limitations of these techniques through their combination suggest an increasingly widespread approach.

**Keywords:** GNSS; GPS; InSAR; Europe; Italy; PS

**Citation:** Del Soldato, M.; Confuorto, P.; Bianchini, S.; Sbarra, P.; Casagli, N. Review of Works Combining GNSS and InSAR in Europe. *Remote Sens.* **2021**, *13*, 1684. https://doi.org/10.3390/rs13091684

Academic Editor: Mimmo Palano

Received: 31 March 2021
Accepted: 24 April 2021
Published: 27 April 2021



## 1. Introduction

The Global Navigation Satellite System (GNSS) and Synthetic Aperture Radar Interferometry (InSAR) can be used for various applications owing to their principal functions. GPS enables the collection of three-dimensional (3D) deformation information, considering the coordinates of surveyed points. Since the 1970s, InSAR [1] has permitted the investigation of the phase difference (interferogram) between two images acquired over the same area at different times.

The GNSS term refers to any satellite constellation that provides information globally about the positioning, timing, and navigation data of elements to a GNSS receiver that then commutes this information to a precise location. The GNSS family includes several regional satellite systems, e.g., GPS (Global Positioning System), Galileo, and GLONASS (GLObal'naya NAvigatsionnaya Sputnikovaya System). GNSS performances are (i) accurate in the determination of real position, velocity and time of travel between the transmitter and receiver; (ii) intact, allowing the insertion of an alarm in case of anomaly positioning; (iii) able to work continuously without interruptions; and (iv) able to fulfil all the criteria of accuracy, integrity, and continuity. This system provides three-dimensional along the East-West, North-South, and up-down direction for ongoing deformation through the precise determination of the receiver position with millimetric accuracy in order to assess the main components. GNSS is very useful for investigating regionally low spatial frequencies, e.g., tectonic or geodynamic deformation fields, but local patterns or movements cannot be appreciated due to the large distance between adjacent GNSS stations. The main limitation of the measurement approach of GNSS is a lack of data in some areas due to the absence of GNSS stations since this technique is station-dependent [2]; however, the European continent is sufficiently covered by more than 300 distributed multi-GNSS reference stations [3].

The InSAR approach is a consolidated remote sensing technique employed for the detection and monitoring of ground deformation that has become widely used over the last decade. The first application dates back to the late 1980s [4] and was based on the DInSAR (Differential InSAR) approach that enabled the collection of cumulative deformation derived from the phase difference of two subsequent SAR acquisitions captured during two consecutive satellite passages over the same area. The transition from a single interferogram (DInSAR) to a multi-interferogram approach permitted a multi-temporal analysis (MTInSAR, Multi Temporal InSAR) [5], which sensibly and rapidly improved the investigation of long-term deformation events. MTInSAR approaches are indeed applicable to investigate the temporal evolution of ground displacement over time by exploiting coherent (i.e., electromagnetically stable) backscattered points corresponding to reflected elements on the ground. The spread and rapid development of different processing algorithms has permitted wide application to various geohazard fields [6], such as subsidence, landslide, tectonic and volcanic phenomena [7,8].

These InSAR techniques also have certain drawbacks strictly related to the (i) geometrical effects between the LOS (Line of Sight) and the slope; (ii) atmospheric contribution and aliasing phase; and (iii) snow, vegetation, or variable land cover of a target area. Nonetheless, these techniques enable the (i) investigation of wide areas with a high measurement precision; (ii) containing of costs with respect to the benefits [9]; (iii) temporal repetition (up to 6 days over the same area with the Sentinel-1 constellation); (iv) all-weather and day/night operation; (v) investigation of remote and inaccessible areas; and (vi) the possibility to back-analyze a certain phenomenon back to the '90s by using archived satellite data, e.g., ERS1/2 data.

GNSS data play a crucial role in different phases of the InSAR processes since changes in atmospheric refraction can result in a misinterpretation of InSAR results [10]. Specifically, such phases include SAR satellite acquisition, InSAR processing and InSAR post-processing. The GNSS pointwise data can be used for InSAR calibration; Reference Point(s) identification, to which the displacement measured by Persistent Scatterers (PS) is referred; tropospheric and/or ionospheric delay correction [10] or unwrapping the interferometric phase [11] during MTInSAR processing; post-calibration and deformation velocity correction assessed by MTInSAR approaches [12–14]; GNSS deformation measurements comparison with the MTInSAR ground deformation [15]; or MTInSAR product validation by comparison with information derivable by the GNSS network [16]. These are the only possibilities available when attempting to combine the SAR and GNSS datasets taking into consideration the singular analysis of ground deformation.

Indeed, many authors have published interesting works about the combination of InSAR and GNSS data for different purposes, such as (i) atmospheric correction (e.g., [17,18]); (ii) subsidence or uplift analysis (e.g., [9,19]); (iii) tectonic and seismic investigation e.g., [20,21]); (iv) landslide back-analysis (e.g., [22,23]); (v) volcanic event studies e.g., [24,25]); (vi) infrastructure investigation (e.g., [26,27]); (vii) glacial analysis and recognition (e.g., [28,29]); and (viii) other more general applications (e.g., [12,30]).

GNSS data and InSAR are also combined to investigate the atmospheric contribution to InSAR processing with consequences to the velocity and cumulative displacement in the ground deformation analysis [31].

Atmospheric investigation is a complex task due to its heterogeneity. It can be simplified by separately considering the troposphere, approximatively from 0 to 40 km above the Earth's surface, and the ionosphere, approximatively from 50 to 1000 km. In addition, the troposphere influences the wet lower portion (approximately 0–11 km) and the dry upper layer (approximately 11 km above the Earth) differently, which can be more difficult and simpler to model, respectively [32]. Considering the wide areas covered by the SAR images, e.g., 250 km with Sentinel-1 [33], the atmospheric effects combined with the reported high precision of the available InSAR approaches could be challenging, especially if the available datasets of SAR images are limited [34,35].

This review focuses on the collection and analysis of scientific contributions that combined GNSS/GPS data and InSAR approaches, i.e., both DInSAR and MTInSAR techniques, for pursuing different perspectives, such as ground deformation recognition and analysis or atmospheric contribution assessment over the entire European continent, excluding Turkey and Russia. In addition, a detailed investigation on Italy, the country most represented by scientific contributions combining GNSS and InSAR data, was conducted. The state-of-the-art of the various applications of these two techniques can improve the knowledge of the scientific community and help in the further development of new approaches or additional applications in different fields.

In this review, the DInSAR and MTInSAR approaches were both considered to cover a wider range of field applications, to both rapid and slow phenomena, but analyzing the scientific contribution separately. In contrast, GNSS, GPS, cGPS (continuous GPS), and DGPS (Differential GPS) are considered unique approaches for avoiding discrimination with low statistical soundness considering that the basic approach is the same.

## 2. Data Collection

Data collection was conducted through Google Scholar's freely accessible web search engine to gather a high number of scientific contributions published that considered both InSAR and GNSS/GPS data.

To collect the scientific articles, book chapters and conference abstracts published on the Google Scholar's search engine were based on the following criteria. The first keyword adopted was the name of each European country to which the following two lists of keywords related to GNSS and SAR were connected. For the SAR group, the keywords used were "InSAR" (and related acronyms as "MT-InSAR", Multi Temporal InSAR, "DInSAR", Differential InSAR, or "A-DInSAR", Advanced DInSAR), "PS" (Persistent Scatterers), and "PSI" (Persistent Scatterers Interferometry). For the global absolute measurements, the terms "GNSS" and "GPS" (with related derived names as cGPS) were used.

The use of Google Scholar for this type of research review is considered an appropriate choice, as this search engine can be considered *"essentially a superset of WoS and Scopus, with substantial extra coverage"* [36]. The same author states that, in Google Scholar, approximately 95% of the available citations correspond with those contained in the Web of Science (WOS) database and approximately 92% correspond with those of Scopus in addition to a relevant number of other citations. For this reason, Google Scholar is considered an adapted search engine and a powerful tool for researching the existing literature considering the topic of the combination of GNSS and InSAR data. However, the source of the contributions highlighted by research on Google Scholar has to be validated since this system is less automatized than others, and its database can be completed with information added by the users [37].

The collected scientific contributions, including articles, reviews, book chapters and congress proceedings, were arranged in different folders based on the various searched countries. Then, the contributions were examined iteratively and catalogued in a table considering the following derivable information:

- Title of the scientific contribution;
- List of authors;
- Type of submission (journal, book chapter, conference paper);
- Year of the publication;
- Country;
- Area of interest (province, municipality, local toponym);
- Scale of the presented work (national, regional local);
- Field of application (i.e., Atmosphere, Glacial; Infrastructure, Landslide, Subsidence; Subsidence/Uplift; Tectonic; Uplift; Vulcanic, Other, and Not Specified);

- Aim of the research;
- Investigated period;
- Type of processing (DInSAR or MTInSAR);
- InSAR approach and algorithm;
- Sensor band;
- Satellite name;
- Data used for the investigation;
- How the two data were used (combination, comparison, validation).

## 3. Temporal and Spatial Distribution of Scientific Production in Europe

### 3.1. Spatial Distribution

A total of 191 scientific contributions, among which 12 were about the theorical combination of GNSS and InSAR data and 179 focused on the combined techniques, were collected and analyzed considering all European mainlands. Spatially, not all countries have been the subject of combined GNSS and InSAR studies (considering DInSAR or MTInSAR approaches). In fact, 15 European nations, in addition to Gibraltar, Andorra, San Marino, and Vatican City, are not represented by publications regarding the GNSS and InSAR data combination (Figure 1). The higher number of publications have focused on Italy, with 47 contributions divided into 36 journal articles (e.g., [17,27,38–43]), 4 book chapters [44–46], and 7 conference proceedings [47–52]. This number of contributions, which is much higher than that collected for all the other countries, can be justified when considering the high number of geohazards affecting the territory, e.g., landslide events, subsidence phenomena, geothermal processes and mining activities, and the higher use of InSAR within the Italian scientific community.

After Italy, the other nations most represented are as follows: (i) Spain, with 14 articles in peer-reviewed journals (e.g., [53–57]) and 3 congress proceedings [58–60] for a total a 17 scientific contributions; (ii) Germany, with a total number of 15 scientific publications equally divided into various scientific journals (e.g., [61–64]) and conference proceedings (e.g., [65,66]); (iii) Portugal, with 8 peer-reviewed scientific articles (e.g., [31,67,68]), 1 book chapter [69], and 3 conference paper (e.g., [70]), and Iceland, represented by 11 contributions in peer-reviewed journals (e.g., [71–73]) and only 1 conference abstract [74], for a total of 12 scientific contribution for both countries; (iv) Norway, with 8 contributions in scientific international journals (e.g., [22,75–77]), 2 book chapters of peer-reviewed journals [78,79], and only 1 proceedings abstract for a conference [80]; and (v) Greece, equally divided in 5 peer-reviewed journal articles (e.g., [81–83]) and 5 abstracts for congresses or meetings (e.g., [84–86]), for a total contribution of 10. It has to be considered that, for the Germany, 11 scientific contributions, i.e., 6 articles published in international peer-reviewed journals and 5 abstracts for congresses refer to the Upper Rhine Graben Area, were focused on the boundary between France and Germany. To avoid repeated counting, it was decided prior to processing to assign all the contributions about the Upper Rhine Graben Area to Germany.

Following the classification, Netherlands is represented by 8 scientific works equally divided into international journal articles and conference papers. Switzerland and Romania provided 7 scientific contributions, 4 peer-reviewed articles, and 3 meeting proceedings, and 5 publications in international journals and 2 extended abstracts, respectively. The United Kingdom is named in 6 scientific works, of which 1 involves the Scotland territory, divided into 4 international peer-reviewed journals and 2 conference proceedings, while France only includes 3 scientific international articles and 1 conference abstract.

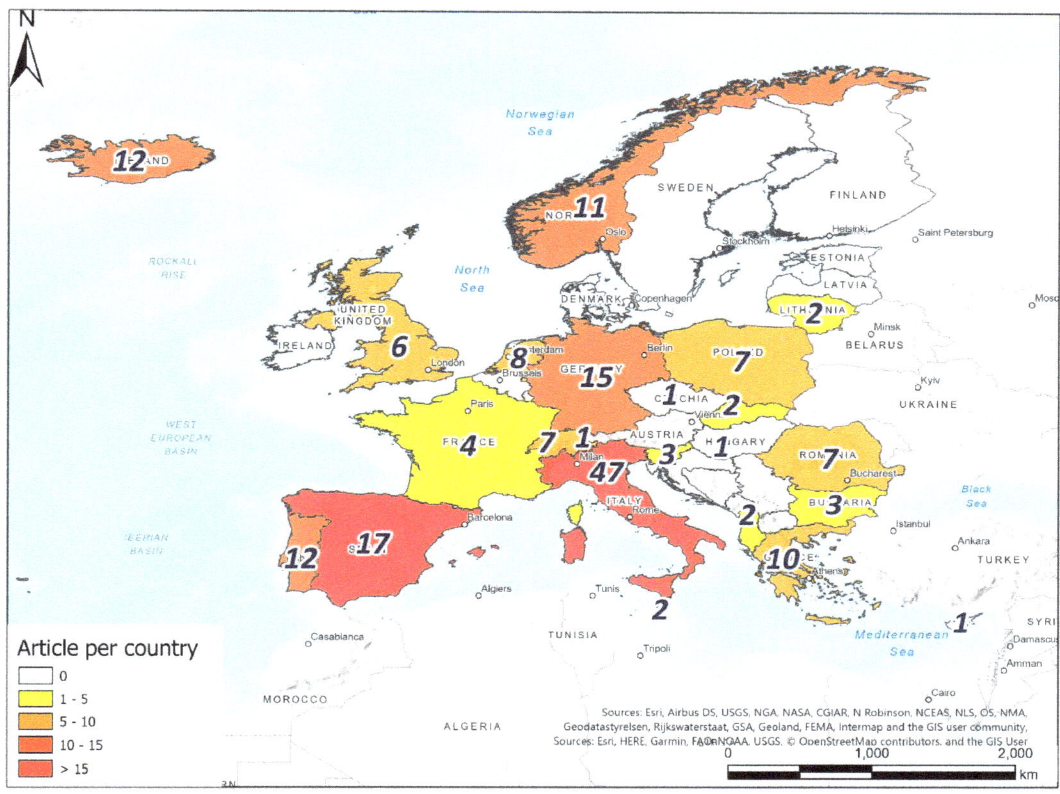



**Figure 1.** Spatial distribution of the scientific contribution of Europe.

Albania (1 article published in international journal and 1 congress abstract), Lithuania (1 publication in scientific per-reviewed journal and 1 congress abstract), and Slovakia and Malta (two scientific peer-reviewed journals) were represented by 2 scientific contributions, while Liechtenstein, Hungary, Czech Republic, and Cyprus were closely ranked, with only one scientific work.

### 3.2. Temporal Distribution

The first scientific contribution outlining the combined use of GNSS and InSAR data dates back to 1999 [42]. This article focused on a model used to estimate the geometry and slip distribution of the fault plane originating from the 26 September 1997, Colfiorito earthquakes (central Italy) by combining SAR interferometry and GPS measurements. In the following years, since 2000, the scientific community maintained a greater interest in data integration, also involving other European countries, such as Iceland, Spain, and Greece. Since 2009, excluding 2008, with a high number of 8 scientific contributions, a slight increment of scientific contributions was identifiable until 2012, with more than 10 articles, books or conference proceedings. The maximum number of scientific contributions was reached in 2015, with 29 contributions from 14 different countries. Next, in 2018 and 2012, a total of 20 and 17 works were published, respectively. It is worth noting that Italian case studies were published annually, ranging from the first application in 1999 to 2019, only excluding 2009, 2016, and 2020 (Figure 2).

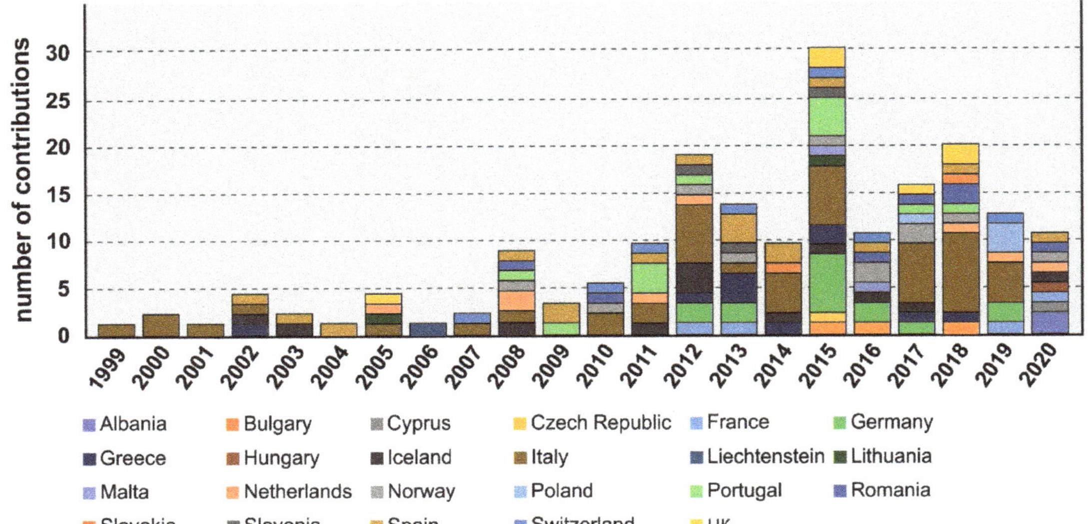

**Figure 2.** Temporal evolution of the scientific contributions combining GNSS and InSAR data in Europe per country.

The analysis of the temporal evolution in relation to the field of application of scientific contributions (Figure 3) revealed that early studies were focused on the analysis of tectonic-induced deformations. The rise of an anticline beneath an urban area in Catania in 2001 [87] and the modeling of the dislocation of the edifices involved in the 1997 Umbria-Marche seismic sequence [88] were analyzed by integrating GNSS and InSAR data. In 2001, the first atmospheric investigation application was presented, showing primary results regarding the calibration of atmospheric effects on SAR interferograms by GPS and local atmosphere models [38]. In 2002, the joint application of GPS and DInSAR for volcano and seismic monitoring was carried out in different cases in Spain for the 1992–2002 period [55]. In 2005, the first application of the combined use of GNSS and SAR data for recognizing subsidence was published. In particular, five different measurement techniques, among which are the DGPS and MTInSAR approaches, were exploited for mapping and monitoring land subsidence in the Venice lagoon [43]. A year later, the first application to landslide studies was published dealing with the Triesenberg-Triesen landslide in Leichteinstein, where GPS monitoring data were compared with 3D models generated using an MTInSAR approach on ERS datasets [89]. Among the less represented fields of application, in 2002, the first application to glacial research was proposed, where an interferogram was generated to produce an ice surface motion map of the Gjàlp volcano in Vatnajokull, Iceland [29]. Regarding infrastructure and uplift case studies, the first key scientific contribution focused on the monitoring of the urban area of Bratislava, Slovakia [26] and on the investigation of an area close to the geothermal site of Landau, Germany [19,63].

In recent years, the combined use of GNSS and SAR data has been adopted for large-scale analyses. Indeed, ground deformation maps have been produced at national scale in the United Kingdom [90] and at continental scale covering all of Europe [91]. In this case, GNSS data were implemented to identify and filter out possible residual atmospheric artefacts that may affect the quality of the employed MTInSAR data.

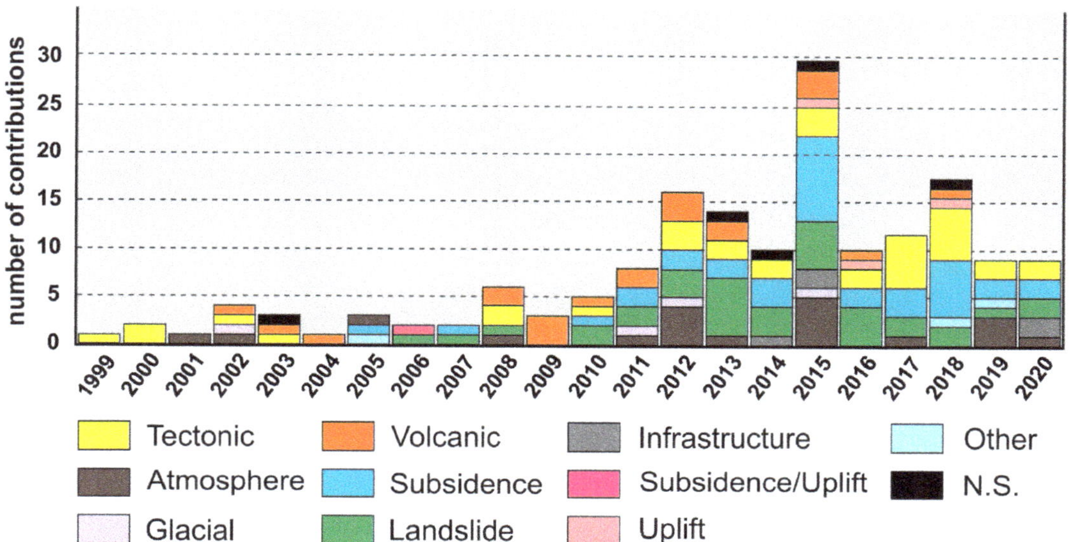

**Figure 3.** Temporal evolution of the scientific contributions combining GNSS and InSAR data in Europe considering the field of application.

### 3.3. Field of Application

The joint application of SAR and GNSS data has been adopted in 9 different fields, classified as follows: (i) Tectonic; (ii) Atmosphere; (iii) Glacial; (iv) Volcanic; (v) Subsidence; (vi) Landslide; (vii) Infrastructure; (viii) Subsidence/Uplift; and ix) Uplift. In addition, another class has been added to include all nonconventional applications, as well as a Not Specified class (N.S.) for applications in which it was not possible to assign a specific field. A higher percentage of publications deal with subsidence phenomena (20.2%), followed by landslide events (19.7%) and tectonic analyses or seismic event back analyses (18.6%) (Figure 4).

Subsidence phenomena are spatially distributed in Europe, with a higher concentration of studies (i) in the Upper Rhine Graben, at the boundary between France and Germany [62,92–94]; (ii) in Italy, over the Pianura Padana Plain (central Italy), in the Emilia Romagna Region [44,47,48] and the areas of the Po Plain [20,45,49,95,96], in the Venice lagoon [41,43,97,98], in the Friuli Venezia Giulia coastal plain (North-East of Italy) [99], the Firenze-Prato-Pistoia basin (Tuscany Region, central Italy) [100], and the Sibari plain (Calabria Region, South Italy) [101]; and (iii) in Spain, the Alto Guadalentín [102,103] (South-East of Spain), in Lorca town [54] and in the Cardona salt mine [60] (North-East of Spain). In Portugal, instead, only two scientific contributions have been reported in connection to the subsiding area of Lisbon [70,104]. Eastern Europe shows a widespread distribution of land subsidence issues, which have been analyzed by combining GNSS and InSAR data, in the coastal portion of Lithuania, close to Kaszuby Lakeland [105], in the agglomeration of Warsaw, Poland [106], in the Bucharest metropolitan area in Romania [107,108], and close to the Tatra Mountains in Slovakia [109]. A specific focus on mining-induced land subsidence events and collapses was provided by research which studied the Upper Silesia Coal basin, covering a part of the territory of Poland [110] and the Czech Republic [111]. North Europe is represented only by the surface motion of peatland in Flow County, Scotland [112], and the localized ground subsidence in the urban settings of London [113] for the United Kingdom and the analysis in the prediction of long-term settlement in the Trondheim harbor, mid-Norway [76].

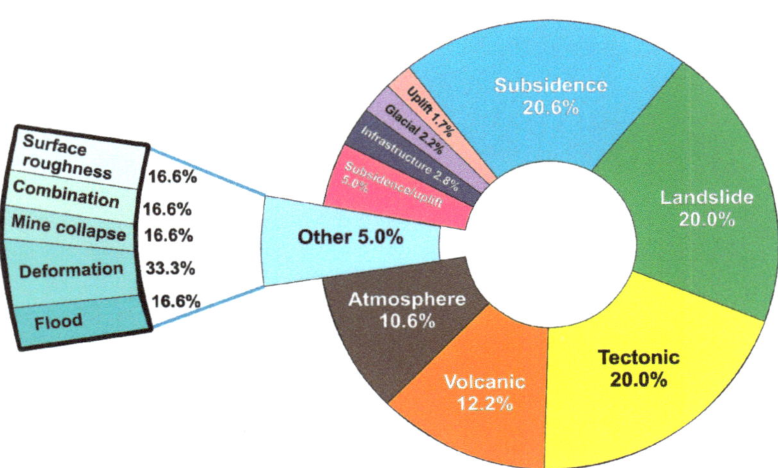


**Figure 4.** Distribution of the scientific works dealing with the combined use of GNSS and SAR data according to the field of application.

A more widespread distribution of scientific contributions from landslide studies with the joint application of GNSS and SAR data can be observed in Europe. A total of 14 nations have at least one reported application. The country with the highest number of contributions is Norway, represented by the cases of the Åknes landslide [114], the Gamanjunni rock slope [78], the Osmundneset landslide in Sognog Fjordane [115], the periglacial landscape activity at Nordnesfjellet [116], and the mapping [79] and monitoring [80] of landslides in North and West Norway. Norway is followed by Switzerland, with 6 contributions, including 3 peer reviewed articles in international journals and 3 proceedings, which focused on rock glacier detection and monitoring [117,118], slope instability monitoring events in the Western Alps [119,120], and deep-seated landslide recognition by Osco [121] and Aletsch Forest [122]. The third most represented country is Italy, with 5 peer-reviewed articles and a proceeding focus on specific case studies of the Bosmatto landslides in the Northwestern Alps [39], investigations over the wider area of the Northern Apennines [123], and of the Assisi landslides [23,52], in addition to a quantitative hazard and risk assessment of slow-moving landslides in the Arno Basin in Tuscany [124]. Next, there are 4 contributions associated with Spain concerning the monitoring of the Vallcebre landslide [125], the Tena Valley [126,127], which was affected by very slow landslides, and the Portalet landslide area [53]. In Slovenia, 3 works have been dedicated to the deformation monitoring of the Potoška Planina [128–130], while 2 peer-reviewed publications set in Malta deal with slow-moving coastal landslides [131] and landslide susceptibility modeling [132] of the Northwestern coast. Scientific contributions are also found investigating the Triesenberg–Triesen landslide in Liechtenstein [89], the Trifon Zarezan landslide in Bulgary [133], the ground deformation affecting the Choirokoitia UNESCO World Heritage site in Cyprus [134], the La Vallette landslide in France [135], the deformation phenomena that occurred in the Tatra Mountains in Slovakia [109], the ground deformation affecting Kulcs village in Hungary [136], and the investigation of the Grande da Pipa river basin in Portugal [67].

The analysis of tectonic movements (including pre-, sin-, and post-event movements) via GNSS and SAR is mainly concentrated in the seismic areas of Europe. The scientific contributions mainly come from Italy, with 16 publications, Greece and Iceland, with 5, and Romania, with 4. Italy constitutes a large representation due to the various seismic sequences that have occurred within the last 20 years, such as the 2009 L'Aquila earthquake [46,137,138], the 2012 Emilia Romagna earthquake [139], and the 2016 Central Italy seismic swarm [21,50,140–143], in addition to the old 1997 Colfiorito earthquake [42,88]. Furthermore, in Italy, scientific publications on the investigation of the geodynamic processes active throughout the whole Italian territory [20] or focused on the Southeastern Po Plain [144] and the geodynamic movements of the Mount Etna [87] or the strain rate of the Hyblean Plateau [145] in Sicily were developed by exploiting a combination of GNSS- and InSAR-derived information. The scientific literature devoted to Greece has reported interesting research on the investigation of pre-earthquake deformation processes by examining GPS, DInSAR, MTInSAR, and ancillary data collected over Zakynthos Island [146]. In addition, the seismological analysis of the critical estimation of a future strong seismic event in the broader area of Cephalonia [82], the post-seismic investigation of a seismic sequence that occurred during the period of January–February 2014 [147] or a specific earthquake, such as the Lefkada involving the Cephalonia Transformation Fault (CTF), close to Cephalonia Island [81], and the post-seismic motion analysis of one of the most active extending grabens, the Corinthian Gulf [85], were collected. Iceland is represented by five contributions. One is the investigation of a $M_w$ (moment magnitude) 6.5 earthquake that occurred in 2000 [71,148]. The others are on the identification of extensional and interseismic deformation recorded in the Northern Vulcanic Zone [149], as well as the investigation of ground deformation close to geothermal power plants in Hengill due to the interaction of regional tectonics and volcanic deformation [150] and the post-seismic viscous relaxation in the Southern Iceland [151]. In the South Carpatians, Romania, crustal deformation was investigated by detecting small-magnitude tectonic processes [152], and GNSS data, combined with SAR interferometry, were adopted for the analysis of the abnormal seismic behaviors of Izvoarele-Galați and for the development of an early-warning system for marine geohazards along the Black Sea coast [153]. Furthermore, Zoran [154] combined geospatial information, i.e., GPS and SAR images, with in situ information to assess the seismic hazard associated with the Vrancea area in Romania. Other interesting research in the field of tectonism utilized earthquake observations with remote sensing data from the $M_w$ 6.4 Durres (Albania) earthquake that occurred in 2019 [155] and the $M_w$ 6.5 Lorca earthquake (Spain) [56]. Studies on seismic [156] and crustal deformation [157], sometimes connected to strain rate analysis [66], or interactions with volcanic systems [68,69], connected to earthquake seismic swarms or active tectonic faults, are more common.

Volcanic (10.9%) applications are mainly concentrated in (i) Greece, focused on the investigation of Santorini volcano [158,159] and Nisyros volcano [83,86]; (ii) Italy, focused on the evaluation of the deformation of Mt. Etna, Sicily Region (Southern Italy) [160,161] and the Campi Flegrei, Campania Region (Southern Italy) [40,51]; (iii) Portugal, focused on the assessment of the geodynamics and displacement mechanisms affecting the Azores islands [13,14,69]; and (iv) Spain [55], focused on the volcanic deformation of the Canary archipelago [58] and Tenerife island [24,57,162,163], inevitably depending on the presence of active volcanoes. Iceland is represented by two scientific articles, one focusing on the lateral dyke grown [72] and the gradual caldera collapse regulated by lateral magma overflow [25] of the Bàrdarbunga stratovolcano in the South-eastern sector of the Iceland island; and a conference paper combining InSAR and GNSS data [74] that focuses on ground deformation measurements of the Northern Volcanic Zone of the island. In addition, Wadge et al. [17] published an article on Mt. Etna focused on the measurement of the tropospheric water value to better consider the contribution of the atmosphere to the differential radar interferograms of ERS1/2.

Atmosphere applications (10.4%), unlike subsidence, landslide activity, and tectonic and volcanic phenomena, generally involve wide areas, ranging from regional to national scales. Atmospheric applications of the combination of GNSS and SAR data encompass the field of water vapor determination [61,164–169], atmospheric model or delay analysis [17,18,31,38,170], ionospheric artifacts detection [171], atmospheric and wet refractivity reconstruction [65,172] and tropospheric correction or delay [77,173–175], with the final goal of assessing a more precise atmospheric correction in ground deformation analysis of SAR datasets in wide areas or peculiar zones, e.g., volcanic.

The Subsidence/Uplift class (4.4%) includes surface deformation analyses performed at a large, regional or national scale, in which both subsidence and uplift phenomena can be identified. Case studies were identified in Netherlands, such as surface deformation analyses performed over the whole national territory [176–179] or monitoring works of the water defence structure of Waddenzee and IJsselmeer [180,181]. Two other scientific contributions focused on the Polish and Lithuanian Baltic coastal areas [105] and the surface displacement pattern affecting the permafrost areas in northern Norway [75].

The categories of Uplift, Infrastructure, and Glacial are poorly represented. In fact, three peer-reviewed articles, focused on the Tower Hamlets Council area of London [182] and on the Geothermal site Landau in Germany [19,63], which only investigated uplift phenomena resulting from the swelling of clayey soil and geothermal consequences, respectively, can be cited. The lower number of studies focused on the investigation of infrastructure can be attributed to several factors, e.g., the resolution of SAR images, the distribution of the GNSS network compared to the infrastructure dimensions, the particularity of the application and the difficulty associated with interpreting the results. An investigation of the nonlinear deformation of infrastructure via SAR data and GNSS comparison was performed in Bratislava, Slovakia [26]. Other infrastructure applications were indirect, including the monitoring of subway construction in Bucharest, Romania [183], ground deformation detection focused on a new geothermal power plant at Reykjanes, Iceland [73], observation of ground deformation associated with the Kozloduy Nuclear Power Plants (NPPs), Bulgary [184], and experimental monitoring of localized deformation on the Roman aqueducts in Rome, Italy [27]. Glacial applications are represented by specific case studies in Iceland, which focused on the 3D surface motion of the glacier surface of the Gjàlp volcano in Vatnajokull [28,29]; France, with a time-series measure of the Argentière glacier on the Mont Blanc Massif [185]; and northern Scotland (United Kingdom) through the investigation of the Glacial Isostatic Adjustment (GIA) [186].

In addition, some very particular studies of the combined use of GNSS and SAR data, categorized as Other (Figure 4), recorded the following: (i) sea level change detection along the coastline of Brest, France [187], along the North Sea and Baltic Sea Coast in Germany [188] and in the Northern Mediterranean Sea [189]; (ii) flood inundation modeling of Northamptonshire [190]; (iii) mine collapse monitoring in Rudna Mine, Poland [191]; (iv) surface and roughness and sediment texture characterization [192]; and (v) a new approach description of precise datum collection [193].

*3.4. Satellite SAR Sensors*

Considering all the aforementioned scientific contributions, whether they are peer-reviewed articles, book chapters or conference papers, 32% of them exploit DInSAR-based data, while the remaining 68% relied on one of the previously described MTInSAR approaches. In detail, more than 100 publications use PSInSAR (Persistent Scatterers Interferometry SAR) data [194], while a minor portion utilized the SPN (Stable Point Network [195]), SqueeSAR [196], SBAS (Small BAseline Subset [197]), iSBAS (Intermittent SBAS [198]), PSP-DINFSAR (Persistent Scatterer Pairs Differential InSAR [199]), CPT (Coherent Pixel Technique [200]), SPINUA (Stable Point Interferometry over Unurbanized Areas [201]), and TSIA (Two-Scale Interferometric Analysis, processing chain that performs a sequence of low-resolution (small-scale) and full-resolution (large-scale) processing [23], based on the SBAS [197] approach) algorithms. It is impossible to establish a precise number of times each algorithm was used since, in some articles, more than one approach was used for different datasets, and in many contributions, the processing algorithm was not clearly specified.

Considering the satellite wavelength of the whole database collected, approximately 150 C-band datasets were implemented, as well as 35 X-band and 15 L-band datasets. For 15 papers, the C- and X-band datasets were combined, while only 7 C- and L-band datasets were used in parallel. None of the scientific studied viewed combined X- and L-bands. Finally, applications analyzing all three bands were implemented 5 times. Publications taken into consideration were also categorized according to the satellite constellation of the SAR data. Most of the paper shows a massive application of ENVISAT data (31.1%), covering the period 2002–2010, and ERS data (27.2%), covering the period from 1992 to 2000, followed by the data of Sentinel-1 (14.2%), acquiring SAR images since April 2014, which was largely diffused over the last few years due to its short and constant revisiting time, due to its scientific contribution aim and free availability [33]. All these satellite constellations acquire in the C-band (5.6 cm wavelength), confirming its more widespread use in more common applications as a good compromise for urban and nonurbanized areas. The minor frequencies of other satellites are registered, listed from the high to low: TerraSAR-X (X-band, approximately 10%), RadarSAT (X-band, approximately 6%), Alos (L-band, approximately 6%), and COSMO-SkyMed (X-band, approximately 4.7%).

Satellite data are also used in combination with others to obtain longer timespans of analysis or to compare the performances of different bands. The combination of different constellation datasets is reported in the satellite graphs of Figure 5. It is interesting to note that, in at least one paper, all constellation satellites were combined with the other three constellations, except for RadarSAT, which did not show application when combined with the other 3 datasets. In addition, it is possible to state that the aforementioned constellation satellites are rarely combined with the other two and commonly with other datasets. Exceptions are also detectable for the COSMO-SkyMed, X-band, and Alos, L-band, whose data were more often used combined with two other constellations than standalone or only with one other dataset.

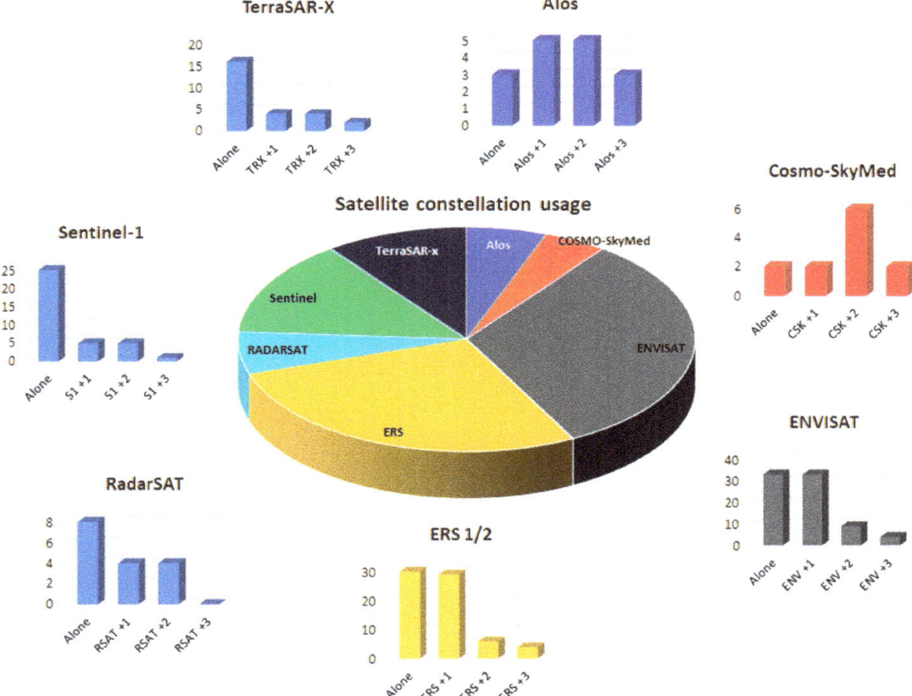


**Figure 5.** Satellite constellation usage considering all the applicative scientific contributions (center) and a more detailed analysis of the use of the dataset of a single satellite (Alone) or with another satellite (+1), with other 2 (+2) or three (+3) constellations (satellite graphs).

## 4. Temporal and Spatial Distribution in Italy

A more specific analysis of the temporal and spatial distribution of previous studies, as well as within the field of application, was conducted focusing on the Italian literature, since Italy is the most represented country, accounting for the 47 published scientific contributions on the integrated use of GNSS and InSAR data.

The highest number of scientific contributions is recorded in Emilia-Romagna and Umbria, which are two regions where tectonic, seismic, subsidence, and landslide events were analyzed (Figure 6).

| Region | Articles |
| --- | --- |
| PIEMONTE | 0 |
| VALLE D'AOSTA | 1 |
| LOMBARDIA | 2 |
| TRENTINO-ALTO ADIGE | 0 |
| VENETO | 6 |
| FRIULI-VENEZIA GIULIA | 2 |
| LIGURIA | 1 |
| EMILIA-ROMAGNA | 11 |
| TOSCANA | 2 |
| UMBRIA | 10 |

| Region | Articles |
| --- | --- |
| MARCHE | 3 |
| LAZIO | 3 |
| ABRUZZO | 6 |
| MOLISE | 0 |
| CAMPANIA | 3 |
| PUGLIA | 0 |
| BASILICATA | 0 |
| CALABRIA | 1 |
| SICILIA | 6 |
| SARDEGNA | 0 |

**Figure 6.** Spatial distribution per region in Italy.

Despite the higher number of articles, books, and conference proceedings with case studies performed over the Italian territory, not all the regions are covered. In fact, six Italian regions do not exhibit any phenomena investigated by either GNSS or SAR data. One article [20] took into consideration the entire Italian territory using both techniques to study Earth surface deformation to provide the displacements with respect to different ground point position components.

In the Valle d'Aosta Region, in Northwestern Italy, a large landslide in a high alpine environment was studied by GNSS, InSAR, and GB-InSAR remote sensing data to reduce uncertainties. The joint combination of these techniques provided a comprehensive view of the deformation field of the landslide [39].

Landslide investigations are also identified in the Umbria Region, central Italy, for the analysis and monitoring of the slow-moving Assisi landslide [23,52]. The geodetic and interferometric products of four deep-seated landslides reactivated by the excavation of a double road tunnel that induced deformation were monitored in the Northern Apennine [123], while the same type of data was used in the Arno River basin, Tuscany Region, to update previous hazards and risk map realized by [124], Catani, et al. [202].

The highest number of applications, 15 in total, is represented by subsidence phenomena investigations. Seven scientific contributions on this theme, i.e., 2 peer-reviewed articles, 2 book chapters, and 3 conference proceedings, target the Emilia-Romagna Region, central Italy, 4 of which are specifically on the Po Plain. Only one is focused on the Tuscany Region, central Italy, more specifically in the Firenze-Prato-Pistoia Plain [100], as one on the Sibari Plain in the Calabria Region, South Italy [101], and on the plain that lies between the Tagliamento and Isonzo Rivers in the Friuli-Venezia Giulia Region, North-East of Italy [99]. The remaining 5 scientific contributions focus on the vertical deformation of the Venice lagoon [41,43,97,98] or, more generally, on the subsidence phenomenon affecting the North Adriatic Sea [45].

The high number of scientific contributions sourced from the Umbria Region, central Italy, resulted from the Assisi landslide analysis and monitoring, as well as the investigation of the earthquakes that occurred in 1997, 2009, and 2016 involving Umbria and the neighboring regions. In total, 16 contributions targeted the investigation of tectonic and seismic investigations.

In contrast to the landslide or subsidence investigations, the seismic analyses typically involved more than one region as earthquakes affect larger regions. For this reason, 4 articles and one conference proceeding focus on more than one region of central Italy. The 2009 L'Aquila earthquake, Abruzzo Region, was also represented in an article and in a book chapter analyzing the coseismic rupture [46] and slip distribution [138] concentrating the study area close to the epicenter. The 2012 Emilia-Romagna Region, instead, did not have relevant effects on the boundary regions. One of these contributions addresses the modeling of the influence of fluids and pore-pressure changes on surface displacements and on the Coulomb Failure Function (CFF) in the co-seismic and post-seismic period related to the 2012 mainshock [139]. Wang et al. [142] looked for the source parameters and triggering link between the earthquake sequences of 2009 and 2016 recorded in central Italy. In addition, 4 works combining InSAR and GNSS data were developed for determining the source parameter of the three main shocks recorded during the 2016 Central Italy earthquake sequence [50,140], for a modern observation of the spatio-temporal evolution of the seismic sequence [143] and for a novel record of near-field co-seismic fault slip measurement [21].

The spatial distribution of 5 volcanic scientific contributions, including 4 journal articles and one symposium proceeding, is strictly related to the location of volcanic cones in Italy. In fact, 3 publications focus on the analysis of nonlinear deformation [51], the evaluation of fault reactivation [203] and the investigation of magma injection [204] of the Campi Flegrei, Campania Region, and two on the ground deformation pattern of the Mt. Etna volcano in Sicily [160,161].

The Mt. Etna volcano is also the subject of two peer-reviewed articles investigating the varied tropospheric compensation expectations between the GPS and SAR interferograms [38] and the dynamic models of atmospheric movement for calculating the delays affecting radar processing [17]. The other area where the atmospheric delay analysis was undertaken was centered over the Como province in Lombardia Region, North Italy, presenting the results in an international scientific journal [18] and in an international symposium [170].

In conclusion, the Italian territory has been a relevant case study since 1999 [42], and it was continuously under investigation by the scientific community until 2019 that analyzed both GNSS and SAR data for different purposes (Figure 7).

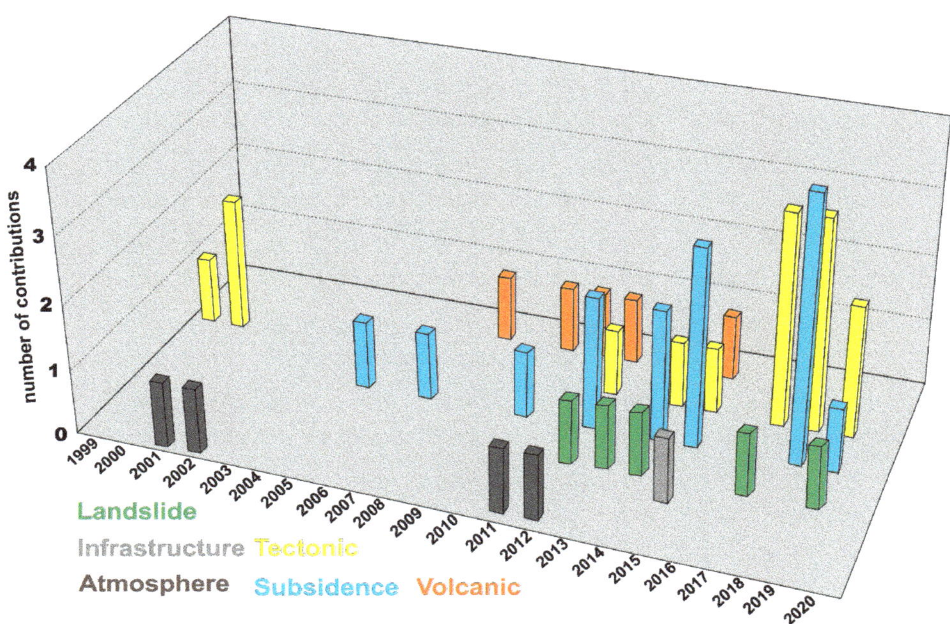


**Figure 7.** Temporal distribution of the scientific contribution recorded in Italy categorized according their field of application. Scientific contributions are missing for 2003, 2004, 2006, 2009, 2016, and 2020.

## 5. Discussion

A substantial number of scientific contributions using remote sensing data derived from GNSS and SAR systems to investigate and analyze ground deformation in European countries, covering a time span of 31 years (from 1999 to 2020), were collected.

The spatial distribution of all the scientific contributions, considering scientific articles, book chapters or conference proceedings, mirrors the distribution of the main relevant geohazards affecting each country. Italy and Spain are well represented due to the high number of geohazards affecting their territory, especially considering the subsidence occurring in both Italy [205] and Spain [9] and the landslides affecting Italy [206]. The limited number of scientific contributions combining GNSS and InSAR data to study landslides is surprising considering the relevance of these natural hazards and the major consequences occurring throughout Europe [207]. Nonetheless, the Italian territory is also affected by frequent seismic swarms and strong earthquakes in addition to the presence of volcanoes, and both phenomena are deeply investigated by the joint use of GNSS and SAR data. In the same way, it is surprising how low the number of applications recorded in the Netherlands is considering the importance of subsidence phenomena in that country (e.g., [208]), which exert social and economic consequences and induce damage to structures and infrastructure [209,210]. Another well-represented nation is Norway, which is strongly affected by landslides, often triggered in quick clay soils [211], and characterized by the presence of permafrost, which can cause four types of landforms [212]: (i) palsas (e.g., [213]),

(ii) rock glaciers (e.g., [214,215]), (iii) ice-cored moraines (e.g., [214,216]), and (iv) ice-wedge polygons (e.g., [217]).

In addition to scientific papers dedicated to the analysis of displacement, general articles employing atmospheric analyses, in which GNSS products are implemented to correct interferograms or remove the geodetic influence on InSAR displacement measurements over wide areas, were collected and analyzed. This approach is even more relevant due to the nature of the wide areas investigated and monitored by InSAR approaches.

The combination of DInSAR and GNSS data has not always been categorical, but various trends and approaches utilized over the years can be illustrated.

The various works analyzed have shown the potential of combining both data for different purposes. For instance, GNSS data were used to improve the accuracy of the interferograms generated [11], topographically correcting the interferograms and the elevation data to measure ice surface motion maps along the Gjalp Volcano. Most of the works jointly used GNSS and DInSAR data to produce a more precise and reliable analysis of the movements detected by either combining or validating the available data. Indeed, in most of the pioneering works, authors worked to validate DInSAR data, since in most of the cases, the latter were used for the first time and their results needed to be validated. For instance, in Raucoules et al. [135], in which 3D data derived from TerraSAR-X processing over the La Vallette landslide (southern France) were validated with GNSS data; in other cases, such as in Bovenga et al. [23] and Sakkas et al. [146], GNSS data, along with geological and geotechnical data, were used to ground truth C- and X-band imagery.

Approximatively half of the scientific works are focused on single case studies or events with a local influence, while 45% involve regional areas. Temporally, more recent papers involve more large areas due to various factors: (i) the possibility to have more and improved algorithms for processing large SAR datasets; (ii) the technological progress of the last few years, allowing us to have available advanced computational resources, e.g. cloud computing by virtual machine system and parallel or distributed computing architectures used to process larger dataset over wide areas; (iii) the willingness of different radar satellite constellations, also in different bands, i.e., X-band, C-band and L-band; and (iv) the ease of obtaining free SAR images at a medium resolution, such as the Sentinel-1 constellation, by the Copernicus Access Hub of the ESA (European Space Agency), ensuring systematic worldwide covering acquisition since April 2014, or the availability of limited imagery datasets at a high resolution, i.e., X-band, by means of dedicated scientific projects in open calls. Only a few articles, 10 of the 174 examined, have as a main aim the analysis of more case studies in the same nation, within very wide areas, or throughout entire countries. In detail, four articles are noteworthy for the extension and the relevance of the work. The first two focused on the ground deformation occurring in Netherlands [176,178], while the remaining focus on the generation of atmosphere Precipitable Water Vapor (PWV) maps over large areas in Portugal [169] and the feasibility assessment of a national InSAR ground deformation map over the United Kingdom [90], which both took place in 2017. In 2020, the first continental-scale processing of Sentinel-1 imagery was described in Lanari et al. [91], which provided a calibrated ground deformation map of Europe with the velocity of deformation and cumulative displacement corrected by the GNSS network data.

The authorship, taking into account the affiliation of the first author of the scientific contributions on ground deformation or atmosphere contribution analyses over European countries, demonstrates that approximately 80% of the total refers to the same nations. For the remaining approximately 20%, 10 authorships are referred to Countries outside of Europa (e.g., 4 from USA, 5 from China, and 1 from Canada), while the other main authorships are of Italian (8), Portugal (6), and French (5) researchers. This can be justified by considering (i) collaborations or (ii) projects from different universities, institutions, or entities and (iii) the PhD abroad periods or (iv) research abroad time of researchers. Italy, with 9 contributions authored by non-Italian investigators, is the most investigated nation, not only by Italian researchers but also those from European (2) and international (5) investigators. Next, Portugal is unique in that it was only analyzed by Portugal researchers that also investigate Romania (6 contributions to 7 are from Portugal researchers), followed by Spain, with 5 contributions from European researchers.

The correlation between the temporal evaluation of European and Italian studies is interesting. For the first few years, this relation is easily comprehensible considering that the scientific contribution collected for the first three years, i.e., 1999–2000 and 2001, involved the Italian territory. It is interesting to note, comparing Figures 2 and 7, that the trend is very similar, with an incremental increase that progressed until the peak contribution in 2015, which was then followed by a reduction. The high number of scientific contributions recorded over the Italian territory severely influences the trend of the temporal distribution that occurred over Europe. In contrast, the studies published in 2020 do not follow this role; in fact, the last analyzed year exhibits a relevant number of scientific contributions of 2019 even if no publications were published involving Italy. A similar correlation can also be identified when considering the European and Italian temporal distributions and the associated field of application.

Combining the information regarding the year of publication of the scientific contributions collected and the investigated SAR satellite, it is possible to state that the later publications aimed to investigate the evolution of a phenomenon as long as possible. Obviously, this cannot be affirmed for publications involving atmospheric interactions and corrections or for those investigating the co-seismic earthquake effects of DInSAR.

The incremental increase in the number of publications reported demonstrates that the remote sensing techniques investigated in this review, i.e., geodetic and InSAR approaches, were being readily employed by the scientific community as a result of their benefits and the consequences for stakeholders and end-users as environmental managers. In recent years, this interest began to be traduced in the publication of national and regional GMS (Ground Motion Services) due to the free availability of InSAR data and, in some cases, the already GNSS-calibrated model. In fact, a higher number of applications, and thus a greater number of papers, was collected and recorded in those countries in which a Ground Motion Service (GMS) already exists or is nearly ready for presentation. In 2002, Italian law allocated funds for a Not-ordinary Plan of Environmental Remote Sensing aiming at monitoring considered at risk hydro-geomorphological areas, providing nationwide ground deformation maps calibrated by GNSS data. ERS and ENVISAT interferometric products, processed by PSInSAR and PSP-DINSFAR approaches, were published in the first phase of the project (2008–2009) and then updated in a second phase (2010–2011), while COSMO-SkyMed data were implemented in the same project in the third and final phases (2013–2015) [218,219]. Next, the first nation that made available updated GMS data over the entire territory was Norway in 2018; this publication was provided by the WebGIS portal Radasat-2 data (2010–2018) and Sentinel-1 (2015–2019), not calibrated by GNSS information [220]. In 2019, a year later, BodenBewegungsdienst Deutschland presented the Germany GMS, making the Sentinel-1 PS data available, GNSS-calibrated, processed thanks to the PSI approach [221] by the DLR German Aerospace Center research institute. In addition, Danish GMS and Dutch GMS are under development and will be published soon. More locally, Tuscany (central Italy) and Veneto Region (North-East Italy) have produced regional GMSs that are providing Sentinel-1 MTInSAR data free of charge processed by the SqueeSAR [196] algorithm. These three regions are the object of continuous monitoring, which is systematically updated and interpreted every 12 days. In addition, the application of a data-mining algorithms also highlight the PS with relevant trend changes [222,223]. Taking into account the usefulness and versatility demonstrated by this review regarding the combination of GNSS and InSAR data in ground deformation analysis, in the future, PS data will become available, and the national and regional GMS should be GNSS-calibrated. Currently, this procedure is sometimes conducted by researchers as a further step for specific investigations with good results [20,100].

Consequently, the promising results obtained by the scientific community and the free availability of data, which permit drastic cost reductions, have drawn increasing interest from the administrative managing office for the mapping and monitoring of the ground deformation issue [223–225].

In addition, since 2016, Europe has been working on an initiative pushing to implement the available data provided by MTInSAR products at continental scale in a project named European Ground Motion Service (EGMS) [226]. The three main products provided by the EGMS are: (i) the ground deformation maps and time-series along the LOS direction; (ii) an advanced ground deformation map in LOS direction combined with the GNSS data of the EUREF permanent network (EPN) [227]; and (iii) the two main deformation components derivable from the combination of the two orbits, i.e., the horizontal East-West and the vertical up-down deformation. All these products will be updated annually and will open several other opportunities for investigation with a consequent advancement in the knowledge in various fields, e.g., ground deformation prediction [228], mapping [224,225,229], and monitoring [230] of natural hazards, and atmospheric delay analysis.

Another use of GNSS data, which has become increasingly widespread over recent years, is aimed at the retrieval of 3D surface displacement, thus overcoming the limitations associated with DInSAR LOS displacement measurements, e.g., References [92,128]. Some limits of the geodetic and InSAR approaches can be overcome by the combination of GNSS and InSAR data, but some limitations cannot be removed. In the latest year, GNSS data were mostly used for completing SAR-derived information and filling in points were data were missing, thus improving the temporal coverage of the deformation time series, as performed in a pervious study [100], or for investigating the "real" ground deformation velocity, thereby remedying the geodetic displacement that affects the ordinary InSAR measurements. Furthermore, the areal coverage of results can be improved by joining GNSS stations and derived-PS points by PSI approaches, which can obtain deformation data over areas invisible to SAR systems, e.g., Reference [39]. It enables the obtainment of information in areas that are not covered by direct measurements or by the use of a single technique. On the other hand, GNSS stations are placed in strategic positions to create an approximatively regular network and cannot completely overpass the InSAR limits, such as the coverage of shadow areas due to the interaction between the topography and the LoS of SAR satellites. To overcome this limitation, a traditional survey with an operator by GPS must be conducted, reducing the easy and fast repeatability and considerably enlarging the time and monetary expenses.

The substantial improvement of both technical and technological approaches in recent decades, as well as the demonstrated useful and versability of the combination of GNSS and InSAR remote sensing techniques for different purposes, by more than 180 scientific contributions suggests that the incremental use of these approaches in the future will be further enhanced and widespread. The free availability of Sentinel-1 images and InSAR products by GMS will help in the increasing application in different fields and the relative spreading and dissemination.

## 6. Conclusions

A review of the published scientific contributions that focused on the European continent, excluding Turkey and Russia, combining GNSS (Global Navigation Satellite System) and InSAR (Interferometric Synthetic Aperture Radar) data, is reported here. A total of 191 scientific contributions were collected and analyzed in detail. The high number of contributions includes peer-reviewed articles and book chapters published in international journals, as well as the abstracts, conference proceedings, and extended abstracts of national and international congresses and symposiums.

An analysis of the spatial and temporal distribution of the scientific contributions was completed for Europe, while a specific focus was dedicated to Italy, since it is the most represented country, with 47 scientific contributions. Italy is also the first nation that reported a study combining GNSS and InSAR data. In fact, the first research work was recorded in 1999 and investigated the geometry and slip distribution of the fault plane originated Colfiorito earthquakes (central Italy), occurred on 26 September 1997.

Most of the joint applications in Europe, as also recognizable for Italy, focus on landslide, subsidence, and seismic/tectonic case studies. The spatial distribution mostly corresponds to the distribution of geohazards in Europe, e.g., subsidence phenomena investigated in Netherlands and volcano case studies in southern Italy, Iceland, and Spanish islands.

Considering the free access of InSAR data by Ground Motion Services (GMS), a good correspondence can be recognized. This research is interesting since it is a good demonstration that the scientific community and administrative managing offices are interested in the remotely sensed analysis of ground deformation. This can be a good starting point for the continued development of improvements in this branch of remote sensing research and in the free accessibility of data policy. Scientific advancement by the scientific community, and consequently the local administrator and the entire population, could receive a great benefit from open access data both from the perspective of research and in confidence in the products.

**Author Contributions:** Conceptualization and writing, M.D.S.; bibliographic review and creation of the database, M.D.S., P.S., and P.C.; methodology and figures preparation, M.D.S. and P.C.; review and homogenization, S.B. and N.C. All authors have read and agreed to the published version of the manuscript.

**Funding:** This research received no external funding.

**Conflicts of Interest:** The authors declare no conflict of interest.

# References

1. Graham, L.C. Synthetic interferometer radar for topographic mapping. *Proc. IEEE* **1974**, *62*, 763–768. [CrossRef]
2. Zulkifli, N.A.; Din, A.H.M.; Som, Z.A.M. Vertical land motion quantification using space-based geodetic methods: A review. In Proceedings of the IOP Conference Series: Earth and Environmental Science, Kuala Lumpur, Malaysia, 24–25 April 2018; p. 012024.
3. Bruyninx, C.; Legrand, J.; Fabian, A.; Pottiaux, E. GNSS metadata and data validation in the EUREF Permanent Network. *GPS Solut.* **2019**, *23*, 1–14. [CrossRef]
4. Gabriel, A.K.; Goldstein, R.M.; Zebker, H.A. Mapping small elevation changes over large areas: Differential radar interferometry. *J. Geophys. Res. Solid Earth* **1989**, *94*, 9183–9191. [CrossRef]
5. Crosetto, M.; Monserrat, O.; Cuevas-González, M.; Devanthéry, N.; Crippa, B. Persistent scatterer interferometry: A review. *ISPRS J. Photogramm. Remote Sens.* **2016**, *115*, 78–89. [CrossRef]
6. Tomás, R.; Li, Z. Earth Observations for Geohazards: Present and Future Challenges. *Remote Sens.* **2017**, *9*, 194. [CrossRef]
7. Zhou, X.; Chang, N.-B.; Li, S. Applications of SAR interferometry in earth and environmental science research. *Sensors* **2009**, *9*, 1876–1912. [CrossRef] [PubMed]
8. Pepe, A.; Calò, F. A review of interferometric synthetic aperture RADAR (InSAR) multi-track approaches for the retrieval of Earth's surface displacements. *Appl. Sci.* **2017**, *7*, 1264. [CrossRef]
9. Tomás, R.; Romero, R.; Mulas, J.; Marturià, J.J.; Mallorquí, J.J.; Lopez-Sanchez, J.M.; Herrera, G.; Gutiérrez, F.; González, P.J.; Fernández, J. Radar interferometry techniques for the study of ground subsidence phenomena: A review of practical issues through cases in Spain. *Environ. Earth Sci.* **2014**, *71*, 163–181. [CrossRef]
10. Hanssen, R.F.; Weckwerth, T.M.; Zebker, H.A.; Klees, R. High-resolution water vapor mapping from interferometric radar measurements. *Sci. Environ.* **1999**, *283*, 1297–1299. [CrossRef]
11. Gudmundsson, S.; Carstensen, J.M.; Sigmundsson, F. Unwrapping ground displacement signals in satellite radar interferograms with aid of GPS data and MRF regularization. *IEEE Trans. Geosci. Remote Sens.* **2002**, *40*, 1743–1754. [CrossRef]
12. Simonetto, E.; Durand, S.; Burdack, J.; Polidori, L.; Morel, L.; Nicolas-Duroy, J. Combination of INSAR and GNSS measurements for ground displacement monitoring. *Procedia Technol.* **2014**, *16*, 192–198. [CrossRef]
13. Catalão, J.; Nico, G.; Hanssen, R.; Catita, C. Integration of InSAR and GPS for vertical deformation monitoring: A case study in Faial and Pico Islands. In Proceedings of the Fringe 2009 Workshop, Frascati, Italy, 30 November–4 December 2009; pp. 1–7.
14. Catalão, J.; Nico, G.; Hanssen, R.; Catita, C. Merging GPS and atmospherically corrected InSAR data to map 3-D terrain displacement velocity. *IEEE Trans. Geosci. Remote Sens.* **2011**, *49*, 2354–2360. [CrossRef]
15. Samsonov, S.; Tiampo, K. Analytical optimization of a DInSAR and GPS dataset for derivation of three-dimensional surface motion. *IEEE Geosci. Remote Sens. Lett.* **2006**, *3*, 107–111. [CrossRef]
16. Lee, I.; Chang, H.-C.; Ge, L. GPS campaigns for validation of InSAR derived DEMs. *J. Glob. Position. Syst.* **2005**, *4*, 82–87. [CrossRef]

17. Wadge, G.; Webley, P.; James, I.; Bingley, R.; Dodson, A.; Waugh, S.; Veneboer, T.; Puglisi, G.; Mattia, M.; Baker, D. Atmospheric models, GPS and InSAR measurements of the tropospheric water vapour field over Mount Etna. *Geophys. Res. Lett.* **2002**, *29*, 1905. [CrossRef]
18. Cheng, S.; Perissin, D.; Lin, H.; Chen, F. Atmospheric delay analysis from GPS meteorology and InSAR APS. *J. Atmos. Sol. Terr. Phys.* **2012**, *86*, 71–82. [CrossRef]
19. Heimlich, C.; Gourmelen, N.; Masson, F.; Schmittbuhl, J.; Kim, S.-W.; Azzola, J. Uplift around the geothermal power plant of Landau (Germany) as observed by InSAR monitoring. *Geotherm. Energy* **2015**, *3*, 1–12. [CrossRef]
20. Farolfi, G.; Piombino, A.; Catani, F. Fusion of GNSS and Satellite Radar Interferometry: Determination of 3D Fine-Scale Map of Present-Day Surface Displacements in Italy as Expressions of Geodynamic Processes. *Remote Sens.* **2019**, *11*, 394. [CrossRef]
21. Wilkinson, M.W.; McCaffrey, K.J.; Jones, R.R.; Roberts, G.P.; Holdsworth, R.E.; Gregory, L.C.; Walters, R.J.; Wedmore, L.; Goodall, H.; Iezzi, F. Near-field fault slip of the 2016 Vettore M w 6.6 earthquake (Central Italy) measured using low-cost GNSS. *Sci. Rep.* **2017**, *7*, 1–7. [CrossRef]
22. Lauknes, T.; Shanker, A.P.; Dehls, J.; Zebker, H.; Henderson, I.; Larsen, Y. Detailed rockslide mapping in northern Norway with small baseline and persistent scatterer interferometric SAR time series methods. *Remote Sens. Environ.* **2010**, *114*, 2097–2109. [CrossRef]
23. Bovenga, F.; Nitti, D.O.; Fornaro, G.; Radicioni, F.; Stoppini, A.; Brigante, R. Using C/X-band SAR interferometry and GNSS measurements for the Assisi landslide analysis. *Int. J. Remote Sens.* **2013**, *34*, 4083–4104. [CrossRef]
24. Fernández, J.; Yu, T.-T.; Rodrıguez-Velasco, G.; González-Matesanz, J.; Romero, R.; Rodrıguez, G.; Quirós, R.; Dalda, A.; Aparicio, A.; Blanco, M. New geodetic monitoring system in the volcanic island of Tenerife, Canaries, Spain. Combination of InSAR and GPS techniques. *J. Volcanol. Geotherm. Res.* **2003**, *124*, 241–253. [CrossRef]
25. Gudmundsson, M.T.; Jónsdóttir, K.; Hooper, A.; Holohan, E.P.; Halldórsson, S.A.; Ófeigsson, B.G.; Cesca, S.; Vogfjörd, K.S.; Sigmundsson, F.; Högnadóttir, T. Gradual caldera collapse at Bárdarbunga volcano, Iceland, regulated by lateral magma outflow. *Science* **2016**, *353*, 6296. [CrossRef]
26. Bakon, M.; Perissin, D.; Lazecky, M.; Papco, J. Infrastructure non-linear deformation monitoring via satellite radar interferometry. *Procedia Technol.* **2014**, *16*, 294–300. [CrossRef]
27. Tapete, D.; Morelli, S.; Fanti, R.; Casagli, N. Localising deformation along the elevation of linear structures: An experiment with space-borne InSAR and RTK GPS on the Roman Aqueducts in Rome, Italy. *Appl. Geogr.* **2015**, *58*, 65–83. [CrossRef]
28. Magnússon, E.; Björnsson, H.; Rott, H.; Roberts, M.J.; Pálsson, F.; Gudmundsson, S.; Bennett, R.A.; Geirsson, H.; Sturkell, E. Localized uplift of Vatnajökull, Iceland: Subglacial water accumulation deduced from InSAR and GPS observations. *J. Glaciol.* **2011**, *57*, 475–484. [CrossRef]
29. Gudmundsson, S.; Gudmundsson, M.T.; Björnsson, H.; Sigmundsson, F.; Rott, H.; Carstensen, J.M. Three-dimensional glacier surface motion maps at the Gjálp eruption site, Iceland, inferred from combining InSAR and other ice-displacement data. *Ann. Glaciol.* **2002**, *34*, 315–322. [CrossRef]
30. Ge, L. Integration of GPS and radar interferometry. *GPS Solut.* **2003**, *7*, 52–54. [CrossRef]
31. Mateus, P.; Nico, G.; Tomé, R.; Catalão, J.; Miranda, P.M. Experimental study on the atmospheric delay based on GPS, SAR interferometry, and numerical weather model data. *IEEE Trans. Geosci. Remote Sens.* **2012**, *51*, 6–11. [CrossRef]
32. Spilker, J., Jr. Tropospheric effects on GPS. *Glob. Posiotioning Syst. Theory Appl.* **1996**, *1*, 517–546.
33. Torres, R.; Snoeij, P.; Geudtner, D.; Bibby, D.; Davidson, M.; Attema, E.; Potin, P.; Rommen, B.; Floury, N.; Brown, M. GMES Sentinel-1 mission. *Remote Sens. Environ.* **2012**, *120*, 9–24. [CrossRef]
34. Gonzalez, F.R.; Parizzi, A.; Brcic, R. Evaluating the impact of geodetic corrections on interferometric deformation measurements. In Proceedings of the EUSAR 2018, 12th European Conference on Synthetic Aperture Radar, Aachen, Germany, 4–7 June 2018; pp. 1–5.
35. Shanker, A.P.; Zebker, H. Edgelist phase unwrapping algorithm for time series InSAR analysis. *JOSA A* **2010**, *27*, 605–612. [CrossRef]
36. Martín-Martín, A.; Orduna-Malea, E.; Thelwall, M.; López-Cózar, E.D. Google Scholar, Web of Science, and Scopus: A systematic comparison of citations in 252 subject categories. *J. Informetr.* **2018**, *12*, 1160–1177. [CrossRef]
37. Halevi, G.; Moed, H.; Bar-Ilan, J. Suitability of Google Scholar as a source of scientific information and as a source of data for scientific evaluation—Review of the literature. *J. Informetr.* **2017**, *11*, 823–834. [CrossRef]
38. Bonforte, A.; Ferretti, A.; Prati, C.; Puglisi, G.; Rocca, F. Calibration of atmospheric effects on SAR interferograms by GPS and local atmosphere models: First results. *J. Atmos. Sol. Terr. Phys.* **2001**, *63*, 1343–1357. [CrossRef]
39. Carlà, T.; Tofani, V.; Lombardi, L.; Raspini, F.; Bianchini, S.; Bertolo, D.; Thuegaz, P.; Casagli, N. Combination of GNSS, satellite InSAR, and GBInSAR remote sensing monitoring to improve the understanding of a large landslide in high alpine environment. *Geomorphology* **2019**, *335*, 62–75. [CrossRef]
40. D'Auria, L.; Pepe, S.; Castaldo, R.; Giudicepietro, F.; Macedonio, G.; Ricciolino, P.; Tizzani, P.; Casu, F.; Lanari, R.; Manzo, M. Magma injection beneath the urban area of Naples: A new mechanism for the 2012–2013 volcanic unrest at Campi Flegrei caldera. *Sci. Rep.* **2015**, *5*, 13100. [CrossRef]
41. Da Lio, C.; Teatini, P.; Strozzi, T.; Tosi, L. Understanding land subsidence in salt marshes of the Venice Lagoon from SAR Interferometry and ground-based investigations. *Remote Sens. Environ.* **2018**, *205*, 56–70. [CrossRef]

42. Stramondo, S.; Tesauro, M.; Briole, P.; Sansosti, E.; Salvi, S.; Lanari, R.; Anzidei, M.; Baldi, P.; Fornaro, G.; Avallone, A. The September 26, 1997 Colfiorito, Italy, earthquakes: Modeled coseismic surface displacement from SAR interferometry and GPS. *Geophys. Res. Lett.* **1999**, *26*, 883–886. [CrossRef]
43. Teatini, P.; Tosi, L.; Strozzi, T.; Carbognin, L.; Wegmüller, U.; Rizzetto, F. Mapping regional land displacements in the Venice coastland by an integrated monitoring system. *Remote Sens. Environ.* **2005**, *98*, 403–413. [CrossRef]
44. Bitelli, G.; Bonsignore, F.; Del Conte, S.; Novali, F.; Pellegrino, I.; Vittuari, L. Integrated use of Advanced InSAR and GPS data for subsidence monitoring. In *Engineering Geology for Society and Territory-Volume 5*; Springer: Berlin/Heidelberg, Germany, 2015; pp. 147–150.
45. Tosi, L.; Teatini, P.; Strozzi, T.; Carbognin, L.; Brancolini, G.; Rizzetto, F. Ground surface dynamics in the northern Adriatic coastland over the last two decades. In *Rendiconti Lincei*; Accademia Nazionale dei Lincei: Roma, Italy, 2010; Volume 21, pp. 115–129.
46. Volpe, M.; Atzori, S.; Piersanti, A.; Melini, D. The 2009 L'Aquila earthquake coseismic rupture: Open issues and new insights from 3D finite element inversion of GPS, InSAR and strong motion data. *Ann. Geophys.* **2015**, *58*, 1–17.
47. Bitelli, G.; Bonsignore, F.; Del Conte, S.; Novali, F.; Pellegrino, I.; Vittuari, L. Subsidence monitoring update for Emilia-Romagna region (Italy) by integrated use of InSAR and GNSS data. In Proceedings of the EGU General Assembly Conference Abstracts, Vienna, Austria, 27 April–2 May 2014; p. 15840.
48. Bitelli, G.; Bonsignore, F.; Pellegrino, I.; Vittuari, L. Evolution of the techniques for subsidence monitoring at regional scale: The case of Emilia-Romagna region (Italy). *Proc. IAHS* **2015**, *372*, 315–321. [CrossRef]
49. Cenni, N.; Loddo, F.; Zucca, F.; Meisina, C.; Baldi, P.; Belardinelli, M.; Bacchetti, M.; Mantovani, E.; Viti, M.; Casula, G. The spatio-temporal pattern of subsidence in the Po basin monitored by different techniques. *Algorithms* **2014**, *99*, 194–214.
50. Cheloni, D.; De Novellis, V.; Albano, M.; Antonioli, A.; Anzidei, M.; Atzori, S.; Avallone, A.; Bignami, C.; Bonano, M.; Calcaterra, S.J. Geodetic model of the 2016 Central Italy earthquake sequence inferred from InSAR and GPS data. *Geophys. Res. Lett.* **2017**, *44*, 6778–6787. [CrossRef]
51. Minet, C.; Goel, K.; Aquino, I.; Avino, R.; Berrino, G.; Caliro, S.; Chiodini, G.; De Martino, P.; Del Gaudio, C.; Ricco, C. Measuring non-linear deformation of the Campi Flegrei caldera (Naples, Italy) using a multi-method insar-geophysical approach. In Proceedings of the 2012 IEEE International Geoscience and Remote Sensing Symposium, Munich, Germany, 22–27 July 2012; pp. 1174–1177.
52. Radicioni, F.; Stoppini, A.; Brigante, R.; Fornaro, G.; Bovenga, F.; NITTI, D.O. Long-term GNSS and SAR data comparison for the deformation monitoring of the Assisi landslide. In Proceedings of the FIG Working Week, Rome, Italy, 6–10 May 2012.
53. Herrera, G.; Notti, D.; García-Davalillo, J.C.; Mora, O.; Cooksley, G.; Sánchez, M.; Arnaud, A.; Crosetto, M. Analysis with C-and X-band satellite SAR data of the Portalet landslide area. *Landslides* **2011**, *8*, 195–206. [CrossRef]
54. Fernandez, J.; Prieto, J.F.; Escayo, J.; Camacho, A.G.; Luzón, F.; Tiampo, K.F.; Palano, M.; Abajo, T.; Pérez, E.; Velasco, J. Modeling the two-and three-dimensional displacement field in Lorca, Spain, subsidence and the global implications. *Sci. Rep.* **2018**, *8*, 14782. [CrossRef] [PubMed]
55. Fernández, J.; Romero, R.; Carrasco, D.; Luzón, F.; Araña, V. InSAR volcano and seismic monitoring in Spain. Results for the period 1992–2000 and possible interpretations. *Opt. Lasers Eng.* **2002**, *37*, 285–297. [CrossRef]
56. De Michele, M.; Briole, P.; Raucoules, D.; Lemoine, A.; Rigo, A. Revisiting the shallow Mw 5.1 Lorca earthquake (southeastern Spain) using C-band InSAR and elastic dislocation modelling. *Remote Sens. Lett.* **2013**, *4*, 863–872. [CrossRef]
57. Prieto, J.F.; Gonzalez, P.; Seco, A.; Rodríguez-Velasco, G.; Tunini, L.; Perlock, P.A.; Arjona, A.; Aparicio, A.; Camacho, A.G.; Rundle, J. Geodetic and Structural Research in La Palma, Canary Islands, Spain: 1992–2007 Results. *Pure Appl. Geophys.* **2009**, *166*, 1461–1484. [CrossRef]
58. Cong, X.; Eineder, M.; Fritz, T. Atmospheric delay compensation in differential SAR Interferometry for volcanic deformation monitoring-Study case: El Hierro. In Proceedings of the 2012 IEEE International Geoscience and Remote Sensing Symposium, Munich, Germany, 22–27 July 2012; pp. 3887–3890.
59. Samsonov, S.; Tiampo, K.; González, P.J.; Prieto, J.; Camacho, A.G.; Fernández, J. Surface deformation studies of Tenerife Island, Spain from joint GPS-DInSAR observations. In Proceedings of the 2008 Second Workshop on Use of Remote Sensing Techniques for Monitoring Volcanoes and Seismogenic Areas, Naples, Italy, 11–14 November 2008; pp. 1–6.
60. Rodriguez-Lloveras, X.; Puig-Polo, C.; Lantada, N.; Gili, J.A.; Marturià, J. Two decades of GPS/GNSS and DInSAR monitoring of Cardona salt mines (NE of Spain)–natural and mining-induced mechanisms and processes. *Proc. IAHS* **2020**, *382*, 167–172. [CrossRef]
61. Alshawaf, F.; Fuhrmann, T.; Heck, B.; Hinz, S.; Knöpfler, A.; Luo, X.; Mayer, M.; Schenk, A.; Thiele, A.; Westerhaus, M. Integration of InSAR and GNSS observations for the determination of atmospheric water vapour. In *Earth Observation of Global Changes (EOGC)*; Springer: Berlin/Heidelberg, Germany, 2013; pp. 147–162.
62. Fuhrmann, T.; Knöpfler, A.; Mayer, M.; Schenk, A.; Westerhaus, M.; Zippelt, K.; Heck, B. An Inventory of Surface Movements in the Upper Rhine Graben Area, Southwest Germany, from SAR-Interferometry, GNSS and Precise Levelling. In *IAG 150 Years*; Springer: Berlin/Heidelberg, Germany, 2015; pp. 419–425.
63. Heimlich, C.; Masson, F.; Schmittbuhl, J. Geodetic analysis of surface deformation at the power plant of Landau (Germany) related to the 2013–2014 event. In Proceedings of the Proc. European Geothermal Congress, Strasbourg, France, 19–23 September 2016.

64. Kalia, A.; Frei, M.; Lege, T. A Copernicus downstream-service for the nationwide monitoring of surface displacements in Germany. *Remote Sens. Environ.* **2017**, *202*, 234–249. [CrossRef]
65. Heublein, M.; Zhu, X.X.; Alshawaf, F.; Mayer, M.; Bamler, R.; Hinz, S. Compressive sensing for neutrospheric water vapor tomography using GNSS and InSAR observations. In Proceedings of the 2015 IEEE International Geoscience and Remote Sens. Symposium (IGARSS), Milan, Italy, 26–31 July 2015; pp. 5268–5271.
66. Westerhaus, M.; Fuhrmann, T.; Mayer, M.; Zippelt, K.; Heck, B. Resolving the velocity and strain fields in the Upper Rhine Graben Area from a Combination of Levelling, GNSS and InSAR. In Proceedings of the EGU General Assembly Conference Abstracts, Vienna, Austria, 17–22 April 2016; EPSC2016-13011.
67. Oliveira, S.C.; Zêzere, J.L.; Catalão, J.; Nico, G. The contribution of PSInSAR interferometry to landslide hazard in weak rock-dominated areas. *Landslides* **2015**, *12*, 703–719. [CrossRef]
68. Marques, F.; Catalão, J.; Hildenbrand, A.; Madureira, P. Ground motion and tectonics in the Terceira Island: Tectonomagmatic interactions in an oceanic rift (Terceira Rift, Azores Triple Junction). *Tectonophysics* **2015**, *651*, 19–34. [CrossRef]
69. Fernandes, R.M.; Catalão, J.; Trota, A.N. The contribution of space-geodetic techniques to the understanding of the present-day geodynamics of the Azores triple junction. In *Volcanoes of the Azores*; Springer: Berlin/Heidelberg, Germany, 2018; pp. 57–69.
70. Henriques, M.J.; Lima, J.N.; Falcão, A.P.; Mancuso, M.; Heleno, S.; Falcao, A.P. Land Subsidence in Lisbon Area: Validation Of PsinSAR Results. In Proceedings of the Proc. Of FIG Working Week, Marrakech, Morocco, 18–22 May 2011.
71. Pedersen, R.; Jónsson, S.; Árnadóttir, T.; Sigmundsson, F.; Feigl, K.L. Fault slip distribution of two June 2000 Mw6. 5 earthquakes in South Iceland estimated from joint inversion of InSAR and GPS measurements. *Earth Planet. Sci. Lett.* **2003**, *213*, 487–502. [CrossRef]
72. Sigmundsson, F.; Hooper, A.; Hreinsdóttir, S.; Vogfjörd, K.; Ófeigsson, B.; Rafn Heimisson, E.; Dumont, S.; Parks, M.; Spaans, K.; Gudmundsson, G.; et al. Segmented lateral dyke growth in a rifting event at Bárðarbunga volcanic system, Iceland. *Nat. Geosci.* **2015**, *517*, 191–195. [CrossRef] [PubMed]
73. Parks, M.; Sigmundsson, F.; Sigurðsson, Ó.; Hooper, A.; Hreinsdóttir, S.; Ófeigsson, B.; Michalczewska, K. Deformation due to geothermal exploitation at Reykjanes, Iceland. *J. Volcanol. Geotherm. Res.* **2020**, *391*, 106438. [CrossRef]
74. Spaans, K.; Sigmundsson, F.; Hreinsdóttir, S.; Öfeigsson, B. High resolution surface deformation measurements in Iceland's Northern Volcanic Zone: Unraveling multiple deformation sources using InSAR and GPS. In Proceedings of the EGU General Assembly Conference Abstracts, Vienna, Austria, 22–27 April 2012; p. 10604.
75. Eriksen, H.Ø.; Lauknes, T.R.; Larsen, Y.; Corner, G.D.; Bergh, S.G.; Dehls, J.; Kierulf, H.P. Visualizing and interpreting surface displacement patterns on unstable slopes using multi-geometry satellite SAR interferometry (2D InSAR). *Remote Sens. Environ.* **2017**, *191*, 297–312. [CrossRef]
76. L'Heureux, J.; Long, M.; Vanneste, M.; Sauvin, G.; Hansen, L.; Polom, U.; Lecomte, I.; Dehls, J.; Janbu, N. On the prediction of settlement from high-resolution shear-wave reflection seismic data: The Trondheim harbour case study, mid Norway. *Eng. Geol.* **2013**, *167*, 72–83. [CrossRef]
77. Shamshiri, R.; Motagh, M.; Nahavandchi, H.; Haghighi, M.H.; Hoseini, M.J.R.S.o.E. Improving tropospheric corrections on large-scale Sentinel-1 interferograms using a machine learning approach for integration with GNSS-derived zenith total delay (ZTD). *Remote Sens. Environ.* **2020**, *239*, 111608. [CrossRef]
78. Böhme, M.; Bunkholt, H.; Oppikofer, T.; Dehls, J.; Hermanns, R.; Eriksen, H.; Lauknes, T.; Eiken, T. Using 2D InSAR, dGNSS and structural field data to understand the deformation mechanism of the unstable rock slope Gamanjunni 3, northern Norway. Landslides and Engineered Slopes. Experience, Theory and Practice. In Proceedings of the 12th International Symposium on Landslides, Napoli, Italy, 12–19 June 2016; p. 443.
79. Dehls, J.; Fischer, L.; Böhme, M.; Saintot, A.; Hermanns, R.; Oppikofer, T.; Lauknes, T.; Larsen, Y.; Blikra, L. *Landslide Monitoring in Western Norway Using High Resolution TerraSAR-X and Radarsat-2 InSAR*; CRC Press: Milton, UK, 2012; pp. 1321–1325.
80. Dehls, J.; Henderson, I.; Lauknes, T.; Larsen, Y. Regional landslide mapping and detailed site characterization using InSAR. In Proceedings of the "GeoEdmonton", Edmonton, AB, Canada, 21–24 September 2008; pp. 21–24.
81. Avallone, A.; Cirella, A.; Cheloni, D.; Tolomei, C.; Theodoulidis, N.; Piatanesi, A.; Briole, P.; Ganas, A. Near-source high-rate GPS, strong motion and InSAR observations to image the 2015 Lefkada (Greece) Earthquake rupture history. *Sci. Rep.* **2017**, *7*, 10358. [CrossRef]
82. Lagios, E.; Papadimitriou, P.; Novali, F.; Sakkas, V.; Fumagalli, A.; Vlachou, K.; Del Conte, S. Combined seismicity pattern analysis, DGPS and PSInSAR studies in the broader area of Cephalonia (Greece). *Tectonophysics* **2012**, *524*, 43–58. [CrossRef]
83. Papoutsis, I.; Papanikolaou, X.; Floyd, M.; Ji, K.; Kontoes, C.; Paradissis, D.; Zacharis, V. Mapping inflation at Santorini volcano, Greece, using GPS and InSAR. *Geophys. Res. Lett.* **2013**, *40*, 267–272. [CrossRef]
84. Elias, P.; Sykioti, O.; Drakatos, G.; Paronis, D.; Chousianitis, K.; Sabatakakis, N.; Anastasopoulos, V.; Briole, P. Landslides modelling and monitoring by exploiting satellite SAR acquisitions, optical imagery, GPS and in-situ measurements in Greece. Preliminary results. In Proceedings of the EGU General Assembly Conference Abstracts, Vienna, Austria, 27 April–2 May 2014; p. 4402.
85. Briole, P.; Avallone, A.; Agatza-Balodimou, E.; Billiris, H.; Charade, O.; Lyon-Caen, H.; Mitsakaki, C.; Papazissi, K.; Paradissis, D.; Veis, G. A ten years analysis of deformation in the Corinthian Gulf via GPS and SAR Interferometry. In Proceedings of the Wegener Meeting, Athens, Greece, 12–14 June 2002.

86. Sakkas, V.; Novali, F.; Lagios, E.; Ferretti, A.; Vassilopoulou, S.; Bellotti, F.; Allievi, J. Combined Squee-SAR TM and GPS ground deformation study of Nisyros-Yali volcanic field (Greece) for period 2002–2012. In Proceedings of the 2015 IEEE International Geoscience and Remote Sensing Symposium (IGARSS), Milan, Italy, 26–31 July 2015; pp. 4672–4675.

87. Borgia, A.; Lanari, R.; Sansosti, E.; Tesauro, M.; Berardino, P.; Fornaro, G.; Neri, M.; Murray, J. Actively growing anticlines beneath Catania from the distal motion of Mount Etna's decollement measured by SAR interferometry and GPS. *Geophys. Res. Lett.* **2000**, *27*, 3409–3412. [CrossRef]

88. Salvi, S.; Stramondo, S.; Cocco, M.; Tesauro, M.; Hunstad, I.; Anzidei, M.; Briole, P.; Baldi, P.; Sansosti, E.; Fornaro, G. Modeling coseismic displacements resulting from SAR interferometry and GPS measurements during the 1997 Umbria-Marche seismic sequence. *J. Seismol.* **2000**, *4*, 479–499. [CrossRef]

89. Colesanti, C.; Wasowski, J. Investigating landslides with space-borne Synthetic Aperture Radar (SAR) interferometry. *Eng. Geol.* **2006**, *88*, 173–199. [CrossRef]

90. Novellino, A.; Cigna, F.; Brahmi, M.; Sowter, A.; Bateson, L.; Marsh, S. Assessing the Feasibility of a National InSAR Ground Deformation Map of Great Britain with Sentinel-1. *Geosciences* **2017**, *7*, 19. [CrossRef]

91. Lanari, R.; Bonano, M.; Casu, F.; Luca, C.D.; Manunta, M.; Manzo, M.; Onorato, G.; Zinno, I. Automatic generation of sentinel-1 continental scale DInSAR deformation time series through an extended P-SBAS processing pipeline in a cloud computing environment. *Remote Sens.* **2020**, *12*, 2961. [CrossRef]

92. Fuhrmann, T.; Caro Cuenca, M.; Knöpfler, A.; Van Leijen, F.; Mayer, M.; Westerhaus, M.; Hanssen, R.; Heck, B. Estimation of small surface displacements in the Upper Rhine Graben area from a combined analysis of PS-InSAR, levelling and GNSS data. *Geophys. J. Int.* **2015**, *203*, 614–631. [CrossRef]

93. Fuhrmann, T.; Knöpfler, A.; Mayer, M.; Schenk, A.; Westerhaus, M.; Zippelt, K.; Heck, B. Towards a fusion of SAR-interferometry, GNSS and precise levelling in the Upper Rhine Graben Area, Southwest Germany. In Proceedings of the ESA Living Planet Symposium, Edinburgh, UK, 9–13 September 2013. SP-722.

94. Haghshenas Haghighi, M.; Motagh, M. Sentinel-1 InSAR over Germany: Large-scale interferometry, atmospheric effects, and ground deformation mapping. *Zeitschrift Geodäsie Geoinformation Landmanagement* **2017**, *2017*, 245–256.

95. Fiaschi, S.; Fabris, M.; Floris, M.; Achilli, V. Estimation of land subsidence in deltaic areas through differential SAR interferometry: The Po River Delta case study (Northeast Italy). *Int. J. Remote Sens.* **2018**, *39*, 8724–8745. [CrossRef]

96. Zerbini, S.; Richter, B.; Rocca, F.; van Dam, T.; Matonti, F. A combination of space and terrestrial geodetic techniques to monitor land subsidence: Case study, the Southeastern Po Plain, Italy. *J. Geophys. Res. Solid Earth* **2007**, *112*, B05401. [CrossRef]

97. Bock, Y.; Wdowinski, S.; Ferretti, A.; Novali, F.; Fumagalli, A. Recent subsidence of the Venice Lagoon from continuous GPS and interferometric synthetic aperture radar. *Geochem. Geophys. Geosyst.* **2012**, *13*. [CrossRef]

98. Teatini, P.; Tosi, L.; Strozzi, T.; Carbognin, L.; Cecconi, G.; Rosselli, R.; Libardo, S. Resolving land subsidence within the Venice Lagoon by persistent scatterer SAR interferometry. *Phys. Chem. Earth Parts A/B/C* **2012**, *40*, 72–79. [CrossRef]

99. Da Lio, C.; Tosi, L. Land subsidence in the Friuli Venezia Giulia coastal plain, Italy: 1992–2010 results from SAR-based interferometry. *Sci. Total Environ.* **2018**, *633*, 752–764. [CrossRef]

100. Del Soldato, M.; Farolfi, G.; Rosi, A.; Raspini, F.; Casagli, N. Subsidence Evolution of the Firenze–Prato–Pistoia Plain (Central Italy) Combining PSI and GNSS Data. *Remote Sens.* **2018**, *10*, 1146. [CrossRef]

101. Cianflone, G.; Tolomei, C.; Brunori, C.A.; Dominici, R. InSAR time series analysis of natural and anthropogenic coastal plain subsidence: The case of Sibari (Southern Italy). *Remote Sens.* **2015**, *7*, 16004–16023. [CrossRef]

102. Béjar-Pizarro, M.; Guardiola-Albert, C.; García-Cárdenas, R.P.; Herrera, G.; Barra, A.; López Molina, A.; Tessitore, S.; Staller, A.; Ortega-Becerril, J.A.; García-García, R.P. Interpolation of GPS and geological data using InSAR deformation maps: Method and application to land subsidence in the alto guadalentín aquifer (SE Spain). *Remote Sens.* **2016**, *8*, 965. [CrossRef]

103. Bonì, R.; Herrera, G.; Meisina, C.; Notti, D.; Béjar-Pizarro, M.; Zucca, F.; González, P.J.; Palano, M.; Tomás, R.; Fernández, J. Twenty-year advanced DInSAR analysis of severe land subsidence: The Alto Guadalentín Basin (Spain) case study. *Eng. Geol.* **2015**, *198*, 40–52. [CrossRef]

104. Heleno, S.I.; Oliveira, L.G.; Henriques, M.J.; Falcão, A.P.; Lima, J.N.; Cooksley, G.; Ferretti, A.; Fonseca, A.M.; Lobo-Ferreira, J.P.; Fonseca, J.F. Persistent scatterers interferometry detects and measures ground subsidence in Lisbon. *Remote Sens. Environ.* **2011**, *115*, 2152–2167. [CrossRef]

105. Graniczny, M.; Cyziene, J.; van Leijen, F.; Minkevicius, W.; Mikulenas, V.; Satkunas, J.; Przylucka, M.; Kowalski, Z.; Uscinowicz, S.; Jeglinski, W. Vertical ground movements in the Polish and Lithuanian Baltic coastal area as measured by satellite interferometry. *Baltica* **2015**, *28*. [CrossRef]

106. Krynski, J.; Zak, L.; Ziolkowski, D.; Cisak, J.; Lagiewska, M. Estimation of height changes of GNSS stations from the solutions of short vectors and PSI measurements. *Geod. Cartogr.* **2017**, *66*, 73–83. [CrossRef]

107. Armaş, I.; Gheorghe, M.; Lendvai, A.M.; Dumitru, P.D.; Bădescu, O.; Călin, A. InSAR validation based on GNSS measurements in Bucharest. *Int. J. Remote Sens.* **2016**, *37*, 5565–5580. [CrossRef]

108. Armaş, I.; Mendes, D.A.; Popa, R.-G.; Gheorghe, M.; Popovici, D. Long-term ground deformation patterns of Bucharest using multi-temporal InSAR and multivariate dynamic analyses: A possible transpressional system? *Sci. Rep.* **2017**, *7*, 43762. [CrossRef]

109. Czikhardt, R.; Papčo, J.; Bakoň, M. Feasibility of the Sentinel-1 Multi-temporal InSAR system based on the SNAP and StaMPS: Case study from the Tatra Mts., Slovakia. *Procedia Comput. Sci.* **2018**, *138*, 366–373. [CrossRef]

110. Tondaś, D.; Pawłuszek, K.; Ilieva, M.; Kapłon, J.; Rohm, W. Investigation for mining-induced deformation in Upper Silesia Coal Basin with multi-GNSS in Near Real-Time. In Proceedings of the 4th Joint International Symposium on Deformation Monitoring (JISDM), Athens, Greece, 15–17 May 2019.

111. Kadlečík, P.; Kajzar, V.; Nekvasilová, Z.; Wegmüller, U.; Doležalová, H. Evaluation of the subsidence based on dInSAR and GPS measurements near Karvina, Czech Republic. *AUC Geogr.* **2015**, *50*, 51–61. [CrossRef]

112. Alshammari, L.; Large, D.J.; Boyd, D.S.; Sowter, A.; Anderson, R.; Andersen, R.; Marsh, S. Long-term peatland condition assessment via surface motion monitoring using the ISBAS DInSAR technique over the Flow Country, Scotland. *Remote Sens.* **2018**, *10*, 1103. [CrossRef]

113. Mason, P.; Ghail, R.; Bischoff, C.; Skipper, J. Detecting and Monitoring Small-Scale Discrete Ground Movements across London, Using Persistent Scatterer InSAR (PSI). 2015. Available online: https://spiral.imperial.ac.uk/handle/10044/1/26693 (accessed on 31 March 2021).

114. Bardi, F.; Raspini, F.; Ciampalini, A.; Kristensen, L.; Rouyet, L.; Lauknes, T.R.; Frauenfelder, R.; Casagli, N. Space-borne and ground-based InSAR data integration: The Åknes test site. *Remote Sens.* **2016**, *8*, 237. [CrossRef]

115. Booth, A.M.; Dehls, J.; Eiken, T.; Fischer, L.; Hermanns, R.L.; Oppikofer, T. Integrating diverse geologic and geodetic observations to determine failure mechanisms and deformation rates across a large bedrock landslide complex: The Osmundneset landslide, Sogn og Fjordane, Norway. *Landslides* **2015**, *12*, 745–756. [CrossRef]

116. Eckerstorfer, M.; Eriksen, H.Ø.; Rouyet, L.; Christiansen, H.H.; Lauknes, T.R.; Blikra, L.H. Comparison of geomorphological field mapping and 2D-InSAR mapping of periglacial landscape activity at Nordnesfjellet, northern Norway. *Earth Surface Process. Landf.* **2018**, *43*, 2147–2156. [CrossRef]

117. Delaloye, R.; Strozzi, T.; Lambiel, C.; Perruchoud, E. Landslide-like development of rockglaciers detected with ERS-1/2 SAR interferometry. In Proceedings of the ESA FRINGE Symposium 2007, Frascati, Italy, 26–30 November 2007.

118. Kenner, R.; Phillips, M.; Beutel, J.; Limpach, P.; Papke, J.; Hasler, A.; Raetzo, H. Investigating the dynamics of a rock glacier using terrestrial laser scanning, time-lapse photography, in situ GPS measurements and satellite SAR inter-ferometry: Ritigraben rock glacier, Switzerland. In Proceedings of the International Conference on Permafrost, Potsdam, Germany, 20–24 June 2016; pp. 20–24.

119. Barboux, C.; Delaloye, R.; Strozzi, T.; Collet, C.; Raetzo, H. TSX InSAR assessment for slope instabilities monitoring in alpine periglacial environment (Western Swiss Alps, Switzerland). In Proceedings of the Proc. ESA FRINGE, Frascati, Italy, 19–23 September 2011; pp. 19–23.

120. Barboux, C.; Strozzi, T.; Delaloye, R.; Wegmüller, U.; Collet, C. Mapping slope movements in Alpine environments using TerraSAR-X interferometric methods. *ISPRS J. Photogramm. Remote Sens.* **2015**, *109*, 178–192. [CrossRef]

121. Strozzi, T.; Ambrosi, C.; Raetzo, H. Interpretation of aerial photographs and satellite SAR interferometry for the inventory of landslides. *Remote Sens.* **2013**, *5*, 2554–2570. [CrossRef]

122. Strozzi, T.; Delaloye, R.; Kääb, A.; Ambrosi, C.; Perruchoud, E.; Wegmüller, U. Combined observations of rock mass movements using satellite SAR interferometry, differential GPS, airborne digital photogrammetry, and airborne photography interpretation. *J. Geophys. Res. Earth Surf.* **2010**, *115*. [CrossRef]

123. Bayer, B.; Simoni, A.; Schmidt, D.; Bertello, L. Using advanced InSAR techniques to monitor landslide deformations induced by tunneling in the Northern Apennines, Italy. *Eng. Geol.* **2017**, *226*, 20–32. [CrossRef]

124. Lu, P.; Catani, F.; Tofani, V.; Casagli, N. Quantitative hazard and risk assessment for slow-moving landslides from Persistent Scatterer Interferometry. *Landslides* **2014**, *11*, 685–696. [CrossRef]

125. Crosetto, M.; Gili, J.; Monserrat, O.; Cuevas-González, M.; Corominas, J.; Serral, D. Interferometric SAR monitoring of the Vallcebre landslide (Spain) using corner reflectors. *Nat. Hazards Earth Syst. Sci.* **2013**, *13*, 923–933. [CrossRef]

126. García-Davalillo, J.C.; Herrera, G.; Notti, D.; Strozzi, T.; Álvarez-Fernández, I. DInSAR analysis of ALOS PALSAR images for the assessment of very slow landslides: The Tena Valley case study. *Landslides* **2014**, *11*, 225–246. [CrossRef]

127. Herrera, G.; Gutiérrez, F.; García-Davalillo, J.; Guerrero, J.; Notti, D.; Galve, J.; Fernández-Merodo, J.; Cooksley, G. Multi-sensor advanced DInSAR monitoring of very slow landslides: The Tena Valley case study (Central Spanish Pyrenees). *Remote Sens. Environ.* **2013**, *128*, 31–43. [CrossRef]

128. Komac, M.; Holley, R.; Mahapatra, P.; van der Marel, H.; Bavec, M. Coupling of GPS/GNSS and radar interferometric data for a 3D surface displacement monitoring of landslides. *Landslides* **2015**, *12*, 241–257. [CrossRef]

129. Mahapatra, P.; van der Marel, H.; Hanssen, R.; Holley, R.; Samiei-Esfahany, S.; Komac, M.; Fromberg, A. Radar transponders and their combination with GNSS for deformation monitoring. In Proceedings of the 2012 the IEEE International Geoscience and Remote Sensing Symposium, Munich, Germany, 22–27 July 2012; pp. 3891–3894.

130. Mahapatra, P.S.; Samiei-Esfahany, S.; van der Marel, H.; Hanssen, R.F. On the use of transponders as coherent radar targets for SAR interferometry. *IEEE Trans. Geosci. Remote Sens.* **2013**, *52*, 1869–1878. [CrossRef]

131. Mantovani, M.; Devoto, S.; Piacentini, D.; Prampolini, M.; Soldati, M.; Pasuto, A. Advanced SAR interferometric analysis to support geomorphological interpretation of slow-moving coastal landslides (Malta, Mediterranean Sea). *Remote Sens.* **2016**, *8*, 443. [CrossRef]

132. Piacentini, D.; Devoto, S.; Mantovani, M.; Pasuto, A.; Prampolini, M.; Soldati, M. Landslide susceptibility modeling assisted by Persistent Scatterers Interferometry (PSI): An example from the northwestern coast of Malta. *Nat. Hazards* **2015**, *78*, 681–697. [CrossRef]

133. Atanasova, M.; Nikolov, H. Ground displacements detection in Trifon Zarezan landslide based on GNSS and SAR data. *MMM Geo Inf.* **2018**, *11*, 7–15.
134. Themistocleous, K.; Danezis, C.; Gikas, V. Monitoring ground deformation of cultural heritage sites using SAR and geodetic techniques: The case study of Choirokoitia, Cyprus. *Appl. Geomat.* **2020**, 1–13. [CrossRef]
135. Raucoules, D.; De Michele, M.; Malet, J.-P.; Ulrich, P. Time-variable 3D ground displacements from high-resolution synthetic aperture radar (SAR). Application to La Valette landslide (South French Alps). *Remote Sens. Environ.* **2013**, *139*, 198–204. [CrossRef]
136. Bozsó, I.; Bányai, L.; Hooper, A.; Sz, E.; Wesztergom, V. Integration of Sentinel-1 Interferometry and GNSS Networks for Derivation of 3-D Surface Changes. *IEEE Geosci. Remote Sens. Lett.* **2021**, *18*, 692–696. [CrossRef]
137. Cheloni, D.; Giuliani, R.; D'Anastasio, E.; Atzori, S.; Walters, R.; Bonci, L.; D'Agostino, N.; Mattone, M.; Calcaterra, S.; Gambino, P. Coseismic and post-seismic slip of the 2009 L'Aquila (central Italy) Mw 6.3 earthquake and implications for seismic potential along the Campotosto fault from joint inversion of high-precision levelling, InSAR and GPS data. *Tectonophysics* **2014**, *622*, 168–185. [CrossRef]
138. Wang, Y.-Z.; Zhu, J.-J.; Ou, Z.-Q.; Li, Z.-W.; Xing, X.-M. Coseismic slip distribution of 2009 L'Aquila earthquake derived from InSAR and GPS data. *J. Cent. South Univ.* **2012**, *19*, 244–251. [CrossRef]
139. Nespoli, M.; Belardinelli, M.E.; Gualandi, A.; Serpelloni, E.; Bonafede, M. Poroelasticity and fluid flow modeling for the 2012 Emilia-Romagna earthquakes: Hints from GPS and InSAR data. *Geofluids* **2018**, 1–15. [CrossRef]
140. Cheloni, D. Geodetic model of the 2016 Central Italy earthquake sequence inferred from InSAR and GPS measurements. In Proceedings of the EGU General Assembly Conference Abstracts, Vienna, Austria, 23–28 April 2017; p. 9191.
141. Cheloni, D.; D'Agostino, N.; Scognamiglio, L.; Tinti, E.; Bignami, C.; Avallone, A.; Giuliani, R.; Calcaterra, S.; Gambino, P.; Mattone, M. Heterogeneous Behavior of the Campotosto Normal Fault (Central Italy) Imaged by InSAR GPS and Strong-Motion Data: Insights from the 18 January 2017 Events. *Remote Sens.* **2019**, *11*, 1482. [CrossRef]
142. Wang, L.; Gao, H.; Feng, G.; Xu, W. Source parameters and triggering links of the earthquake sequence in central Italy from 2009 to 2016 analyzed with GPS and InSAR data. *Tectonophysics* **2018**, *744*, 285–295. [CrossRef]
143. Walters, R.J.; Gregory, L.C.; Wedmore, L.N.; Craig, T.J.; McCaffrey, K.; Wilkinson, M.; Chen, J.; Li, Z.; Elliott, J.R.; Goodall, H. Dual control of fault intersections on stop-start rupture in the 2016 Central Italy seismic sequence. *Earth Planet. Sci. Lett.* **2018**, *500*, 1–14. [CrossRef]
144. Farolfi, G.; Bianchini, S.; Casagli, N. Integration of GNSS and satellite InSAR data: Derivation of fine-scale vertical surface motion maps of Po Plain, Northern Apennines, and Southern Alps, Italy. *IEEE Trans. Geosci. Remote Sens.* **2018**, *57*, 319–328. [CrossRef]
145. Vollrath, A.; Zucca, F.; Bekaert, D.; Bonforte, A.; Guglielmino, F.; Hooper, A.J.; Stramondo, S. Decomposing DInSAR time-series into 3-D in combination with GPS in the case of low strain rates: An application to the Hyblean Plateau, Sicily, Italy. *Remote Sens.* **2017**, *9*, 33. [CrossRef]
146. Sakkas, V.; Novali, F.; Vassilopoulou, S.; Damiata, B.N.; Fumagalli, A.; Lagios, E. Combined PSI And Differential GPS Study Of Zakynthos Island (W. Greece) For The Period 1992–2012. In Proceedings of the ESA Living Planet Symposium, Edinburgh, UK, 9–13 September 2013; p. 214.
147. Briole, P.; Elias, P.; Parcharidis, I.; Bignami, C.; Benekos, G.; Samsonov, S.; Kyriakopoulos, C.; Stramondo, S.; Chamot-Rooke, N.; Drakatou, M. The seismic sequence of January–February 2014 at Cephalonia Island (Greece): Constraints from SAR interferometry and GPS. *Geophys. Suppl. Mon. Not. R. Astron. Soc.* **2015**, *203*, 1528–1540. [CrossRef]
148. Decriem, J.; Árnadóttir, T.J.T. Transient crustal deformation in the South Iceland Seismic Zone observed by GPS and InSAR during 2000–2008. *Tectonophysics* **2012**, *581*, 6–18. [CrossRef]
149. Metzger, S.; Jónsson, S. Plate boundary deformation in North Iceland during 1992–2009 revealed by InSAR time-series analysis and GPS. *Tectonophysics* **2014**, *634*, 127–138. [CrossRef]
150. Juncu, D.; Árnadóttir, T.; Hooper, A.; Gunnarsson, G. Anthropogenic and natural ground deformation in the Hengill geothermal area, Iceland. *J. Geophys. Res. Solid Earth* **2017**, *122*, 692–709. [CrossRef]
151. Jonsson, S. Importance of post-seismic viscous relaxation in southern Iceland. *Nat. Geosci.* **2008**, *1*, 136–139. [CrossRef]
152. Szűcs, E.; Bozsó, I.; Kovács, I.J.; Bányai, L.; Gál, Á.; Szakács, A.; Wesztergom, V. Probing tectonic processes with space geodesy in the south Carpathians: Insights from archive SAR data. *Acta Geod. Geophys.* **2018**, *53*, 331–345. [CrossRef]
153. Gheorghe, M.; Armaș, I.; Năstase, E.-I.; Munteanu, A. *Potential of InSAR Monitoring for Seismic Areas in Romanian*; Center for Risk Studies, Spatial Modelling, Terrestrial and Coastal System Dynamics: Bucharest, Romania, 2018; pp. 23–31.
154. Zoran, M. Use of geospatial and in situ information for seismic hazard assessment in Vrancea area, Romania. In Proceedings of the 2008 Second Workshop on Use of Remote Sensing Techniques for Monitoring Volcanoes and Seismogenic Areas, Naples, Italy, 11–14 November 2008; pp. 1–5.
155. Ganas, A.; Elias, P.; Briole, P.; Cannavo, F.; Valkaniotis, S.; Tsironi, V.; Partheniou, E. Ground deformation and seismic fault model of the M6. 4 Durres (Albania) Nov. 26, 2019 earthquake, based on GNSS/INSAR observations. *Geosciences* **2020**, *10*, 210. [CrossRef]
156. Grassi, F.; Cenni, N.; Mancini, F. Combination of satellite SAR and GNSS data of co-seismic deformation after the November 26, 2019 Albania earthquake: First results. In Proceedings of the EGU General Assembly Conference Abstracts, Vienna, Austria, 4–8 May 2020; p. 4553.

157. Atanasova, M.; Nikolov, H. Detection of the Earth crust deformation in Provadia area using InSAR technique. In *XXVI International Symposium on Modern Technologies, Education and Professional Practice in Geodesy and Related Fields*; Sofia, Bulgary, 2016. Available online: https://www.researchgate.net/publication/334964442_XXIV_-th_INTERNATIONAL_SYMPOSIUM_MODERN_TECHNOLOGIES_EDUCATION_AND_PROFESSIONAL_PRACTICE_IN_GEODESY_AND_RELATED_FIELDS_Study_of_the_earth_crust_movements_on_the_territory_of_Bulgaria_with_GPS (accessed on 31 March 2021).
158. Lagios, E.; Sakkas, V.; Novali, F.; Bellotti, F.; Ferretti, A.; Vlachou, K.; Dietrich, V. SqueeSAR™ and GPS ground deformation monitoring of Santorini Volcano (1992–2012): Tectonic implications. *Tectonophysics* **2013**, *594*, 38–59. [CrossRef]
159. Lagios, E.; Sakkas, V.; Novali, F.; Ferreti, A.; Damiata, B.; Dietrich, V.J. Reviewing and updating (1996–2012) ground deformation in Nisyros Volcano (Greece) determined by GPS and SAR Interferometric Techniques (1996–2012). In *Nisyros Volcano*; Springer: Berlin/Heidelberg, Germany, 2018; pp. 285–301.
160. Currenti, G.; Napoli, R.; Del Negro, C. Toward a realistic deformation model of the 2008 magmatic intrusion at Etna from combined DInSAR and GPS observations. *Earth Planet. Sci. Lett.* **2011**, *312*, 22–27. [CrossRef]
161. Palano, M.; Puglisi, G.; Gresta, S. Ground deformation patterns at Mt. Etna from 1993 to 2000 from joint use of InSAR and GPS techniques. *J. Volcanol. Geotherm. Res.* **2008**, *169*, 99–120. [CrossRef]
162. Fernández, J.; González-Matesanz, F.; Prieto, J.; Rodríguez-Velasco, G.; Staller, A.; Alonso-Medina, A.; Charco, M. GPS monitoring in the NW part of the volcanic island of Tenerife, Canaries, Spain: Strategy and results. *Pure Appl. Geophys.* **2004**, *161*, 1359–1377. [CrossRef]
163. Fernández, J.; Tizzani, P.; Manzo, M.; Borgia, A.; González, P.; Martí, J.; Pepe, A.; Camacho, A.; Casu, F.; Berardino, P. Gravity-driven deformation of Tenerife measured by InSAR time series analysis. *Geophys. Res. Lett.* **2009**, *36*. [CrossRef]
164. Alshawaf, F.; Fersch, B.; Hinz, S.; Kunstmann, H.; Mayer, M.; Meyer, F. Water vapor mapping by fusing InSAR and GNSS remote sensing data and atmospheric simulations. *Hydrology Earth Syst. Sci.* **2015**, *19*, 4747–4764. [CrossRef]
165. Alshawaf, F.; Fuhrmann, T.; Heck, B.; Hinz, S.; Knoepfler, A.; Luo, X.; Mayer, M.; Schenk, A.; Thiele, A.; Westerhaus, M. Atmospheric water vapour determination by the integration of INSAR and GNSS observations. In Proceedings of the Fringe 2011 Workshop, Frascati, Italy, 9–23 September 2011. ESA SP-697.
166. Alshawaf, F.; Hinz, S.; Mayer, M.; Meyer, F.J. Constructing accurate maps of atmospheric water vapor by combining interferometric synthetic aperture radar and GNSS observations. *J. Geophys. Res. Atmos.* **2015**, *120*, 1391–1403. [CrossRef]
167. Benevides, P.; Nico, G.; Catalao, J.; Miranda, P. Merging SAR interferometry and GPS tomography for high-resolution mapping of 3D tropospheric water vapour. In Proceedings of the 2015 IEEE International Geoscience and Remote Sensing Symposium (IGARSS), Milan, Italy, 26–31 July 2015; pp. 3607–3610.
168. Benevides, P.; Nico, G.; Catalão, J.; Miranda, P.M. Bridging InSAR and GPS tomography: A new differential geometrical constraint. *IEEE Trans. Geosci. Remote Sens.* **2015**, *54*, 697–702. [CrossRef]
169. Mateus, P.; Catalão, J.; Nico, G. Sentinel-1 interferometric SAR mapping of precipitable water vapor over a country-spanning area. *IEEE Trans. Geosci. Remote Sens.* **2017**, *55*, 2993–2999. [CrossRef]
170. Cheng, S.; Perissin, D.; Chen, F.; Lin, H. Atmospheric delay analysis from GPS and InSAR. In Proceedings of the 2011 IEEE International Geoscience and Remote Sensing Symposium, Vancouver, BC, Canada, 24–29 July 2011; pp. 1650–1653.
171. Chen, G.; Hay, G.J.; Carvalho, L.M.; Wulder, M.A. Object-based change detection. *Int. J. Remote Sens.* **2012**, *33*, 4434–4457. [CrossRef]
172. Heublein, M.; Alshawaf, F.; Erdnüß, B.; Zhu, X.X.; Hinz, S. Compressive sensing reconstruction of 3D wet refractivity based on GNSS and InSAR observations. *J. Geod.* **2019**, *93*, 197–217. [CrossRef]
173. Milczarek, W.; Kopeć, A.; Głąbicki, D. Estimation of tropospheric and ionospheric delay in DInSAR calculations: Case study of areas showing (natural and induced) seismic activity. *Remote Sens.* **2019**, *11*, 621. [CrossRef]
174. Roque, D.; Simonetto, E.; Falcão, A.; Perissin, D.; Durand, F.; Morel, L.; Fonseca, A.; Polidori, L. An analysis of displacement measurements for Lisbon, Portugal, using combined InSAR and GNSS data. *Orbit* **2008**, *34209*, 28.
175. Wilgan, K.; Geiger, A. High-resolution models of tropospheric delays and refractivity based on GNSS and numerical weather prediction data for alpine regions in Switzerland. *J. Geod.* **2019**, *93*, 819–835. [CrossRef]
176. Cuenca, M.C.; Hanssen, R.; Hooper, A.; Arikan, M. Surface deformation of the whole Netherlands after PSI analysis. In Proceedings of the Fringe 2011 Workshop, Frascati, Italy, 9–23 September 2011; pp. 19–23.
177. Gee, D.; Sowter, A.; Grebby, S.; de Lange, G.; Athab, A.; Marsh, S. National geohazards mapping in Europe: Interferometric analysis of the Netherlands. *Eng. Geol.* **2019**, *256*, 1–22. [CrossRef]
178. Hanssen, R.; Caro Cuenca, M.; Klees, R.; Van der Marel, H. Decadal vertical deformation of the Netherlands via the geodetic integration of gravimetry, GNSS, leveling and SAR interferometry. In Proceedings of the AGU Fall Meeting Abstracts, San Francisco, CA, USA, 3–7 December 2012.
179. Parizzi, A.; Rodriguez Gonzalez, F.; Brcic, R. A covariance-based approach to merging InSAR and GNSS displacement rate measurements. *Remote Sens.* **2020**, *12*, 300. [CrossRef]
180. Hanssen, R.F.; Van Leijen, F.J. Monitoring deformation of water defense structures using satellite radar interferometry. In Proceedings of the 13th FIG Symposium on Deformation Measurement and Analysis, Lisbon, Portugal, 12–15 May 2008.
181. Hanssen, R.F.; van Leijen, F.J. Monitoring water defense structures using radar interferometry. In Proceedings of the 2008 the IEEE Radar Conference, Rome, Italy, 26–30 May 2008; pp. 1–4.

182. Bonì, R.; Bosino, A.; Meisina, C.; Novellino, A.; Bateson, L.; McCormack, H. A methodology to detect and characterize uplift phenomena in urban areas using Sentinel-1 data. *Remote Sens.* **2018**, *10*, 607. [CrossRef]

183. Gheorghe, M.; Armaș, I.; Dumitru, P.; Călin, A.; Bădescu, O.; Necsoiu, M. Monitoring subway construction using Sentinel-1 data: A case study in Bucharest, Romania. *Int. J. Remote Sens.* **2020**, *41*, 2644–2663. [CrossRef]

184. Bignami, C.; Stramondo, S. Ground deformation observed at Kozloduy (Bulgaria) and Akkuyu (Turkey) NPPS by means of multitemporal SAR inteferometry. Πανελλήνια και Διεθνή Γεωγραφικά Συνέδρια, Συλλογή Πρακτικών **2015**, 1337–1355.

185. Ponton, F.; Walpersdorf, A.; Gay, M.; Trouvé, E.; Mugnier, J.-L.; Fallourd, R.; Cotte, N.; Ott, L.; Serafini, J. GPS and TerraSAR-X time series measure temperate glacier flow in the Mont Blanc massif (France): The Argentière glacier test site. In Proceedings of the EGU General Assembly Conference Abstracts, Vienna, Austria, 2–27 April 2012; p. 9525.

186. Stockamp, J.; Li, Z.; Bishop, P.; Hansom, J.; Rennie, A.; Petrie, E.; Tanaka, A.; Bingley, R.; Hansen, D.; Ouwehand, L. Investigating glacial isostatic adjustment in Scotland with InSAR and GPS observations. In Proceedings of the FRINGE, Frascati, Italy, 23–27 March 2015; pp. 23–27.

187. Poitevin, C.; Wöppelmann, G.; Raucoules, D.; Le Cozannet, G.; Marcos, M.; Testut, L. Vertical land motion and relative sea level changes along the coastline of Brest (France) from combined space-borne geodetic methods. *Remote Sens. Environ.* **2019**, *222*, 275–285. [CrossRef]

188. Riedel, A.; Riedel, B.; Tengen, D.; Gerke, M. Investigations on vertical land movements along the North Sea and Baltic Sea coast in Germany with PS Interferometry. *Int. Arch. Photogramm. Remote Sens. Spat. Inf. Sci.* **2019**, *42*. [CrossRef]

189. Zerbini, S.; Raicich, F.; Prati, C.M.; Bruni, S.; Del Conte, S.; Errico, M.; Santi, E. Sea-level change in the Northern Mediterranean Sea from long-period tide gauge time series. *Earth Sci. Rev.* **2017**, *167*, 72–87. [CrossRef]

190. Wilson, M.; Atkinson, P. The use of elevation data in flood inundation modelling: A comparison of ERS interferometric SAR and combined contour and differential GPS data. *Int. J. River Basin Manag.* **2005**, *3*, 3–20. [CrossRef]

191. Ilieva, M.; Rudziński, Ł.; Pawłuszek-Filipiak, K.; Lizurek, G.; Kudłacik, I.; Tondaś, D.; Olszewska, D. Combined Study of a Significant Mine Collapse Based on Seismological and Geodetic Data—29 January 2019, Rudna Mine, Poland. *Remote Sens.* **2020**, *12*, 1570. [CrossRef]

192. Van Der Wal, D.; Herman, P.M.; Wielemaker-van Den Dool, A. Characterisation of surface roughness and sediment texture of intertidal flats using ERS SAR imagery. *Remote Sens. Environ.* **2005**, *98*, 96–109. [CrossRef]

193. Mahapatra, P.; van der Marel, H.; van Leijen, F.; Samiei-Esfahany, S.; Klees, R.; Hanssen, R. InSAR datum connection using GNSS-augmented radar transponders. *J. Geod.* **2018**, *92*, 21–32. [CrossRef]

194. Ferretti, A.; Prati, C.; Rocca, F. Permanent scatterers in SAR interferometry. *IEEE Trans. Geosci. Remote Sens.* **2001**, *39*, 8–20. [CrossRef]

195. Arnaud, A.; Adam, N.; Hanssen, R.; Inglada, J.; Duro, J.; Closa, J.; Eineder, M. ASAR ERS interferometric phase continuity. In Proceedings of the IGARSS 2003. 2003 IEEE International Geoscience and Remote Sensing Symposium. Proceedings (IEEE Cat. No. 03CH37477), Toulouse, France, 21–25 July 2003; pp. 1133–1135.

196. Ferretti, A.; Fumagalli, A.; Novali, F.; Prati, C.; Rocca, F.; Rucci, A. A new algorithm for processing interferometric data-stacks: SqueeSAR. *IEEE Trans. Geosci. Remote Sens.* **2011**, *49*, 3460–3470. [CrossRef]

197. Berardino, P.; Fornaro, G.; Lanari, R.; Sansosti, E. A new algorithm for surface deformation monitoring based on small baseline differential SAR interferograms. *Geosci. Remote Sens. IEEE Trans.* **2002**, *40*, 2375–2383. [CrossRef]

198. Sowter, A.; Bateson, L.; Strange, P.; Ambrose, K.; Syafiudin, M.F. DInSAR estimation of land motion using intermittent coherence with application to the South Derbyshire and Leicestershire coalfields. *Remote Sens. Lett.* **2013**, *4*, 979–987. [CrossRef]

199. Costantini, M.; Falco, S.; Malvarosa, F.; Minati, F. A new method for identification and analysis of persistent scatterers in series of SAR images. In Proceedings of the Geoscience and Remote Sensing Symposium, 2008, IGARSS 2008, IEEE International, Boston, MA, USA, 7–11 July 2008; pp. II-449–II-452.

200. Blanco-Sanchez, P.; Mallorquí, J.J.; Duque, S.; Monells, D. The coherent pixels technique (CPT): An advanced DInSAR technique for nonlinear deformation monitoring. *Pure Appl. Geophys.* **2008**, *165*, 1167–1193. [CrossRef]

201. Bovenga, F.; Nutricato, R.; Guerriero, A.R.L.; Chiaradia, M. SPINUA: A flexible processing chain for ERS/ENVISAT long term interferometry. In Proceedings of the Envisat & ERS Symposium, Salzburg, Austria, 6–10 September 2004. ESA SP-572.

202. Catani, F.; Casagli, N.; Ermini, L.; Righini, G.; Menduni, G. Landslide hazard and risk mapping at catchment scale in the Arno River basin. *Landslides* **2005**, *2*, 329–342. [CrossRef]

203. Vilardo, G.; Isaia, R.; Ventura, G.; De Martino, P.; Terranova, C. InSAR Permanent Scatterer analysis reveals fault re-activation during inflation and deflation episodes at Campi Flegrei caldera. *Remote Sens. Environ.* **2010**, *114*, 2373–2383. [CrossRef]

204. Vilardo, G.; Ventura, G.; Terranova, C.; Matano, F.; Nardò, S. Ground deformation due to tectonic, hydrothermal, gravity, hydrogeological, and anthropic processes in the Campania Region (Southern Italy) from Permanent Scatterers Synthetic Aperture Radar Interferometry. *Remote Sens. Environ.* **2009**, *113*, 197–212. [CrossRef]

205. Solari, L.; Del Soldato, M.; Bianchini, S.; Ciampalini, A.; Ezquerro, P.; Montalti, R.; Raspini, F.; Moretti, S. From ERS 1/2 to Sentinel-1: Subsidence monitoring in Italy in the last two decades. *Front. Earth Sci.* **2018**, *6*, 149. [CrossRef]

206. Solari, L.; Del Soldato, M.; Raspini, F.; Barra, A.; Bianchini, S.; Confuorto, P.; Casagli, N.; Crosetto, M. Review of satellite interferometry for landslide detection in Italy. *Remote Sens.* **2020**, *12*, 1351. [CrossRef]

207. Herrera, G.; Mateos, R.M.; García-Davalillo, J.C.; Grandjean, G.; Poyiadji, E.; Maftei, R.; Filipciuc, T.-C.; Auflič, M.J.; Jež, J.; Podolszki, L. Landslide databases in the Geological Surveys of Europe. *Landslides* **2018**, *15*, 359–379. [CrossRef]

208. Zijerveld, L.; Stephenson, R.; Cloetingh, S.; Duin, E.; Van den Berg, M. Subsidence analysis and modelling of the Roer Valley Graben (SE Netherlands). *Tectonophysics* **1992**, *208*, 159–171. [CrossRef]
209. Costa, A.L.; Kok, S.; Korff, M. Systematic assessment of damage to buildings due to groundwater lowering-induced subsidence: Methodology for large scale application in the Netherlands. *Proc. IAHS* **2020**, *382*, 577–582. [CrossRef]
210. Daalen, T.M.v.; Fokker, P.A.; Bogaard, P.J.; Meulen, M. Why we urgently need a public subsidence information service in the Netherlands. *Proc. IAHS* **2020**, *382*, 821–823. [CrossRef]
211. L'Heureux. A study of the retrogressive behaviour and mobility of Norwegian quick clay landslides. The Landslides Engineered Slopes: Protecting Society through Improved Understanding. In Proceedings of the 11th International & 2nd North American Symposium on Landslides, Banff, AB, Canada, 2–8 June 2012; Taylor, Francis Group: London, UK, 2012; pp. 981–988.
212. Subcommittee, P. *Glossary of Permafrost and Related Ground-Ice Terms*; Associate Committee on Geotechnical Research, National Research Council of Canada: Ottawa, ON, Canada, 1988; p. 156.
213. Sollid, J.; Sørbel, L. Palsa bogs at Haugtjørnin, Dovrefjell, South Norway. *Norsk Geografisk Tidsskrift Nor. J. Geogr.* **1974**, *28*, 53–60. [CrossRef]
214. Lilleøren, K.S.; Etzelmüller, B. A regional inventory of rock glaciers and ice-cored moraines in Norway. *Geogr. Ann. Ser. A Phys. Geogr.* **2011**, *93*, 175–191. [CrossRef]
215. Shakesby, R.A.; Dawson, A.G.; Matthews, J.A. Rock glaciers, protalus ramparts and related phenomena, Rondane, Norway: A continuum of large-scale talus-derived landforms. *Boreas* **1987**, *16*, 305–317. [CrossRef]
216. Østrem, G. Ice-cored moraines in Scandinavia. *Geogr. Ann. Ser. A Phys. Geogr.* **1964**, *46*, 282–337.
217. Svensson, H. Ice-wedge casts and relict polygonal patterns in Scandinavia. *J. Quat. Sci.* **1988**, *3*, 57–67. [CrossRef]
218. Costantini, M.; Ferretti, A.; Minati, F.; Falco, S.; Trillo, F.; Colombo, D.; Novali, F.; Malvarosa, F.; Mammone, C.; Vecchioli, F. Analysis of surface deformations over the whole Italian territory by interferometric processing of ERS, Envisat and COSMO-SkyMed radar data. *Remote Sens. Environ.* **2017**, *202*, 250–275. [CrossRef]
219. Di Martire, D.; Paci, M.; Confuorto, P.; Costabile, S.; Guastaferro, F.; Verta, A.; Calcaterra, D. A nation-wide system for landslide mapping and risk management in Italy: The second Not-ordinary Plan of Environmental Remote Sensing. *Int. J. Appl. Earth Obs. Geoinf.* **2017**, *63*, 143–157. [CrossRef]
220. Dehls, J.F.; Larsen, Y.; Marinkovic, P.; Lauknes, T.R.; Stødle, D.; Moldestad, D.A. INSAR. No: A National Insar Deformation Mapping/Monitoring Service In Norway–From Concept To Operations. In Proceedings of the IGARSS 2019–2019 IEEE International Geoscience and Remote Sensing Symposium, Yokohama, Japan, 28 July–2 August 2019; pp. 5461–5464.
221. Goel, K.; Adam, N.; Shau, R.; Rodriguez-Gonzalez, F. Improving the reference network in wide-area Persistent Scatterer Interferometry for non-urban areas. In Proceedings of the 2016 IEEE International Geoscience and Remote Sensing Symposium (IGARSS), Beijing, China, 10–15 July 2016; pp. 1448–1451.
222. Raspini, F.; Bianchini, S.; Ciampalini, A.; Del Soldato, M.; Solari, L.; Novali, F.; Del Conte, S.; Rucci, A.; Ferretti, A.; Casagli, N. Continuous, semi-automatic monitoring of ground deformation using Sentinel-1 satellites. *Sci. Rep.* **2018**, *8*, 1–11. [CrossRef]
223. Del Soldato, M.; Solari, L.; Raspini, F.; Bianchini, S.; Ciampalini, A.; Montalti, R.; Ferretti, A.; Pellegrineschi, V.; Casagli, N. Monitoring ground instabilities using SAR satellite data: A practical approach. *ISPRS Int. J. Geo-Inf.* **2019**, *8*, 307. [CrossRef]
224. Montalti, R.; Solari, L.; Bianchini, S.; Del Soldato, M.; Raspini, F.; Casagli, N. A Sentinel-1-based clustering analysis for geo-hazards mitigation at regional scale: A case study in Central Italy. *Geomat. Nat. Hazards Risk* **2019**, *10*, 2257–2275. [CrossRef]
225. Solari, L.; Del Soldato, M.; Montalti, R.; Bianchini, S.; Raspini, F.; Thuegaz, P.; Bertolo, D.; Tofani, V.; Casagli, N. A Sentinel-1 based hot-spot analysis: Landslide mapping in north-western Italy. *Int. J. Remote Sens.* **2019**, *40*, 7898–7921. [CrossRef]
226. Crosetto, M.; Solari, L.; Mróz, M.; Balasis-Levinsen, J.; Casagli, N.; Frei, M.; Oyen, A.; Moldestad, D.A.; Bateson, L.; Guerrieri, L. The evolution of wide-area dinsar: From regional and national services to the European ground motion service. *Remote Sens.* **2020**, *12*, 2043. [CrossRef]
227. Bruyninx, C.; Habrich, H.; Söhne, W.; Kenyeres, A.; Stangl, G.; Völksen, C. Enhancement of the EUREF permanent network services and products. In *Geodesy for Planet Earth*; Springer: Berlin/Heidelberg, Germany, 2012; pp. 27–34.
228. Intrieri, E.; Raspini, F.; Fumagalli, A.; Lu, P.; Del Conte, S.; Farina, P.; Allievi, J.; Ferretti, A.; Casagli, N. The Maoxian landslide as seen from space: Detecting precursors of failure with Sentinel-1 data. *Landslides* **2018**, *15*, 123–133. [CrossRef]
229. Rosi, A.; Tofani, V.; Agostini, A.; Tanteri, L.; Stefanelli, C.T.; Catani, F.; Casagli, N. Subsidence mapping at regional scale using persistent scatters interferometry (PSI): The case of Tuscany region (Italy). *Int. J. Appl. Earth Obs. Geoinf.* **2016**, *52*, 328–337. [CrossRef]
230. Raspini, F.; Bianchini, S.; Ciampalini, A.; Del Soldato, M.; Montalti, R.; Solari, L.; Tofani, V.; Casagli, N. Persistent Scatterers continuous streaming for landslide monitoring and mapping: The case of the Tuscany region (Italy). *Landslides* **2019**, *16*, 2033–2044. [CrossRef]

*Article*

# 3D Displacement Field of Wenchuan Earthquake Based on Iterative Least Squares for Virtual Observation and GPS/InSAR Observations

**Luyun Xiong [1], Caijun Xu [1,2,3,*], Yang Liu [1,2,3], Yangmao Wen [1,2,] and Jin Fang [1]**

[1]   School of Geodesy and Geomatics, Wuhan University, Wuhan 430079, China; lyxiong@whu.edu.cn (L.X.); yang.liu@sgg.whu.edu.cn (Y.L.); ymwen@sgg.whu.edu.cn (Y.W.); jfang@whu.edu.cn (J.F.)
[2]   Key Laboratory of Geospace Environment and Geodesy, Ministry of Education, Wuhan University, Wuhan 430079, China
[3]   Key Laboratory of Geophysical Geodesy, Ministry of Natural Resources, Wuhan 430079, China
*   Correspondence: cjxu@sgg.whu.edu.cn; Tel.: +86-27-6877-8805

Received: 5 January 2020; Accepted: 16 March 2020; Published: 18 March 2020

**Abstract:** The acquisition of a 3D displacement field can help to understand the crustal deformation pattern of seismogenic faults and deepen the understanding of the earthquake nucleation. The data for 3D displacement field extraction are usually from GPS/interferometric synthetic aperture radar (InSAR) observations, and the direct solution method is usually adopted. We proposed an iterative least squares for virtual observation (VOILS) based on the maximum a posteriori estimation criterion of Bayesian theorem to correct the errors caused by the GPS displacement interpolation process. Firstly, in the simulation examples, both uniform and non-uniform sampling schemes for GPS observation were used to extract 3D displacement. On the basis of the experimental results of the reverse fault, the normal fault with a strike-slip component, and the strike-slip fault with a reverse component, we found that the VOILS method is better than the direct solution method in both horizontal and vertical directions. When a uniform sampling scheme was adopted, the percentages of improvement for the reverse fault ranged from 3% to 9% and up to 70%, for the normal fault with a strike-slip component ranging from 4% to 8% and up to 68%, and for the strike-slip fault with a reverse component ranging from 1% to 8% and up to 22%. After this, the VOILS method was applied to extract the 3D displacement field of the 2008 Mw 7.9 Wenchuan earthquake. In the East–West (E) direction, the maximum displacement of the hanging wall was 1.69 m and 2.15 m in the footwall. As for the North–South (N) direction, the maximum displacement of the hanging wall was 0.82 m for the southwestern, 0.95 m for the northeastern, while that of the footwall was 0.77 m. In the vertical (U) direction, the maximum uplift was 1.19 m and 0.95 m for the subsidence, which was significantly different from the direct solution method. Finally, the derived vertical displacements were also compared with the ruptures from field investigations, indicating that the VOILS method can reduce the impact of the interpolated errors on parameter estimations to some extent. The simulation experiments and the case study of the 3D displacement field for the 2008 Wenchuan earthquake suggest that the VOILS method proposed in this study is feasible and effective, and the degree of improvement in the vertical direction is particularly significant.

**Keywords:** Wenchuan earthquake; 3D displacement field; iterative least squares for virtual observation; GPS; InSAR

## 1. Introduction

Interferometric synthetic aperture radar (InSAR) technology has become one of the methods for monitoring surface deformation due to its advantages of large spatial coverage, day-and-night observations, and short revisit interval. Since the co-seismic displacement field of the 1992 Landers earthquake was obtained by European Remote Sensing-1 (ERS-1) satellite SAR images [1], this technique has been applied to measure the surface deformation related to the occurrence of earthquakes [2–8], volcanic activity [9–12], and natural and/or anthropic land subsidence [13–15]. However, the 3D displacement field of the deformation area cannot be obtained from the single geometry InSAR line of sight (LOS) displacements, which can be used to reflect the movement change of the surface. Moreover, the global positioning system (GPS) technique is widely used in the inversion problems in many fields, such as co- and post-seismic earthquakes [16–22], intrusion, and inflation/deflation at active volcanoes [23,24]. GPS observations can provide 3D surface deformation, but the GPS stations are relatively sparse and have low spatial resolution, which is insufficient to show detailed surface movement. Consequently, many researchers fused these two data sets to infer the 3D surface deformation because InSAR and GPS data can complement each other in spatial and temporal resolutions [25–30].

There are two main categories of methods to infer the 3D displacement field from combing GPS and InSAR data. The first is the traditional direct solution method, used to interpolate the GPS data into the same spatial resolution of InSAR data that have been downsampled, and use the least squares estimation to solve the equation. It has been verified that the optimal inversion result of the target energy function can be obtained by least squares without designing the global optimal algorithm [31]. A second method, such as the simultaneous and integrated strain tensor estimation from geodetic and satellite deformation measurements (SISTEM) method [32], does not require the interpolation of the GPS data, and simultaneously provides solutions of the strain tensor, the displacement field, and the rigid body rotation tensor. Moreover, such a method is based on the elastic dislocation model [33] and takes into account some prior constraints of the smooth change between adjacent points [34].

Since the interpolated displacement field coming from the traditional direct solution method is generally affected by errors, to overcome such a problem, we propose here an iterative least squares for virtual observation based on the maximum a posteriori estimation criterion of Bayesian theorem [35]. Firstly, the simulation examples are given to verify the validity of the method, and then it is applied to estimate the 3D displacement field for the 2008 Wenchuan earthquake. The main objectives of this method are: (1) to aim at the insensitivity of the North–South deformation, the interpolated displacement field with GPS data is used as a priori initial 3D displacement field to make up for the defect of extracting 3D deformation from single geometry InSAR one-dimensional (1D) LOS observations; (2) to use InSAR 1D LOS observations to correct errors caused by the GPS 3D deformation interpolation process; (3) to realize the reasonable fusion of GPS/InSAR observations in 3D deformation extraction based on the iteration method.

Some previous studies have derived the 3D displacement field for the 2008 Wenchuan earthquake [34,36–38]. Their basic processes are to use GPS data to correct the InSAR observations before interpolating them, and to solve 3D displacement field jointly with InSAR data. There is almost no related study to construct the 3D displacement field based on a similar iterative process currently. Thus, the proposed method was applied to retrieve the 3D displacement field of the 2008 Wenchuan earthquake and was verified accordingly.

## 2. Iterative Least Squares for Virtual Observation

### 2.1. Mathematical Background

The LOS deformation obtained by InSAR technology does not represent the actual 3D surface deformation, but the projection of the East–West (E), North–South (N), and vertical (U) deformation in the LOS direction [39]. As shown in Figure 1, H is the projection of satellite flight direction on the ground, $\phi$ is the azimuth of satellite flight direction (positive clockwise from the North), and $\theta$ is the radar incidence angle at the reflection point. $U_e$, $U_n$, and $U_u$ are the displacements of three directions of E, N, and U respectively. The relationship between the observed values $L_{InSAR}$ and the surface deformation in these three directions can be expressed as [40]

$$L_{InSAR} = \begin{bmatrix} -\sin\theta\sin(\phi - 3\pi/2) & -\sin\theta\cos(\phi - 3\pi/2) & \cos\theta \end{bmatrix}\begin{bmatrix} U_e & U_n & U_u \end{bmatrix}^T \qquad (1)$$

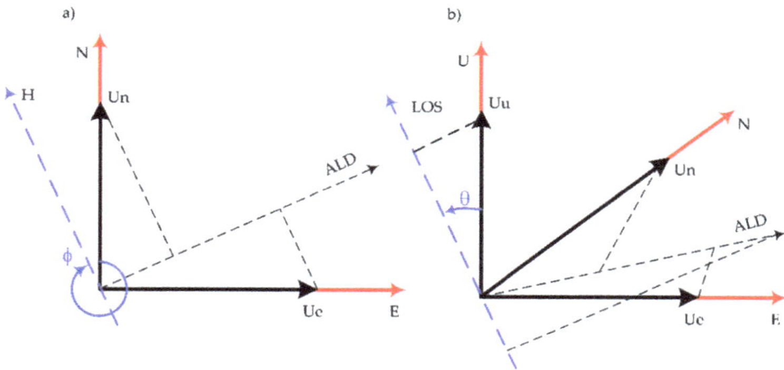


**Figure 1.** Sketch map of the relationship between interferometric synthetic aperture radar (InSAR) observation and surface displacements [41]. (**a**) Top-view showing the North–South (N) and East–West (E) components projection on the azimuth look direction (ALD). (**b**) 3D sketch of surface deformation components and InSAR observations.

Equation (1) is the theoretical equation, when $\mathbf{L}_{InSAR}$ are the observations, and their measurement errors are included at the same time. Let $S_e = -\sin\theta\sin(\phi - 3\pi/2)$, $S_n = -\sin\theta\cos(\phi - 3\pi/2)$ and $S_u = \cos\theta$, then Equation (1) can be written as

$$\mathbf{L}_{InSAR} = \mathbf{BX} + \mathbf{\Delta}_{InSAR} \qquad (2)$$

where $\mathbf{X} = \begin{bmatrix} U_e^1 & U_n^1 & U_u^1 & U_e^2 & U_n^2 & U_u^2 & \cdots & U_e^i & U_n^i & U_u^i \end{bmatrix}^T$, $\mathbf{L}_{InSAR} = \begin{bmatrix} L_1 & L_2 & \cdots & L_i \end{bmatrix}^T$, $\mathbf{B} = I_n \otimes \begin{bmatrix} S_e & S_n & S_u \end{bmatrix}$, $i$ is the $i$th number of the points ($i = 1, 2, \cdots n$). $\otimes$ denotes the Kronecker product, and the projection vector $\begin{bmatrix} S_e & S_n & S_u \end{bmatrix}$ denotes the average of the all points.

The likelihood function between the InSAR LOS displacements $\mathbf{L}_{InSAR}$ and the 3D displacement $\mathbf{X}$ can be described as follows:

$$p(\mathbf{L}_{InSAR}|\mathbf{X}) = (2\pi)^{-n/2}|\mathbf{D}_{InSAR}|^{-1/2} \times \exp\left[-\frac{1}{2}(\mathbf{BX} - \mathbf{L}_{InSAR})^T\mathbf{D}_{InSAR}^{-1}(\mathbf{BX} - \mathbf{L}_{InSAR})\right] \qquad (3)$$

where $|\mathbf{D}_{InSAR}|$ is the absolute value of the determinant of the variance matrix $\mathbf{D}_{InSAR}$.

The GPS stations are interpolated into the spatial density of the LOS observations, and they are regarded as the virtual observations to constrain the 3D displacement field. The expression is showed as

$$\mathbf{L}_{GPS} = \mathbf{X} + \mathbf{\Delta}_{GPS} \qquad (4)$$

Thus, the prior information constrained on the 3D displacement can be presented by a probability density function (PDF) as

$$p(\mathbf{X}) = (2\pi)^{-3n/2}|\mathbf{D}_{GPS}|^{-1/2} \times \exp\left[-\frac{1}{2}(\mathbf{X} - \mathbf{L}_{GPS})^T\mathbf{D}_{GPS}^{-1}(\mathbf{X} - \mathbf{L}_{GPS})\right] \qquad (5)$$

where $\mathbf{L}_{\text{GPS}} = \begin{bmatrix} X_{\text{GPS}}^i & Y_{\text{GPS}}^i & Z_{\text{GPS}}^i \end{bmatrix}^{\text{T}}$, and $|\mathbf{D}_{\text{GPS}}|$ is the absolute value of the determinant of the variance matrix $\mathbf{D}_{\text{GPS}}$.

According to the Bayesian theorem [35], the posterior PDF for the 3D displacement is calculated as follows:

$$p(\mathbf{X}|\mathbf{L}_{\text{InSAR}}) = \frac{p(\mathbf{L}_{\text{InSAR}}|\mathbf{X})p(\mathbf{X})}{p(\mathbf{L}_{\text{InSAR}})} \tag{6}$$

where the denominator is a normalizing constant independent of $\mathbf{X}$, which is set as nc. Thus, substituting Equations (4) and (5) into Equation (6):

$$p(\mathbf{X}|\mathbf{L}_{\text{InSAR}}) = nc(2\pi)^{-2n}|\mathbf{D}_{\text{InSAR}}|^{-1/2} \times |\mathbf{D}_{\text{GPS}}|^{-1/2} \times \exp\left[-\frac{1}{2}V(\mathbf{X})\right] \tag{7}$$

where $V(\mathbf{X}) = (\mathbf{BX} - \mathbf{L}_{\text{InSAR}})^{\text{T}}\mathbf{D}_{\text{InSAR}}^{-1}(\mathbf{BX} - \mathbf{L}_{\text{InSAR}}) + (\mathbf{X} - \mathbf{L}_{\text{GPS}})^{\text{T}}\mathbf{D}_{\text{GPS}}^{-1}(\mathbf{X} - \mathbf{L}_{\text{GPS}})$. Based on the maximum a posteriori estimation criterion $p(\mathbf{X}|\mathbf{L}_{\text{InSAR}}) = \min$, which is equivalent to

$$(\mathbf{B\hat{X}} - \mathbf{L}_{\text{InSAR}})^{\text{T}}\mathbf{D}_{\text{InSAR}}^{-1}(\mathbf{B\hat{X}} - \mathbf{L}_{\text{InSAR}}) + (\mathbf{\hat{X}} - \mathbf{L}_{\text{GPS}})^{\text{T}}\mathbf{D}_{\text{GPS}}^{-1}(\mathbf{\hat{X}} - \mathbf{L}_{\text{GPS}}) = \min \tag{8}$$

where $\mathbf{\hat{X}}$ is the estimation of the $\mathbf{X}$. According to the principle of generalized least squares adjustment [42], we can obtain the expression of the 3D displacement $\mathbf{\hat{X}}$

$$\mathbf{\hat{X}} = \mathbf{L}_{\text{GPS}} + \mathbf{D}_{\text{GPS}}\mathbf{B}^{\text{T}}\left(\mathbf{BD}_{\text{GPS}}\mathbf{B}^{\text{T}} + \mathbf{D}_{\text{InSAR}}\right)^{-1}(\mathbf{L}_{\text{InSAR}} - \mathbf{BL}_{\text{GPS}}) \tag{9}$$

*2.2. The VOILS Method and its Algorithmic Flow*

Although GPS points can coincide with some observation points of InSAR, the spatial density of GPS data points is still not enough, therefore it is necessary to obtain the spatial density consistent with InSAR data points by interpolation. Consequently, this paper uses the ordinary Kriging interpolation method [43] to interpolate GPS points and obtain the prior initial 3D displacement field at the observation points of InSAR, which is as shown in Equation (5). The 3D displacement field obtained by interpolation inevitably contains errors, therefore, this study designs an iterative least squares for virtual observation, which has the advantages of correcting errors in interpolated GPS displacement and reasonably fusing two types of data. The iteration process is the key step that is different from the traditional method, which is expected to realize the main objectives as described above.

Figure 2 shows the flow chart of the VOILS method. It can be described as: (1) Downsampling InSAR data to obtain sparse LOS displacements; (2) Interpolating GPS data to the spatial density of InSAR downsampled data by the Kriging method as virtual observations; (3) Calculating E, N, and U component's displacement by the least squares for virtual observation (i.e., Equation (9)); (4) If the maximum absolute value of the difference between the front and back of the 3D displacement is less than the given threshold value $\delta$, the parameter $\mathbf{\hat{X}}^i$ is the output and the iteration will be terminated, otherwise the virtual observation values E, N, and U will be updated, and the steps 2~4 will repeat.

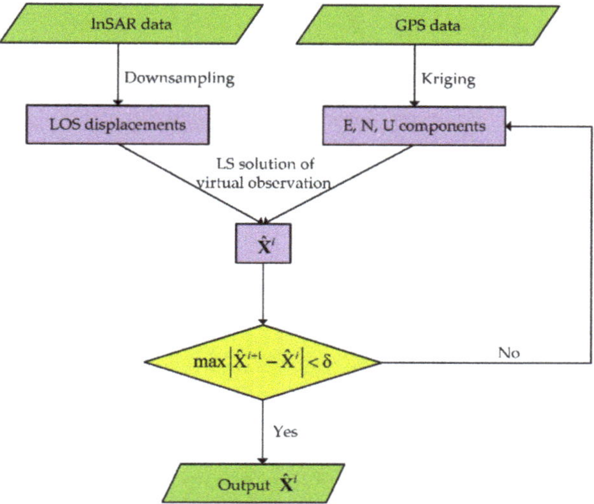


**Figure 2.** Flow chart of the iterative least squares for virtual observation.

## 3. Simulation Experiment

We adopt the following strategies (i.e., uniform and non-uniform) to select GPS observation points. In Scheme 1, we sample the GPS observation points with the uniform mode. To investigate the impact of the different density on the extracted 3D displacement field, 9, 25, 49, and 121 GPS points are selected respectively, which correspond to the spatial resolutions of 30 km, 20 km, 14 km, and 10 km of GPS observation points. In Scheme 2, we employ the non-uniform way to sample the GPS observation points in the displacement field. We use the randperm function in matrix laboratory (MATLAB) to select 9, 25, 49, and 121 points randomly according to the number of uniformly sampled GPS points.

In the comparative analysis of the results, two indicators, i.e., the root mean square error (RMSE) and the percentage of improvement of VOILS method compared with the direct solution method (Per), are used to evaluate the performances of the VOILS method.

(1) RMSE is calculated by the difference $d$ between the fitting value at each grid point and the original simulated true value. The expression is as follows:

$$RMSE = \sqrt{\frac{1}{n}\sum_{i=1}^{n}(d_i)^2} \tag{10}$$

where $n$ means the number of the grid points.

(2) Per, the percentage value of improvement, is presented as follows

$$Per = \frac{|RMSE_{New} - RMSE_{Old}|}{RMSE_{Old}} \times 100\% \tag{11}$$

where $RMSE_{New}$ and $RMSE_{Old}$ represent the RMSE obtained by the VOILS method and the direct solution method, respectively.

The uniform elastic half-space dislocation theory [44] is used to simulate the 3D surface displacement field. The deformation area is 100 km × 100 km. The model parameters of reverse fault, normal fault with a strike-slip component, and strike-slip fault with a reverse component are listed in Table 1. The simulated 3D displacement field is calculated by FORTRAN programs EDGRN/EDCMP [45].

**Table 1.** Model parameters of reverse fault, normal fault with a strike-slip component and strike-slip fault with a reverse component.

| Type | X (km) | Y (km) | Depth (km) | Length (km) | Width (km) | Slip (m) | Strike (°) | Dip (°) | Rake (°) |
|---|---|---|---|---|---|---|---|---|---|
| reverse fault | 0 | −15 | 0 | 30 | 10 | 5 | 15 | 60 | 90 |
| normal fault with a strike-slip component | 0 | −15 | 0 | 30 | 10 | 5 | 15 | 60 | −70 |
| strike-slip fault with a reverse component | 0 | 0 | 0 | 30 | 10 | 5 | 150 | 85 | 20 |

The simulation field is divided into $51 \times 51$ grid cells with a 2km interval (see Figure 3, in which both horizontal and vertical coordinates range from −50 km to 50 km, and the projection coefficient is [Se Sn Su]= [0.340 −0.095 0.935] [26]). The random errors adhering to normal distribution whose standard deviations are 3 mm, 5 mm, and 30 mm are added to the horizontal component, vertical component, and LOS deformation, respectively. Twenty-five selected points are displayed in the upper right corner subfigure of Figure 3.

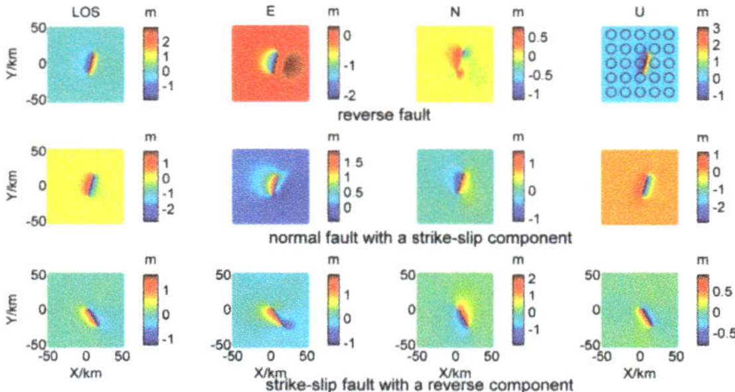


**Figure 3.** Simulated line of sight (LOS) displacement field and 3D displacement field for three types of faults. (The first line is a reverse fault. The second one is a normal fault with a strike-slip component. The last one is a strike-slip fault with a reverse component.).

## 3.1. Experiment with the Reverse Fault

In Scheme 1, we carried out 200 simulation experiments. The RMSE values of the corresponding 3D displacement field are calculated with these two methods, and the results are counted to get the histogram of Figure 4. It shows the RMSE values of 3D displacement field with different GPS spatial resolutions. It can be seen that the histogram obtained by the VOILS method is closer to the origin of coordinate than that computed by the direct solution method, which implies that the VOILS method exhibits some improvement relative to the direct solution method and can retrieve the 3D displacement field better.

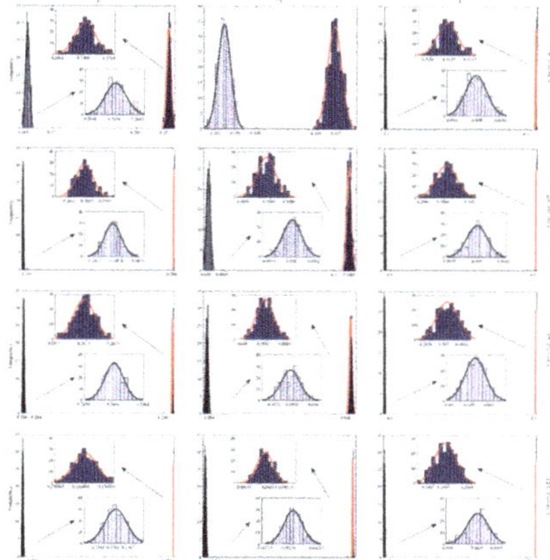

**Figure 4.** Root mean square error (RMSE) values with a normal density curve histogram for the reverse fault of Scheme 1. (The black normal density curve is the result of the iterative least squares for virtual observation (VOILS) method. The red one is the direct solution method, and the small figure in each subfigure is the local enlarged image of the original image).

In order to quantitatively illustrate the advantages of the VOILS method and form a more intuitive comparison with the direct solution method, mean improvement percentages for Scheme 1 are listed in Table 2, which is calculated by Formula (11). It shows that the improvements in all directions are comparable under different GPS spatial resolutions. The improvement degrees of the three directions from high to low are U, E, and N, which may be related to the satellite projection coefficient. The projection coefficients of U, E, and N directions are 0.935, 0.340, and −0.095, respectively. Thus, the vertical direction contributes the most to the LOS deformation, followed by the E and N directions.

The experiments under different GPS spatial resolutions were also carried out. The improved percentages are listed in Table 1. In addition, the corresponding number of GPS points are listed in brackets. It can be seen that the percentages of improvement display a relatively stable state with the increasing number of GPS points, which indicates that the proposed method has the advantage of not being subject to GPS spatial resolution. Compared with the direct solution method, the percentages of improvement for horizontal directions (i.e., E and N directions) range from 3% to 9%. Unexpectedly, the vertical direction (i.e., U direction) improvement percentage exceeds 68%.

**Table 2.** The Per in the reverse fault for two schemes.

| Scheme | E (%) | N (%) | U (%) | GPS spatial resolution (No.) |
|---|---|---|---|---|
| Scheme 1 | 8.68 | 5.16 | 68.86 | 30 km × 30 km (9) |
| | 7.79 | 4.41 | 69.48 | 20 km × 20 km (25) |
| | 7.93 | 4.33 | 69.55 | 14 km × 14 km (49) |
| | 7.89 | 4.03 | 69.21 | 10 km × 10 km (121) |
| | 8.17 | 4.74 | 69.56 | 30 km × 20 km (15) |
| | 8.44 | 4.90 | 69.88 | 30 km × 10 km (33) |
| | 8.53 | 4.95 | 70.13 | 30 km × 6 km (91) |
| | 7.92 | 4.38 | 70.17 | 20 km × 10 km (55) |
| | 7.93 | 4.34 | 70.68 | 20 km × 6 km (105) |
| | 7.80 | 4.18 | 69.82 | 14 km × 10 km (77) |
| | 7.78 | 4.10 | 70.34 | 14 km × 6 km (119) |
| | 7.91 | 3.88 | 69.72 | 10 km × 6 km (187) |
| Scheme 2 | 7.13 | 4.05 | 67.17 | 9 |
| | 7.41 | 3.95 | 69.65 | 25 |
| | 8.10 | 4.17 | 70.93 | 49 |
| | 10.96 | 6.65 | 71.76 | 121 |

Two hundred simulation experiments were also carried out in Scheme 2, and Per averages are shown in Table 2. It can be seen that the experiment results from Scheme 2 are similar to those from Scheme 1. Moreover, we executed several random experiments with non-uniformly sampled GPS points and calculated the Per. As shown in Figure 5, the proposed method in three directions exhibits different degrees of improvement relative to the direct solution method. Evidently, the vertical direction displays obvious improvement. It can be observed that the experimental results show a relatively stable state with the variations of the random sampling GPS points.

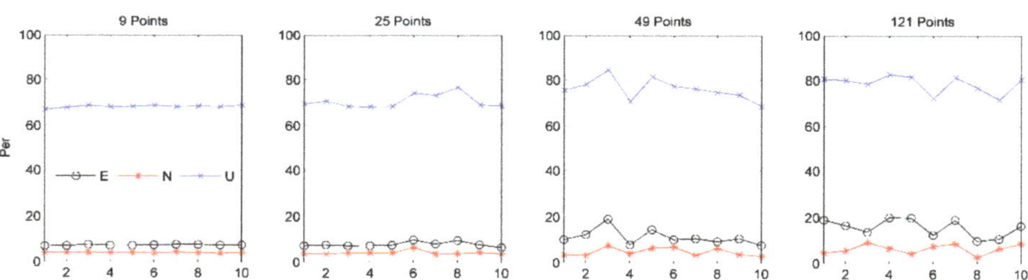


**Figure 5.** The variation of improvement percentages for Scheme 2 under different global positioning system (GPS) points.

### 3.2. Experiments with the Normal Fault with a Strike-Slip Component and Strike-Slip Fault with a Reverse Component

In the above experiments, the fault slip is assumed to be the reverse fault. The normal fault with a strike-slip component and strike-slip fault with a reverse component will be considered in this section further.

Firstly, we used Scheme 1 and Equation (11) to compute the Per, as shown in Table 3. It can be observed that for different types of fault movement, the improvement ratios in different directions under different GPS spatial resolutions are similar. The improvement degrees of three directions from high to low are vertical, E, and N. The results of Table 3 are similar to the ones of Table 2. In the case study of the normal fault with a strike-slip component, the improvement ratios of the horizontal direction range from 4% to 8%. Additionally, those of the vertical direction are up to 68%. As for the strike-slip fault with a reverse component, the results show that different directions in improvement ratios demonstrate significant differences. It can be seen that vertical direction has larger improvement ratios exceeding 21%, while lower improvement ratios below 2% occur in the N direction. Despite that, the VOILS method is better than the direct solution method.

**Table 3.** The Per in the normal fault with a strike-slip component and strike-slip fault with a reverse component for two schemes.

| Scheme | Type | E (%) | N (%) | U (%) | GPS spatial resolution (No.) |
|---|---|---|---|---|---|
| Scheme 1 | normal fault with a strike-slip component | 7.22 | 4.79 | 68.01 | 30 km × 30 km (9) |
| | | 5.99 | 4.80 | 68.63 | 20 km × 20 km (25) |
| | | 6.18 | 4.86 | 68.56 | 14 km × 14 km (49) |
| | | 6.15 | 4.80 | 68.29 | 10 km × 10 km (121) |
| | strike-slip fault with a reverse component | 7.78 | 1.51 | 21.15 | 30 km × 30 km (9) |
| | | 7.93 | 1.53 | 22.12 | 20 km × 20 km (25) |
| | | 7.99 | 1.55 | 22.36 | 14 km × 14 km (49) |
| | | 7.97 | 1.54 | 22.55 | 10 km × 10 km (121) |
| Scheme 2 | normal fault with a strike-slip component | 5.81 | 4.70 | 67.77 | 9 |
| | | 6.32 | 5.09 | 72.61 | 25 |
| | | 11.30 | 5.07 | 76.81 | 49 |
| | | 9.74 | 4.60 | 76.07 | 121 |
| | strike-slip fault with a reverse component | 7.90 | 1.58 | 16.83 | 9 |
| | | 7.84 | 1.42 | 23.33 | 25 |
| | | 7.78 | 1.55 | 18.17 | 49 |
| | | 9.41 | 1.81 | 25.96 | 121 |

Consequently, in Scheme 2, we performed 200 simulated experiments. The Per averages are shown in Table 3. We found that the results are similar to Table 2. Figure 6 portrays the histogram of RMSE values with a normal density curve at nine non-uniformly sampled GPS points. In addition, Figure 7 shows the improvement ratio of the VOILS method compared with the direct solution method. Obviously, RMSE values calculated by the VOILS method are smaller than those computed by the direct solution method, which indicates that the VOILS method is better than the direct solution method.

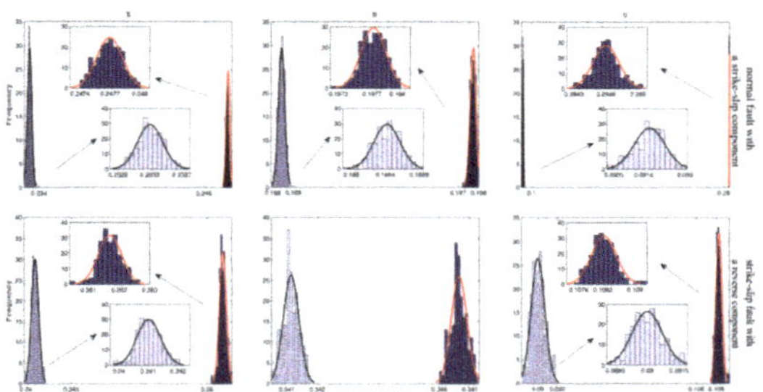

**Figure 6.** Root mean square error (RMSE) values with a normal density curve histogram under nine global positioning system (GPS) points for Scheme 2.

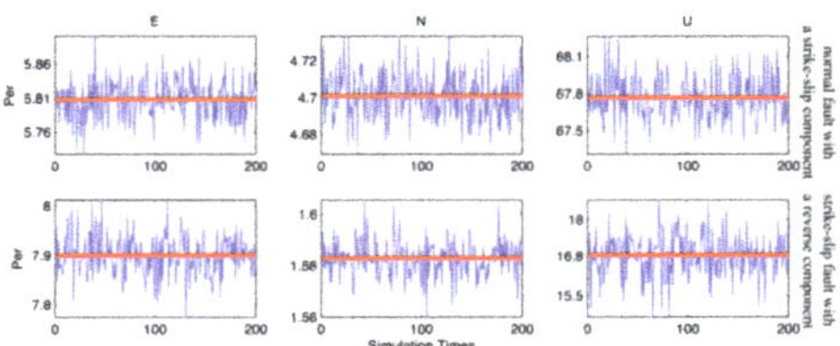

**Figure 7.** The improvement ratios of the VOILS method compared with the direct solution method under nine global positioning system (GPS) points for Scheme 2.

As for the results shown in Figure 7, the mean Per of the E, N, and U directions for the normal fault with a strike-slip component are 5.81%, 4.70%, and 67.77%, respectively. While those of E,

N, and U directions for the strike-slip fault with reverse component are 7.90%, 1.58%, and 16.83%, respectively. From Tables 2 and 3, it can be found that the improvement ratios of the U direction for the strike-slip fault with a reverse component are smaller than those of the reverse fault and normal fault with a strike-slip component. This may be because the slip vector of a strike-slip accompanied by reverse components is mainly in a horizontal direction, while those of a normal fault with a strike-slip component and reverse fault are in a vertical direction. This correlates with the largest contribution of U direction to LOS deformation.

In summary, it is concluded that the improvement degree in the vertical direction is better than that in the horizontal direction. The magnitude of vertical deformation calculated by the VOILS method is closer to the true value than that computed from the direct solution method. This confirms that the VOILS method can make full use of InSAR observations. Our simulation results fully show that the VOILS method is feasible and effective. Furthermore, the proposed method displays obvious advantages in retrieving 3D displacement fields compared to the direct solution method, especially for vertical deformation.

## 4. 3D Displacement Field Extraction of the Wenchuan Earthquake

### 4.1. GPS and InSAR Data

The Wenchuan earthquake occurred on May 12, 2008, on the Longmenshan fault zone in Sichuan Province (Figure 8). The earthquake caused surface rupture in the Yingxiu–Beichuan fault of the main central fault and the Guanxian–Jiangyou fault of the Qianshan fault [46]. Previous studies have illustrated that this event is a complex slip mechanism as a variable combination of a reverse and right-lateral component [46–50]. After the earthquake, the co-seismic deformation observation data, including GPS data and InSAR data from advanced land observation satellite (ALOS) satellite phased array type L-band synthetic aperture radar (PALSAR) images, were obtained. In this paper, we employed the downsampled 3792 LOS displacements from Xu et al. [47], and 473 GPS co-seismic displacement data calculated by Wang et al. [5]. It is noted that 297 GPS points (i.e., 284 campaigned stations and 13 continuous stations) are used in our study, with the spatial distribution of the data shown in Figure 8.

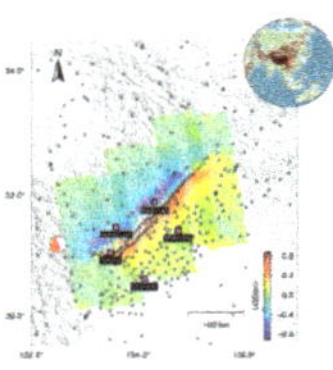

**Figure 8.** The tectonic setting map of the Wenchuan earthquake. The color-shaded dots denote the LOS displacements from the advanced land observation satellite (ALOS) satellite phased array type L-band synthetic aperture radar (PALSAR) images. Open circles denote the GPS stations. The yellow star depicts the epicenter of the 2008 Wenchuan earthquake, and the red bench ball denotes the focal mechanism solution from the U.S. Geological Survey (USGS). The gray lines denote the rupture traces of this event. The magenta rectangles represent the locations of surrounding cities.

### 4.2. Results and Comparative Analysis

The 3D displacement field of the Wenchuan earthquake obtained by the VOILS method is shown in Figure 9a–c. Meanwhile, the LOS displacement field and its residual errors are calculated, as shown in Figure 9d–e.

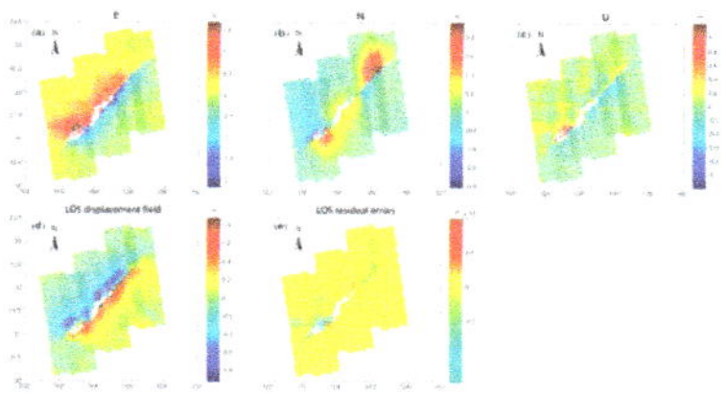

**Figure 9.** The results obtained by the VOILS method, (**a**), (**b**), and (**c**) are the deformations of E, N, and vertical (U) directions, respectively. (**d**) is the modelled LOS displacement field. (**e**) are the residual errors between the modelled and observed LOS displacements.

Figure 9a shows the E direction deformation, and it can be observed that the hanging wall moves eastward with a maximum displacement of 1.69 m, while the footwall moves westward with a maximum displacement of 2.15 m, showing a right-lateral trend. As for the N direction deformation (Figure 9b), the southwest of the hanging wall moves southward with a maximum displacement of 0.82 m and the northeast section moves northward with a maximum displacement of 0.95 m. The whole footwall moves northward with a maximum displacement of 0.77 m, and the deformation decreases in the northeast direction. The relatively large displacements of the hanging wall and footwall near the epicenter indicate that the reverse component in this area is the largest. Finally, the U direction deformation is shown in Figure 9c. The hanging wall and footwall near the fault zone appear uplift and subsidence. In addition, the maximum uplift and the maximum subsidence near the epicenter are 1.19 m and 0.95 m respectively, which means that reverse movement existed in the southwest of the fault. The 3D deformation characteristics clearly show the local movement characteristics of the fault and reflect the seismogenic fault's complexity and heterogeneity.

From Figure 9d–e, we know that the LOS displacement field is close to that of Figure 8. Meanwhile, the unit for fitting residual errors after iteration calculation is millimeters, which is at the same order magnitude of the threshold $\delta$ (2 mm) adopted in the iteration process. It is useful to compare the derived coseismic displacement field with independent GPS observations. The InSAR points with a distance of approximately 3 km around the GPS points were chosen for horizontal comparison. Figure 10 shows the horizontal displacement vectors, indicating that the agreement between the GPS observations and the derived displacements is quite good. The RMSE values between them are 6.1 cm, 2.3 cm, and 5.8 cm in the E, N, and U directions, respectively.

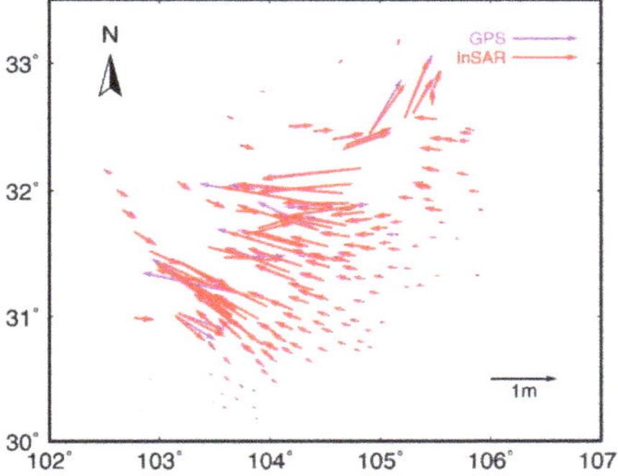

**Figure 10.** The coseismic horizontal displacements. The purple arrows show the horizontal displacement vectors derived from the GPS observations, and the red arrows denote the vectors from the derived horizontal displacement.

In order to compare with the above results, the 3D displacement field (Figure 11a–c) was calculated using the direct solution method. The LOS deformation and its residual errors are shown in Figure 11d–e.

Figure 11a–c show similar features as Figure 9a–c, while the numerical values are different. It can be seen from Figure 11e that the residuals along both sides of the seismogenic fault are larger, especially in the southwest of the hanging wall. This can be ascribed to the fault being close to the epicenter, which caused larger deformation, as well as poor interpolation accuracy with less near-field GPS data. This is similar to the characteristic distribution of the LOS deformation residual map produced by Luo et al. [34]. However, the numerical values are different, which may be related to the data and the method used.

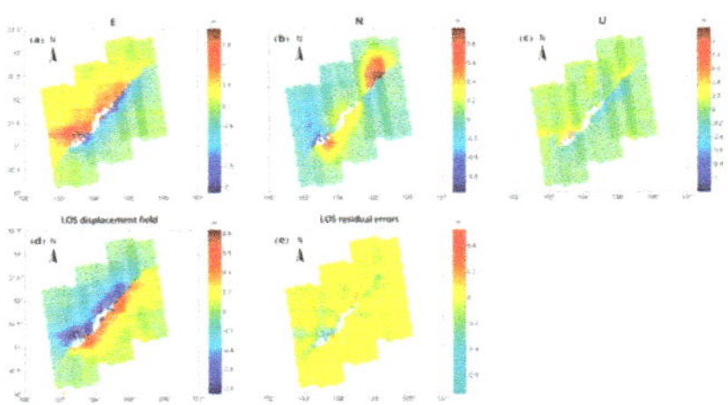

**Figure 11.** The results obtained by the direct solution method, (**a**), (**b**), and (**c**) are the deformations of E, N, and U directions, respectively. (**d**) is the modelled LOS displacement field. (**e**) are the residual errors between the modelled and observed LOS displacements.

The results indicate that the deformation directions along both sides of the fault are basically opposite, and the larger displacements in the three directions are nearer to the source. Near the epicenter of the E direction, the movement of the hanging wall to the east, relative to the footwall, is dominant, and along the NE direction, the deformation of the hanging wall and footwall is opposite. In the N direction, the northward movement of the northeastern part of the hanging wall is dominant, with a small amount of reverse component. Both the hanging wall and footwall of the U direction have subsidence near the fault zone, and the uplift of the hanging wall is greater than the subsidence of the footwall. In summary, we can understand that the main fault near the epicenter is a reverse fault with a right-lateral strike-slip component, while the northeastern segment is a right-lateral strike-slip with a small amount of reverse component, which is in line with the results of other research [46–50].

In order to further verify the degree of improvement in the vertical direction, we compared the derived displacements to the surface rupture measurements from Xu et al. (2009) [50]. Figure 12 shows the comparison of these data. Note that the average displacements of InSAR points with a distance of about 5 km around the field points are used for comparison. It shows that the vertical displacements from the VOILS method are more consistent with the field observations, which indicates that the VOILS method can effectively reduce the impacts of the interpolated errors on parameter estimation to some extent. Meanwhile, these results indicate that the VOILS method performs better than the direct solution method in this event.

Figure 11e shows that the residual magnitude of InSAR is different from that of Figure 9e. The residual magnitude of InSAR fitted by VOILS method is mm-level, which is significantly smaller than that of the direct solution method. Meanwhile, the results of this study are also different from those of the existing researches [28,36–38]. Therefore, the results obtained by the VOILS method are relatively stable and the fitting is relatively good. On the other hand, the RMSE value of the direct solution method is 0.07m, representing a precision of cm-level. The above analyses show that the VOILS method can make full use of the InSAR observations, and has the ability to exert benefits from GPS and InSAR data.

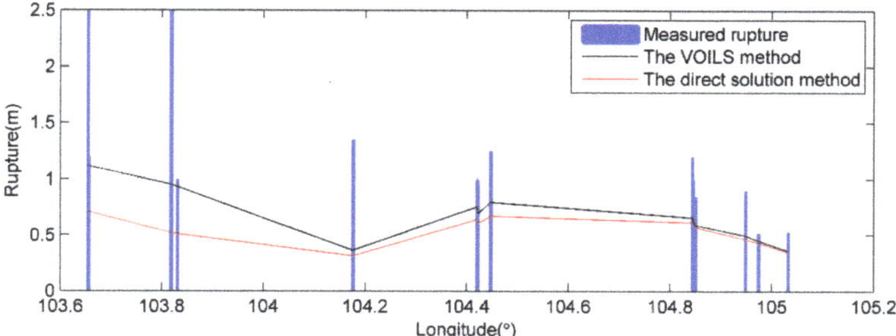

**Figure 12.** Comparison of the derived vertical displacements with the surface rupture measurements (The blue bar represents the measured rupture from field investigations (Xu et al., 2009) [50]. The black and red lines represent the displacements by making subtraction between the derived deformation on hanging wall and footwall from the VOILS method and the direct solution method, respectively.).

## 5. Conclusions

In the conventional direct solution method, GPS data need to be interpolated first, which will undoubtedly introduce errors. Therefore, this paper proposes the VOILS method, which can correct the errors in GPS 3D deformation interpolation, and realize the reasonable fusion of GPS/InSAR observations in 3D displacement extraction based on the iteration process, making better use of InSAR data and calculating relatively stable 3D displacement fields.

In the simulation examples, uniform and non-uniform schemes are adopted to select GPS points. In the uniform sampling scheme, the spatial resolutions of GPS points of 30 km, 20 km, 14 km, and 10 km were considered. Accompanied with the comprehensive analyses of the reverse fault, normal fault with a strike-slip component, and strike-slip fault with a reverse component, it demonstrates that the VOILS method is better than the direct solution method in both horizontal and vertical directions. From the results of the uniform sampling scheme, we see that the percentages of improvement for the reverse fault are approximately 3~9% and 70%, for the normal fault with a strike-slip component approximately 4~8% and 68%, and for the strike-slip fault with a reverse component approximately 1~8% and 22%. Moreover, the degree of improvement in the vertical direction is more obvious. The simulation experiments show that the VOILS method can make better use of the InSAR observations, and has feasibility and validity.

The VOILS method was applied to get the 3D displacement field of the 2008 Wenchuan earthquake. The LOS residual magnitude obtained by the VOILS method is mm-level, while the magnitude of the direct solution method is cm-level. Furthermore, the results of the VOILS method are relatively stable. Based on the analyses of 3D deformation, it can be seen that the mechanism of the main shock was mainly a reverse motion with a right-lateral strike-slip component. In the northeast of the fault, the mechanism is dominated by the right-lateral strike-slip. In particular, the comparison with the rupture displacements from the field investigations shows that the VOILS method is better than the direct solution method.

**Author Contributions:** Conception and the design of the experiments, C.X. and Y.L.; the performance of the experiments, L.X.; formal analysis, L.X., Y.L., C.X. Y.W. and J.F.; writing—original draft preparation, L.X.; writing—review and editing, L.X., C.X., Y.L., Y.W. and J.F.; supervision, C.X. All authors have read and agreed to the published version of the manuscript.

**Funding:** This work is co-supported by the National Key Research Development Program of China under Grant No. 2018YFC1503603, No. 2018YFC1503604, and the National Natural Science Foundation of China under Grant No. 41721003, No. 41874011, and No. 41774011.

**Acknowledgments:** The authors thank academic editor and anonymous reviewers for their comments and constructive suggestions, which highly improved the quality of the manuscript. Thanks are also given to Guangyu

Xu for helping drawing and Hanqing Chen for helping polish this manuscript. The ALOS SAR data were provided by JAXA. Some figures were plotted by the Generic Mapping Tools software [51].

**Conflicts of Interest:** The authors declare no conflict of interest.

## References

1. Massonnet, D.; Rossi, M.; Carmona, C.; Adragna, F.; Peltzer, G.; Feigl, K.L.; Rabaute, T. The displacement field of the Landers earthquake mapped by radar interferometry. *Nature* **1993**, *364*, 138–142. [CrossRef]
2. Wright, T.J. Toward mapping surface deformation in three dimensions using InSAR. *Geophys. Res. Lett.* **2004**, *31*, 169–178. [CrossRef]
3. Xu, G.; Xu, C.; Wen, Y. Sentinel-1 observation of the 2017 Sangsefid earthquake, northeastern Iran: Rupture of a blind reserve-slip fault near the Eastern Kopeh Dagh. *Tectonophysics* **2018**, *731*, 131–138. [CrossRef]
4. Shen, Z.-K.; Sun, J.; Zhang, P.; Wan, Y.; Wang, M.; Bürgmann, R.; Zeng, Y.; Gan, W.; Liao, H.; Wang, Q. Slip maxima at fault junctions and rupturing of barriers during the 2008 Wenchuan earthquake. *Nat. Geosci.* **2009**, *2*, 718–724. [CrossRef]
5. Wang, Q.; Qiao, X.J.; Lan, Q.G.; Freymueller, J.T. Rupture of deep faults in the 2008 Wenchuan earthquake and uplift of the Longmen Shan. *Nat. Geosci.* **2011**, *4*, 634–640.
6. Fang, J.; Xu, C.; Wen, Y.; Wang, S.; Xu, G.; Zhao, Y.; Yi, L. The 2018 Mw 7.5 Palu earthquake: A supershear rupture event constrained by InSAR and broadband regional seismograms. *Remote Sens.* **2019**, *11*, 1330. [CrossRef]
7. Wang, S.; Xu, C.; Wen, Y.; Yin, Z.; Jiang, G.; Fang, L. Slip model for the 25 November 2016 Mw 6.6 Aketao earthquake, western China, revealed by Sentinel-1 and ALOS-2 observations. *Remote Sens.* **2017**, *9*, 325. [CrossRef]
8. He, P.; Wen, Y.; Xu, C.; Chen, Y. Complete three-dimensional near-field surface displacements from imaging geodesy techniques applied to the 2016 Kumamoto earthquake. *Remote Sens. Environ.* **2019**, *232*, 111321. [CrossRef]
9. Muller, C.; Del Potro, R.; Biggs, J.; Gottsmann, J.; Ebmeier, S.K.; Guillaume, S.; Cattin, P.-H.; Van Der Laat, R. Integrated velocity field from ground and satellite geodetic techniques: Application to Arenal volcano. *Geophys. J. Int.* **2015**, *200*, 861–877. [CrossRef]
10. Palano, M.; Puglisi, G.; Gresta, S. Ground deformation patterns at Mt. Etna from 1993 to 2000 from joint use of InSAR and GPS techniques. *J. Volcanol. Geoth. Res.* **2008**, *169*, 99–120. [CrossRef]
11. Guo, Q.; Xu, C.; Wen, Y.; Liu, Y.; Xu, G. The 2017 noneruptive unrest at the Caldera of Cerro Azul Volcano (Galápagos Islands) revealed by InSAR observations and geodetic modelling. *Remote Sens.* **2019**, *11*, 1992. [CrossRef]
12. Pritchard, M.; Biggs, J.; Wauthier, C.; Sansosti, E.; Arnold, D.W.D.; Delgado, F.; Ebmeier, S.K.; Henderson, S.T.; Stephens, K.; Cooper, C.; et al. Towards coordinated regional multi-satellite InSAR volcano observations: Results from the Latin America pilot project. *J. Appl. Volcanol.* **2018**, *7*, 5. [CrossRef]
13. Bonì, R.; Meisina, C.; Cigna, F.; Herrera, G.; Notti, D.; Bricker, S.; McCormack, H.; Tomás, R.; Béjar-Pizarro, M.; Mulas, J.; et al. Exploitation of satellite A-DInSAR time series for detection, characterization and modelling of land subsidence. *Geosciences* **2017**, *7*, 25.
14. Bonì, R.; Herrera, G.; Meisina, C.; Notti, D.; Béjar-Pizarro, M.; Zucca, F.; Gonzalez, P.J.; Palano, M.; Tomás, R.; Fernandez, J.; et al. Twenty-year advanced DInSAR analysis of severe land subsidence: The Alto Guadalentín Basin (Spain) case study. *Eng. Geol.* **2015**, *198*, 40–52. [CrossRef]
15. Caló, F.; Notti, D.; Galve, J.; Abdikan, S.; Görüm, T.; Pepe, A.; Balik Şanli, F. DInSAR-based eetection of land subsidence and correlation with groundwater depletion in Konya Plain, Turkey. *Remote Sens.* **2017**, *9*, 83. [CrossRef]
16. Cheloni, D.; Serpelloni, E.; Devoti, R.; D'Agostino, N.; Pietrantonio, G.; Riguzzi, F.; Anzidei, M.; Avallone, A.; Cavaliere, A.; Cecere, G.; et al. GPS observations of coseismic deformation following the 2016, August 24, Mw 6 Amatrice earthquake (central Italy): Data, analysis and preliminary fault model. *Ann. Geophys.* **2016**, *59*. [CrossRef]
17. Jiang, Z.; Huang, D.; Yuan, L.; Hassan, A.; Zhang, L.; Yang, Z. Coseismic and postseismic deformation associated with the 2016 Mw 7.8 Kaikoura earthquake, New Zealand: Fault movement investigation and seismic hazard analysis. *Earth Planets Space* **2018**, *70*, 62. [CrossRef]
18. Houlié, N.; Dreger, D.; Kim, A. GPS source solution of the 2004 Parkfield earthquake. *Sci. Rep.* **2014**, *4*, 3646. [CrossRef]
19. Johanson, I.A.; Fielding, E.J.; Rolandone, F.; Buμrgmann, R. Coseismic and postseismic slip of the 2004 Parkfield earthquake from space-geodetic data. *Bull. Seismol. Soc. Am.* **2006**, *96*, S269–S282. [CrossRef]

20. Murray, J. Slip on the San Andreas Fault at Parkfield, California, over two earthquake cycles, and the implications for seismic hazard. *Bull. Seismol. Soc. Am.* **2006**, *96*, S283–S303. [CrossRef]

21. Liu, P.; Custodio, S.; Archuleta, R.J. Kinematic Inversion of the 2004 M 6.0 Parkfield Earthquake Including an Approximation to Site Effects. *Bull. Seismol. Soc. Am.* **2006**, *96*, S143–S158. [CrossRef]

22. Delouis, B.; Nocquet, J.-M.; Vallée, M. Slip distribution of the February 27, 2010 Mw = 8.8 Maule Earthquake, central Chile, from static and high-rate GPS, InSAR, and broadband teleseismic data. *Geophys. Res. Lett.* **2010**, *37*. [CrossRef]

23. Palano, M.; Viccaro, M.; Zuccarello, F.; Gresta, S. Magma transport and storage at Mt. Etna (Italy): A review of geodetic and petrological data for the 2002–03, 2004 and 2006 eruptions. *J. Volcanol. Geoth. Res.* **2017**, *347*, 149–164. [CrossRef]

24. Owen, S.; Segall, P.; Lisowski, M.; Miklius, A.; Murray, M.; Bevis, M.; Foster, J. 30 January 1997 eruptive event on Kilauea Volcano, Hawaii, as monitored by continuous GPS. *Geophys. Res. Lett.* **2000**, *27*, 2757–2760. [CrossRef]

25. Gudmundsson, S.; Gudmundsson, M.T.; Björnsson, H.; Sigmundsson, F.; Rott, H.; Carstensen, J.M. Three-dimensional glacier surface motion maps at the Gja´lp eruption site, Iceland, inferred from combining InSAR and other ice-displacement data. *Ann. Glaciol.* **2002**, *34*, 315–322. [CrossRef]

26. Samsonov, S.; Tiampo, K. Analytical optimization of a DInSAR and GPS dataset for derivation of three-dimensional surface motion. *IEEE Geosci. Remote Sens. Lett.* **2006**, *3*, 107–111. [CrossRef]

27. Samsonov, S.; Tiampo, K.; Rundle, J.; Li, Z. Application of DInSAR-GPS optimization for derivation of fine-scale surface motion maps of southern California. *IEEE Trans. Geosci. Remote Sens.* **2007**, *45*, 512–521. [CrossRef]

28. Song, X.G.; Shen, X.; Jiang, Y.; Wan, J.-H. Coseismic 3D deformation field acquisition of the Wenchuan earthquake based on InSAR and GPS data. *Seismol. Geol.* **2015**, *37*, 222–231. (In Chinese)

29. Luo, H.; Chen, T. Three-dimensional surface displacement field associated with the 25 April 2015 Gorkha, Nepal, earthquake: Solution from integrated InSAR and GPS measurements with an extended SISTEM approach. *Remote Sens.* **2016**, *8*, 559. [CrossRef]

30. Guo, Z.; Wen, Y.; Xu, G.; Wang, S.; Wang, X.; Liu, Y.; Xu, C. Fault slip model of the 2018 Mw 6.6 Hokkaido eastern Iburi, Japan, earthquake estimated from satellite Radar and GPS measurements. *Remote Sens.* **2019**, *11*, 1667. [CrossRef]

31. Hu, J. Theory and Method of Estimating Three-Dimensional Displacement with InSAR Based on the Modern Surveying Adjustment. Ph.D. Thesis, Central South University, Changsha, China, 2013.

32. Guglielmino, F.; Nunnari, G.; Puglisi, G.; Spata, A. Simultaneous and integrated strain tensor estimation from geodetic and satellite deformation measurements to obtain three-dimensional displacement maps. *IEEE Trans. Geosci. Remote Sens.* **2011**, *49*, 1815–1826. [CrossRef]

33. Song, X.; Jiang, Y.; Shan, X.; Qu, C. Deriving 3D coseismic deformation field by combining GPS and InSAR data based on the elastic dislocation model. *Int. J. Appl. Earth Obs. Geoinf.* **2017**, *57*, 104–112. [CrossRef]

34. Luo, H.; Liu, Y.; Chen, T.; Xu, C.; Wen, Y. Derivation of 3-D surface deformation from an integration of InSAR and GNSS measurements based on Akaike's Bayesian Information Criterion. *Geophys. J. Int.* **2016**, *204*, 292–310. [CrossRef]

35. Bagnardi, M.; Hooper, A. inversion of surface deformation data for rapid estimates of source parameters and uncertainties: A Bayesian approach. *Geochem. Geophy. Geosy.* **2018**, *19*, 2194–2211. [CrossRef]

36. Ban, B.S.; Wu, J.C.; Chen, Y.Q.; Feng, G.C.; Hu, S.C. Calculation of three-dimensional terrain deformation of Wenchuan earthquake with GPS and InSAR data. *J. Geod. Geodyn.* **2010**, *30*, 25–28. (In Chinese)

37. Xu, K.K.; Niu, Y.F.; Wu, J.C. Establishment of 3-D coseismic terrain displacement field with GPS and InSAR. *J. Geod. Geodyn* **2014**, *34*, 15–18. (In Chinese)

38. Shan, X.J.; Qu, C.Y.; Guo, L.M.; Zhang, G.-H.; Song, X.; Zhang, G.; Wen, S.-Y.; Wang, C.; Xu, X.-B.; Liu, Y.-H. The vertical coseismic deformation field of the Wenchuan earthquake based on the combination of GPS and InSAR measurements. *Seismol. Geol.* **2014**, *36*, 718–730. (In Chinese)

39. He, X.F.; He, M. *InSAR Earth Observation Data Processing and Comprehensive Measuring*; Science Press: Beijing, China, 2012; p. 210.

40. Fuhrmann, T.; Garthwaite, M.C. Resolving three-dimensional surface motion with InSAR: Constraints from multi-geometry data fusion. *Remote Sens.* **2019**, *11*, 241. [CrossRef]

41. Wen, Y.M. Coseismic and Postseismic Deformation Using Synthetic Aperture Radar Interferometry. Ph.D. Thesis, Wuhan University, Wuhan, China, 2009.

42. Cui, X.Z.; Yu, Z.C.; Tao, B.Z.; Liu, D.; Yu, Z.; Sun, H.; Wang, X. *Generalized Surveying Adjustment*, 2nd ed.; Wuhan University Press: Wuhan, China, 2009; p. 21.

43. Gudmundsson, S.; Sigmundsson, F.; Carstensen, J.M. Three-dimensional surface motion maps estimated from combined interferometric synthetic aperture radar and GPS data. *J. Geophys. Res.* **2002**, *107*, 2250. [CrossRef]

44. Okada, Y. Surface deformation due to shear and tensile faults in a half-space. *Bull. Seismol. Soc. Am.* **1985**, *75*, 1135–1154.

45. Wang, R.; Martín, F.L.; Roth, F. Computation of deformation induced by earthquakes in a multi-layered elastic crust—FORTRAN programs EDGRN/EDCMP. *Comp. Geosci.* **2003**, *29*, 195–207. [CrossRef]

46. Xu, X.W.; Wen, X.Z.; Ye, J.Q.; Ma, B.Q. The Ms 8.0 Wenchuan earthquake surface ruptures and its seismogenic structure. *Seismol. Geol* **2008**, *30*, 597–629. (In Chinese)

47. Xu, C.; Liu, Y.; Wen, Y.; Wang, R.-Q. Coseismic slip distribution of the 2008 Mw 7.9 Wenchuan earthquake from joint inversion of GPS and InSAR data. *Bull. Seismol. Soc. Am.* **2010**, *100*, 2736–2749. [CrossRef]

48. Zhang, G.H.; Qu, C.Y.; Song, X.G. Slip distribution and source parameters inverted from co-seismic deformation derived by InSAR technology of Wenchuan Mw 7.9 earthquake. *Chin. J. Geophys.* **2010**, *53*, 269–279. (In Chinese)

49. Shan, X.J.; Qu, C.Y.; Song, X.G.; Zhang, G.F. Coseismic surface deformation caused by the Wenchuan Ms 8.0 earthquake from InSAR data analysis. *Chin. J. Geophys.* **2009**, *52*, 496–504. (In Chinese)

50. Xu, X.; Wen, X.; Yu, G.; Chen, G.; Klinger, Y.; Hubbard, J.; Shaw, J. Coseismic reverse- and oblique-slip surface faulting generated by the 2008 Mw 7.9 Wenchuan earthquake, China. *Geology* **2009**, *37*, 515–518. [CrossRef]

51. Wessel, P.; Smith, W.H.F. New, improved version of generic mapping tools released. *Eos Trans. Am. Geophys. Union* **1998**, *79*, 579. [CrossRef]

 *remote sensing*

Article

# A Covariance-Based Approach to Merging InSAR and GNSS Displacement Rate Measurements

**Alessandro Parizzi *, Fernando Rodriguez Gonzalez and Ramon Brcic**

Remote Sensing Technology Institute German Aerospace Center (DLR) Münchenerstraße 20, 82234 Weßling, Germany; fernando.rodriguezgonzalez@dlr.de (F.R.G.); ramon.brcic@dlr.de (R.B.)
* Correspondence: alessandro.parizzi@dlr.de

Received: 4 December 2019; Accepted: 2 January 2020; Published: 16 January 2020

**Abstract:** This paper deals with the integration of deformation rates derived from Synthetic Aperture Radar Interferometry (InSAR) and Global Navigation Satellite System (GNSS) data. The proposed approach relies on knowledge of the variance/covariance of both InSAR and GNSS measurements so that they may be combined accounting for the spectral properties of their errors, hence preserving all spatial frequencies of the deformation detected by the two techniques. The variance/covariance description of the output product is also provided. A performance analysis is carried out on realistic simulated scenarios in order to show the boundaries of the technique. The proposed approach is finally applied to real data. Five Sentinel-1A/B stacks acquired over two different areas of interest are processed and discussed. The first example is a merged deformation map of the northern part of the Netherlands for both ascending and descending geometries. The second example shows the deformation at the junction between the North and East Anatolian Fault using three consecutive descending stacks.

**Keywords:** InSAR; GNSS; deformation

---

## 1. Introduction and Motivation

Synthetic Aperture Radar Interferometry (InSAR) deformation rate measurements support a wide range of applications in the fields of geology, geophysics, and geohazards [1]. Since the nature of the interferometric measurements is inherently relative, the additive delays—like those of the troposphere and ionosphere—make the accuracy of such measurements strongly dependent on the distance [2–4]. Hence, InSAR performance varies according to the particular application. Typically, interferometry works well for applications such as infrastructure monitoring or urban subsidence, since they involve relatively short scales (10–20 km). However, there is also scientific interest in using InSAR to measure the strain accumulation of tectonic faults. This type of study requires very accurate measurements (1 mm/year at distances larger than 100 km) [5].

The new generation of Synthetic Aperture Radar (SAR) sensors—like Sentinel-1—provides systematically acquired data with a swath width of 250 km [6]; future missions plan to further extend this to 350 km [7]. Due to the atmospheric errors at such scales, the requirement of 1 mm/y could be a challenging goal for InSAR, especially if the available time series have a reduced observation time [8]. Therefore, in order to fully exploit the coverage capabilities of these missions, we are motivated to develop techniques that merge interferometric deformation rates with other geodetic measurements, not only to improve the performance of applications particularly affected by atmospheric effects (e.g., inter-seismic deformations), but also to remove the reference point, hence facilitating an easier integration with the other geometries or techniques.

The best candidate for this integration with SAR interferometry are the deformation rates estimated using a Global Navigation Satellite System (GNSS). This is due firstly to their well-known complementarity where InSAR coverage is combined with GNSS accuracy. Indeed, there is a long tradition of research in this area as evidenced by studies performed by geo-scientists; for example, in supporting interferometric phase unwrapping [9,10]. Combining InSAR and GNSS deformation rates has also been studied in order to calibrate the InSAR-estimated velocities [11–13], and to help the limited geometric sensitivity of the SAR system in measuring deformation [14–16]. The integration of InSAR with other instrumentation such as active transponders has been investigated in order to design optimal geodetic networks that enhance the spatial distribution of the deformation measurements [17–19].

However, it is important to highlight that the main argument of this work is not to claim the complementary InSAR/GNSS—which is well known, as previously mentioned—but rather to discuss the use of their error statistics in the combination, as well as its propagation through to the final product. In [8], the effect of the correction of the interferometric phase for tropospheric delays and solid earth tides using external models was shown. After such corrections, the mean variograms of the interferometric measurements error show a stationary behavior that can be well approximated by a covariance function. An optimal combination of the two measurement techniques is hence possible, since the spectral properties of their own errors can be taken into account. Moreover, no assumptions about the displacement pattern characteristics [19] and GNSS measurement density are necessary as in [11] and the data need not be filtered. Only the InSAR/GNSS differences, which can be statistically characterized, are estimated and used to compensate the original InSAR measurements. The error can then be propagated from the data to the results, allowing full characterization of the output uncertainty.

The measurement of strain accumulation along hundreds of kilometers is a potential application for this framework. The integration of InSAR and GNSS does not remove any deformation component, but performs a weighted merging of the different spatial frequencies of the deformation detected by the two techniques according to their reliability. Moreover, the mathematical modeling, often applied on final measurements, requires a proper weighting of the different input data [20]. Other possible applications include the InSAR-based National Ground Motion Services [21]. Such projects are often required to provide a product that merges the InSAR-derived results and the results derived by the GNSS networks already deployed on the territory. Since these products are part of a Service, a consistent description and traceability of the uncertainties is strictly required.

In Section 2 the proposed methodology is described, separating the capability of retrieving the absolute motion from the calibration of the residual atmospheric errors. An analytical description of the error of the merged product is also provided. In Section 3, simulations are performed in order to provide reference numbers in terms of coverage and quality versus performance. Finally, the results of using real data in significant test cases are presented and discussed. The first example covers the northern part of the Netherlands and demonstrates merging on a large scale in order to provide a consistent product like those in national ground motion services. The second example is aimed at addressing geophysical applications, and shows the calibration of interferometric measurements over North Anatolian Fault to the Eurasian plate.

## 2. Methodology

Let us consider $N$ locations in the processed area of interest, where two deformation rate measurements $v_D$ and $\underline{v}_G$ are performed using InSAR and GNSS, respectively. With $\underline{s}$ as the radar line of sight (LoS) it is possible to state the problem modeling the difference between the two velocities at the $i$th position $\Delta_i$ as follows:

$$\Delta_i = v_{D,i} - \underline{v}_{G,i}^T \underline{s} = \delta(r_i, a_i) + v_{ref} + n(r_i, a_i), \tag{1}$$

where $v_{ref}$ is the velocity of the reference point used in InSAR processing, $\delta(r_i, a_i)$ is a space-variant error screen in the interferometric data, and $n(r_i, a_i)$ is the random error. The implicit hypothesis behind this definition is that the spatially correlated noise, basically related to the residual atmospheric delay, is present in the interferometric measurement only. This characteristic allows extraction, through the subtraction of the two velocities, of the parameters $\delta(r,a)$ and $v_{ref}$, which have to be estimated in order to properly merge the data. It is worth noticing that this operation can also be seen as a calibration of the InSAR data: it exploits GNSS reference rates so that the interferometric measurement is no longer relative to the reference point while also removing residual systematic effects. Figure 1 displays an overview of the steps of the proposed approach. In the first step, the error statistics are derived and the covariance matrix $\mathbf{R}$ is computed. The reference point velocity $v_{ref}$ is then estimated and subtracted from the offset vector $\underline{\Delta}$. Finally, the residual error screen is interpolated at each InSAR measurement point, exploiting knowledge of the error statistics.

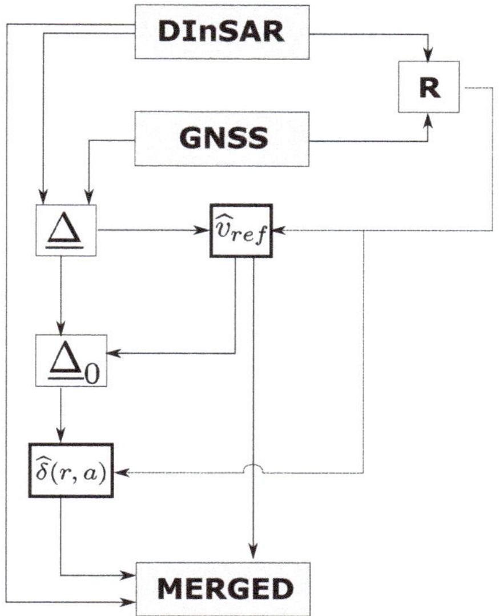


**Figure 1.** Flow chart of the proposed method.

*2.1. Error Description of the Input Data*

A proper handling of the relation defined in Equation (1) requires a statistical description of the vector $\underline{\Delta}$. Therefore, the covariance matrix $\mathbf{R}$ of the difference measurements has to be derived. It is possible to distinguish between three contributions: The random noise of the GNSS measurements $\sigma_{G,i}^2$ obtained by projecting the variance of the different components of the estimated rates onto the radar LoS, the random noise of the InSAR measurements $\sigma_{D,i}^2$, and spatially correlated noise due to residual atmospheric effects. The full covariance matrix can be defined as:

$$\mathbf{R} = \mathbb{E}[\underline{\Delta}\underline{\Delta}^T] = \mathrm{diag}(\sigma_{G,0}^2, ...\sigma_{G,N-1}^2) + \\ + \mathrm{diag}(\sigma_{D,0}^2, ...\sigma_{D,N-1}^2) + \mathbf{C}_\alpha. \tag{2}$$

The first two contributions are diagonal matrices, since they represent the spatially uncorrelated errors of the independent GNSS and InSAR measurement processes, respectively. On the other hand, the derivation of the covariance matrix $\mathbf{C}_\alpha$ that represents the residual atmospheric error is particularly interesting [22]. A covariance function $\Gamma(d)$ representing the residual error has to be robustly estimated and evaluated at the $N$ positions where the $\Delta_i$ are located; with $d_{i,j}$ as the distance between the $i$th and the $j$th measurement, this is:

$$\mathbf{C}_\alpha(i,j) = \Gamma(d_{i,j}). \tag{3}$$

The first two components of Equation (2) can be estimated using the supplied accuracy of the GNSS data and the interferometric temporal coherences, respectively. However, estimation of $\mathbf{C}_\alpha$ requires a more detailed discussion.

*2.2. Estimation of the Residual Atmospheric Error Covariance Matrix*

The covariance of the residual atmospheric error can be estimated from interferograms where the residual atmospheric delay is the dominant measured delay. The fast revisit time of the Sentinel-1 mission allows the computation of short time interferograms with temporal baselines $\delta t$ down to 6 days. Following the approach in [8], the short temporal baseline interferograms are used to compute variograms merely representing the residual atmospheric errors. Since such errors are expected to be on the order of magnitude of a centimeter, deformation rates of several tens of cm/y are necessary in order to bias the estimation of atmospheric errors. Such rates can be reached in landslides or mining areas that are typically restricted in coverage. Since the variograms are computed by averaging many different measurements over the whole scene, the effect of such areas should not strongly impact the estimation. It should be noted that since variogram estimation requires the unwrapped phase, this step must be performed at the end of the processing. At this stage, areas of very high deformation are apparent and can eventually be masked out. The effect of solid earth tides has been corrected in the interferometric phase using models [8]. However, the projection of tectonic plate motion onto the line of sight is a large-scale effect whose spatial characteristics cause a bias in variogram estimation. Such movements are mainly horizontal and can reach 6–7 cm/y. How large the horizontal motion must be in order to be comparable with the troposphere in a short temporal baseline interferogram can be easily estimated. For a 1 cm gradient along the radar swath, a horizontal motion of $1/\left(sin(\theta_{near}) - sin(\theta_{far})\right)$ is necessary where $\theta_{near}$ is the incidence angle at near range and $\theta_{far}$ the incidence angle at far range, leading to a horizontal motion of $\approx$ 5 cm or 152 cm/y for a revisit time of 12 days. Therefore, keeping the maximum $\delta t < 30$ days should result in negligible impact. It should be mentioned that the eventual presence of seismic events in the time series should also be assessed, and co-seismic interferometric pairs not be used to generate the variograms.

Given that such interferograms are almost deformation-free, it is possible to assume that the average of the variograms $\mathbb{E}[V]$ is a good estimator $\widehat{V}$ (the ˆsymbol indicates an estimated parameter) of the covariance characteristics of the residual atmospheric delays. Since velocity estimation implies a linear regression on the phase, a scaling factor accounting for acquisitions' time span and number must be applied to convert the single-phase measurement accuracy into deformation rate accuracy:

$$\widehat{V}_{rate}(d) = \frac{\lambda^2}{16\pi^2} \frac{\widehat{V}(d)}{2} \frac{M}{M \sum_k t_k^2 - \left(\sum_k t_k\right)^2}, \tag{4}$$

where $d$ is the distance, $\lambda$ the radar wavelength, $t_k$ the acquisition times, and $M$ the number of interferograms used for the linear regression. The analysis in [8] showed that if atmospheric phase corrections based on European Centre for Medium-Range Weather Forecasts (ECMWF) models are performed [23,24], the residual atmospheric effects after processing are well-modeled as stationary.

$\widehat{V}_{rate}(d)$ can then be fitted using a covariance model in order to compute the model parameters, converted to $\widehat{\Gamma}(d)$, and to estimate the covariance matrix $\mathbf{C}_\alpha$ in Equation (3). In this study, the exponential covariance model was used, as seen in Figure 2.

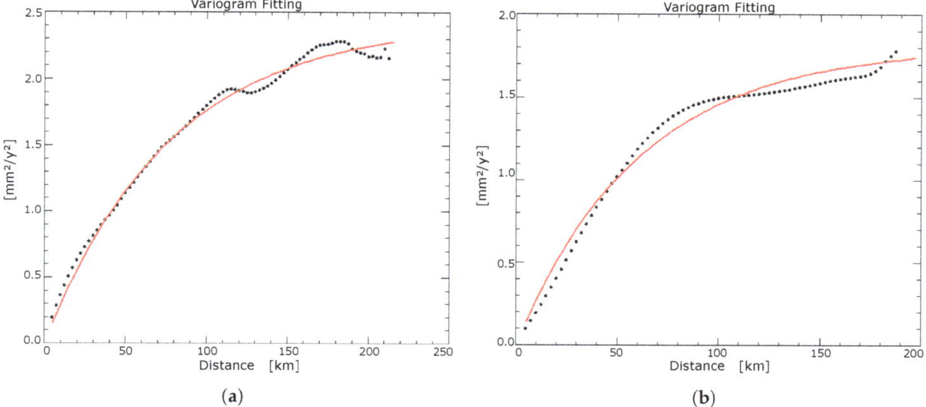

**Figure 2.** Examples of an exponential fitting of the deformation rate variograms. The black dots represent the estimated variogram and the bold red line represents the estimated model. The variograms in (**a**) and (**b**) refer to the stacks 123/2 and 123/3 of the North Anatolian Fault data set respectively , see Section 4.2.

## 2.3. Estimation of the Reference Point Motion

Equation (1) shows how $\underline{\Delta}$ can be seen as an observation of $v_{ref}$ with superimposed random and colored noise. Since the noise statistics of $\underline{\Delta}$ exhibit stationary behavior, the derivation of the reference point velocity can be carried out by averaging the observed $\underline{\Delta}$. Therefore, according to the considerations in the previous Section 2.1, $\widehat{v}_{ref}$ can be found by taking into account the computed covariance matrix $\mathbf{R}$:

$$\widehat{v}_{ref} = (\underline{u}\mathbf{R}^{-1}\underline{u}^T)^{-1}\underline{u}^T\mathbf{R}^{-1}\underline{\Delta}, \tag{5}$$

where $\underline{u}$ is a unitary vector that maps, one to one, the measurements $\underline{\Delta}$ with the unknown scalar $v_{ref}$. The estimated $\widehat{v}_{ref}$ represents the motion of the reference point in its local LoS and has to be added to the InSAR velocities in order to make them absolute.

## 2.4. Estimation of the 0-Mean Calibration Screen

In the previous Section 2.3, the overall offset between InSAR and GNSS deformation rates was estimated while accounting for the covariance matrix. Now, the space-variant error screen between the GNSS and InSAR velocities can be estimated by performing a covariance-based interpolation (Kriging) of the residual offsets $\underline{\Delta}_0 = \underline{\Delta} - \widehat{v}_{ref}$. A set of coefficients $\underline{c}$ that, combined with the vector $\underline{\Delta}_0$, allows the reconstruction of $\delta(r,a)$ everywhere must be estimated.

$$\widehat{\delta}(r,a) = \underline{c}^T\underline{\Delta}_0 \tag{6}$$

According to theory, this can be obtained by imposing the condition that the interpolation/prediction error be uncorrelated:

$$\epsilon_n = \widehat{\delta}(r_n, a_n) - \delta(r_n, a_n) \tag{7}$$

with the data $\underline{\Delta}_0$ [25]. Substituting Equation (6) into Equation (7) and imposing the uncorrelatedness condition with $\underline{\Delta}_0$ gives:

$$\underline{c} = \mathbf{R}^{-1}\underline{\rho}, \tag{8}$$

where $\underline{\rho} = \mathbb{E}[\delta(r, a)\underline{\Delta}_0]$ is the vector representing the correlation between the data vector $\underline{\Delta}_0$ and the error screen $\delta$ at the current position.

## 2.5. Variance and Covariance of the Results

To enhance usability, the error of the final product should be characterized. The final result is obtained by compensating the InSAR velocities for the estimated $\widehat{v}_{ref}$ and $\widehat{\delta}(r, a)$. The reference point deformation rate is a bias added to all points. Its error is therefore a constant value in the final product. Its contribution can be derived by computing the variance of the linear system inversion:

$$\sigma_{\widehat{v}_{ref}}^2 = (\underline{u}\mathbf{R}^{-1}\underline{u}^T)^{-1}. \tag{9}$$

The error of the estimated $\widehat{\delta}$ will be space variant. The variance and covariance of the estimated error screen $\widehat{\delta}(r, a)$ can be derived as in [25].

$$\mathbb{E}[\epsilon_n^2] = \mathbb{E}[(\delta(r_n, a_n) - \underline{c}^T\underline{\Delta}_0)^2] = \Gamma(0) - \underline{\rho_n}^T\mathbf{R}^{-1}\underline{\rho_n} \tag{10}$$

where $\Gamma(0)$ represents the original error due to the spatially correlated signal, and the second part of the equation represents a "mitigation factor" that reduces such variance according to the distance from the GNSS data. Analogously, the covariance is:

$$\mathbb{E}[\epsilon_n\epsilon_m] = \Gamma(d_{m,n}) - \underline{\rho_m}^T\mathbf{R}^{-1}\underline{\rho_n}. \tag{11}$$

Some conclusions concerning the final product can now be drawn. It provides absolute measurements that can be characterized by a variance directly derived from the applied methodology. It still contains a residual spatial correlation, though mitigated by the removal of $\widehat{\delta}(r, a)$. However, this residual error is no longer stationary. Its covariance depends on the two considered points, as highlighted by Equation (11). The variogram or covariance representation of the error—as for the input InSAR velocities—is hence no longer possible, but must must to be computed locally, accounting for the position $(r, a)$ with respect to the GNSS stations.

## 3. Simulations

Simulations were performed in order to assess the validity of the method and evaluate its performance. The scope of these simulations is to understand and quantify the effects of the spatial distribution (density) and the quality of the GNSS measurements in different processing scenarios, and to provide some numbers that summarize the achievable accuracies in the retrieval of $\widehat{v}_{ref}$ and $\widehat{\delta}$.

A set of randomly distributed points with zero deformation was simulated over a surface of $175 \times 250$ km$^2$ (Sentinel-1 slice size) in order to represent the InSAR measurements. Random noise (clutter) and spatially correlated noise (atmospheric residuals) were added to the points. For the generation of spatially correlated noise, an exponential covariance $\Gamma(d) = \sigma_\alpha^2 exp(-d/L_c)$ was used. A reduced set of GNSS zero-deformation data were also simulated at random positions within the scene. According to the model only random noise was added to the simulated GNSS velocities. For the sake of simplicity, positions were uniformly distributed within the scene. The reconstruction depends on how well the available samples are able to represent the error spectrum. In practice, the more high-pass the error, the more samples will be needed. Of course, a regular sampling is desirable. If the data are concentrated in an area, the re-construction will be good in this area. When moving away from data points, the estimator will extrapolate.

The simulations were performed while varying the two main parameters, the number of reference GNSS measurements, and their accuracy. Two different scenarios were tested with low and high atmospheric residual power $\sigma_\alpha^2$. In the first scenario, $\sigma_\alpha^2 = 2$ mm$^2$/y$^2$; for the Sentinel-1 mission, this is comparable to having a long time series of $T_{obs} \approx 3$ y in the case of applied atmospheric corrections [8]. In the second scenario, $\sigma_\alpha^2 = 9$ mm$^2$/y$^2$, representing the case of a short time series of $T_{obs} \approx 1$ y [8].

A realistic correlation length, according to real data, was used for the simulations—$L_{corr} = 60$ km in both cases. This comes from direct experience with the data. We observed that the average variograms after tropospheric corrections exhibited values of 40–100 km when fitted with an exponential model. An example of the simulation framework is shown in Figure 3.

Observing the simulation results in Figures 4 and 5 it can be concluded that, due to the spatial correlation of the error, higher densities of GNSS measurements improves the estimation of $\delta$ since it provides a better sampling of the error field, but does not help much in retrieving the absolute velocity $v_{ref}$, due to the "data redundancy" introduced by the spatial covariance.

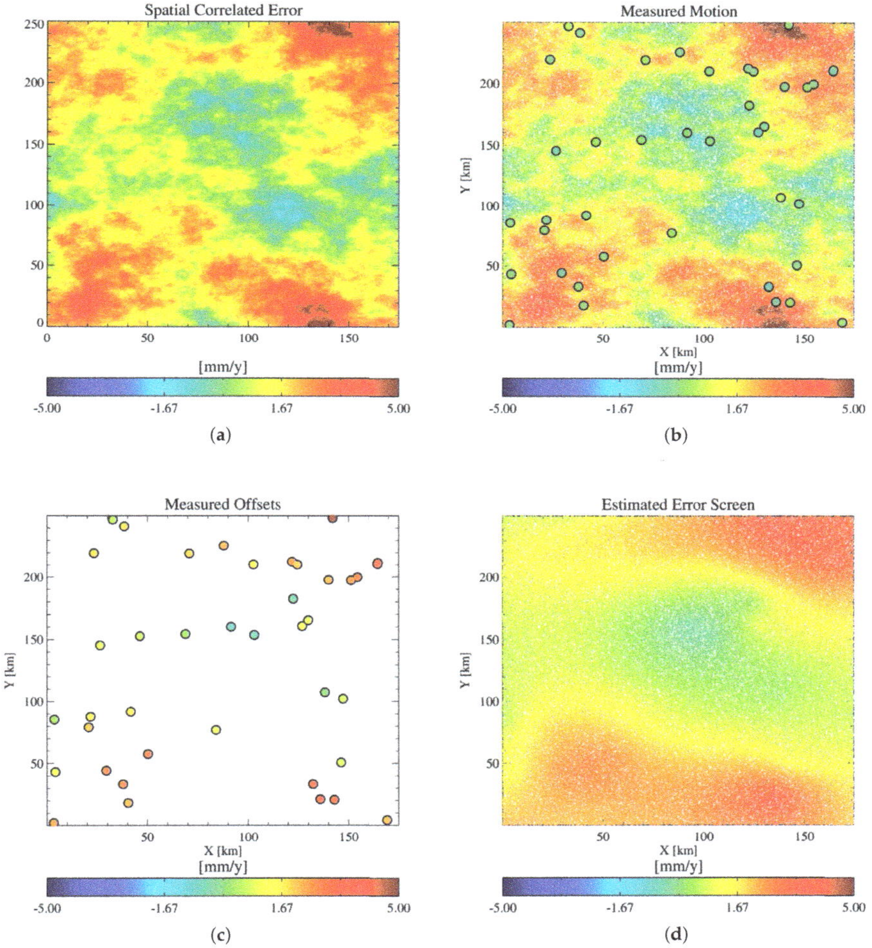

**Figure 3.** *Cont.*

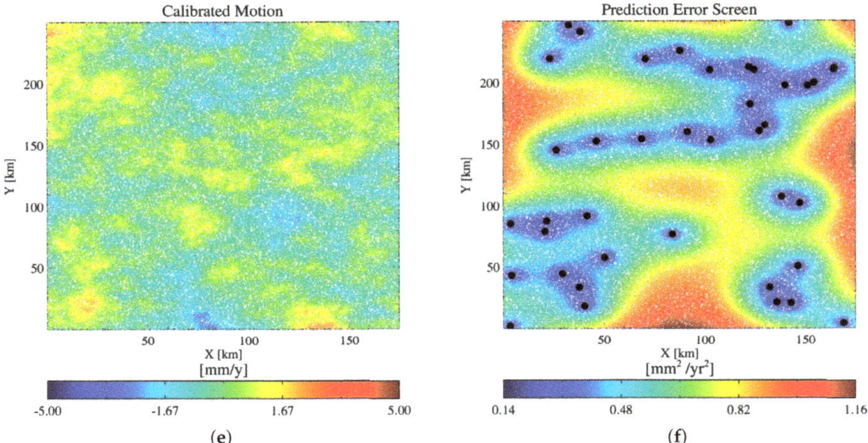


**Figure 3.** The Figure shows an example of the framework used for the simulations. The simulated motion is 0 over the entire scene. In (**a**), the simulated spatially correlated noise is depicted. In (**b**), the dense Synthetic Aperture Radar Interferometry (InSAR) measurements and the coarse Global Navigation Satellite System (GNSS; large dots, atmosphere-free) are depicted. In (**c**), the measured offsets InSAR/GNSS ($\underline{\Delta}_0$) are depicted. In (**d**), the estimated error screen to be removed from (**b**) is depicted. In (**e**), the final results obtained by calibrating (**b**) with (**d**) are depicted. In (**f**), the estimated error for the merged product is depicted.

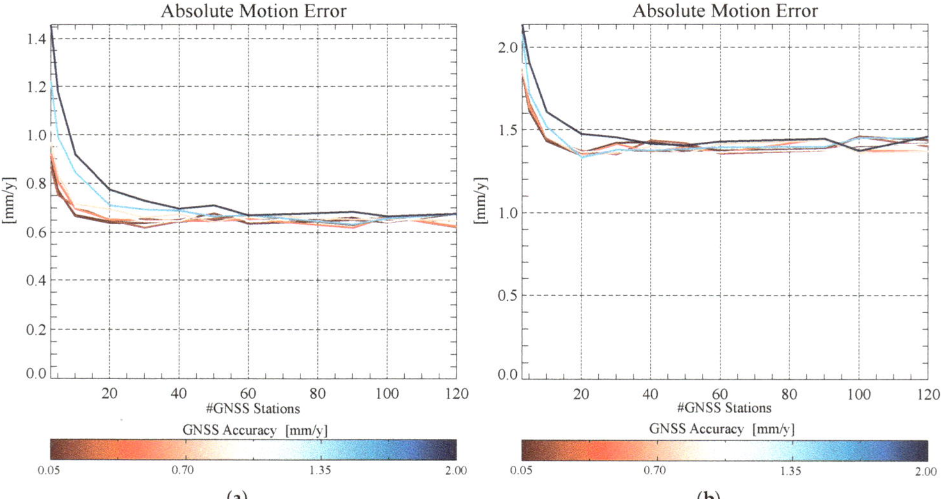


**Figure 4.** Performance simulations in retrieving $\hat{v}_{ref}$ carried out with different atmospheric noise sills: (**a**) $\sigma_\alpha^2 = 2$ mm²/y² and (**b**) $\sigma_\alpha^2 = 9$ mm²/y².

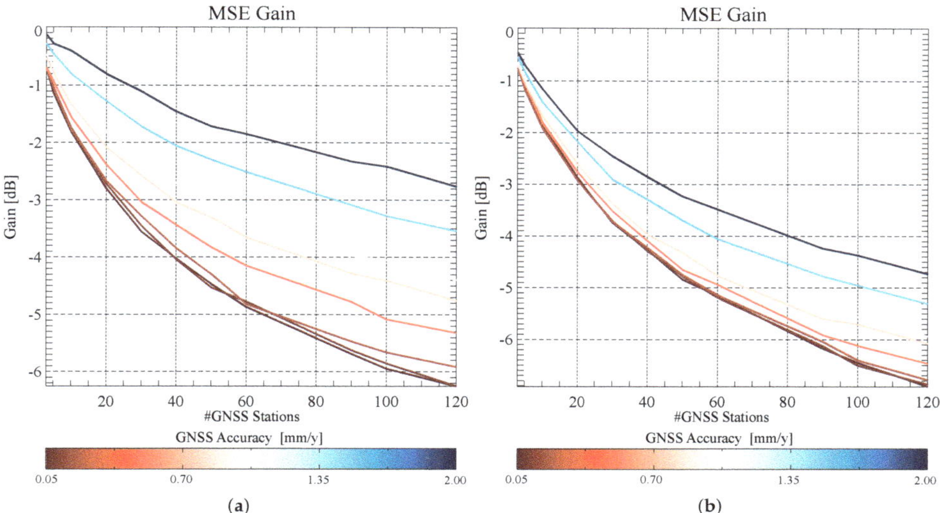

**Figure 5.** Performance simulations in retrieving $\widehat{\delta}$ carried out with different atmospheric noise sills: (a) $\sigma_\alpha^2 = 2\,\text{mm}^2/\text{y}^2$ and (b) $\sigma_\alpha^2 = 9\,\text{mm}^2/\text{y}^2$.

### 3.1. Performance in Retrieving $\widehat{v}_{ref}$

The accuracy of $\widehat{v}_{ref}$ is that of the mean computed from a set of correlated samples. The level of dependence of the given dataset is controlled by the spatial density of the latter and the correlation length $L_{corr}$ of the superimposed noise. We observed that it is logical to expect that, given a $L_{corr}$, increasing the number of data improves estimation of $\widehat{v}_{ref}$ up to a certain level, since as the density of measurements increases, so to does the amount of correlation between measurement, hence limiting its impact on final performance. In a typical scenario ( $T_{obs} \approx 3\,\text{y}$ , $\sigma_\alpha^2 = 2\,\text{mm}^2/\text{y}^2$), the accuracy can easily be brought below 1 mm/y, even with a limited set of GNSS stations (Figure 4).

### 3.2. Performance in Retrieving $\widehat{\delta}$

Evaluating the performance of $\widehat{\delta}$ is not a simple task. The most natural way would be to compare the variograms of the results before and after application of the technique. This approach would deliver a deeper insight into the achieved gain, since it would be possible to show this as a function of the scale. However, as mentioned in Section 2.5, the error of the merged results is no longer stationary making a variogram representation questionable. In order to avoid this issue, a simpler but more robust approach was followed. The performances were evaluated in terms of mean square error (MSE) in *dB*, over the whole scene. In practice, the mean power of the residual deformation signal after removal of the estimated error screen $\widehat{\delta}$ from the measured rates $v_{meas}$ is:

$$MSE_{dB} = 10Log_{10}\left( \mathbb{E}[(v_{meas} - \widehat{\delta})^2] \right). \tag{12}$$

The results displayed in Figure 5a,b show that, in both cases, the performance cannot improve beyond a certain level by improving the quality of the GNSS measurements only. An improvement in GNSS coverage is also necessary in order to better compensate the higher wave-numbers of the error screen.

## 4. Results

For the study, two Sentinel-1A/B datasets covering the northern part of the Netherlands (two stacks, ascending and descending) and the junction between the North Anatolian Fault and East Anatolian Fault (three stacks, descending) were used. The interferograms were computed, corrected for tropospheric delays using ECMWF ERA-5 (ECMWF Re-Analysis) data [8] and processed using the PSInSAR technique [26]. In order to preserve all wave-numbers of the deformation signal, no spatial high-pass filtering or polynomial detrending was performed on the final data. The technique described above was applied to the estimated deformation rates. As visible in Figures 6–11, the method is able to retrieve the absolute motion while also mitigating the undesired residual atmospheric error, hence displaying all available spectral components in the deformation signal. Figures 7a, 9a, and 11a show that the residual zero-mean error $\Delta_0$ is quite small, varying between $\pm 2$ mm/y. This can already be considered as a kind of validation of the InSAR data, whose accuracy is close to 1 mm/y.

For the reference measurements, the GNSS data processed by Nevada Geodetic Laboratories [27,28] were used. The correspondence GNSS/PSs for the calculation of vector $\underline{\Delta}$ was implemented by averaging all of the PS rates within a radius of 250 m from the GNSS station.

A further elucidation needs to be made. The reference system that describes SAR geometry (state vectors) does not account for continental drift making plate movement visible in the interferometric measurements, projected along the LoS. The GNSS data used for the calibration are also not continental drift compensated. Hence, the two data-sets can be considered "compatible". After merging (also adding $v_{ref}$), the data contains continental drift that could dominant the visualialization. Therefore, plate movement was compensated using the model in [29] to show movement relative to a specific plate (Eurasian). The removed motion is basically an overall offset of several mm/y, plus a light ramp due to the LoS projection of the horizontal motion. This is generally not necessary since it is only a representation issue, but it facilitates the interpretation of the results. For the sake of completeness, the removed model is also shown in Figures 7b, 9b, and 11b.

### 4.1. Netherlands Datasets

The systematic generation of deformation maps on a national scale using SAR interferometry has become more feasible, as missions like Sentinel-1 provide global coverage on a regular basis. Often, these kinds of services must be combined with existing geodetic measurements in order to make them integrated and comparable [21]. In this work, a typical example is provided where InSAR velocities covering a large area—acquired in descending and ascending passes—are combined with a consistent number of GNSS-derived velocities. The considered datasets are two Sentinel-1A/B stacks covering the northern part of the Netherlands. The area includes many examples of man-induced subsidences, such as in Groeningen [30]. The details of the dataset are described by Table 1. Observing the raw result of the PSI-processing, some spatially correlated deformation signals are visible when observing between $\pm 4$ mm/y. Since no large-scale deformation phenomena are expected in this area, such patterns must therefore be related to the residual atmospheric signal. In the area of interest, a considerably high number of GNSS data are available (see Table 1). The previously described methodology was applied to the data using the available GNSS measurements [27]. The results in Figures 6 and 8 show how the technique permits the further reduction of the residual atmospheric error and the relation of every single PS measurement to the Eurasian plate only. Further information about the measured $\Delta_0$ and the removed continental drift model is also available in Figures 7 and 9. Since the ascending/descending data are now absolute, it would be possible to combine them directly using the error description provided in Equations (9)–(11).

**Table 1.** Netherlands Dataset.

| Track | Orbit | Acquisitions | Time [years] | GNSS Stations |
|-------|-------|--------------|--------------|---------------|
| 088 | ASCE | 140 | 3.4 | 47 |
| 037 | DESCE | 122 | 3.0 | 37 |

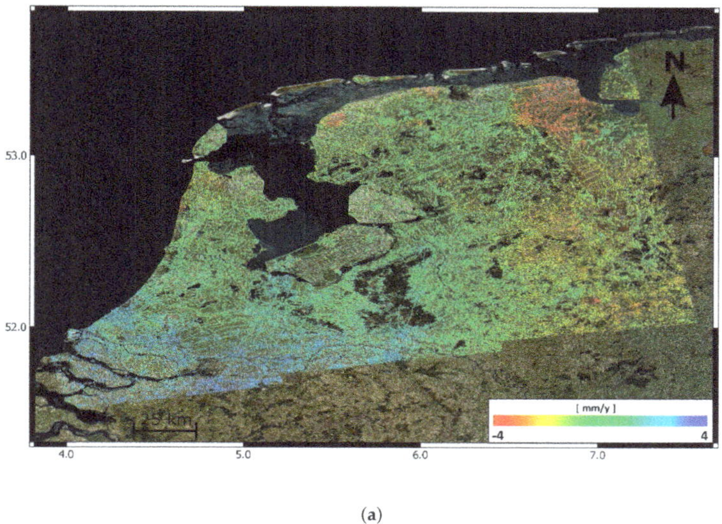

(**a**)

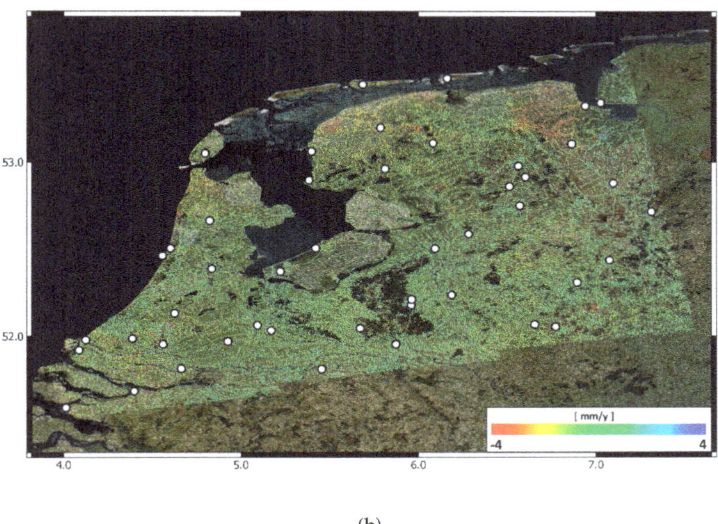

(**b**)

**Figure 6.** Results for the ascending Holland stack (**a**) before and (**b**) after performing merging with GNSS data. In (**b**), the GNSS stations used are displayed with black–white circles.

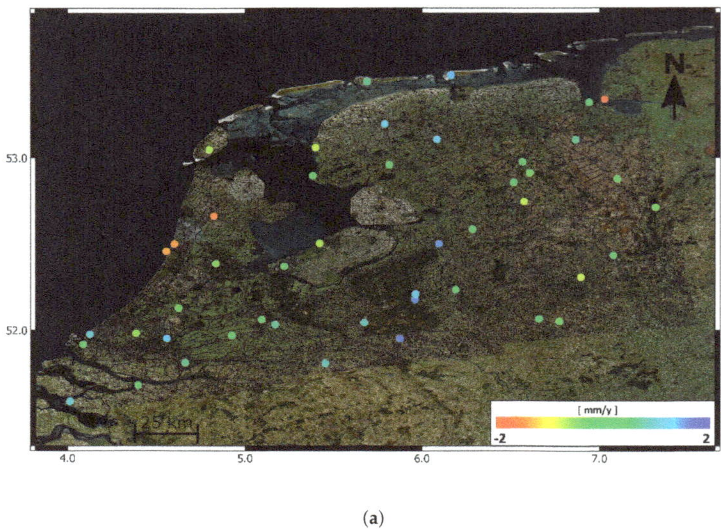



(**a**)

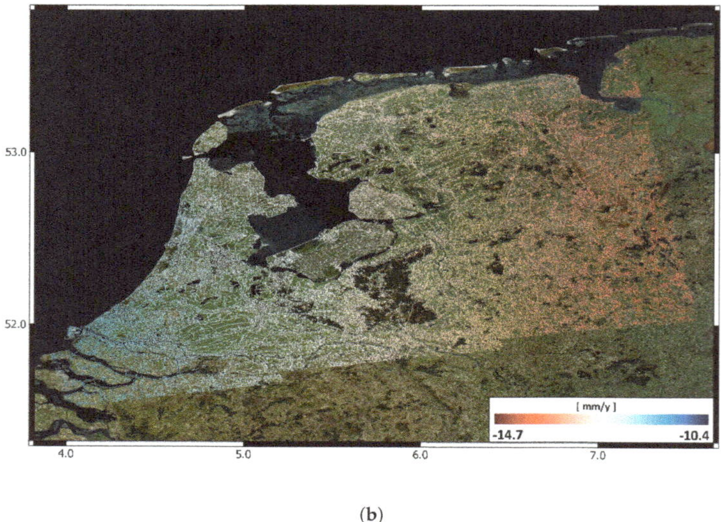



(**b**)

**Figure 7.** Results for the ascending Holland stack, showing the measured and color-coded $\Delta_0$ in (**a**) and the removed continental drift screen in (**b**).

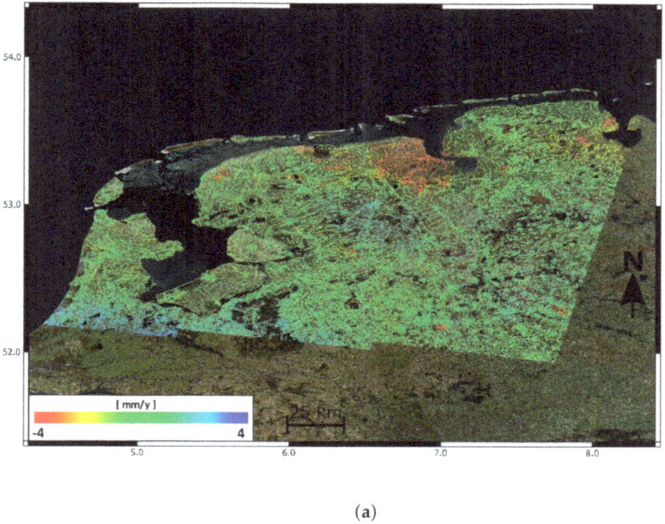

(**a**)

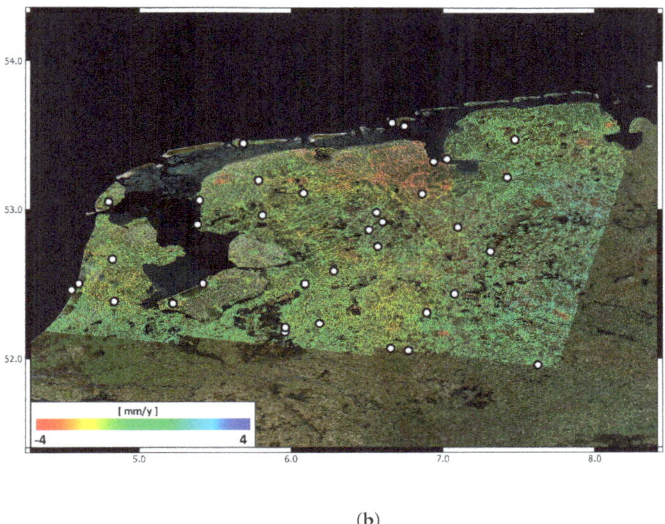

(**b**)

**Figure 8.** Results for the descending Holland stack, showing (**a**) before and (**b**) after performing merging with GNSS data. In (**b**), the GNSS stations used are displayed with black–white circles.

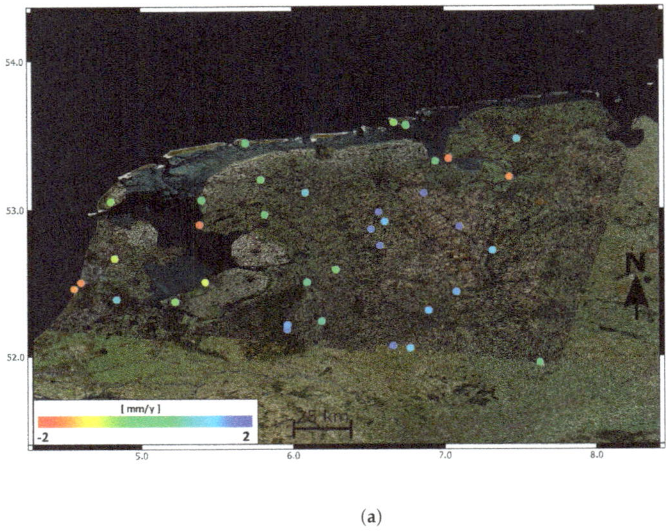

(**a**)

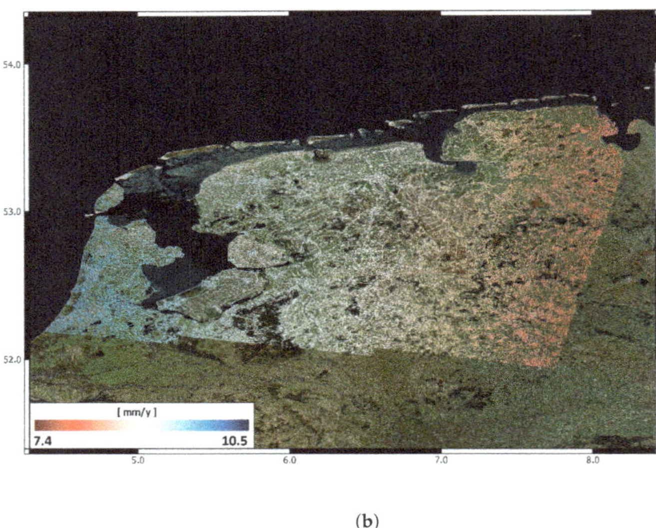

(**b**)

**Figure 9.** Results for the descending Holland stack, showing (**a**) the measured and color-coded $\Delta_0$ and (**b**) the removed continental drift screen.

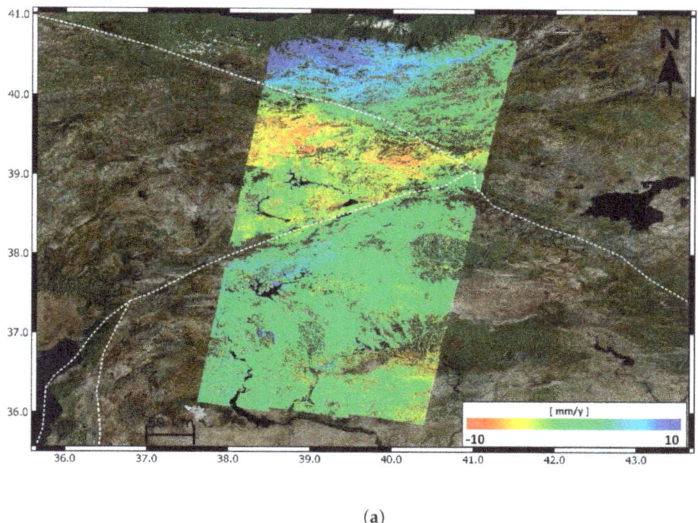

(**a**)

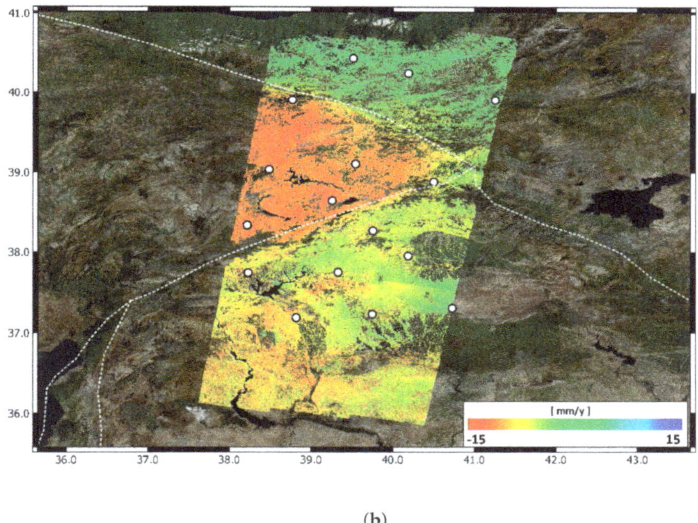

(**b**)

**Figure 10.** Results for the North Anatolian Fault (**a**) before and (**b**) after performing merging with GNSS data. In (**b**), the GNSS stations used are displayed with black–white circles and the fault lines are displayed using a white dashed line.

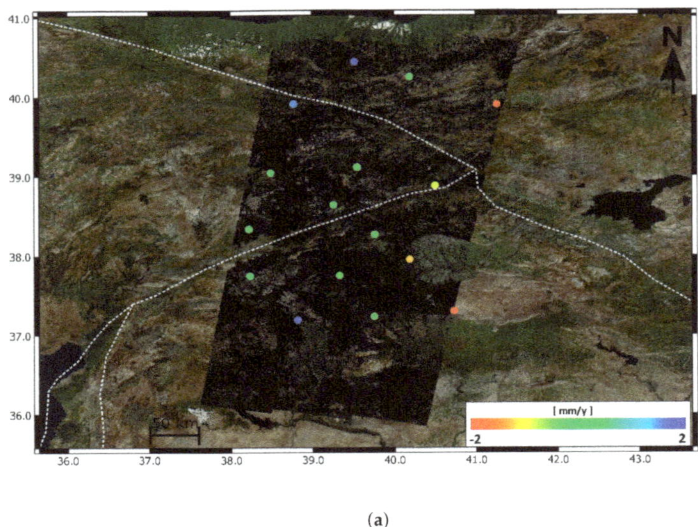

(**a**)

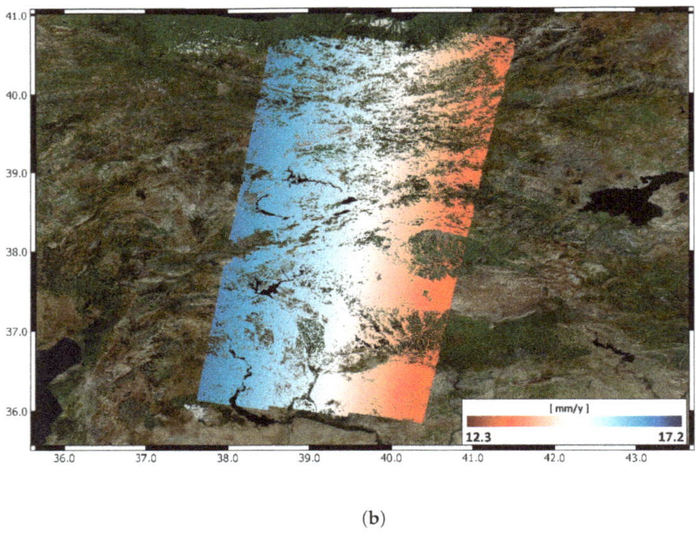

(**b**)

**Figure 11.** Results for the North Anatolian Fault, showing (**a**) the measured and color-coded $\Delta_0$ and (**b**) the removed continental drift screen.

*4.2. North Anatolian Fault Dataset*

Applications related to the measurement of tectonic movements are the most challenging for InSAR. The requirement of an accuracy of 1 mm/y at more that 100 km (<10 nstrain) pushes the technique to its limits [5]. Notwithstanding, it has been demonstrated that, since actual SAR missions are characterized by very stable oscillators [31] and very good orbit knowledge [32,33], such numbers can be considered achievable. The major limitation is atmospheric effects. If the interferometric time series is not long enough, integration with other data would be necessary in order to fulfill the requirements at very large distances. Moreover, the possibility of referencing the measurements to standard reference systems (e.g., Eurasian Plate, etc.) would extend the usability of the final results.

The technique must then also be demonstrated on an appropriate example for tectonics. The North Anatolian Fault was chosen, as it is a typical case study investigated by many geo-scientists using SAR interferometry [34,35]. The area of interest is covered by a Sentinel-1A/B stripe extending for more than 600 km in the along-track direction, see Table 2. In order to make the processing feasible, the stripe was divided into three frames. The three frames were processed independently using the PSInSAR technique. The area is very tectonically active and it is very important to preserve all scales of the measured deformation signal. Figure 10 displays the results together with a mapping of the main path of the North and East Anatolian Fault (dashed white line).

**Table 2.** North Anatolian Fault Dataset.

| Track | Orbit | Acquisitions | Time [years] | GNSS Stations |
|-------|-------|--------------|--------------|---------------|
| 123/2 | DESCE | 134 | 3.3 | 5 |
| 123/3 | DESCE | 134 | 3.3 | 9 |
| 123/4 | DESCE | 134 | 3.3 | 4 |

## 5. Conclusions

The results show how the technique is able to both remove dependence on the reference point and mitigate residual atmospheric errors present in the final InSAR results. The described methodology works under some straightforward assumptions on the input data that have to be considered:

- The GNSS/InSAR time series must overlapped in space and time,
- the 3D GNSS velocities and their variances are known, and
- the motion of the GNSS station should be representative of the motion of the area.

The method was studied on both simulated and real data. Simulations show that even with a reduced set of GNSS stations it is possible to retrieve the absolute motion with good accuracy. On the other hand, the capability of the approach to mitigate residual atmospheric errors depends both on the GNSS station density and the spatial correlation of the errors. In order to provide some numbers, the error was modeled as an $AR(1)$ process with a correlation length of 60 km and variable overall variance. Such numbers were used because they match with the experiments. Two real test cases were then processed to demonstrate the approach by trying to address applications that should benefit from the technique. The results are promising and open the door to an accurate cross-validation between GNSS/InSAR. The extension of the approach to time series would also be interesting, merging InSAR phases with continuous GNSS measurements. The approach could be conceptually equivalent, since the deformation rates are just a scaling of the single-phase measurements, but the spatial covariance of each interferogram would be needed.

For the sake of precision, it is noted that estimation of $\hat{v}_{ref}$ is also generally possible using only InSAR data. The key parameter to be considered here is indeed the length of the time series $T_{obs}$. Theoretically speaking, with a long time series, it is possible to robustly estimate the motion of the reference point by comparing the interferometric deformation rates with the rates estimated from group delays, using co-registration shifts or the PS positions [36,37]. Moreover, if $T_{obs}$ is large enough, the residual atmospheric effects are also sufficiently small to fulfill even the strict requirements [8] set by applications aimed at measuring tectonic movements [5]. Notwithstanding, if $T_{obs}$ is not large enough to fulfill the accuracy requirements at large distances, then merging with GNSS is necessary.

In general, it is possible to conclude that the developed method helps to increase, when necessary, the accuracy of InSAR over large distances [8]. The final product has the spatial coverage of InSAR, and is an optimal combination of InSAR and GNSS information based on the knowledge of the error statistics. This preserves all scales of the deformation signal and also allows characterization of the final product error.

**Author Contributions:** Conceptualization, A.P., F.R.G. and R.B.; Methodology, A.P.; Software, A.P. and F.R.G.; Investigation, A.P.; Writing–Original Draft Preparation, A.P.; Writing–Review and Editing, A.P. and R.B. All authors have read and agreed to the published version of the manuscript.

**Funding:** This research received no external funding.

**Acknowledgments:** The authors would like to thank Ekaterina Tymofyeyeva for the suggestions that helped improving the manuscript.

**Conflicts of Interest:** The authors declare no conflict of interest.

## Abbreviations

The following abbreviations are used in this manuscript:

| | |
|---|---|
| SAR | Synthetic Aperture Radar |
| InSAR | Synthetic Aperture Radar Interferometry |
| PS | Persistent Scatterer |
| PSI | Persistent Scatterers Interferometry |
| LoS | Line of Sight |
| GNSS | Global Navigation Satellite System |
| ECMWF | European Centre for Medium-Range Weather Forecasts |
| ERA | ECMWF Re-Analysis |

## References


1. Rosen, P.; Hensley, S.; Joughin, I.; K.Li, F.; Madsen, S.; Rodriguez, E.; Goldstein, R.M. Synthetic aperture radar interferometry. *Proc. IEEE* **2000**, *88*, 333–382. [CrossRef]
2. Rosen, P.A.; Hensley, S.; Zebker, H.A.; Webb, F.H.; Fielding, E.J. Surface deformation and coherence measurements of Kilauea Volcano, Hawaii, from SIR-C radar interferometry. *J. Geophys. Res. Planets* **1996**, *101*, 23109–23125. [CrossRef]
3. Gray, A.L.; Mattar, K.E.; Sofko, G. Influence of ionospheric electron density fluctuations on satellite radar interferometry. *Geophys. Res. Lett.* **2000**, *27*, 1451–1454. [CrossRef]
4. Hanssen, R. *Radar Interferometry: Data Interpretation and Error Analysis*; Springer: Dordrecht, The Netherlands, 2001.
5. Wright, T.J. The earthquake deformation cycle. *Astron. Geophys.* **2016**, *57*, 4.20–4.26. atw148. [CrossRef]
6. Torres, R.; Snoeij, P.; Geudtner, D.; Bibby, D.; Davidson, M.; Attema, E.; Potin, P.; Rommen, B.; Floury, N.; Brown, M.; et al. GMES Sentinel-1 mission. *Remote Sens. Environ.* **2012**, *120*, 9–24. [CrossRef]
7. Moreira, A.; Krieger, G.; Hajnsek, I.; Papathanassiou, K.; Younis, M.; Lopez-Dekker, P.; Huber, S.; Villano, M.; Pardini, M.; Eineder, M.; et al. Tandem-L: A Highly Innovative Bistatic SAR Mission for Global Observation of Dynamic Processes on the Earth's Surface. *IEEE Geosci. Remote Sens. Mag.* **2015**, *3*, 8–23. [CrossRef]
8. Gonzalez, F.R.; Parizzi, A.; Brcic, R. Evaluating the impact of geodetic corrections on interferometric deformation measurements. In Proceedings of the EUSAR 2018: 12th European Conference on Synthetic Aperture Radar, Aachen, Germany, 4–7 June 2018; pp. 1–5.
9. Gudmundsson, S.; Carstensen, J.M.; Sigmundsson, F. Unwrapping ground displacement signals in satellite radar interferograms with aid of GPS data and MRF regularization. *IEEE Trans. Geosci. Remote Sens.* **2002**, *40*, 1743–1754. [CrossRef]
10. Shanker, A.P.; Zebker, H. Edgelist phase unwrapping algorithm for time series InSAR analysis. *JOSA A* **2010**, *27*, 605–612. [CrossRef]
11. Wei, M.; Sandwell, D.; Smith-Konter, B. Optimal combination of InSAR and GPS for measuring interseismic crustal deformation. *Adv. Space Res.* **2010**, *46*, 236–249. [CrossRef]
12. Tong, X.; Sandwell, D.; Smith-Konter, B. High-resolution interseismic velocity data along the San Andreas fault from GPS and InSAR. *J. Geophys. Res. Solid Earth* **2013**, *118*, 369–389. [CrossRef]

13. Farolfi, G.; Bianchini, S.; Casagli, N. Integration of GNSS and Satellite InSAR Data: Derivation of Fine-Scale Vertical Surface Motion Maps of Po Plain, Northern Apennines, and Southern Alps, Italy. *IEEE Trans. Geosci. Remote Sens.* **2019**, *57*, 319–328. [CrossRef]

14. Wang, H.; Wright, T. Satellite geodetic imaging reveals internal deformation of western Tibet. *Geophys. Res. Lett.* **2012**, *39*, doi:10.1029/2012GL051222. [CrossRef]

15. Catalão, J.; Nico, G.; Hanssen, R.; Catita, C. Merging GPS and atmospherically corrected InSAR data to map 3-D terrain displacement velocity. *IEEE Trans. Geosci. Remote Sens.* **2011**, *49*, 2354–2360. [CrossRef]

16. Bekaert, D.; Segall, P.; Wright, T.; Hooper, A. A network inversion filter combining GNSS and InSAR for tectonic slip modeling. *J. Geophys. Res. Solid Earth* **2016**, *121*, 2069–2086. [CrossRef]

17. Mahapatra, P.S.; Samiei-Esfahany, S.; van der Marel, H.; Hanssen, R.F. On the use of transponders as coherent radar targets for SAR interferometry. *IEEE Trans. Geosci. Remote Sens.* **2014**, *52*, 1869–1878. [CrossRef]

18. Mahapatra, P.; van der Marel, H.; van Leijen, F.; Samiei-Esfahany, S.; Klees, R.; Hanssen, R. InSAR datum connection using GNSS-augmented radar transponders. *J. Geod.* **2018**, *92*, 21–32. [CrossRef]

19. Mahapatra, P.S.; Samiei-Esfahany, S.; Hanssen, R.F. Geodetic Network Design for InSAR. *IEEE Trans. Geosci. Remote Sens.* **2015**, *53*, 3669–3680. [CrossRef]

20. Jonsson, S.; Zebker, H.; Segall, P.; Amelung, F. Fault Slip Distribution of the 1999 Mw 7.1 Hector Mine, California, Earthquake, Estimated from Satellite Radar and GPS Measurements. *Bull. Seismol. Soc. Am.* **2002**, *92*, 1377–1389. [CrossRef]

21. Kalia, A.; Frei, M.; Lege, T. A Copernicus downstream-service for the nationwide monitoring of surface displacements in Germany. *Remote Sens. Environ.* **2017**, *202*, 234–249. [CrossRef]

22. Fattahi, H.; Amelung, F. InSAR bias and uncertainty due to the systematic and stochastic tropospheric delay. *J. Geophys. Res. Solid Earth* **2015**, *120*, 8758–8773. [CrossRef]

23. Cong, X.; Balss, U.; Eineder, M.; Fritz, T. Imaging geodesy—Centimeter-level ranging accuracy with TerraSAR-X: An update. *IEEE Geosci. Remote Sens. Lett.* **2012**, *9*, 948–952. [CrossRef]

24. Cong, X. SAR Interferometry for Volcano Monitoring: 3D-PSI Analysis and Mitigation of Atmospheric Refractivity. Ph.D. Thesis, Technische Universität München, München, Germany, 2014.

25. Proakis, J.; Manolakis, D. *Digital Signal Processing*; Prentice Hall: Upper Saddle River, NJ, USA, 2006.

26. Ferretti, A.; Prati, C.; Rocca, F. Permanent scatterers in SAR interferometry. *IEEE Trans. Geosci. Remote Sens.* **2001**, *39*, 8–20. [CrossRef]

27. MIDAS Velocity Fields. Available online: http://geodesy.unr.edu/ (accessed on 1 October 2018).

28. Blewitt, G.; Kreemer, C.; Hammond, W.C.; Gazeaux, J. MIDAS robust trend estimator for accurate GPS station velocities without step detection. *J. Geophys. Res. Solid Earth* **2016**, *121*, 2054–2068. [CrossRef]

29. Rouby, H.; Métivier, L.; Rebischung, P.; Altamimi, Z.; Collilieux, X. ITRF2014 plate motion model. *Geophys. J. Int.* **2017**, *209*, 1906–1912. [CrossRef]

30. Ketelaar, V.G. *Satellite Radar Interferometry: Subsidence Monitoring Techniques*; Springer Science & Business Media: Berlin/Heidelberg, Germany, 2009; Volume 14.

31. Larsen, Y.; Marinkovic, P.; Dehls, J.F.; Perski, Z.; Hooper, A.J.; Wright, T.J. The Sentinel-1 constellation for InSAR applications: Experiences from the InSARAP project. In Proceedings of the 2017 IEEE International Geoscience and Remote Sensing Symposium (IGARSS), Fort Worth, TX, USA, 23–28 July 2017; pp. 5545–5548.

32. Yoon, Y.T.; Eineder, M.; Yague-Martinez, N.; Montenbruck, O. TerraSAR-X Precise Trajectory Estimation and Quality Assessment. *IEEE Trans. Geosci. Remote Sens.* **2009**, *47*, 1859–1868. [CrossRef]

33. Peter, H.; Jäggi, A.; Fernández, J.; Escobar, D.; Ayuga, F.; Arnold, D.; Wermuth, M.; Hackel, S.; Otten, M.; Simons, W.; et al. Sentinel-1A–First precise orbit determination results. *Adv. Space Res.* **2017**, *60*, 879–892. [CrossRef]

34. Cakir, Z.; Ergintav, S.; Akoğlu, A.M.; Çakmak, R.; Tatar, O.; Meghraoui, M. InSAR velocity field across the North Anatolian Fault (eastern Turkey): Implications for the loading and release of interseismic strain accumulation. *J. Geophys. Res. Solid Earth* **2014**, *119*, 7934–7943. [CrossRef]

35. Hussain, E.; Wright, T.J.; Walters, R.J.; Bekaert, D.P.; Lloyd, R.; Hooper, A. Constant strain accumulation rate between major earthquakes on the North Anatolian Fault. *Nat. Commun.* **2018**, *9*, 1392. [CrossRef]

36.  Madsen, S.N.; Zebker, H.A.; Martin, J. Topographic mapping using radar interferometry: Processing techniques. *IEEE Trans. Geosci. Remote Sens.* **1993**, *31*, 246–256. [CrossRef]
37.  Eineder, M.; Minet, C.; Steigenberger, P.; Cong, X.; Fritz, T. Imaging Geodesy—Toward Centimeter-Level Ranging Accuracy With TerraSAR-X. *IEEE Trans. Geosci. Remote Sens.* **2011**, *49*, 661–671. TGRS.2010.2060264. [CrossRef]

*remote sensing*

*Article*

# Geodetic Model of the March 2021 Thessaly Seismic Sequence Inferred from Seismological and InSAR Data

Vincenzo De Novellis [1,2], Diego Reale [1], Guido Maria Adinolfi [3], Eugenio Sansosti [1,*]
and Vincenzo Convertito [2]

[1]  National Research Council (CNR) of Italy, Istituto per il Rilevamento Elettromagnetico dell'Ambiente (IREA), 80124 Napoli, Italy; vincenzo.denovellis@cnr.it (V.D.N.); diego.reale@cnr.it (D.R.)
[2]  Istituto Nazionale di Geofisica e Vulcanologia, Osservatorio Vesuviano, 80124 Napoli, Italy; vincenzo.convertito@ingv.it
[3]  Dipartimento di Scienze e Tecnologie, Università degli Studi del Sannio, 82100 Benevento, Italy; gmadinolfi@unisannio.it
*   Correspondence: eugenio.sansosti@cnr.it

**Abstract:** In this work, we propose a geodetic model for the March 2021 Thessaly seismic sequence (TSS). We used the interferometric synthetic aperture radar (InSAR) technique and exploited a dataset of Sentinel-1 images to successfully detect the surface deformation caused by three major events of the sequence and constrain their kinematics, further strengthened by seismic data analysis. Our geodetic inversions are consistent with the activation of distinct blind faults previously unknown in this region: three belonging to the NE-dipping normal fault associated with the Mw 6.3 and Mw 6.0 events, and one belonging to the SW-dipping normal fault associated with the Mw 5.6, the last TSS major event. We performed a Coulomb stress transfer analysis and a 1D pore pressure diffusivity modeling to investigate the space–time evolution of the sequence; our results indicate that the seismic sequence developed in a sort of domino effect. The combination of the lack of historical records of large earthquakes in this area and the absence of mapped surface features produced by past faulting make seismic hazard estimation difficult for this area: InSAR data analysis and modeling have proven to be an extremely useful tool in helping to constrain the rupture characteristics.

**Keywords:** 2021 Thessaly seismic sequence; 3D complex fault geometry; analytical modeling; InSAR measurements

**Citation:** De Novellis, V.; Reale, D.; Adinolfi, G.M.; Sansosti, E.; Convertito, V. Geodetic Model of the March 2021 Thessaly Seismic Sequence Inferred from Seismological and InSAR Data. *Remote Sens.* **2021**, *13*, 3410. https://doi.org/10.3390/rs13173410

Academic Editor: Fumio Yamazaki

Received: 7 July 2021
Accepted: 24 August 2021
Published: 27 August 2021

**Publisher's Note:** MDPI stays neutral with regard to jurisdictional claims in published maps and institutional affiliations.



## 1. Introduction

The Thessaly seismic sequence (TSS), Central Greece, started on 3 March 2021 (10:16:08 UTC) with a Mw 6.3 event and struck an area about 25 km WNW of the town of Larissa (Figure 1). Fortunately, it caused no fatalities, while only three people were slightly injured due to the partial collapse of buildings with load-bearing masonry walls in the village of Damasi; moreover, widespread liquefaction phenomena were observed in the fields located close to local rivers [1]. Based on various seismological observatories and institutes (AUTH, NOA, GCMT, USGS, OCA, INGV, GFZ), the first event was characterized by a normal fault mechanism (Figure 1a). Several aftershocks (Mw > 5) concentrated to the southeast during the first few hours after the event within a distance of 10 km (Figure 1a).

In the following days, TSS was characterized by other two major events: a Mw 6.0 on March 4 (18:38:19 UTC), which has been localized about 7 km northwest from the first event; and a Mw 5.6 on March 12 (12:57:50 UTC), localized 12 km further toward the NW. A large number of smaller events have also been recorded until mid-April when the sequence decreased in frequency and magnitude.

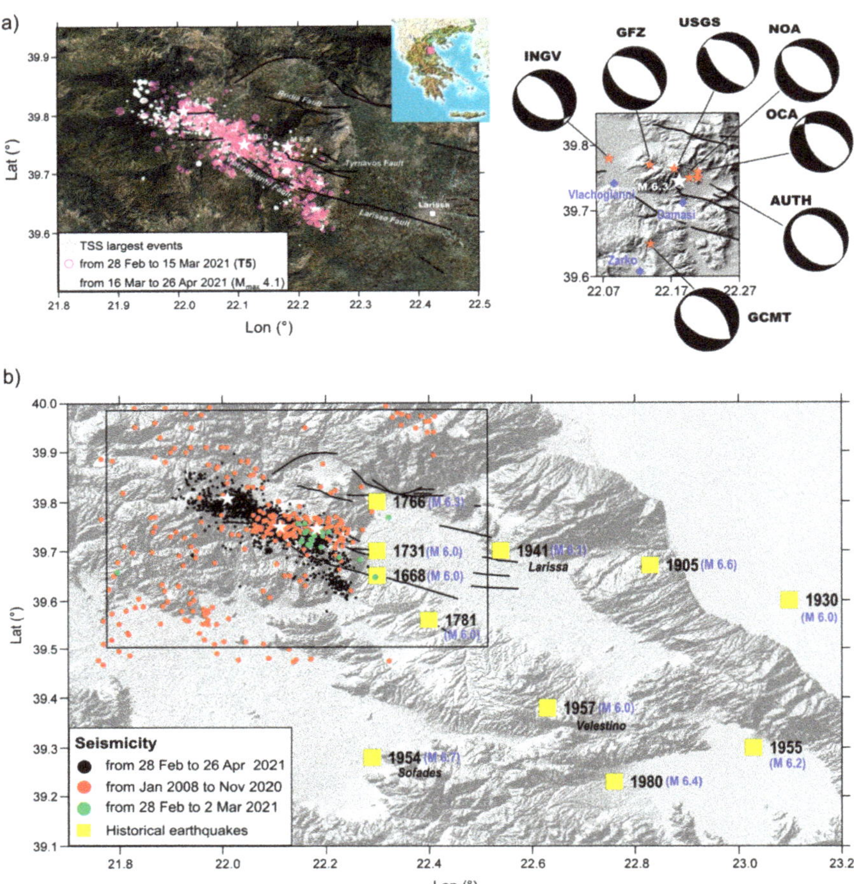

**Figure 1.** Study area and seismological data. (**a**; left) Main tectonic lineaments redrawn after [2] and represented by black lines; relocated earthquakes from 28 February 2021 to 15 March 2021 and from 16 March 2021 to 26 April 2021, represented by magenta and white circles, respectively. (**a**; right) Mw 6.3 epicentral area with focal mechanisms provided by various institutes; most damaged villages are also shown. (**b**) Recent recorded earthquakes in the Tyrnavos area. Historical earthquakes that occurred in Thessaly area are also shown (data by [3]). Recent earthquakes occurred in an area not covered by past events, thus filling a seismic gap.

The Thessaly region is located at the back-arc area of the Aegean microplate and is one of the most seismically active areas of Greece [2,4]. Since 1 January 2008, more than 320 revised events have been detected from the Institute of Geodynamics—National Observatory of Athens database (IG-NOA) in the extended fault area near Tyrnavos: this background seismicity at the end of 2020 had a magnitude value ranging between 0.7 to 3.6. The seismic activity resumed on 9 January 2021, with events with a magnitude ranging between 1.5 to 2. A comprehensive description of the historical and instrumental seismicity of the area is provided in detail in [3,5].

This region is mainly characterized by NW–SE-striking normal faults, testifying an extensional stress regime with low strain rate. Several active fault zones have been detected into the central and northern part of Greece and some of them have been associated with past destructive earthquakes [6] (Figure 1b). In particular, the seismicity of the Thessaly Basin originates mainly along two fault zones: the northern one, concentrated in the Larissa plain, and a southern fault zone [7]. These two zones suggest that there has been a seismic gap in the Larissa plain [8]. Overall, the 2021 TSS sequence affected the area surrounded by past events, thus filling the gap between previous ruptures.

Concerning the structures involved in TSS, a sporadic series of NW–SE striking surface breaks were found on the mountains to the north of Zarko village [1]. Unfortunately, no tectonic ruptures were found corresponding to the Mw 6.0 and Mw 5.6 events. Therefore, the geodetic data play a crucial role in locating the faults and in better constraining their geometry.

Here, we exploit the complete dataset of Sentinel-1 (S1) InSAR measurements from both ascending and descending orbits to investigate the ground displacement field. The S1 short repeat time made it possible to isolate the three major events that characterized the sequence (Figure 2a). In addition, given the rather low magnitude of the aftershocks (Mw $\leq$ 5.0) compared to the main events, we assumed that they only marginally contributed to the overall surface deformation. We modeled the available InSAR deformation maps to retrieve the parameters characterizing some finite dislocation sources. Then we performed a Coulomb stress transfer and a 1D diffusivity analysis in order to investigate possible fault interactions. Finally, we highlight the key role played by the configuration of the Thessaly Basin characterized by blind faults interconnected at depth, particularly interesting from both the neotectonic and active tectonics points of view.

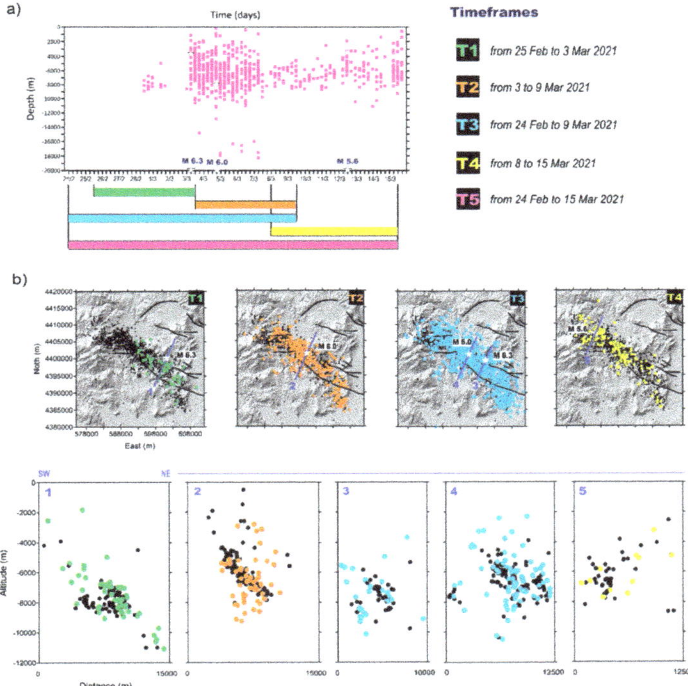

**Figure 2.** Seismicity evolution. (**a**) Depth vs. time plot of the seismicity occurred between 28 February 2021 and 15 March 2021; some timeframes used in the subsequent discussion are also shown in different colors (T). (**b**) Maps and sections of the relocated seismicity plotted according to their origin time. Black dots are the relocated earthquakes recorded until 26 April 2021; subsequent earthquakes are shown with colors reflecting the introduced timeframes (from green to yellow circles). The three largest events are represented by white stars in the upper panels.

## 2. Exploited Data

### 2.1. Seismological Data

In order to constrain the geometry and location of the main fault structures involved during the TSS, we considered 1853 earthquakes that occurred in the area from 28 February 2021 to 26 April 2021 with magnitudes ranging between 0.2 and 6.3 (Figure 2). The analyzed data were collected by the Hellenic Seismic Network operated by the Institute of Geodynamics of the National Observatory of Athens (NOA-IG). For the absolute earthquake location, we used the phase readings provided by NOA-IG at 116 seismic stations (Figure S2). Our database is composed of 23,371 P- and 21,455 S-arrival times. Phase arrival times were re-weighted according to the epicentral distance provided by a preliminary location. As a result, most of the data (about 69% of P-waves and about 71% S-wave) used for absolute earthquake locations were recorded by stations located within an epicentral distance of 100 km.

Absolute earthquake locations were calculated according to a probabilistic approach using the NonLinLoc code [9] and a local 1-D velocity model [10] (see Table 1).

**Table 1.** Structural model used to relocate the TSS sequence and to compute Coulomb stress change. VP = P-wave velocity, VS = S-wave velocity, $\rho$ = density. The density was computed from vs. assuming a shear modulus of 33 GPa.

| Depth (km) | VP (km/s) | VS (km/s) | $\rho$ (kg/m$^3$) |
|---|---|---|---|
| 0.0–4.0 | 4.8 | 2.7 | 4437 |
| 4.0–7.0 | 5.7 | 3.3 | 3146 |
| 7.0–11.50 | 6.1 | 4.5 | 2747 |
| 11.50–16.50 | 6.3 | 3.6 | 2575 |
| 16.50–35.0 | 6.5 | 3.7 | 2419 |
| 35.0 $\infty$ | 7.8 | 4.4 | 1680 |

The VP/VS ratio used for the analysis was 1.78: this choice corresponds to the best value that allows to minimize the rms and location errors obtained after re-locating the data subset of the main earthquakes ($M_W \geq 5.0$). We obtained the mean error values, vertical, and horizontal equal to 0.67 km, 0.81 km, and a mean rms of 0.26 s, respectively. Then, we applied a double difference earthquake location technique (HypoDD [11]) to compute the earthquake relative locations that minimize the residuals between the observed and theoretical travel time differences (or double differences) for pairs of earthquakes at each station. We obtained the relative locations of 1760 earthquakes with mean errors, vertical, and horizontal, equal to 242 m and 246 m, respectively. The mean rms was 0.08 s (Figure S3). To achieve proper least squares error estimates, we followed the method proposed in [11]. The double difference earthquake locations showed a clustered seismicity, tighter along the NE–SW direction with clear earthquake alignments in vertical sections, which was not evident in the absolute catalog locations. For more details about the use of double difference technique (see the Supplementary Materials).

Earthquake locations (Figure 1 and Figure S4) suggest a 40 km NW–SE alignment, which is consistent with the regional trend of the mapped surface faults, although the earthquake clusters of 5–10 km length revealed secondary fault structures. This result is supported by the vertical section plots, which clearly show NE-dipping planes, in agreement with the retrieved focal mechanism solutions (Figure 2b). We pointed out that the continuity of these alignments seems to be interrupted laterally. As shown in Sections 2 and 4, earthquake locations seem to show NE- and SW-dipping planes, replicating an antithetic-synthetic fault set, typical of areas involved in normal faulting. In the northern area, the earthquake location highlights a S-dipping fault plane (Section 5). Our results show a segmented fault zone with predominantly NE-trending structures and complex geometries, typical of normal fault systems as those activated by Mw 6.1 L'Aquila and Mw 6.5 Central Italy earthquakes [12,13].

### 2.2. InSAR Measurements

The TSS was imaged by Sentinel-1 (S1) with a regular six day repeat interval along two ascending (tracks 102 and 175) and two descending (tracks 7 and 80) orbits; we considered the interval from 24 February 2021 to 15 March 2021, which spanned the whole seismic sequence and included four acquisitions per track. The complete list of Sentinel-1 images used in this study is reported in Table S1 and are shown graphically in Figure S1, Supplementary Materials. Sentinel-1 data were processed by using our own internally developed InSAR processing chain [14–16].

Notably, almost simultaneous acquisitions from ascending and descending orbits were available, which may support the joint inversion of these two components of the displacement vector. A peculiarity of this S1 dataset is that the second image of track 102 was acquired just in between the first two largest earthquakes (Mw 6.3 and Mw 6.0); in this way, although they occurred only 32 h apart, we could separate the two contributions. This was not the case for the other three orbits, of which the interferograms between the first and second acquisition contained no signal and, therefore, were useless for the following analysis.

To ease the discussion, we defined several "timeframes" according to the Sentinel-1 passes and to the events they include (see colored bars in Figures 2a and S1). In particular, we defined T1 and T2 as the two intervals covering the sole Mw 6.3 and Mw 6.0, respectively (Figures S5 and S6). In addition, each interferogram was uniquely labelled with a letter according to Table S1 and Figure S1.

Furthermore, we defined T3 as the timeframe interval including both the first and second main events; in this case, we had seven interferograms (blue bars in Figures 2a and S1): four "long" (from end of February to 8–9 March) and three "short" interferograms (from 2–3 to 8–9 March); clearly, the short ones did not include track 102 since it covered one event only. Figure 2 shows the four T3 "long" interferograms (wrapped and unwrapped) coupled by the same acquisition times. In particular, track 102 (ascending, IW1) and track 07 (descending, IW3) are shown in the first row, while track 175 (ascending, IW3) and track 80 (descending, IW1) in the second one. These maps show three NNW–SSE striking deformation lobes located in the Larissa Basin; to the south, the two largest lobes, associated with the Mw 6.3, were evident, particularly on the descending data; this displacement pattern reached a maximum value of ~−38 cm in LOS (line-of-sight) on ascending data (negative LOS values represent increasing distances from the satellite). To the NW of this pattern, a smaller lobe relevant to the Mw 6.0 was observed. We also note that the T3 ascending pattern was quite consistent with the superimposition of T1 and T2 (Figures S5 and S6) once the different look angles were taken into account. The corresponding descending interferograms presented similar characteristics, considering the effect of opposite side views and possible horizontal components. Furthermore, we defined T4 as the interval covering the Mw 5.6 only, which was imaged once by all four tracks (yellow bars in Figures 2a and S1). The corresponding wrapped and unwrapped interferograms are presented in Figure 2, showing a prevalent negative LOS displacement pattern extending along the EW direction, with a minimum of ~−10 cm, consistent with the lower energy of this event with respect to the previous ones.

Finally, we introduce T5 as the overall timeframe that includes all three main events; similarly to T3, this was formed by seven interferograms (magenta bars in Figures 2a and S1), which are presented in Figures S5 and S6. Note that multiple interferograms within a single timeframe (Figures 3, S5 and S6) basically comprise the same displacement signal since the four considered tracks were shifted in time by no more than one day and there was no significant activity between the end of February and the beginning of the sequence.

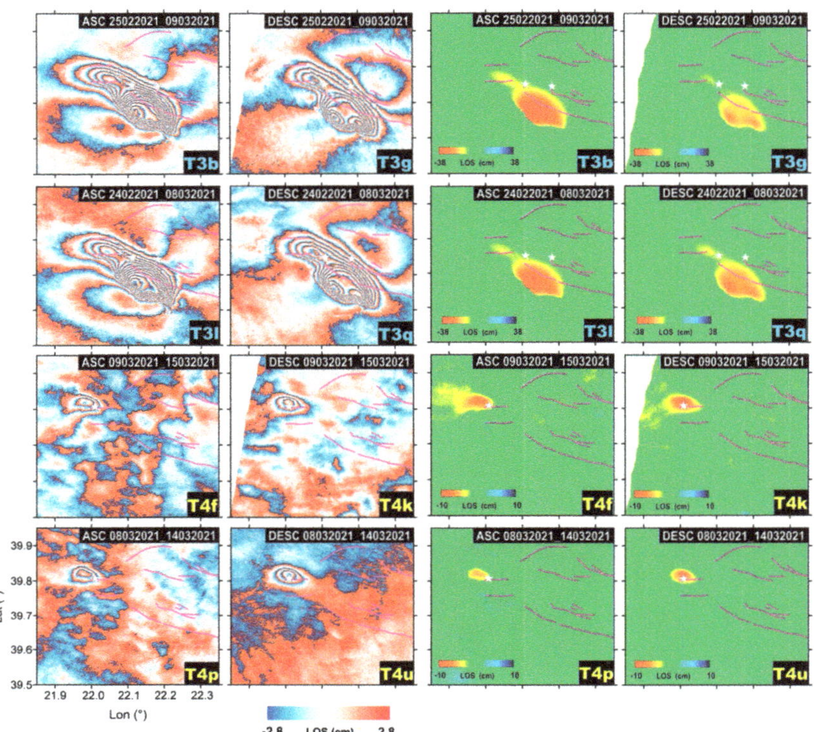

**Figure 3.** InSAR interferograms. A selection of Sentinel-1 InSAR interferograms used in this study (**first two columns**) and corresponding unwrapped data (**last two columns**). Fault lineaments and epicenters of the three largest events are indicated with magenta lines and white stars, respectively. Displacements are in satellite line-of-sight (LOS).

As a further element for the interpretation, we also estimated the vertical and east–west (EW) components of the displacement [17] for each of the considered timeframes; for brevity, they are reported in Figure S7.

## 3. Modeling and Results

### 3.1. Analytical Modeling

We jointly modeled the LOS displacements retrieved from both ascending and descending InSAR data with a finite dislocation fault in an elastic and homogeneous half-space [18,19]. Our source modeling approach was based on a consolidated two-step approach, also implemented in SARscape® software, capable of computing the distribution of the (non-uniform) slip over the fault plan. The first step was implemented by fitting (via a non-linear inversion) a uniform slip model to the data to constrain the geometry of the faulting plane; in particular, it found the east, north, and depth position of the plane center, along with its strike (azimuth) and dip angles.

Clearly, this first step also provides us with an average slip value (and its rake) and with the fault plane dimensions (length and width); however, these parameters are refined afterward. In fact, the second step computes (via a linear inversion) the distribution of the slip values over the defined fault plane by partitioning it into a number of small patches; at this point, the distribution of the patches showing a significant nonzero slip defines the extension of the actual faulting portion of the plane.

We started with a non-linear inversion scheme [20,21] to identify the fault parameters and mechanism (strike, dip, rake, slip, fault location, length, depth, width) using some of the results derived from the seismic waveform analysis (e.g., strike direction or fault depth) for some inversion cases (see Figures S8, S9, S11 and S12). The optimization starts from a random fault configuration within the given parameter ranges and keeps minimizing the cost function $\Phi$, based on the weighted squares of the residuals between the observed and the predicted data:

$$\Phi = \frac{1}{N} \sum_i^N \frac{\left(d_{i,obs} - d_{i,mod}\right)^2}{\sigma_i}$$

where $d_{i,obs}$ and $d_{i,mod}$ are the observed and modeled displacements of the *i-th* data point, respectively; $\sigma_i$ is the weighting parameter (here we assume $\sigma_i$ to be the same for all points); and $N$ is the total number of points used for the inversion. The downhill algorithm is implemented with multiple restarts to guarantee the convergence to the global cost function absolute minimum.

In order to refer the source depth to the actual ground surface, we also applied a topography compensation [22] and assessed possible residual offsets and ramps in the InSAR measurements caused, for instance, by residual orbital signals [23]. Moreover, to reduce the computation load of the inversion process, the InSAR data were preliminarily down-sampled to a 140 m regular grid.

After defining the fault geometry, we applied a linear inversion to obtain the slip distribution, extending the fault length and width to let the slip vanish to zero and sub-dividing the fault plane into sub-faults of $0.5 \times 0.5$ km$^2$. To avoid back-slip, we used a trial-and-error approach to define the system damping, which is the empirical parameter balancing the data fit and the slip distribution roughness.

The adopted linear inversion scheme is described by the equation [24,25]:

$$\begin{bmatrix} \mathbf{d} \\ 0 \end{bmatrix} = \begin{bmatrix} G \\ \varepsilon \nabla^2 \end{bmatrix} \cdot \mathbf{m}$$

where $\mathbf{d}$ is the InSAR data vector; G is the Green's functions matrix; and $\nabla^2$ is the Laplacian operator, which is tuned with the damping factor $\varepsilon$ obtained by trial-and-error [26] to obtain a reliable slip distribution, described with the m vector of parameters. A further constrain of parameter positivity was adopted to prevent back-slip.

The two timeframes that minimize the number of included events while allowing us to study the TSS deformation jointly from both ascending and descending orbits are clearly T3 and T4. We inverted for the two events (Mw 6.3 and Mw 6.0) included in T3 and, separately, for the one (Mw 5.6) included in T4 (Figures S8–S10).

The best-fitting fault model with uniform slip for T3 consists of three distinct ~58°, ~33°, and ~42° NE-dipping fault segments located beneath the three observed lobes of deformation shown in Figure 4 (say Figures 1a,b and 2, respectively). They both extend to a maximum depth of 10 km, the F1a showing a small, left lateral strike-slip component (rake: $-118°$). Although we cannot exclude some effects on the first event, the F2 fault is interpreted as associated with the second event, as confirmed also by modeling the second event (T2) alone (see Figure S11); this showed a $13 \times 3.7$ km$^2$ source plain. Conversely, data in T4 were best-fitted by a uniform slip model with a single ~47° SE-dipping fault (say F3).

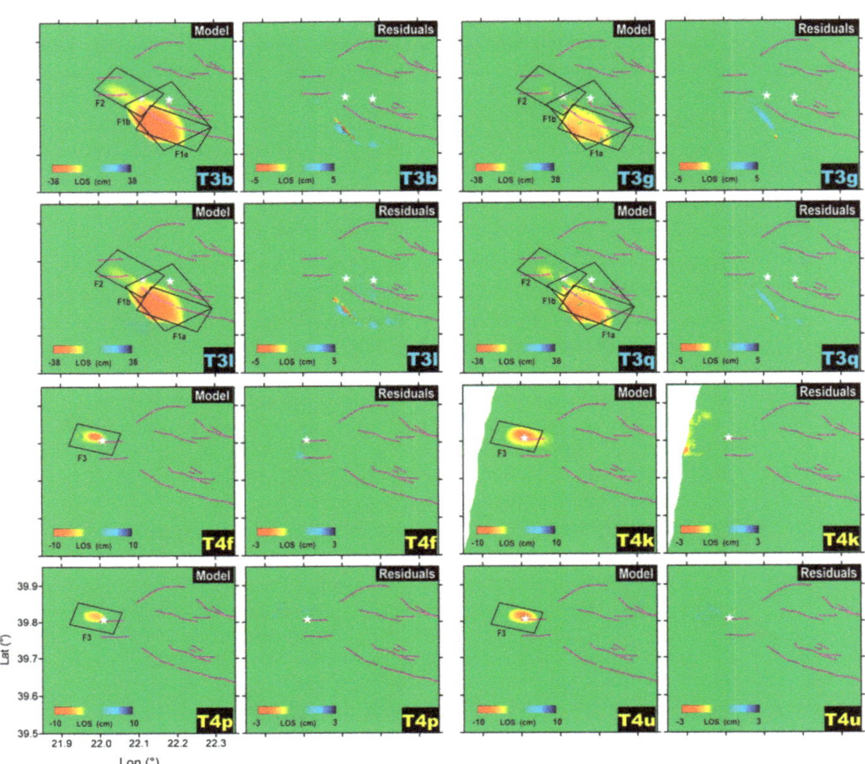

**Figure 4.** Line-of-sight projected displacement maps computed from the distributed slip modeling (T3 and T4 timeframes) and corresponding residual maps for the S1 interferograms shown in Figure 3. Fault lineaments and the retrieved plain faults are indicated by magenta and black lines, respectively; white stars indicate the epicenters of the three largest events. All the uniform slip modeling results are presented in the Supplementary Materials as Figures S8–S12.

We performed uniform slip modeling for all the timeframes defined earlier; for sake of completeness, we report all these results in the Supplementary Materials in Figures S8–S12 (including inversion parameters and statistics). Our final result was achieved by constraining a small number of the model parameters either to a preliminary geodetic model or to seismic data, as explained in the following.

First, to explore the sources responsible for Mw 6.3, we chose to invert the InSAR data for the shortest allowed timeframe (i.e., T1). We noted that a single rectangular fault cannot reproduce the observed bi-lobed fringe pattern associated with this event; as a test, we performed a free inversion of T1 and found that a good fit with the data was reached only when using a two-fault model (Figure S11).

Second, we note that, while the LOS displacement profiles for the Mw 6.3 and Mw 5.6 events were very abrupt, thus clearly indicating the corresponding dipping directions, this was not the case for the Mw 6.0 event, as shown in Figure 5.

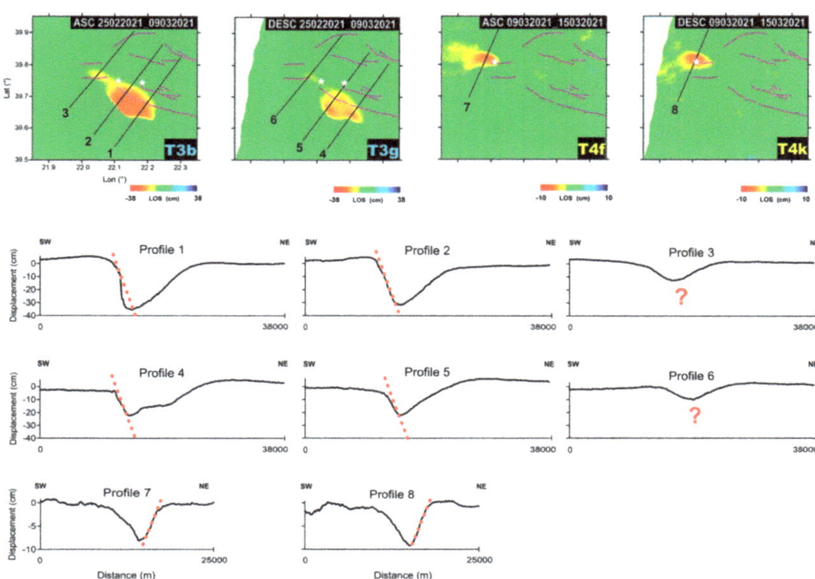

**Figure 5.** Displacement maps and profiles. (**Upper** panels) LOS displacement maps for the indicated timeframes together with the selected profiles. (**Lower** panels) Displacement profiles for Mw 6.3, 6.0, and 5.6 InSAR measurement for T3 and T4; dashed red lines indicate their dipping direction. The question marks in right panels indicate that the displacement in the corresponding profiles did not allow us to discriminate the dipping of the fault plane (Mw 6.0) (see main text for the details).

It is clear, therefore, that the use of the displacement information alone makes it difficult to resolve for the F2 strike value. To overcome this problem, we thus inverted T2, constraining its depth and limiting the strike search interval according to the seismic analysis (Figure S11). Among the dislocation model parameters, the seismological analysis can provide strike and dip values, along with information about the depth. Here, we only used seismic-derived strike and depth information to guide the inversion of T2, while all the other inversions were kept free; therefore, seismic analysis data were used to independently validate the location and mechanism of the retrieved faults other than F2 (Figure 2b).

As an additional analysis, we also inverted for T5, which contained the whole sequence, by applying the same constraints used for T3 and T4; basically, we obtained the same source parameters when using a different combination of all the available interferograms covering the period (Figure S13). Subsequently, we selected the geometric settings derived from our preferred non-linear inversions (Figure S8) for the slip distribution calculation. To this end, we subdivided the fault planes into small patches ($0.5 \times 0.5$ km$^2$) and solved for the slip values on each patch (Figure 6).

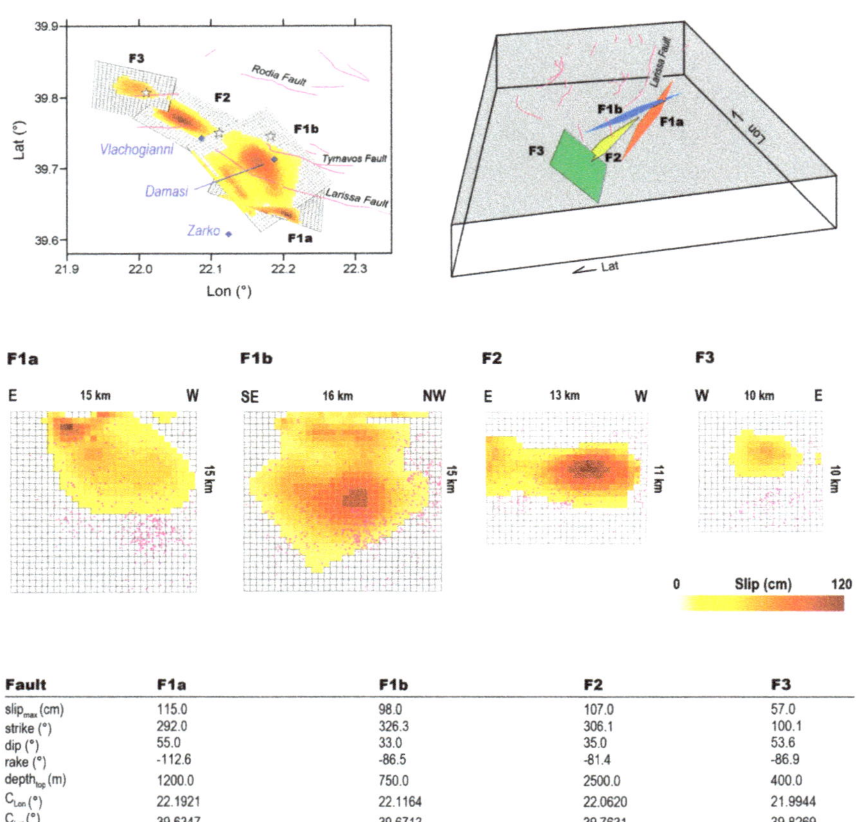



| Fault | F1a | F1b | F2 | F3 |
|---|---|---|---|---|
| $slip_{max}$ (cm) | 115.0 | 98.0 | 107.0 | 57.0 |
| strike (°) | 292.0 | 326.3 | 306.1 | 100.1 |
| dip (°) | 55.0 | 33.0 | 35.0 | 53.6 |
| rake (°) | -112.6 | -86.5 | -81.4 | -86.9 |
| $depth_{top}$ (m) | 1200.0 | 750.0 | 2500.0 | 400.0 |
| $C_{Lon}$ (°) | 22.1921 | 22.1164 | 22.0620 | 21.9944 |
| $C_{Lat}$ (°) | 39.6347 | 39.6713 | 39.7631 | 39.8269 |

**Figure 6.** Linear Inversion results. The four sources and relative retrieved parameters of the linear solutions for T3 + T4 in 2-D and 3-D views. Fault lineaments and epicenter of the three largest events are represented by magenta lines and white stars, respectively. The purple points represent all the events occurring on the fault planes.

For the Mw 6.3, the slip was concentrated between 2 km and 4 km depth: the maximum slip for F1a and F1b faults was about 1.1 m at 2 km and 4 km depths, respectively. Both fault plains present small portions of slip at shallow depths, in the area where ruptures are observed on the ground (Figure 1a). The Mw 6.0 event ruptured a western ~15 km-long segment of the F2 fault system with a maximum slip value reaching ~1.0 m at 5 km depth. Finally, the $M_W$ 5.6 ruptured the ~10 km-long segment that had remained unbroken after the previous events. We found a main patch of slip (with maximum of 0.57 m) located at the center of the fault plane at a 2.5 km depth. The overall geodetic moment of the F1 fault system was $2.88 \cdot 10^{18}$ Nm corresponding to a Mw 6.3 earthquake, while for F2 and F3, we obtained $1.1 \cdot 10^{18}$ Nm and $3.1 \cdot 10^{17}$ Nm, corresponding to Mw 6.0 and 5.6 events, respectively. Looking at Figure 6, most of the earthquakes occurred outside or at the borders of the slipped areas on each of the fault planes. This feature suggests that some of the shear stress was redistributed outside the main patches of the slip after each event.

### 3.2. Coulomb Stress Transfer Analysis

To study the interaction between the three main events of the sequence, we calculated the change in the stress field on each fault plane in terms of Coulomb failure function (ΔCFF) by using an approach similar to that proposed in [27]. We calculated the co- and post-seismic deformation and the associated Coulomb stress change $\Delta CFF = \Delta \tau + \mu(\Delta \sigma + \Delta p)$, where $\Delta\tau$ and $\Delta\sigma$ are the shear and normal stress change, respectively; $\mu$ is the friction coefficient; and $\Delta p$ is the pore pressure change. We used a computer code based on the viscoelastic-gravitational dislocation theory [28] and assumed that $\Delta p = 0$, corresponding to drained conditions, and used $\mu = 0.4$. The method allowed for the use of finite source fault models with heterogeneous slip distribution. It uses the standard linear solid rheology defined by three parameters: the unrelaxed shear modulus $\mu_0$, the viscosity $\eta$, and the parameter $\alpha$, which is the ratio of the fully relaxed modulus to the unrelaxed modulus. As a point of difference with usual analyses that consider the location of the nucleation of the following earthquakes relative to the induced stress variation, here, we investigated the heterogeneity of the stress field on the whole fault plane and the time evolution of the earthquake preparatory process.

As undertaken for the earthquake location, we also used here the layered model proposed in [10] (see Table 1), while the detailed slip distributions and the fault geometry were those inferred from the InSAR modeling. We computed Green functions for a 10-day time window to include the effect of three main events contributing to the stress field. We used 60 equally spaced horizontal points at a distance range of 0–60 km, and 40 points in depth, ranging between 0 and 20 km. Stress field variation was computed on 16 different layers with depths ranging between 0 and 15 km.

We first considered the effect of the Mw 6.3 event on the fault planes of Mw 6.0 and Mw 5.6. Next, we considered the additional effect of the Mw 6.0 event on the fault plane of the Mw 5.6 earthquake. The first interaction is shown in Figure 7a,b while the second is shown in Figure 7c. The results reported in Figure 7a indicate that the Mw 6.3 earthquake strongly affected the Mw 6.0 by producing a significant ΔCFF increase both east and west of the hypocenter and a slighter increase in the upper right corner that likely promoted its next breaking. Similarly, the Mw 6.3 also increased the ΔCFF in the lower part of the fault plane of the Mw 5.6. The occurrence of the Mw 6.0 increased the ΔCFF in the central part of the fault, leading to the following breaking; notably, the earthquakes that occurred before the Mw 6.0 (reported as square in Figure 7a) seemed to encircle the patches that will slip. Moreover, all the aftershocks (reported as crosses in Figure 7a) occurred in the area that was initially inhibited by the ΔCFF and where remaining stress was released after the main event. Although less evident, a similar picture is depicted in Figure 7c for the Mw 5.6.

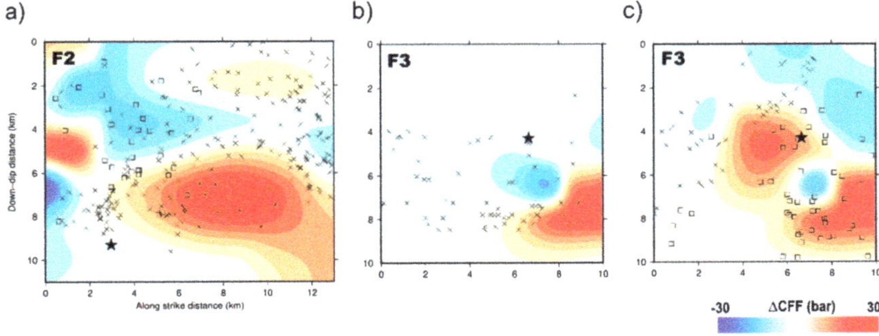


**Figure 7.** Coulomb stress transfer analysis. (**a**) ΔCFF on the fault plane of the Mw 6.0 caused by the Mw 6.3. The squares represent all the events occurring on the fault plane (at distance of 1 km) of the Mw 6.0 before its origin time; the crosses represent all the events following its origin time. (**b**) ΔCFF contribution of the Mw 6.3 on the fault plane of the Mw 5.6. (**c**) Cumulative ΔCFF of both the Mw 6.3 and the Mw 6.0 earthquakes on the fault plane of the Mw 5.6. In each panel, the star identifies the hypocenter of the corresponding earthquake.

### 3.3. Diffusivity Analysis

The results reported in the previous section suggest a clear static stress contribution to the evolution of the seismic sequence. However, in order to further investigate the delayed triggering of the main events, we tested the hypothesis of pore pressure diffusion from the region enclosing the TSS. To this aim, we modeled the pore pressure perturbation caused by a point source in an isotropic fluid saturated medium. We considered the Mw 6.3 as the source and retrieved the best 1D isotropic diffusivity value (*Diso*) following the approach proposed in [29]. According to [30], the triggering front was modeled as $r = \sqrt{4\pi Dt}$, where $D$ is the scalar diffusivity corresponding to *Diso*/4. We only analyzed events with magnitude larger than the minimum magnitude of completeness that corresponds to Mc 1.6 $\pm$ 0.2 and was obtained by using the maximum curvature approach [31].

The resulting $r - t$ plot is shown in Figure 8 and the inferred best *Diso* was $60 \pm 37$ m$^2$/s. A similar value has been obtained for the 2009 L'Aquila sequence in Italy [32].

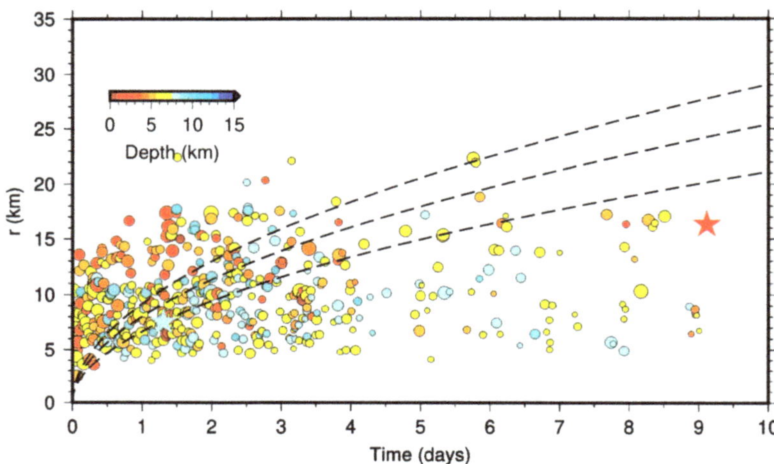


**Figure 8.** r-t plot for the relocated TSS seismic sequence. Only earthquakes with magnitude larger than the minimum magnitude of completeness (Mc 1.6) were considered. The dashed lines correspond to the triggering front evaluated by using the inferred diffusivity value $60 \pm 37$ m$^2$/s and plus/minus one standard deviation. Symbols dimension is proportional to the event magnitude and color coded according to the depth. The two stars indicate the Mw 6.3 (light blue) and Mw 5.6 (orange), respectively.

We note that, except for the very early days after the main event where aftershocks are expected to occur mainly due a redistribution of the stress on the fault planes, the evolution of the whole sequence is compatible with a pore pressure diffusion mechanism likely due to the volume deformation produced by the two main events. In particular, the Mw 6.0 and the Mw 5.6 occurred in correspondence with a triggering front characterized by the lower limit of the inferred diffusivity values (Figure 8).

## 4. Discussion

The TSS represents the largest seismic sequence affecting a continental extensional domain in Greece that has been widely monitored by modern geodetic techniques; thanks to the short satellite revisit time, InSAR measurements allowed for isolating the contribution from the major earthquakes of the sequence to study their interaction. Available geological data indicate that the northern sector of Thessaly represents a large seismic gap. This may be a direct consequence of the limited size of the faults (less than 20 km) and their intrinsic capability to originate earthquakes of small-to-moderate magnitude only [8]. TSS, which finally filled the gap, confirmed this hypothesis.

Our model shows that the seismic activity distributed along distinct segments of a previously unknown fault system composed of three fault plains (F1a, F1b, and F2) dipping NE for the Mw 6.3 and Mw 6.0 earthquakes, and of a small fault (F3) dipping SW for the Mw 5.6 (Figure 6). However, the activation of some portion of the known Tyrnavos and Larissa fault during the aftershocks cannot be completely ruled out. The orientation of the retrieved fault plains is in good agreement with the published focal mechanisms by Greek and other international Institutes (Figure 1a), and is in accordance with the extensional tectonic regime of the area.

For the first two events, the vertical deformation component highlights strong subsidence phenomena reaching more than ~40 cm. Moreover, to make the whole picture more complicated, the E–W component showed significant westward motion (~20 cm) (Figure S7). This latter motion can be related to the movement of the F1a fault, which features a slight left lateral strike-slip component.

Extensive and deep ground cracks generated by the Mw 6.3 were observed close to river beds along with ejected liquefied material; other cracks were parallel to the Vlachogianni fault. Some field reports showed scattered surface alignments only in the Zarko region [1], which can hardly be associated with faults. However, the inferred slip distribution showed that the highest values were confined in the S–E area (Vlachogianni, Damasi, Zarko villages) and this could explain the observed damage distribution (Figures 1a and 6).

Our ΔCFF results (Figure 7) indicate that the seismic sequence developed in a sort of domino effect. Regarding the temporal evolution of the sequence, the delayed triggering of the Mw 6.0 earthquake can be explained by the distribution of the events that occurred earlier; indeed, that distribution envelops the patches of the fault that will break a feature, described in [33] as the encircling maneuver. A discrimination between the three main conceptual models—cascade-up, pre-slip and progressive localization [34,35]—for earthquake initiation would require a specific spatial-temporal analysis of the earthquakes sequence and stress perturbations between different events, which is beyond the scope of the present manuscript. We thus suggest that the most plausible initiation mechanism could be the recently proposed mixed stress loading process involving the cascade-triggering and aseismic-slip transient mechanisms [36].

As for the Mw 5.6 earthquake, the results depicted in Figure 7b,c show that the most relevant effect, in terms of static stress contribution, was due to the Mw 6.0 event. Furthermore, the delay in the triggering (about eight days from the Mw 6.3 and Mw 6.0 events) suggests that mechanisms other than the static stress change—in addition to a dynamic triggering contribution—took place. This is the case, for example, of pore fluid pressure diffusion as observed in other seismic sequences [37–39]. Indeed, a pore pressure increase can reduce the normal stress promoting the rupture. Moreover, fluids can also contribute to decreasing the fault friction [40]. The performed 1D diffusivity analysis indicates that almost all the sequence, and in particular the occurrence of the Mw 6.0 and Mw 5.6, are indeed compatible with a mechanism of pore pressure diffusion that was triggered by the deformation pattern produced by the previous events.

The combination of the lack of historical records of large earthquakes in this area, and the absence of mapped surface features produced by past faulting, make seismic hazard estimation difficult. Therefore, the identification of blind faults may significantly contribute to improve the seismic hazard analysis, particularly when dealing with moderate dip angles. In this context, the InSAR products have proven to be an extremely useful tool in helping to constrain the rupture characteristics.

Considering the possible presence of a multi-fault rupturing structure and the complexity of the tectonic framework of the studied area, we believe that the study of the seismogenic mechanism of the blind faulting deserves further investigation through additional data.

## 5. Conclusions

In this work, we studied the seismic sequence occurred in Thessaly (Greece) during the early days of March 2021. To this end, we used seismic and InSAR data to set up a geodetic model of the causative faulting system. Moreover, we performed Coulomb stress transfer and 1D pore pressure diffusivity analyses to investigate the space–time evolution of the sequence.

Our main findings are summarized as follows:

- the Thessaly seismic sequence nucleated at shallow depths (<12 km) and is related to the activation of several blind, previously unknown faults;
- the seismic sequence developed in a sort of domino effect involving a complex interaction among the normal faults within the activated crustal volume;
- InSAR data and modeling are also extremely useful to constrain the rupture characteristics in the case of blind faults; and
- the used approach can help improve our knowledge of the seismic potential of the Thessaly region and refine the associated seismic hazard.

**Supplementary Materials:** The following are available online at https://www.mdpi.com/article/10.3390/rs13173410/s1, Table S1: InSAR data parameters, Figure S1: Magnitude and time span for seismic activity and coseismic InSAR data pairs, Figure S2: Seismic stations used for earthquake locations, Figure S3: Error histograms for earthquake relocations, Figure S4: Map of the relocated epicenters, Figure S5: Additional Sentinel-1 interferograms (wrapped), Figure S6: Additional Sentinel-1 interferograms (unwrapped), Figure S7: E–W and vertical displacement maps computed from InSAR data, Figure S8: Uniform slip modeling for T3 (b + g + l + q, jointly) and T4 (f + k + p + u, jointly); Figure S9: Uniform slip modeling for T3 (b + g and l + q); Figure S10: Uniform slip modeling for T4 (f + k and p + u); Figure S11: Uniform slip modeling for T1 (a) and T2 (d); Figure S12: Uniform slip modeling for T5 (c + h and m + r).

**Author Contributions:** V.D.N., V.C. and E.S. conceived the paper; D.R. and E.S. processed InSAR data; G.M.A. processed the seismological data; V.D.N. performed the geodetic modeling; V.C. performed the Coulomb stress transfer and diffusivity analyses; V.D.N., V.C. and E.S. wrote the manuscript. All authors have read and agreed to the published version of the manuscript.

**Funding:** This study was partially funded by Pianeta Dinamico—Working Earth INGV-MUR Project.

**Institutional Review Board Statement:** Not applicable.

**Informed Consent Statement:** Not applicable.

**Data Availability Statement:** All the datasets generated and/or analyzed during the present study are available in the Zenodo repository (doi:10.5281/zenodo.5016812).

**Acknowledgments:** Sentinel-1 data were provided through the Copernicus Program of the European Union. We used the SRTM 1 arc-second DEM. Figure 7a–c were generated with Generic Mapping Tools [41].

**Conflicts of Interest:** The authors declare no conflict of interest.

## References

1. Lekkas, E.; Agorastos, K.; Mavroulis, S.; Kranis, C.; Skourtsos, E.; Carydis, P.; Gogou, M.; Katsetsiadou, K.-N.; Papadopoulos, G. The early March 2021 Thessaly earthquake sequence. In *Newsletter of Environmental, Disaster and Crises Management Strategies*; National and Kapodistrian University of Athens: Athens, Greece, 2021; p. 195. [CrossRef]
2. Caputo, R.; Pavlides, S. Late Cainozoic geodynamic evolution of Thessaly and surroundings (central-northern Greece). *Tectonophys* **1993**, *223*, 339–362. [CrossRef]
3. Caputo, R.; Helly, B.; Pavlides, S.; Papadopoulos, G. Archaeo- and palaeoseismological investigationsin Northern Thessaly (Greece): Insights for the seismic potential of the region. *Nat. Haz.* **2006**, *39*, 195–212. [CrossRef]
4. Caputo, R.; Bravard, J.-P.; Helly, B. The Pliocene-quaternary tecto-sedimentary evolution of the Larissa Plain (Eastern Thessaly, Greece). *Geodin. Acta* **1994**, *7*, 57–85. [CrossRef]
5. Caputo, R.; Helly, B. The Holocene activity of the Rodià Fault, Central Greece. *J. Geodyn.* **2005**, *40*, 153–169. [CrossRef]
6. Caputo, R.; Pavlides, S. *The Greek Database of Seismogenic Sources (GreDaSS), Version 2.0.0: A Compilation of Potential Seismogenic Sources (Mw > 5.5) in the Aegean Region*; CINECA IRIS: Bologna, Italy, 2013. [CrossRef]
7. Caputo, R.; Piscitelli, S.; Oliveto, A.; Rizzo, E.; Lapenna, V. The use of electrical resistivity tomography in active tectonics. Examples from the Tyrnavos Basin, Greece. *J. Geodyn.* **2003**, *36*, 19–35. [CrossRef]
8. Caputo, R. Inference of a seismic gap from geological data: Thessaly (Central Greece) as a case study. *Ann. Geofis.* **1995**, *38*. [CrossRef]
9. Lomax, A.; Virieux, J.; Volant, P.; Berge-Thierry, C. Probabilistic earthquake location in 3D and layered models. In *Advances in Seismic Event Location*; Springer: Dordrecht, The Netherlands, 2000; pp. 101–134.
10. Papadimitriou, P.; Kaviris, G.; Makropoulos, K. The Cornet seismological network: 10 years of operation, recorded seismicity and significant applications. *Hell. J. Geosci.* **2010**, *45*, 193–208.

11. Waldhauser, F.; Ellsworth, W.L. A double-difference earthquake location algorithm: Method and application to the northern Hayward fault, California. *Bull. Seismol. Soc. Am.* **2000**, *90*, 1353–1368. [CrossRef]

12. Chiaraluce, L.; Valoroso, L.; Piccinini, D.; Di Stefano, R.; De Gori, P. The anatomy of the 2009 L'Aquila normal fault system (central Italy) imaged by high resolution foreshock and aftershock locations. *J. Geophys. Res. Solid Earth* **2011**, *116*. [CrossRef]

13. Chiaraluce, L.; Di Stefano, R.; Tinti, E.; Scognamiglio, L.; Michele, M.; Casarotti, E.; Marzorati, S. The 2016 central Italy seismic sequence: A first look at the mainshocks, aftershocks, and source models. *Seismol. Res. Lett.* **2017**, *88*, 757–771. [CrossRef]

14. Sansosti, E.; Berardino, P.; Manunta, M.; Serafino, F.; Fornaro, G. Geometrical SAR image registration. *IEEE Trans. Geosci. Remote Sens.* **2006**, *44*, 2861–2870. [CrossRef]

15. Fornaro, G.; Sansosti, E. A two-dimensional region growing least squares phase unwrapping algorithm for interferometric SAR processing. *IEEE Trans. Geosci. Remote Sens.* **1999**, *37*, 2215–2226. [CrossRef]

16. Fornaro, G.; Pauciullo, A.; Reale, D. A Null-Space Method for the Phase Unwrapping of Multitemporal SAR Interferometric Stacks. *IEEE Trans. Geosci. Remote Sens.* **2011**, *49*, 2323–2334. [CrossRef]

17. Wright, T.J.; Parsons, B.E.; Lu, Z. Toward mapping surface deformation in three dimensions using InSAR. *Geophys. Res. Lett.* **2004**, *31*, L01607. [CrossRef]

18. Okada, Y. Surface deformation due to shear and tensile faults in a half-space. *Bull. Seismol. Soc. Am.* **1985**, *75*, 1135–1154. [CrossRef]

19. Okada, Y. Internal deformation due to shear and tensile faults in a half-space. *Bull. Seismol. Soc. Am.* **1992**, *82*, 1018–1040.

20. Levenberg, K. A method for the solution of certain problems in least squares. *Q. Appl. Math.* **1944**, *2*, 164–168. [CrossRef]

21. Marquardt, D. An algorithm for least-squares estimation of nonlinear parameters. *SIAM J. Soc. Ind. Appl. Math.* **1963**, *11*, 431–441. [CrossRef]

22. Williams, C.A.; Wadge, G. An accurate and efficient method for including the effects of topography in three-dimensional elastic models of ground deformation with applications to radar interferometry. *J. Geophys. Res.* **2000**, *105*, 8103–8120. [CrossRef]

23. Pepe, A.; Berardino, P.; Bonano, M.; Euillades, L.D.; Lanari, R.; Sansosti, E. SBAS-based satellite orbit correction for the generation of DInSAR time-series: Application to RADARSAT-1 data. *IEEE Trans. Geosci. Remote Sens.* **2011**, *49*, 5150–5165. [CrossRef]

24. Atzori, S.; Hunstad, I.; Chini, M.; Salvi, S.; Tolomei, C.; Bignami, C.; Stramondo, S.; Trasatti, E.; Antonioli, A.; Boschi, E. Finite fault inversionof DInSAR coseismic displacement of the 2009 L'Aquila earthquake (central Italy). *Geophys. Res. Lett.* **2009**, *36*, L15305. [CrossRef]

25. De Novellis, V.; Atzori, S.; De Luca, C.; Manzo, M.; Valerio, E.; Bonano, M.; Cardaci, C.; Castaldo, R.; Di Bucci, D.; Manunta, M.; et al. DInSAR analysis and analytical modeling of Mount Etna displacements: The December 2018 volcano-tectonic crisis. *Geophys. Res. Lett.* **2019**, *46*, 5817–5827. [CrossRef]

26. Menke, W. *Geophysical Data Analysis: Discrete Inverse Theory*; Academic Press: Cambridge, MA, USA, 1989.

27. Pino, N.A.; Convertito, V.; Madariaga, R. Clock advance and magnitude limitation through fault interaction: The case of the 2016 central Italy earthquake sequence. *Sci. Rep.* **2019**, *9*, 5005. [CrossRef] [PubMed]

28. Wang, R.; Lorenzo-Martin, F.; Roth, F. PSGRN/PSCMP—A new code for calculating co- and post-seismic deformation, geoid and gravity changes based on the viscoelastic-gravitational dislocation theory. *Comput. Geosci.* **2006**, *32*, 527–541. [CrossRef]

29. Noir, J.; Jacques, E.; Bekri, S.; Adler, P.M.; Tapponier, P.; King, G.C.P. Fluid flow triggered migration of events in the 1989 Dobi earthquake sequence of central Afar. *Geophys. Res. Lett.* **1997**, *24*, 2335–2338. [CrossRef]

30. Shapiro, S.; Patzig, R.; Rothert, E.; Rindschwentner, J. Triggering of Seismicity by Pore-pressure Perturbations: Permeability-related Signatures of the Phenomenon. *Pure Appl. Geophys.* **2003**, *456*, 1051–1066. [CrossRef]

31. Wiemer, S.; Wyss, M. Minimum magnitude of complete reporting in earthquake catalogs: Examples from Alaska, the western United States, and Japan. *Bull. Seismol. Soc. Am.* **2000**, *90*, 859–869. [CrossRef]

32. Malagnini, L.; Lucente, F.P.; De Gori, P.; Akinci, A.; Munafo', I. Control of pore fluid pressure diffusion on fault failure mode: Insights from the 2009 L'Aquila seismic sequence. *J. Geophys. Res.* **2012**, *117*, B05302. [CrossRef]

33. Das, S.; Kostrov, B.V. Breaking of a single asperity: Rupture process and seismic radiation. *J. Geophys. Res.* **1983**, *88*, 4277–4288. [CrossRef]

34. Gomberg, J. Unsettled earthquake nucleation. *Nat. Geosci.* **2018**, *11*, 463–464. [CrossRef]

35. Abercrombie, R.E. Similar starts for small and large earthquakes. *Nature* **2019**, *573*, 42–43. [CrossRef] [PubMed]

36. Kato, A.; Ben-Zion, Y. The generation of large earthquakes. *Nat. Rev. Earth Environ.* **2021**, *2*, 26–39. [CrossRef]

37. Gavrilenko, P. Hydromechanical coupling in response to earthquakes: On the possible consequences for aftershocks. *Geophys. J. Int.* **2005**, *161*, 113–129. [CrossRef]

38. Tung, S.; Masterlark, T. Delayed poroelastic triggering of the 2016 October Visso earthquake by the August Amatrice earthquake, Italy. *Geophys. Res. Lett.* **2018**, *45*, 2221–2229. [CrossRef]

39. Convertito, V.; De Matteis, R.; Improta, L.; Pino, N.A. Fluid-Triggered Aftershocks in an Anisotropic Hydraulic Conductivity Geological Complex: The Case of the 2016 Amatrice Sequence, Italy. *Front. Earth Sci.* **2020**, *8*, 541323. [CrossRef]

40. Scholz, C.H. *The Mechanics of Earthquakes and Faulting*, 2nd ed.; Cambridge University Press: Cambridge, UK, 2002.

41. Wessel, P.; Smith, W.H.F. Free software helps map and display data. *EOS Trans. AGU* **1991**, *72*, 445–446. [CrossRef]

*Letter*

# The November 2019 Seismic Sequence in Albania: Geodetic Constraints and Fault Interaction

**Alessandro Caporali [1,2,*], Mario Floris [1], Xue Chen [1], Bilbil Nurce [3], Mauro Bertocco [2] and Joaquin Zurutuza [2]**

[1] Department of Geosciences, University of Padova, 35131 Padova, Italy; mario.floris@unipd.it (M.F.); chenxue@cugb.edu.cn (X.C.)
[2] Center for Space Activities CISAS 'G. Colombo', University of Padova, 35131 Padova, Italy; mauro.bertocco@unipd.it (M.B.); jzurutuza@gmail.com (J.Z.)
[3] Department of Geodesy, Universiteti Politeknik i Tiranës, Tirana 1001, Albania; billnurce@gmail.com
* Correspondence: alessandro.caporali@unipd.it; Tel.: +39-049-827-9122

Received: 29 January 2020; Accepted: 3 March 2020; Published: 5 March 2020

**Abstract:** The seismic sequence of November 2019 in Albania culminating with the Mw = 6.4 event of 26 November 2019 was examined from the geodetic (InSAR and GNSS), structural, and historical viewpoints, with some ideas on possible areas of greater hazard. We present accurate estimates of the coseismic displacements using permanent GNSS stations active before and after the sequence, as well as SAR interferograms with Sentinel-1 in ascending and descending mode. When compared with the displacements predicted by a dislocation model on an elastic half space using the moment tensor information of a reverse fault mechanism, the InSAR and GNSS data fit at the mm level provided the hypocentral depth is set to 8 ± 2 km. Next, we examined the elastic stress generated by the Mw = 7.2 Montenegro earthquake of 1979, with the Albania 2019 event as receiver fault, to conclude that the Coulomb stress transfer, at least for the elastic component, was too small to have influenced the 2019 Albania event. A somewhat different picture emerges from the combined elastic deformation resulting after the two (1979 and 2019) events: we investigated the fault geometries where the Coulomb stress is maximized and concluded that the geometry with highest induced Coulomb stress, of the order of ca. 2–3 bar (0.2–0.3 MPa), is that of a vertical, dextral strike slip fault, striking SW to NE. This optimal receiver fault is located between the faults activated in 1979 and 2019, and very closely resembles the Lezhe fault, which marks the transition between the Dinarides and the Albanides.

**Keywords:** InSAR; GNSS; coseismic displacements; Coulomb stress transfer; seismic hazard

---

## 1. Introduction and Tectonic Setting

The Mw = 6.4 earthquake with epicenter near Durazzo in Albania (Lat = 41.46°N; Long = 19.58°E) at an estimated depth of 26 km (http://geofon.gfz-potsdam.de/eqinfo/event.php?id=gfz2019xdig) took place on 26 November 2019 at 2:54 UTC. The event was preceded by weaker shocks (max Mw = 5.6 on 21 September 2019) and followed by a swarm of aftershocks within some 10s km, with the largest event being of Mw = 5.5 on the same day of 26 November 2019 at 06:08:24 UTC. The epicenters of the aftershocks appear to migrate in the north direction (https://www.emsc-csem.org/Earthquake/262/M6-4-ADRIATIC-SEA-on-November-26th-2019-at-02-54-UTC#aftershocks). The preliminary fault plane solutions indicate a reverse fault with a high dip fault plane (strike = 151°, dip= 72°) with SW vergence and a lower dip fault plane (strike = 335°, dip = 18°) with NE vergence (Figure 1 and Table 1).

**Figure 1.** Location of the 2019 main event of Mw = 6.4 and seismicity of known and unknown (colored dots) fault plane solution since 2003 (Source: https://geofon.gfz-potsdam.de/eqinfo/form.php). The yellow lines indicate tectonic lineaments according to the DISS-Share Database [1,2]. The presumable fault plane of the 1979 Montenegro earthquake of Mw = 7.2 is also shown (red rectangles).

**Table 1.** Fault plane solutions of the events of magnitude greater than 5, published by GFZ (http://geofon.gfz-potsdam.de/eqinfo/). The hypocentral depth of the main event is debated (see the next sections).

| Date (UTC) | Lat (deg) | Long (deg) | Depth (km) | $M_W$ | Strike1 (deg) | Dip1 (deg) | Rake1 (deg) | Strike2 (deg) | Dip2 (deg) | Rake2 (deg) |
|---|---|---|---|---|---|---|---|---|---|---|
| 26 November 2019 2:54 | 41.46 | 19.58 | 26 | 6.4 | 151 | 72 | 89 | 335 | 18 | 98 |
| 26 November 2019 6:08 | 41.54 | 19.42 | 26 | 5.5 | 139 | 65 | 85 | 332 | 22 | 102 |
| 27 November 2019 14:45 | 41.54 | 19.42 | 26 | 5.3 | 155 | 63 | 89 | 337 | 27 | 91 |

The Composite Seismogenic Source is identified in the Database of Individual Seismogenic Sources (DISS) of INGV as ALCS (DISS Working Group, 2018) (http://diss.rm.ingv.it/dissnet/CadmoDriver?_action_do_single=1&_state=find&_token=NULLNULLNULLNULL&_tabber=1&_page=pSASources_d&IDSource=ALCS002). This structure belongs to the External Albanides, a compressive belt that is bordered to the N by the Dinarides, a ca. 700 km long belt overriding the Adria microplate. The two belts differ in strike by some 30°. The change in strike roughly corresponds to the Shkoder–Peje lineament (Figure 1), a major SW to NE striking structure [3] caught in between two thrust zones. Near to its SW termination, the Lezhe fault was described as a strike slip fault by Aliaj et al. [4]. Additional right lateral strike slip faults decouple the NE motion associated to the counterclockwise rotation of the Italian peninsula from the counterclockwise rotation of the Balkan–Hellenides towards SW [5]. The most notable example is the Kefalonia–Lefkada Transform Fault, which is characterized by repeated seismicity [6] (Figure 1). While the Dinarides/Albanian front moves SW, relative to a stable Europe, essentially as a rigid block [7], there is a clear indication that the Apulia region of the Italian peninsula is aseismic and moves NE converging towards the Dinarides. One expects a remarkable right lateral shear region to be accommodated in the mountain chains striking Albania and Western Greece NW to SE, just north of the complicated and intense deformation pattern affecting the West Hellenic arc [8–10]. The regional scale dextral shear field is well described by the pattern of velocities of GNSS stations (Figure 2). In conclusion, the thrusts in the Lower Dinarides, Albanides, and Western Greece need to be interleaved with dextral strike slip faults to accommodate the opposite (i.e., in the NE direction) motion of the Italian peninsula.

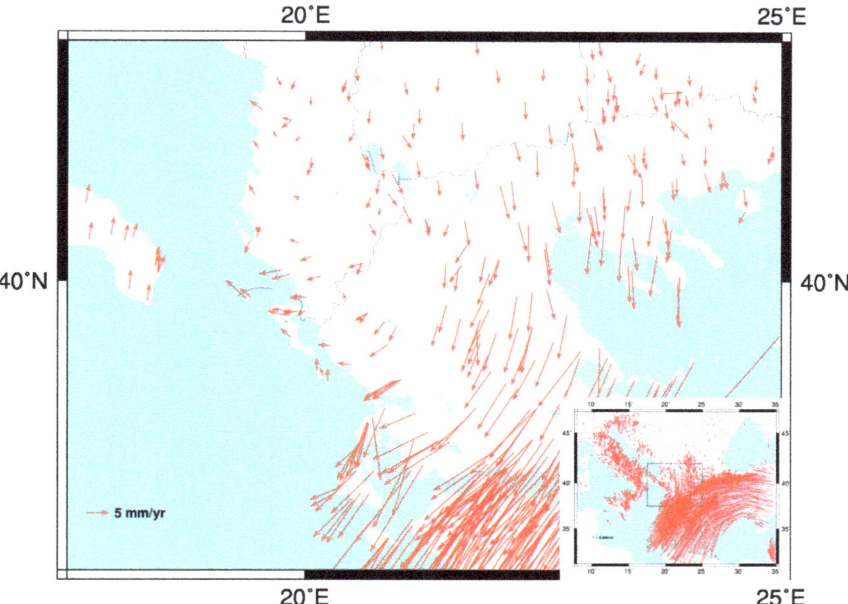


**Figure 2.** Velocities of permanent GNSS Stations in the ETRF2000 "European fixed" reference frame. The velocities on the western side, mostly in Italy, are directed northwards, and those on the central eastern side are directed southwards, implying a dextral shear field. The black rectangle in the insert indicates the area of study. Source: http://pnac.swisstopo.admin.ch/divers/dens_vel/combvel_se_all_cmb_grd_east.jpg. See http://geolabpasaia.org/gnss/agi/maps/EU-DenseVelocities.html#6/42.855/21.955 for an animated map.

## 2. Geodetic Data: InSAR and GNSS

Sentinel-1 data for both ascending (20 November 2019–26 November 2019) and descending (25 November 2019–1 December 2019) passes were analyzed in conjunction with GNSS data from permanent sites. The InSAR data indicate a maximum displacement along the line of sight of approximately 10.5 cm, with an uncertainty that can be conservatively set to 10% or 1 cm (one sigma). The uncertainty is mostly controlled by the tropospheric delay and several other causes [11–13]. The interferograms for the ascending and descending passes are mutually consistent and indicate a very similar deformation pattern. This leads to the conclusion that the atmospheric delay was small enough to maintain correlation between the two interferograms. The direction of flight is ca. 346 degrees for the ascending pass, close to the strike of the fault, and ca. 194 degrees for the descending orbit, with a look angle of ca. 34 degrees of off-nadir angle (Figure 3).

A preliminary calculation based on a model of dislocation in an elastic half space [14,15], with Young's modulus E = 80 GPa and Poisson ratio ν = 0.25, uses the measured (Table 1) moment tensor data to constrain fault geometry (strike and dip of the causative fault) and slip and fault dimensions. The empirical Wells and Coppersmith formulas [16] were used to infer slip and slip area from the momentum magnitude. A model based on a rectangular fault with uniform slip was assumed. This calculation resulted in a maximum surface displacement along the line of sight not larger than 4 cm.

To resolve the discrepancy with the InSAR data, the hypocentral depth of the main event was relocated at a shallower depth. The tests indicated that a depth of 8 ± 2 km was able to result in the observed surface displacement, while keeping the same moment magnitude and orientations of the fault planes (Figure 4). The model uncertainty was estimated as that range of depths such that the resulting line of sight displacements are within 10% of the measured 10.5 cm displacement.

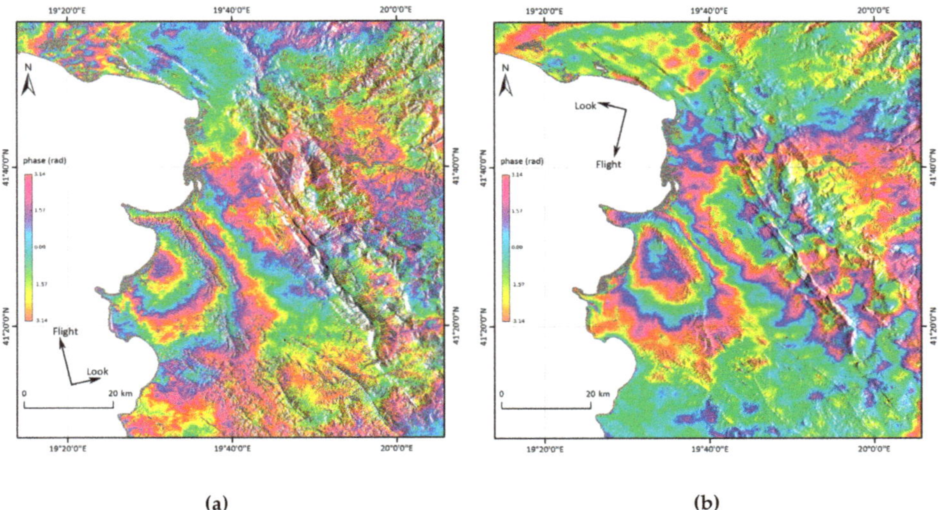



(a)                                                                 (b)

**Figure 3.** Interferograms for the (**a**) ascending and (**b**) descending passes of Sentinel 1.

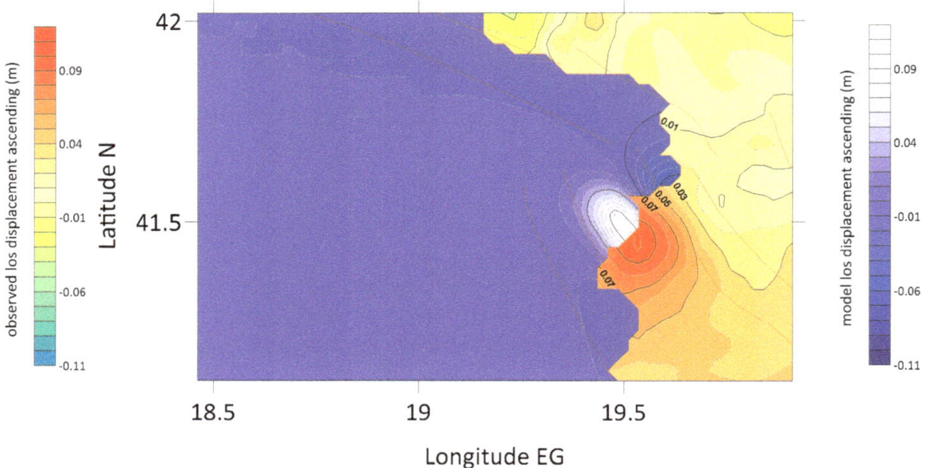



**Figure 4.** Displacement map in the direction of the line of sight (LOS) of the ascending pass of the Sentinel-1 satellite. In the foreground is the observed displacement, truncated at the shoreline, and in the background is the modeled displacement assuming a hypocentral depth of 8 ± 2 km. Brown lines are the fault structures from the project SHARE [1,2]. The plot for the descending pass has negligible differences.

Several permanent GNSS stations belonging to the Albanian permanent networks are routinely processed as part of the Densification of the European Permanent Network [17] and of the Central European GNSS Research Network CEGRN [18–20] using state of the art processing standards [21]. The time series of the horizontal and vertical coordinates available at http://147.162.183.197/ALBANIA/ show sudden discontinuities across the date of the 2019 event with higher amplitudes for the stations (DUR2 and TIR2) near the epicenter, indicating that the stations were displaced by the main event (Figure 5 is an example for DUR2). The previous and subsequent events, being of magnitude 5.5 or less, leave no appreciable signature in the time series, but were nevertheless included in the deformation model. Table 2 gives the coseismic displacements obtained by comparing the horizontal coordinates (vertical coordinates have insufficient accuracy) of the stations before and after the main shock, together with the displacements predicted by the elastic model used for the InSAR data (Table 1, with the hypocentral depth of the main event moved from 26 to 8 km). Table 2 indicates that Durazzo (DUR2) was the station with the highest horizontal displacement (ca. 2 cm). Figure 6 describes the measured horizontal displacements of the processed GNSS sites and the expected vertical motion. The low dip angle plane was assumed as likely principal plane, based on the fact that reverse faults at low dips are activated by a smaller deviatoric stress than at higher dip angle [22]. The green segments in the SW corner indicate the intersection of the assumed fault planes with the Earth surface, at sea. Permanent displacements recorded by a high-resolution digital accelerometer installed at an epicentral distance of ~15 km and probably on the hanging wall of the causative fault are comparable with our GNSS values [23]. Duni and Theodoulidis [23] reported a horizontal Peak Ground Acceleration (PGA) of 1.92 ms$^{-2}$, in good agreement with the average predicted value by a regional model, and a spectral acceleration of at least 5 ms$^{-2}$, for a wide range of periods between 0.3 and 1.0 s.

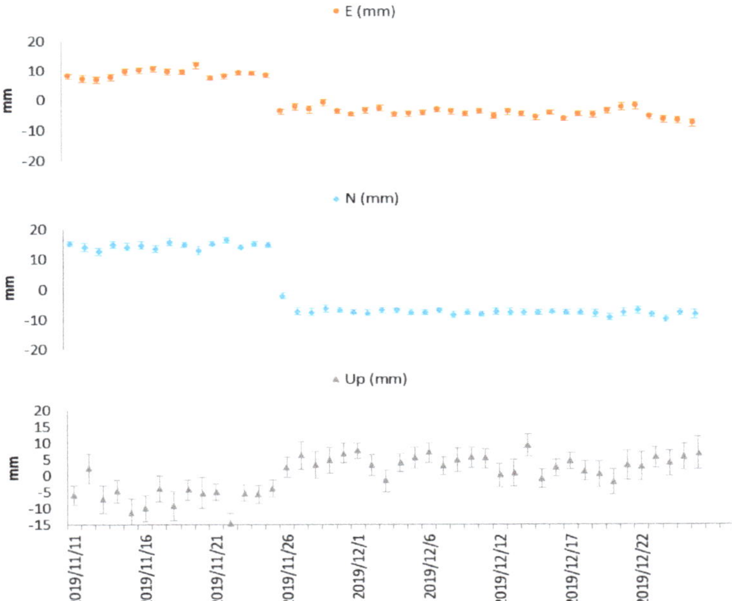

**Figure 5.** Daily time series of the displacement of the GNSS station DUR2 in the east, north, and up directions. The coseismic displacement of 26 November 2019 is in the SW direction and upwards. Error bars correspond to one sigma formal uncertainty.

**Table 2.** Coseismic displacements of selected GNSS stations across the epoch 26 November 2019, 2:54 UTC vs. displacements computed with a dislocation model of an elastic half space based on the moment tensor solution of the main earthquake. The differences between observed and computed displacement in the east and north components have a mean of 0.003 and 0.000 m, respectively, and a root mean square of 0.004 m for both components. Computed displacements are based on Table 1 data with the depth of the main event at 8 km.

| Longitude (deg) | Latitude (deg) | ObsEasting (m) | ObsNorthing (m) | CalcEasting (m) | CalcNorthing (m) | Station Name |
|---|---|---|---|---|---|---|
| 19.945 | 40.708 | 0.001 | −0.002 | 0.000 | 0.000 | BERA |
| 19.451 | 41.316 | −0.012 | −0.018 | -0.018 | −0.025 | DUR2 |
| 19.758 | 40.089 | 0.001 | −0.002 | 0.000 | 0.000 | HIMA |
| 20.698 | 40.707 | 0.001 | −0.001 | 0.000 | 0.000 | KOR2 |
| 20.773 | 40.624 | −0.005 | 0.001 | 0.000 | 0.000 | MALQ |
| 20.440 | 41.685 | −0.002 | −0.003 | −0.004 | −0.002 | PESH |
| 19.875 | 41.768 | −0.003 | −0.003 | −0.009 | −0.010 | RRES |
| 19.496 | 42.051 | 0.002 | −0.003 | 0.000 | −0.001 | SHKO |
| 19.810 | 41.336 | −0.004 | −0.004 | −0.011 | 0.003 | TIR2 |

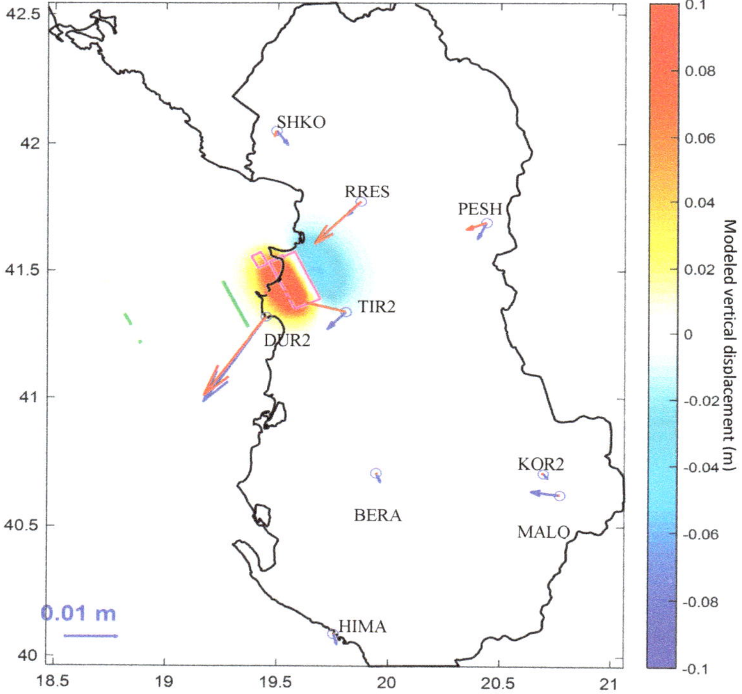


**Figure 6.** Measured (blue arrows) vs. modeled (red arrows) coseismic displacements at permanent GNSS stations in Albania. Purple rectangles represent the fault planes and the green segments the intersections of the fault planes with the Earth surface, under the hypothesis that the low dip faults were activated. The modeled vertical motion is shown with filled contours.

An important question is whether the November 2019 event was in some way affected by the Mw = 7.2 event which took place in Ulcinj (Montenegro) in 1979, ca. 80 km NW of the epicenter of the 2019 event. To this purpose, the change in elastic stress generated by the Montenegro earthquake was projected on planes parallel to the 2019 Albania earthquake using again the deformation of an elastic half space. The Montenegro fault plane was inferred by the pertinent structure MECS001 described in the DISS (Strike = 300°, dip = 28°, and rake = 90°) at a depth of 8 km. The map shows that the transferred stress at the depth of 8 km, the estimated hypocentral depth of the 2019 Albania earthquake, was in the order of 0.1 bar or less, making it unlikely that the 2019 event was triggered by the 1979 earthquake (Figure 7). A friction coefficient of 0.4 was assumed in the calculation of the Coulomb stress.

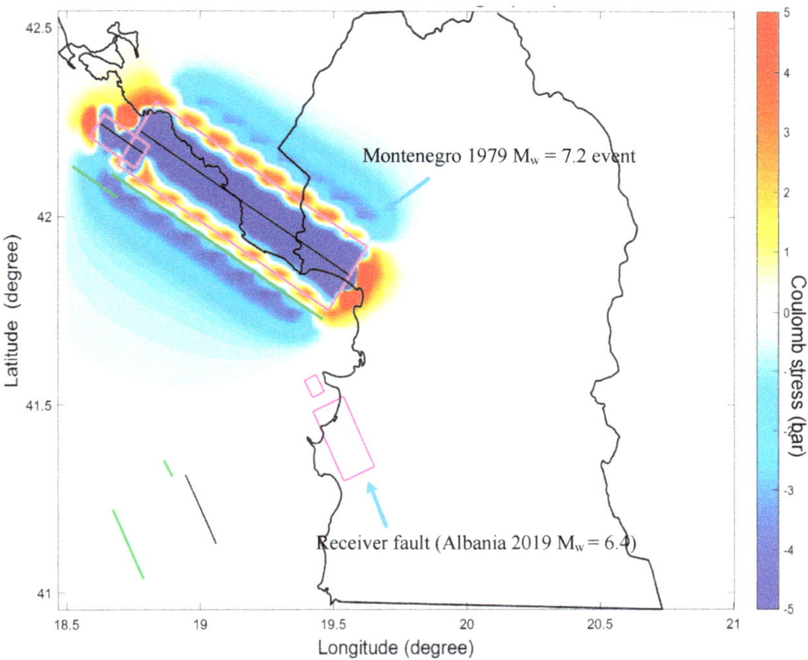


**Figure 7.** Coulomb stress change generated by the 1979 Montenegro event of Mw = 7.2 is projected onto planes parallel to the 2019 Albania earthquake. The amount of transferred elastic stress is negligible. Units: 1 bar = 0.1 MPa.

The maximum stress change resulting from both the 1979 and 2019 events is likely to illuminate a specific area and fault geometry in between the two epicenters. The location and fault angles with the highest Coulomb stress change were therefore investigated, to identify a fault location and geometry with optimal characteristics from the point of view of alignment of the fault to the principal axes of the transferred stress. The analysis indicates that a relatively high shear stress is transferred along a dextral strike in the range 220–240 degrees, dipping nearly 90 degrees and with a rake of ca. 180 degrees. The depth of computation is 8 km, based on the hypocentral depth of the 2019 event. Thus, a likely candidate to accommodate this shear stress is the Lezhe fault (Figure 8). As a consequence of the 1979 and 2019 events, the offshore part of this fault received an extra stress of 2–3 bars, or 0.2–0.3 MPa. This stress change adds to a regional stress of unknown magnitude, but well represented by a plane stress with compressional axis in the same direction SW–NE. If the strain rate is some 30 nstrain/year (1 nstrain = $10^{-9}$) [18], and assuming a shear modulus $\mu$ = 30 GPa, the regional stress rate is on the order of ca. 0.9 kPa/year. Consequently, a seismically induced stress change of 0.2–0.3 MPa (2–3 bar) would imply a time advance of some 2–3 centuries in the stress buildup process at or near the Lezhe fault.

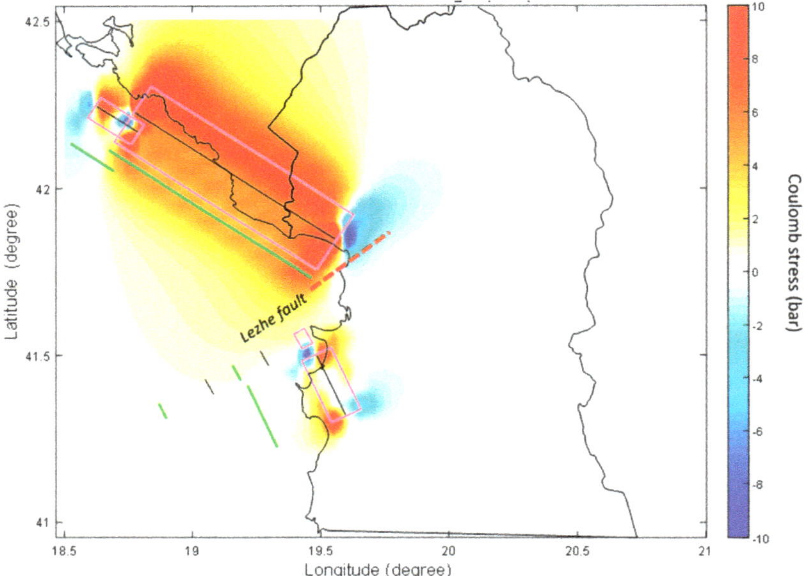


**Figure 8.** Coulomb stress change resulting from the combined action of the 1979 (Montenegro Mw = 7.2) and 2019 (Albania, Mw=6.4) earthquakes, mapped to a plane of strike = 240, dip = 89, and rake = 180 degrees. The Lezhe fault area appears to be positively loaded. The black segments represent the computation depth relative to the mean depth of the fault (purple rectangle). Units 1 bar = 0.1 MPa.

## 3. Conclusions

The November 2019 Albania earthquakes fit very well into the picture of an Adriatic crust subducting under the Albanides/Dinarides. While the epicenters and the fault angles are well constrained, the literature reports a relatively wide range of hypocentral depths, ranging from 9 km (https://earthquakes.ga.gov.au/event/ga2019xgdrrf) to 26 km (http://geofon.gfz-potsdam.de/eqinfo/event.php?id=gfz2019xdig). In most cases, these differences in estimated hypocentral depths reflect the uncertainty in the adopted velocity model. We point out that, based on the InSAR data of Sentinel-1 and on the permanent GNSS stations in Albania, the observed ground displacement requires a shallow hypocenter, of the order of 8 km. We do however expect that a locked zone extends down to some 26 km based on the change of direction of the regional velocity field of GNSS stations and a simple 'arctangent' model [18].

With the improved re-location of the hypocentral depth of the 2019 event, we have further examined the relationship between the 1979 Montenegro earthquake (Mw = 7.2) and the 2019 Albania earthquake (Mw = 6.4) within a model of dislocation in an elastic half space. The model calculations suggest that the 1979 Montenegro event generated a negligible Coulomb stress on the receiver fault likely to have been activated in 2019. Nevertheless, when the two events are considered together again in a purely elastic half space, a non-negligible stress load is visible in the area in-between the two (1979 and 2019) earthquakes. We speculate that the Lezhe fault, a strike slip fault with nearly optimal orientation, could be a likely candidate to receive such an extra load.

**Author Contributions:** M.F. and X.C. analyzed the InSAR data; B.N. provided the GNSS data; M.B. and J.Z. analyzed the GNSS data; and A.C. did the Coulomb stress calculations and wrote the paper. All authors have read and approved the final manuscript.

**Funding:** This research was supported by the Grant "Attività di monitoraggio del territorio veneto a supporto degli strumenti di pianificazione territoriale regionale attraverso il ricorso ai dati forniti dai sistemi GNSS" (Monitoring the territory of the Veneto region for improved mapping using GNSS data) financed by the Regione del Veneto.

**Acknowledgments:** The authors would like to thank Elmar Brockmann of Swisstopo (Wabern, CH) for the dense velocity data set for Europe, and the anonymous reviewers for their constructive and valuable suggestions on the earlier drafts of this manuscript.

**Conflicts of Interest:** The authors declare no conflict of interest.

## References


1. Basili, R.; Kastelic, V.; Demircioglu, M.B.; Garcia Moreno, D.; Nemser, E.S.; Petricca, P.; Sboras, S.P.; Besana-Ostman, G.M.; Cabral, J.; Camelbeeck, T.; et al. The European Database of Seismogenic Faults (EDSF) compiled in the framework of the Project SHARE. 2013. Available online: http://diss.rm.ingv.it/share-edsf/ (accessed on 27 February 2020).
2. DISS Working Group. Database of Individual Seismogenic Sources (DISS), Version 3.2.1: A Compilation of Potential Sources for Earthquakes larger than M 5.5 in Italy and Surrounding Areas. 2018. Available online: http://diss.rm.ingv.it/diss/ (accessed on 27 February 2020).
3. Nieuwland, D.A.; Oudmayer, B.C.; Valbona, U. The tectonic development of Albania: Explanation and prediction of structural styles. *Mar. Pet. Geol.* **2001**, *18*, 161–177. [CrossRef]
4. Aliaj, S.; Adams, J.; Halchuk, S.; Sulstarova, E.; Peci, V.; Muco, B. Probabilistic seismic hazard maps for Albania. In Proceedings of the 13th World conference on earthquake engineering, Vancouver, BC, Canada, 1–6 August 2004; pp. 1–6.
5. Jouanne, F.; Mugnier, J.L.; Koci, R.; Bushati, S.; Matev, K.; Kuka, N.; Shinko, I.; Kociu, S.; Duni, L. GPS constraints on current tectonics of Albania. *Tectonophysics* **2012**, *554*, 50–62. [CrossRef]
6. Caporali, A.; Bruyninx, C.; Fernandes, R.; Ganas, A.; Kenyeres, A.; Lidberg, M.; Stangl, G.; Steffen, H.; Zurutuza, J. Stress drop at the Kephalonia Transform Zone estimated from the 2014 seismic sequence. *Tectonophysics* **2016**, *666*, 164–172. [CrossRef]
7. Burchfiel, B.C.; King, R.W.; Todosov, A.; Kotzev, V.; Durmurdzanov, N.; Serafimovski, T.; Nurce, B. GPS results for Macedonia and its importance for the tectonics of the Southern Balkan extensional regime. *Tectonophysics* **2006**, *413*, 239–248. [CrossRef]
8. Kahle, H.G.; Straub, C.; Reilinger, R.; McClusky, S.; King, R.; Hurst, K.; Veis, G.; Kastens, K.; Cross, P. The strain rate field in the eastern Mediterranean region, estimated by repeated GPS measurements. *Tectonophysics* **1998**, *294*, 237–252. [CrossRef]
9. Peter, Y.; Kahle, H.G.; Cocard, M.; Veis, G.; Felekis, S.; Paradissis, D. Establishment of a continuous GPS network across the Kephalonia Fault Zone, Ionian islands, Greece. *Tectonophysics* **1998**, *294*, 253–260. [CrossRef]
10. Serpelloni, E.; Anzidei, M.; Baldi, P.; Casula, G.; Galvani, A. Crustal velocity and strain-rate fields in Italy and surrounding regions: New results from the analysis of permanent and non-permanent GPS networks. *Geophys. J. Int.* **2005**, *161*, 861–880. [CrossRef]
11. Hanssen, R.F. *Radar Interferometry: Data Interpretation and Error Analysis*; Springer Science & Business Media: Berlin, Germany, 2001; p. 308.
12. Pepe, A.; Calò, F. A Review of Interferometric Synthetic Aperture RADAR (InSAR) Multi-Track Approaches for the Retrieval of Earth's Surface Displacements. *Appl. Sci.* **2017**, *7*, 1264. [CrossRef]
13. Wang, Y.; Guo, Y.; Hu, S.; Li, Y.; Wang, J.; Liu, X.; Wang, L. Ground Deformation Analysis Using InSAR and Backpropagation Prediction with Influencing Factors in Erhai Region, China. *Sustainability* **2019**, *11*, 2853. [CrossRef]
14. Okada, Y. Internal deformation due to shear and tensile faults in a half-space. *Bull. Seismol. Soc. Am.* **1992**, *82*, 1018–1040.
15. Toda, S.; Stein, R.S.; Reasenberg, P.A.; Dieterich, J.H. Stress transferred by the 1995 Mw = 6.9 Kobe, Japan, shock: Effect on aftershocks and future earthquake probabilities. *J. Geophys. Res. Solid Earth* **1998**, *103*, 24543–24565. [CrossRef]
16. Wells, D.L.; Coppersmith, K.J. New empirical relationships among magnitude, rupture length, rupture width, rupture area and surface displacement. *Bull. Seismol. Soc. Am.* **1994**, *84*, 974–1002.

17. Kenyeres, A.; Bellet, J.G.; Bruyninx, C.; Caporali, A.; de Doncker, F.; Droscak, B.; Duret, A.; Franke, P.; Georgiev, I.; Bingley, R.; et al. Regional integration of long-term national dense GNSS network solutions. *GPS Solut.* **2019**, *23*, 122. [CrossRef]
18. Caporali, A.; Aichhorn, C.; Barlik, M.; Becker, M.; Fejes, I.; Gerhatova, L.; Ghitau, D.; Grenerczy, G.; Hefty, J.; Krauss, S.; et al. Surface kinematics in the Alpine–Carpathian–Dinaric and Balkan region inferred from a new multi-network GPS combination solution. *Tectonophysics* **2009**, *474*, 295–321. [CrossRef]
19. Caporali, A.; Zurutuza, J.; Bertocco, M.; Ishchenko, M.; Khoda, O. Present day geokinematics of Central Europe. *J. Geodyn.* **2019**, *132*, 101652. [CrossRef]
20. Zurutuza, J.; Caporali, A.; Bertocco, M.; Ishchenko, M.; Khoda, O.; Steffen, H.; Figurski, M.; Parseliunas, E.; Berk, S.; Nykiel, G. The Central European GNSS Research Network (CEGRN) dataset. *Data Brief* **2019**, *27*, 104762. [CrossRef] [PubMed]
21. Bruyninx, C.; Altamimi, Z.; Caporali, A.; Kenyeres, A.; Lidberg, M.; Stangl, G.; Torres, J.A. Guidelines for EUREF Densifications. Available online: http://epncb.oma.be/_documentation/guidelines/Guidelines_for_EUREF_Densifications.pdf (accessed on 30 November 2019).
22. Turcotte, D.L.; Schubert, G. *Geodynamics*, 3rd ed.; Cambridge University Press: Cambridge, UK, 2014; p. 623.
23. Duni, L.; Theodoulidis, N. Short Note on the 26 November 2019, Durres (Albania) M6.4 Earthquake: Strong Ground Motion with Emphasis in Durres city. Available online: http://www.emsc-csem.org/Files/news/Earthquakes_reports/Short-Note_EMSC_31122019.docx (accessed on 26 November 2019).

*Article*

# Ground Deformation Modelling of the 2020 Mw6.9 Samos Earthquake (Greece) Based on InSAR and GNSS Data

**Vassilis Sakkas**

Department of Geophysics-Geothermics, National Kapodistrian University of Athens, Panepistimiopolis Zografou, 15784 Athens, Greece; vsakkas@geol.uoa.gr; Tel.: +30-2107274914

**Abstract:** Modelling of combined Global Navigation Satellite System (GNSS) and Interferometric Synthetic Aperture Radar (InSAR) data was performed to characterize the source of the Mw6.9 earthquake that occurred to the north of Samos Island (Aegean Sea) on 30 October 2020. Pre-seismic analysis revealed an NNE–SSW extensional regime with normal faults along an E–W direction. Co-seismic analysis showed opening of the epicentral region with horizontal and vertical displacements of ~350 mm and ~90 mm, respectively. Line-of-sight (LOS) interferometric vectors were geodetically corrected using the GNSS data and decomposed into E–W and vertical displacement components. Compiled interferometric maps reveal that relatively large ground displacements had occurred in the western part of Samos but had attenuated towards the eastern and southern parts. Alternating motions occurred along and across the main geotectonic units of the island. The best-fit fault model has a two-segment listric fault plane (average slip 1.76 m) of normal type that lies adjacent to the northern coastline of Samos. This fault plane is 35 km long, extends to 15 km depth, and dips to the north at 60° and 40° angles for the upper and lower parts, respectively. A predominant dip-slip component and a substantial lateral one were modelled.

**Keywords:** 2020 Samos earthquake; SAR interferometry; GNSS; fault modelling; slip distribution

**Citation:** Sakkas, V. Ground Deformation Modelling of the 2020 Mw6.9 Samos Earthquake (Greece) Based on InSAR and GNSS Data. *Remote Sens.* **2021**, *13*, 1665. https://doi.org/10.3390/rs13091665

Academic Editors: Magaly Koch and Mimmo Palano

Received: 19 March 2021
Accepted: 22 April 2021
Published: 24 April 2021

**Publisher's Note:** MDPI stays neutral with regard to jurisdictional claims in published maps and institutional affiliations.

## 1. Introduction

The Northern Aegean Sea (Greece) is characterized by a complex geotectonic setting with intense seismic activity (Figure 1). The area that is bordered by the Northern Anatolian Fault (NAF) zone to the north and the Hellenic Trench to the south exhibits a strong extensional regime consistent with major continental extension [1,2]. The kinematic and dynamic models of the area are compatible with plate tectonic motions [3,4] and highlight the occurrence of strong ground deformation and intense seismicity. The NE–SW trending, dextral, strike-slip faulting in the northern Aegean Sea is associated with and linked to the NAF, which trends parallel to the North Aegean Trough [5–7]. However, the eastern part of the Aegean Sea has diverse fault trends and character. This region accommodates several E–W trending fault zones that exhibit normal-type motions, which are consistent with the extensional stress field [3,8].

The study area of Samos Island—located in the eastern part of the northern Aegean Sea—is subject to extensional forces. It is situated to the south of a NE–SW trending, dextral, strike-slip transfer zone called the Izmir Balikesir Transfer Zone (IBTZ) [1,9], as well as a zone of E–W normal faulting near Izmir Bay (Turkey). East of the IBTZ is the Sakarya Tectonic Unit (STU), which includes Chios Island [10], while in the south the islands of Samos and Ikaria are part of the tectono-metamorphic belt of the Hellenides Tectonic Unit (HTU) [10]. This broad region represents the transition between western Turkey and the Aegean domains [10–12].

The geology of the island consists of four main tectono-metamorphic units that are mainly covered by Mio-Pliocene sedimentary basins [11–14]. Based on geological and seismological studies, there are five active fault zones that have the potential to generate large-magnitude earthquakes [15]. These zones are oriented approximately E–W. Two of them are located in the northern coastal part, and two in the southern central part of Samos. Their dip angles vary between 50 and 75°. The longest fault zone (~27 km) occurs to the north and offshore of the island; it is a normal fault that dips to the north at an angle of 40°. This fault zone is located in the southern part of the Samos Basin (SB) [14]. This basin extends westward towards the deeper (1600 m depth) North Ikarian Basin (NIB).

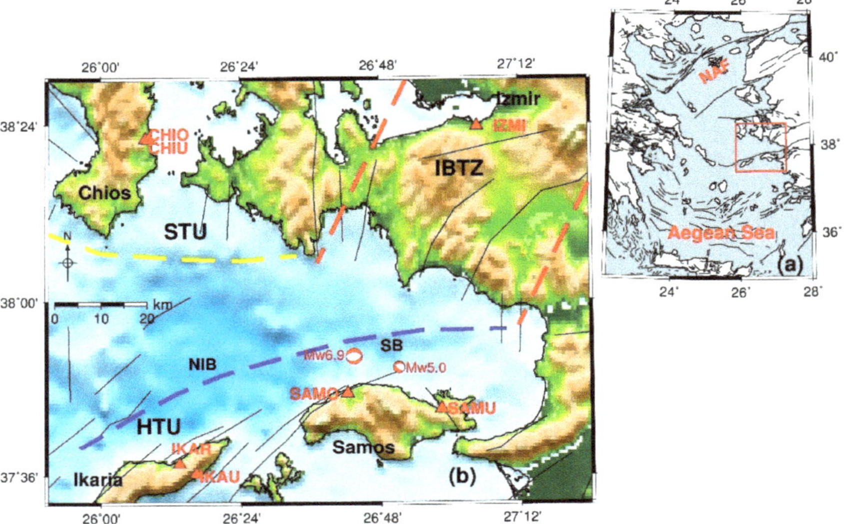


**Figure 1.** (**a**) Index map and (**b**) relief map of the broader area of Samos, which shows the location of continuous Global Navigation Satellite System (GNSS) sites that were used in this study (red triangles), together with the focal mechanism solutions of the Mw6.9 earthquake and its major aftershock. Faults (black lines) were taken from [15–18]. Red dashed lines define the Izmir Balikesir Transfer Zone (IBTZ); yellow dashed line marks the Sakarya Tectonic Unit (STU); blue dashed line defines the Hellenides Tectonic Unit (HTU); NIB: North Ikaria Basin; SB: Samos Basin; NAF: North Anatolian Fault zone.

Several large-magnitude earthquakes have occurred over the last few years in this part of the Aegean [19,20]. The most recent one occurred to the north of Samos on 30 October 2020. This earthquake caused extensive damage on the island and the surrounding areas, as well as significant co-seismic displacements. Combined GNSS and InSAR techniques have been proved effective in precisely measuring the ground deformation. The GNSS technique can provide an absolute 3D vector of ground displacement (estimated errors ~2–3 mm and ~5–8 mm for the horizontal and vertical component, respectively), but it is limited to point-wise coverage. The InSAR provides spatial coverage, but the information is in the line-of-sight (LOS) direction, and further multi-geometrical analysis is required to obtain the true ground motion components. Moreover, the conventional differential interferometry may detect displacements of at least ~28 mm, which is half the wavelength of the radar signal. Thus, joint application of GNSS and InSAR data can effectively determine the real ground displacement field.

The purpose of the present work is to study the ground deformation associated with the Mw6.9 earthquake by combining GNSS and InSAR data for analysis. The methodology has been previously applied to other tectonically active areas of Greece [21–23]. This study aims to quantitatively determine the pre- and co-seismic displacements in the Samos area and produce a model of the activated fault that describes its geometrical and kinematic characteristics.

## 2. Seismological Data

The Mw6.9 earthquake that occurred ~10 km to the north and offshore of Samos (Figure 2) devastated the island and the broader area of Izmir; there were several fatalities. The main event was followed by intense post-seismic activity with more than 200 aftershocks of M $\geq$ 3 over the next forty days. The strongest aftershock occurred about three hours after the earthquake [24].

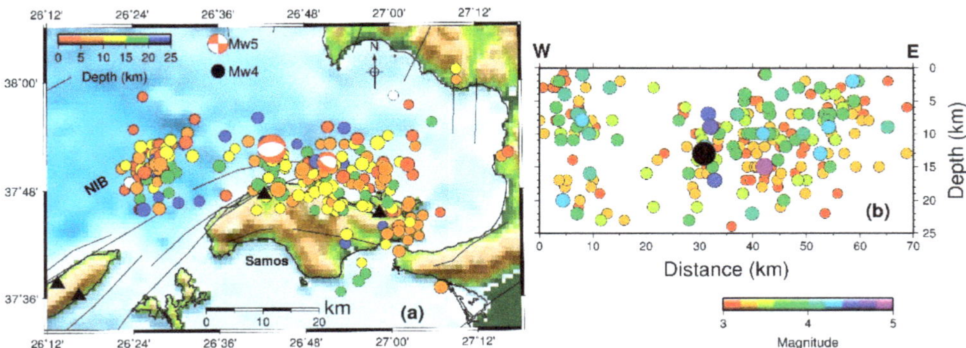


**Figure 2.** (**a**) Post-seismic activity (M $\geq$ 3) following the Mw6.9 earthquake (black circle) and its spatial distribution for the period 30 October 2020 to 28 February 2021. (**b**) Depth distribution of the aftershocks (M $\geq$ 3) along an E–W striking profile. Only earthquakes with M $\geq$ 3 are presented, since events of smaller magnitude have negligible or even no contribution to the ground displacement.

Real-time waveform data from the Hellenic Unified Seismological network were used for the seismological analysis, together with data from other available seismological networks in the area [25–27]. The hypocenters were initially located by using the HypoInverse code [28] and a custom velocity model that was formed for this sequence, which started with a 1-D model for the region of Karaburun (Erythres), Turkey [24,29]. The hypocenters presented herein are relocated events that were extracted using the double-difference method HypoDD [30]. Further details of the processing procedure are found in [19].

The analysis of the earthquake and the strongest aftershock reveals a nearly E–W oriented, dip-slip, normal fault plane (Table 1). The most prominent features of post-seismic activity were the formation of two distinctive clusters. One large group of aftershocks extended eastward from the epicenter, mainly along the northern coast of the island. A smaller cluster occurred west of the epicenter and terminated at the SE margin of the NIB, which itself is characterized by en echelon faults of strike-slip character [17]. The spatial evolution of aftershocks revealed that the seismicity extended over a broader area of ~60 km by ~20 km in the east–west and north–south directions, respectively. The hypocenter depth of the strongest aftershocks (M $\geq$ 4.5) was located mainly within a zone of 10–15 km depth, which also encompasses the hypocenter depth of the earthquake and the main aftershock. It is noted that all of the post-seismic events with M $\geq$ 4.5 occurred up to 31 October, and that the seismicity was significantly decreased about eight days after the main event (i.e., less than three events of M $\geq$ 3 per day). The last significant seismic activity with ~M4 occurred between late January (M4.3 near the epicenter) and early February (M4 near the western-formed cluster) 2021. Furthermore, less than twenty events of 3.0 < M $\leq$ 4.3 occurred over the broader region between January and February 2021.

**Table 1.** Focal mechanism parameters of earthquake and strongest aftershock [25].

|  | Mw6.9 | Mw5.0 |
| --- | --- | --- |
| Date | 30 October 2020 | 30 October 2020 |
| Time (UTC) | 11:51 | 15:14 |
| Latitude (°N) | 37.8759 | 37.8507 |
| Longitude (°E) | 26.7235 | 26.8522 |
| Depth (km) | 13 | 15 |
| Strike (°) CWN | 270 | 264 |
| Dip (°) | 50 | 37 |
| Rake (°) | $-81$ | $-126$ |
| Seismic Moment (N-m) | $2.81 \times 10^{19}$ | $3.9 \times 10^{16}$ |

## 3. GNSS Data

Over the past few decades, the Global Navigation Satellite System (GNSS) technique has been used to study the crustal velocity field, as well as ground deformations due to seismic, volcanic, geologic, or anthropogenic activity [31–35]. Several continuous GNSS stations have been installed in the vicinity of Samos, mainly by the commercial sector (Figure 1). Daily data covering the period before and after the earthquake from four continuous stations were available and processed, namely SAMO, IKAR, and CHIO, which belong to the METRICA S.A commercial network [36], and IZMI, which is an International GNSS Service (IGS) station (https://www.igs.org; accessed on 23 April 2021) (Figure 3).

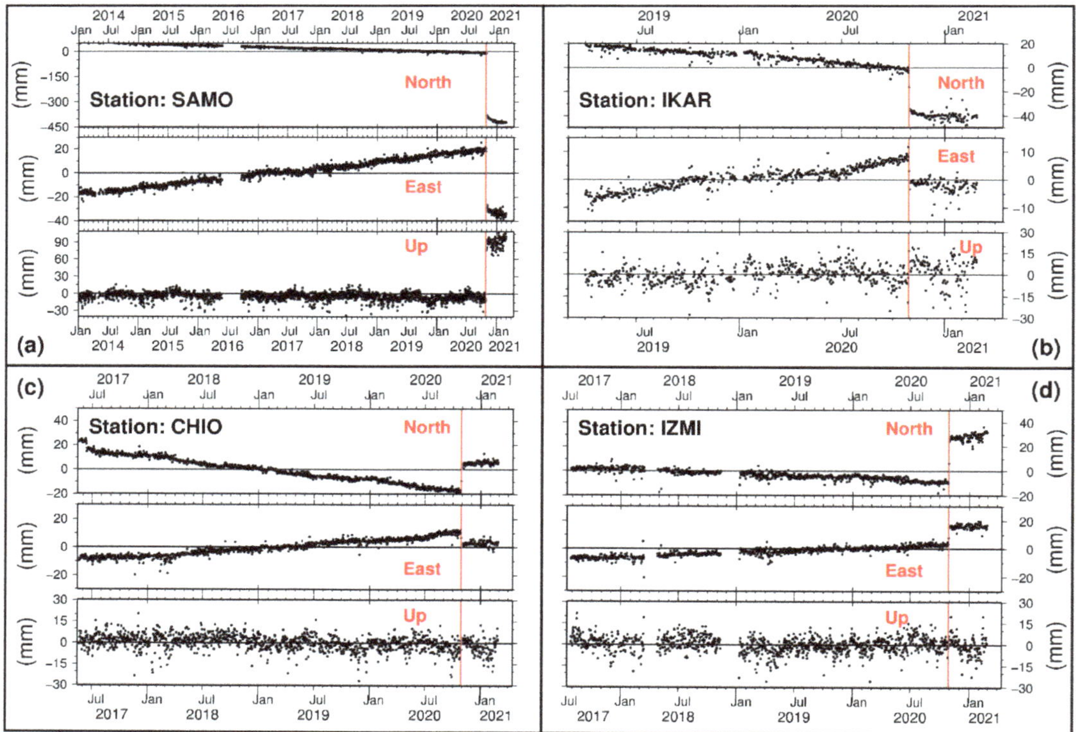


**Figure 3.** Time series of four GNSS stations: (**a**) SAMO, (**b**) IKAR, (**c**) CHIO, and (**d**) IZMI. Red line marks the day of the earthquake. The reference frame is ITRF2014.

Daily data from three more stations—SAMU, IKAU, and CHIU—were available from the Uranus commercial network for the co-seismic period (26 October to 1 November) [37]. The stations SAMO and SAMU are located in the northern part of Samos at ~9 km and ~25 km, respectively, from the earthquake's epicenter. The stations IKAR and IKAU are located on the nearby island of Ikaria, which lies to the west of Samos and ~45 km to the WSW of the epicenter. The stations CHIO and CHIU operate on the island of Chios at ~80 km to the NW of Samos. Finally, station IZMI is located near Izmir (Turkey) at ~65 km to the NE of the epicenter. Data were also processed from GNSS stations that are located on the islands of Lesvos (~140 km to the north of Samos), Leros, and Kalymnos (~65 km and ~86 km to the south of Samos, respectively) [36,37] (Figures S1 and S2). Inspection of the GNSS stations that are located close to the epicenter did not reveal any structural damage after the earthquake [38–40].

The raw GNSS data were processed using Bernese v5.2 GNSS s/w (Astronomical Institute of the University of Bern, Bern, Switzerland) [41]. The coordinates for the local continuous GNSS stations were estimated on the global ITRF2014 reference frame. The processing yielded high-precision datasets of station coordinates, time series of daily coordinates, annual velocities, and co-seismic displacements. Further processing details are provided in Appendix A.

### 3.1. Pre-Seismic Period

Baselines changes between the four local GNSS stations were compiled to assess local deformation prior to the earthquake (Figure 4a). Moreover, differential velocities were calculated relative to station IZMI, and the regional strain field was estimated (Figure 4b). It is noted that station IKAR was only briefly operating before the earthquake (being established on 1 April 2019). However, its calculated velocity vector is in agreement with another station on Ikaria (belonging to Hellenic Cadaster), where data were available for the period from 2013 to 2017. The pre-earthquake deformational field confirms the extensional dynamic conditions of the area [1,8,42].

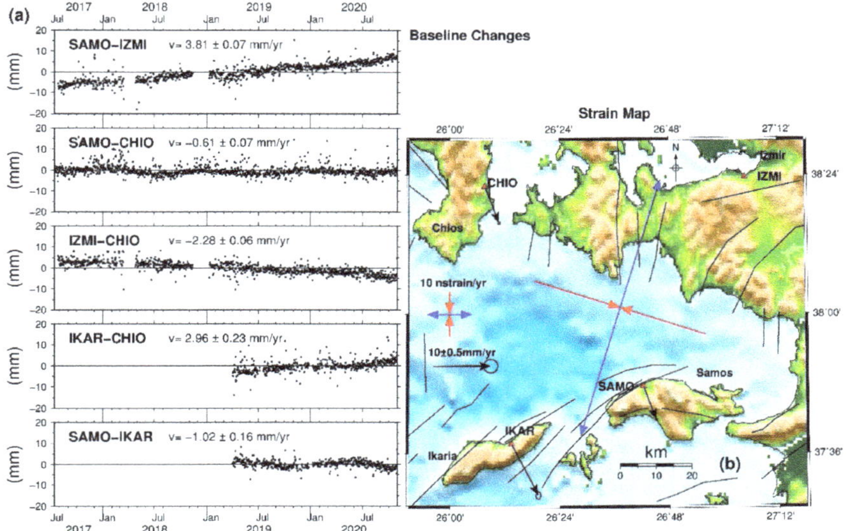


**Figure 4.** (**a**) The daily baseline changes between the four local continuous GNSS stations for the period before the Mw6.9 earthquake, normalized to an average value. (**b**) Map of the strain tensor (blue arrow for extension, red arrow for compression) and the horizontal velocity vectors (black arrows) for the GNSS stations relative to station IZMI. Error ellipses in all figures have been drawn to the 95% confidence interval.

The baseline change between two stations was estimated based on the daily coordinates of the selected sites. Baselines were formed only when data were available for both sites and for a period exceeding 24 h. From the triangle formed by the stations SAMO, IZMI, and CHIO, it is evident that a significant lengthening occurred between SAMO and IZMI, which had a baseline change velocity of 3.81 ± 0.07 mm/yr. Conversely, the distance between IZMI and CHIO was shortened by a smaller value (v = −2.28 ± 0.06 mm/yr). The distance between CHIO and SAMO remained almost unchanged (v = −0.61 ± 0.07 mm/yr). Station IKAR's kinematic behavior indicates lengthening relative to CHIO (v = 2.96 ± 0.23 mm/yr) but shortening with respect to SAMO (v = −1.02 ± 0.16 mm/yr).

The strain tensor was calculated for the pre-seismic period based on the local GNSS station's horizontal differential velocities relative to station IZMI. The strain field represents the dynamic forces in active tectonic areas; it is independent of the reference frame and reveals the changes of dimensions or shape of a deformed area. Due to the large average distance between the stations, the strain tensor was computed on a central point of the network using the algorithm of [43]. The strain tensor of the area has a principal strain direction of 17.5° CWN and an extensional behavior eigenvalue of 60.2 ± 4.8 nstrain/yr, while the minimum eigenvalue was computed to be −39.3 ± 2.5 nstrain/yr (negative for compression). These values are consistent with the aforementioned baseline changes.

### 3.2. Co-Seismic Period

The normal-fault character of the earthquake—as shown by the focal mechanism solution—dominates the pattern and the amplitude of the co-seismic deformational field, as recorded at the GNSS stations (Figure 5). As previously noted, data from three more GNSS stations were included in the processing for the co-seismic deformation. For the computation of the pre-seismic location of the stations, the coordinates from four days prior to the earthquake were averaged up to 11:00 UTC 30 October. The co-seismic location was determined by averaging the estimated coordinates from 12:00 UTC 30 to 31 October, thus avoiding inclusion of more days after the earthquake. Post-seismic motions did occur, as will be presented later.

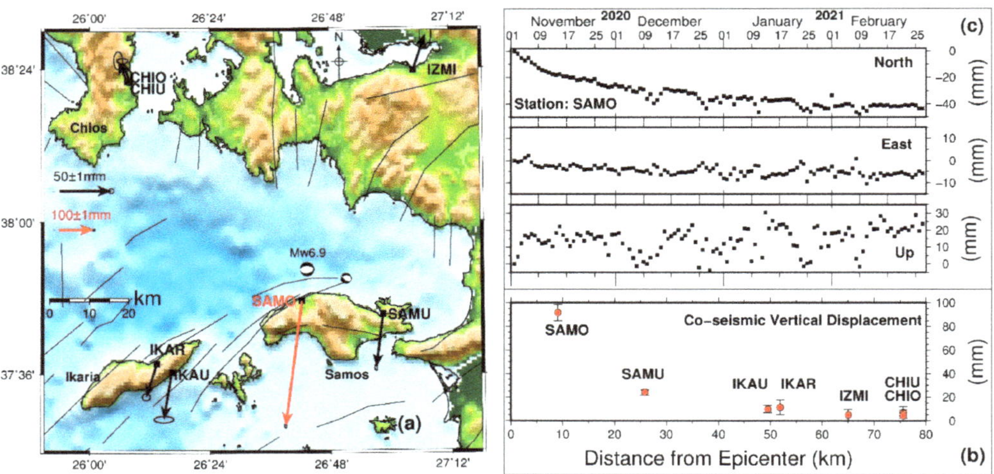


**Figure 5.** (**a**) Map showing the focal mechanism of the earthquake together with the co-seismic horizontal displacement vectors of seven GNSS stations located near the epicenter (station SAMO's vector is at a different scale compared to the others). (**b**) Diagram showing the co-seismic vertical displacements (and error estimates) as a function of distance from the epicenter. (**c**) Time series of the daily coordinates of station SAMO for the period following the earthquake (2 November 2020 to 28 February 2021).

For station SAMO, the post-seismic location was estimated for the days 2 and 3 November, since the station stopped working during the earthquake and re-operated on 2 November, 11:50 UTC. Thus, the co-seismic deformation for this station, as deduced from the analysis, is slightly overestimated since it encompasses motions for about two days after the earthquake. For the days 2–6 November, the station exhibited a near linear post-seismic motion for both horizontal and vertical components (Figure 5c). For this short period, velocity vectors of ~1.5 $\pm$ 0.2 mm/day to the south and ~4.5 $\pm$ 0.8 mm/day for the vertical were estimated. Presuming a linear motion (at least for the first few days), the co-seismic motion at the station was overestimated by ~5 mm and ~10 mm for the horizontal and vertical components, respectively.

The stations located to the south and closer to the epicenter exhibited higher co-seismic horizontal and vertical displacements than the northern stations (CHIO, CHIU, and IZMI) (Table 2). Co-seismic displacements were also observed at the stations on the islands of Lesvos to the north and Leros and Kalymnos on the south (Figure S1). Two stations on Lesvos showed displacements of ~9 mm to the north, while stations on Leros and Kalymnos exhibited displacements of ~20 mm and ~10 mm to the south, respectively.

**Table 2.** Co-Seismic displacements from the GNSS stations (the reference frame is ITRF2014). Notation: D, displacement; STDV, standard deviation; E, east–west; N, north–south; and U, vertical/up.

| Station | Longitude (°) | Latitude (°) | $D_E$ (mm) | $D_N$ (mm) | $D_U$ (mm) | $STDV_E$ (mm) | $STDV_N$ (mm) | $STDV_U$ (mm) |
|---|---|---|---|---|---|---|---|---|
| SAMO | 26.7053 | 37.7928 | −48.94 | −371.27 | 91.70 | 1.15 | 2.75 | 6.71 |
| SAMU | 26.9735 | 37.7575 | −6.95 | −53.27 | 24.10 | 0.52 | 1.29 | 2.27 |
| IKAR | 26.2242 | 37.6282 | −10.22 | −33.16 | 11.35 | 1.58 | 1.37 | 6.28 |
| IKAU | 26.2733 | 37.6054 | −7.69 | −46.65 | 10.03 | 3.87 | 1.30 | 3.19 |
| CHIO | 26.1272 | 38.3679 | −8.75 | 20.59 | 7.10 | 1.72 | 3.53 | 4.71 |
| CHIU | 26.1360 | 38.3665 | −7.30 | 20.83 | 4.44 | 2.19 | 1.01 | 4.92 |
| IZMI | 27.0818 | 38.3948 | 12.21 | 35.20 | 5.03 | 2.20 | 6.05 | 4.71 |

The most prominent feature of the co-seismic displacement is the large horizontal (~370 mm) and vertical (~90 mm) displacements at station SAMO when compared to the other station on the island, SAMU. The two stations on Ikaria had similar vectors with small differences attributed to local tectonic characteristics. The stations on Chios also had similar co-seismic motions. Lastly, station IZMI had a displacement of >35 mm to the NE, which was associated with the severe damage in the urban area of Izmir.

The relaxation period for the region was expected to last for at least 4–6 months after the earthquake. Processing of the GNSS data for almost four months after the earthquake (up to February 2021) showed continued strong deformation at least at station SAMO (Figure 5c) and adjacent stations where data are available (Figure 3). Station SAMO showed southward horizontal motion at ~1.5 mm/day for about 15 days after the earthquake, which reduced to ~0.5 mm/day up to 12 December. The vertical component exhibited an uplift rate of ~3.5 mm/day till 7 November (8 days after the earthquake), then flattened afterwards. The curve for the east component is almost flat, showing no motion in this direction. It is noted that most of the post-seismic activity (number of events and magnitude) occurred during the first fortnight after the main tremor.

## 4. InSAR Data

During the period of the earthquake, co-seismic deformation over the whole island was captured by the Sentinel-1 Synthetic Aperture Radar (SAR) satellites that are maintained by the European Space Agency. Interferometric processing of the acquired SAR images was performed using the Geohazards Exploitation Platform (https://geohazards-tep.eu; accessed on 23 April 2021) and the provided Diapason module, which is being commercially developed by TRE ALTMiRA (Milano, Italy). The interferograms that are presented and post-processed in this study are the final products from the online platform. The SAR images were of ascending and descending orbital trajectory, which yielded three differential interferograms of the deformed region (Table 3).

**Table 3.** Synthetic Aperture Radar (SAR) interferometric pairs.

| Pairs | 24 October–30 October 2020 | 24 October–5 November 2020 | 24 October–5 November 2020 |
|---|---|---|---|
| Sensor | S1B-S1A | S1B-S1B | S1A-S1A |
| Orbit | Ascending | Ascending | Descending |
| Track | 131 | 131 | 36 |
| $B_\perp$ (m) * | 36.03 | 22.83 | 45.47 |
| $B^T$ (time) ** | 4 h 13 min | 6 days | 6 days |
| Incidence Angle, $\theta$ (°) | 36.8513 | 33.7412 | 41.5633 |
| Azimuth Angle, $\varphi$ (°) CWN | 349 | 349 | 191 |

* $B_\perp$ represents the perpendicular baseline between the orbits; ** $B^T$ is the time interval between the occurrences of the earthquake and acquisition of the repeat image.

The coherence of the formed differential interferograms depends mainly on the temporal and spatial decorrelation between the reference and the repeat SAR scenes [44]. The small time separation between the acquired images (only few days), and the use of Sentinel-1 data that have a small and well-controlled orbital tube (ensuing in small spatial baselines), result in the composition of differential interferograms of good coherence, and consequently obtain precise co-seismic ground deformation.

The ascending interferogram that was produced from the 24 and 30 October images (Figure 6a) is of good quality with ~34% coherence at $\geq$0.6. The largest aftershock occurred three hours after the earthquake and is not expected to have contributed to the displacement that is depicted by the interferogram. This assumption is supported by the GNSS data from station SAMU (24-h, 30s interval, continuous data for 30 October), which reveals the total observed co-seismic static displacement occurred during the earthquake and not afterwards (Figure S3. Therefore, this interferogram solely describes the co-seismic deformation given that the repeat scene was acquired just 4 h after the earthquake.

The descending interferogram (24 October and 5 November) exhibited the highest coherence when compared to the others (40% at $\geq$0.6), as well as low tropospheric disturbances (Figure 6c). The strong aftershocks (~M4) that occurred up to 31 October are not expected to have produced measurable deformation. Note that the post-seismic displacement at station SAMO was ~23 mm during the six days after the earthquake, which is less than the sensitivity of the InSAR method (i.e., ~28 mm, which is half the wavelength of the radar signal).

The second ascending interferogram (24 October and 5 November) had similar coherence as the first ascending one (Figure 6b). It was generated to be directly compared to the descending interferogram, since both have the same time span from the earthquake's occurrence. Other interferograms that were created for longer periods and dates, and overlapping the three presented pairs, proved to be of lower quality and were not considered for further interpretation.

The interferograms obtained from all the SAR pairs showed discrete fringes. Significant deformation occurred along the northern coast of Samos, and less towards its southern part. A lack of information due to low coherence was observed in all interferograms in the northern central part of the island. The deterioration is attributed to poor decorrelation caused by dense vegetation of the area. The most intense fringes (i.e., similar number, type (color sequence linked to phase differences), and shape) on all three interferograms were located in a narrow zone in the northern central coastal area. The phase differences in this zone are compatible to increased line-of-sight (LOS) distance between the repeat and the reference scenes. In the western and eastern parts of Samos, both ascending interferograms showed similar shape and type of deformational fringes and were significantly different than the descending interferogram. This difference is due to the horizontal kinematic character of the ground motions depicted from the two orbital geometries, which are associated with changes in range along the LOS direction.

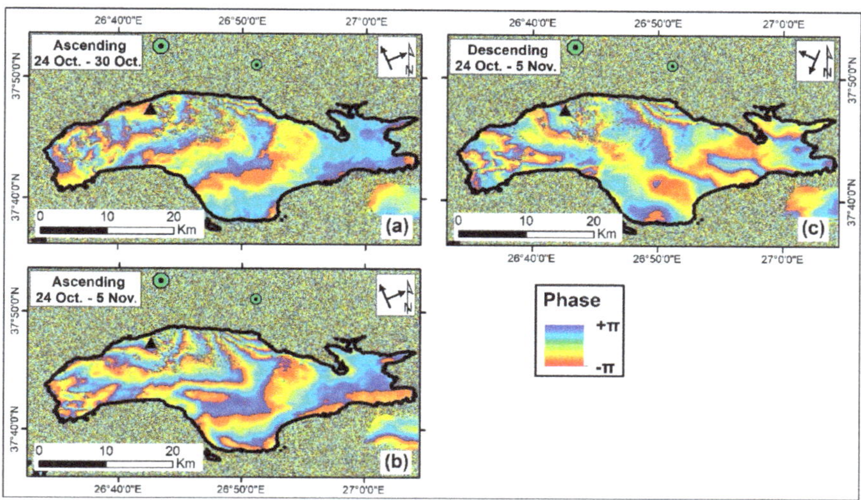

**Figure 6.** Wrapped phase InSAR maps derived from the Sentinel 1 satellites on (**a,b**) ascending and (**c**) descending orbital trajectory. Color scale ($-\pi$ to $\pi$) describes one cycle of phase difference between the reference and repeat SAR images in the line-of-sight direction (LOS), which represents 28 mm of ground displacement along this direction. Black triangles mark the two GNSS stations on Samos (SAMO to the west and SAMU to the east). Green polygons mark the earthquake and the strongest aftershock. Inset arrows indicate the heading azimuth and look direction of the radar satellites.

### 4.1. LOS Displacement Vector

The phase interferograms were unwrapped and the LOS deformational vectors were estimated. They revealed that large co-seismic displacements occurred mainly in the northern and western parts of Samos. Spatial unwrapping provides "relative" information about deformations. To obtain "absolute" values of ground displacements, the GNSS data from the two stations on the island could be used to calibrate the LOS displacement vectors. The differences between the deformation recorded by InSAR and GNSS data are aligned, which creates a smooth surface-of-displacement correction by means of a set of correction for each GNSS station [45–47]. However, the GNSS stations should not be used where intense geodynamic deformation has occurred (as is the present case) [47]. Thus, station SAMO was excluded from the InSAR geodetic correction because of its close proximity to the epicenter; it was inoperable for about two days after the earthquake. Conversely, the station SAMU is located away from the epicenter. It operated continuously after the earthquake, and the produced interferograms exhibit good coherence ($\geq$0.6) in the adjacent area. Therefore, this station was used to define the offset calibration of the InSAR LOS displacement vector. The 3D GNSS derived vector for station SAMU was projected on the LOS direction for ascending and descending acquisition orbits (see Appendix B for details). A cloud of InSAR points around the station that was distributed to a radius of ~200 m of good coherence ($\geq$0.6) was used to calculate the interferometric LOS displacement ($d_{SAR}$) of the area and to be compared with the projected on LOS direction GNSS vector ($d_{GNSS}$). Then the offset calibration ($d_{SAROC}$) of the InSAR data based on the GNSS observations was performed as the difference of the LOS displacement determined by the two techniques: $d_{SAROC} = d_{GNSS} - d_{SAR}$ The resultant geodetically corrected LOS displacement maps of the three interferometric pairs are presented in Figure 7.

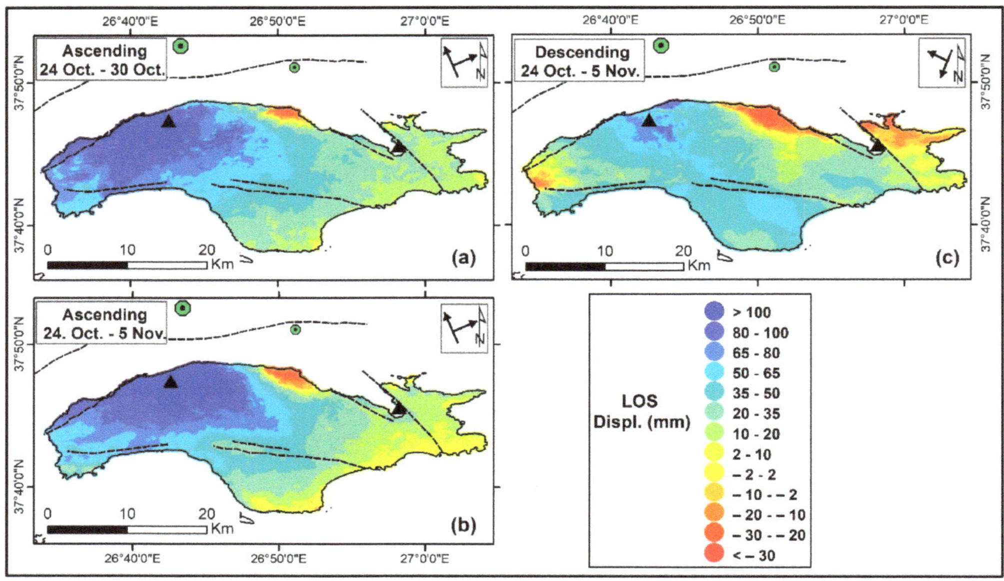



**Figure 7.** LOS displacement maps that were geodetically corrected by GNSS station SAMU for ascending (**a**,**b**) and descending (**c**) orbital acquisition geometry. Black triangles mark the two GNSS stations on Samos (SAMO to the west and SAMU to the east). Dashed lines represent active fault zones from [15]. Green polygons mark the earthquake (Mw6.9) and main aftershock (Mw5.0).

### 4.2. Decomposition of LOS Vector

The LOS displacement vector is composed of projection vectors of the real E–W, N–S, and vertical displacements of a ground target in the direction of the radar wave. Based on the geometry of the SAR imagery, the projection of the LOS displacement vector on the 3D-motion components is defined by the incidence angle $\theta$ and the azimuth angle $\varphi$ of the satellite heading (measured clockwise from North) [48,49] (Appendix B). The LOS vector has different sensitivity for each component. For the present case, the sensitivities of the ascending and descending tracks were 74–83%, 36–65%, and 10–12% for the vertical, E–W, and N–S components, respectively. It is evident that the InSAR technique was more sensitive in detecting vertical displacement rather than E–W motions and was limited in the N–S direction.

The aim of multi-geometry SAR processing is to combine LOS displacement measurements from two or more acquisition geometries in order to estimate the horizontal and vertical components of the recorded displacement signal [49]. Accordingly, displacement data from the first ascending (24–30 October 2020) and the descending interferograms were decomposed to estimate displacement motion for the E–W and vertical directions. The variation of incidence angle $\theta$ along the range was considered in the calculation. The variation of $\theta$ was about ±2° for the ascending pair and about ±1.5° for the descending; these variations were introduced to the fuse procedure. The azimuth angle $\varphi$ of the satellite heading was treated as constant for each track, since it is usually about 1° over the extent of a radar image.

Efforts to use all of the three formed interferograms to retrieve the N–S displacement component were unsuccessful. The estimated N–S components in target points close to GNSS stations (which were acting as control points) were excessively high and of different sign (northward or southward) concerning the GNSS results. Odd and inconsistent behavior was observed also on several other neighboring targets, where both amplitude and sign of the N–S component were unrealistically changing. A plausible explanation could be the limited sensitivity of the InSAR method on the N–S component, especially on Samos where this is the prominent deformational component, as shown by the GNSS data. A more accurate way to illustrate the poorly constrained N–S component is the calculation of the condition number of the coefficient matrix (*cond(A)*) that is used in the decompositional process [49]. The resultant *cond(A)* for the three produced interferograms has a high value (146.9 $\gg$ 1), indicating an ill-conditioned linear system (i.e., *cond(A)* value close to 1 is considered as well-conditioned), whereas, the condition number of the coefficient matrix has a small value (1.25) when neglecting the N–S component from the linear equation system, thus demonstrating a well-constrained linear system. Based on the above issue, the N–S component was excluded from the decompositional process.

The downsampling procedure of the calibrated unwrapped interferograms was performed spatially for the whole island (Figure 8). The results were evaluated for points in the vicinity of the two GNSS stations. There was an overestimation of about 20% and 10% on the vertical and the E–W components, respectively, for the InSAR points close to station SAMO, while the points in the vicinity of station SAMU yielded almost identical results as those of the GNSS observations.

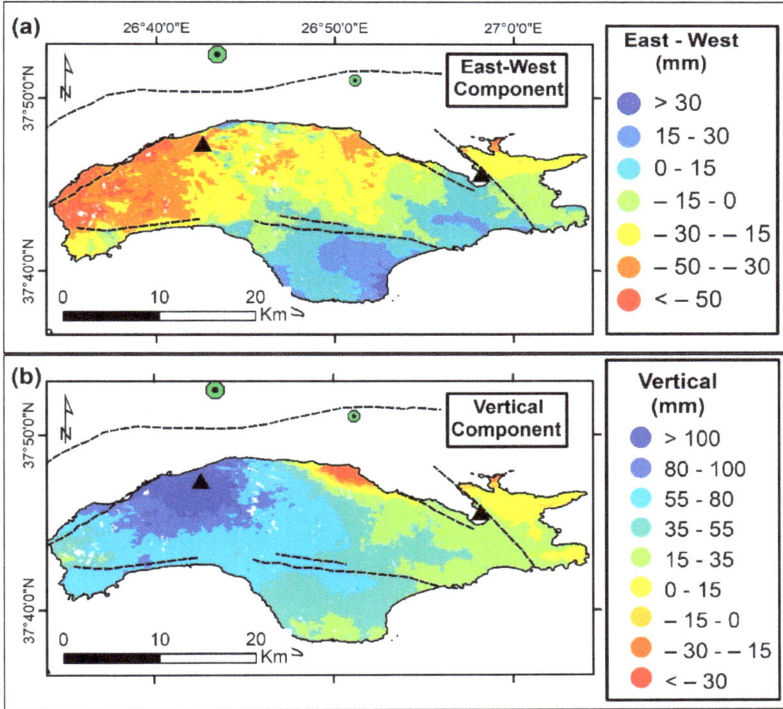


**Figure 8.** Multi-geometry data decomposition of the calibrated unwrapped interferometric displacements maps for the (**a**) E–W and (**b**) vertical co-seismic components. Negative values on the E–W component represent westward motion. Black triangles mark the two GNSS stations on Samos (SAMO to the west and SAMU to the east).

The deformational map of the E–W component (Figure 8a) shows significant westward motion on the western part of Samos, while the vertical component describes intense uplift (Figure 8b). In the central southern part of the island, a differential type of motion (westward to the north and eastward to the south) is evident from the E–W component map. Significant subsidence (up to 80 mm) was observed on the narrow-deformed zone in the central northern coastal region, as was concluded from both wrapped and unwrapped interferograms. Subsidence (~15 mm) was also observed at the northeastern part of the island. The overall maps of the E–W and vertical components highlight the gradual decrease of the co-seismic displacements towards the eastern and southern parts of Samos, which is more clearly depicted in the two ascending interferograms.

## 5. Fault Modelling

Forward and inverse modelling of a rectangular fault surface was performed to define the source, the magnitude, and the type of earthquake that caused the displacements on Samos, as were recorded by the GNSS and SAR data. Displacements that were embedded in a non-uniform slip model were calculated using an inversion algorithm [50–52]. The displacement vectors (together with their error estimates) that were derived from all of the available GNSS stations in the area were used in modelling.

Several of the highly correlated data points in the unwrapped calibrated interferograms were down-sampled using an approach similar to [53] to facilitate the modelling. For each orbital trajectory, LOS displacement values from the calibrated interferograms were numerically extracted by manually picking selected pixels. These pixels had high coherence ($\geq$0.6), were located along formed fringes, and uniformly covered the island. Points/targets were selected that were close to the two GNSS stations for quality control. Some targets were picked from areas of low coherence (<0.6) in order to gain insight about the displacement in these areas and to attain better spatial coverage of the island. Approximately 160 and 190 points/targets from the ascending and descending interferograms, respectively, were chosen that met the above criteria. Finally, 110 points/targets that were common to both tracks were selected from the two data sets and used in modelling. An error of 10% was added to the LOS components. This error was considered sufficient to describe the sensitivity of the LOS vector to the true ground motion along the horizontal component, and to account for the differences between the GNSS and the InSAR data for the selected points in the vicinity of the two stations.

The two data sets were weighted unequally in the joint inversion. The GNSS data points were weighted at 1.0, while a weighting factor that ranged from 0.6 to 1.0 was applied to the InSAR data points based on the coherence levels of the selected points.

### 5.1. Fault Geometry

Determining simultaneously the fault geometry and the slip distribution on the fault plane is computationally expensive [32]. For this reason, inversion was performed only to define the slip distribution on the fault plane. The location, geometry, and characteristics of the model fault plane were derived from (i) the fault traces proposed by [15], (ii) the amplitude and the direction of the GNSS displacement vectors, (iii) the spatial deformation from the interferometric maps, (iv) the information derived from and the constraints imposed by the focal mechanism solution of the earthquake (Table 1) and the spatial distribution of the aftershocks (Figure 2), and (v) the iterative attempts to fit the observed data. The geometrical parameters of the best-fit model are presented in Table 4.

**Table 4.** Fault parameters of the best-fit model.

| Location [1] | Length (km) | Width (km) | Strike (°) (CWN) | Dip (°) | Burial Depth (km) | Locking Depth (km) | Geodetic Moment (N-m) |
|---|---|---|---|---|---|---|---|
| 26.9435° E 37.7956° N | 35 ± 3 | 19.7 | 277 ± 3 | | 1.0 ± 0.3 | 15.0 ± 3.0 | 3.67 × $10^{19}$ |
| Upper Segment | | 5.7 | | 60 ± 3 | 1.0 ± 0.3 | 6.0 ± 0.5 | 1.01 × $10^{19}$ |
| Lower Segment | | 14.0 | | 40 ± 5 | 6.0 ± 0.5 | 15.0 ± 3.0 | 2.66 × $10^{19}$ |

[1] The location (longitude and latitude) indicates the eastern top-most point of the fault plane. The rigidity was assumed to be 30·GPa.

The observed ground deformation indicates that the source of the earthquake was a fault plane to the north and offshore of the island. The gradual attenuation of ground displacements eastward and to the southern part of the island indicated a near E–W striking plane. This explains the spatial distribution of the post-seismic activity along the northern coast of Samos. Intense uplift occurred mainly in the northern and western parts with a large southward displacement component. This describes a normal fault with the activated part dipping to the north such that the mainland of Samos rests on the footwall block. These observations clarify which one of the two possible nodal plane solutions for the centroid moment tensor is the actual one.

The length of the fault plane along the strike direction was set to ~35 km, which extends to the areas of the most intense displacements and the majority of aftershocks. The width of the plane was estimated from the expected ruptured surface for a ~M7 earthquake [54]. The depth distribution of the earthquake and aftershocks was considered when assigning the width and the locking depth of the fault plane. The locking depth was set to 15 km, which takes into account that most of the strong aftershocks (M $\geq$ 4.5) happened in this zone. The tsunami that formed after the earthquake [55] indicates that the fault plane likely reaches close to the sea bottom. Thus, a 1 km vertical burial depth was assumed. Nevertheless, the burial depth was altered several times to assess the sensitivity of the model to this parameter.

The dip angle of the fault plane was examined by fixing the location and the extent of the fault plane and then searching for the optimal angle that minimized the data misfit between observed and predicted surface displacements. Treating the fault as a single plane produced the general pattern and amplitude of ground displacement. However, this approach could not account for the observed intense subsidence along a narrow zone in the northern central part of the island. The motions in this area indicate a hanging-wall block of the fault. Assuming a two-segment fault plane along the width yielded a better fit to the observed vertical displacement.

The characteristics of the two segments along the width (i.e., extension and dip angles) were determined by trial and error inversion. Ultimately, dip angles of $60°$ and $40°$ were defined for the upper and lower segments, respectively. The horizontal boundary between the two parts was estimated to 6 km vertical depth, which marks the transition from an area of moderate seismicity (<6 km depth) to a zone where the main seismic activity was recorded (6–15 km depth). The dip angle for the upper segment coincides with the characteristics of tectonic faults in the area, which are described as high-angle normal faults [11]. The modelled dip angle of the lower segment is consistent with the focal mechanism solution and errors on its estimation.

### 5.2. Slip Distribution

After the geometry of the activated fault was determined, the plane was discretized to horizontal (rows) and vertical (columns) subsegments and then inverted to define the slip distribution along the plane. The average spacing of the data points close to the epicentral area (i.e., northern Samos) was ~3 km. Therefore, a grid of 5 rows × 12 columns was deemed adequate. Two rows along the dip direction were assigned to the upper segment and three rows to the lower one. The grid has cells of ~3 km along the strike direction and ~2.5 km and ~3.0 km along the dip direction for the upper and lower segments, respectively.

A checkerboard resolution test was performed to assess the ability of the discretized plane to resolve the observed data (Appendix C). The test was applied to a fault plane with the same geometrical characteristics as described above. For forward modelling, alternating dip-slip values of 0 m and 1 m were assigned to the cells, and then the 3D displacement vector was calculated to produce synthetic observational data. Next, these data were used to resolve the slip of a uniform slip model. It was found that the model recovered the slip on all of the cells of the upper three rows but lost resolution for the deeper cells (fully resolved at ~60%).

A smoothing factor, *k*, was introduced to constrain irregular oscillations of the slip distribution along adjacent cells [49] and to control the roughness of the model. Increasing *k* results in a smoother distribution of slip motion but increases the misfit. Thus, there is no unique solution. A wide range of values for *k* was applied during the inversion process. The preferred model for slip distribution has *k* = 3000 in the inflation corner with the trade-off curve between roughness and misfit.

Concerning the limitations on slip, the modelling allowed all cells to include a lateral strike-slip component (range from −1.0 m to 1.0 m, where negative sign indicates left-lateral motion). However, the modelling focused on determining the magnitude of dip-slip motion. Accordingly, the range of dip-slip was set free but with an absolute magnitude of less than 2.5 m, which was based on the focal mechanism of the earthquake and studies by [54]. The slip distribution of the best-fit model is presented in Figure 9.

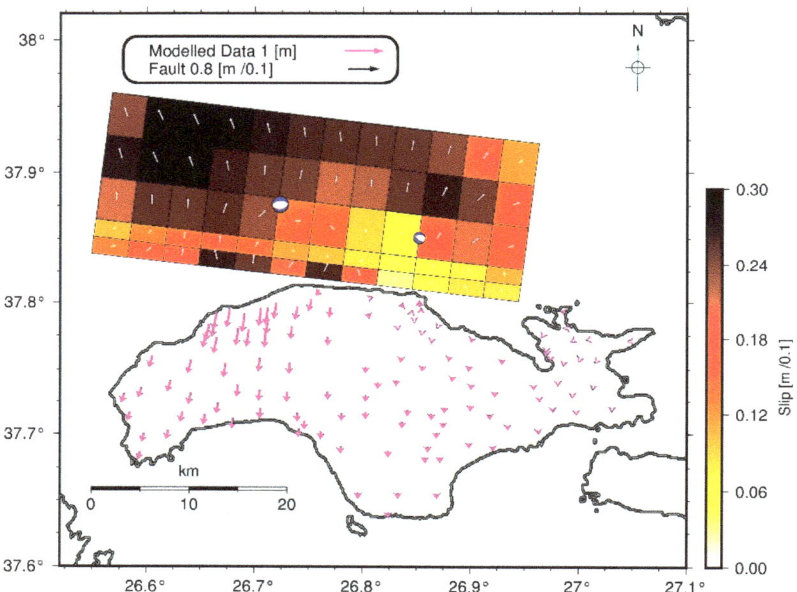


**Figure 9.** Slip distribution along the fault plane for the preferred model. The upper segment (2 rows) has a dipping angle of 60° and the lower one (3 rows) of 40°. White arrows on the fault plane describe the unit slip vector. Pink arrows on the data points are the modelled horizontal displacement vectors. The focal mechanisms of the earthquake, and the main aftershock are also presented.

The preferred slip-distribution model shows an average slip of 1.76 m along the entire fault plane. The slip on the upper and lower segments of the plane was computed to 0.48 m and 1.28 m, respectively. The average slip value is significantly higher than the expected value of ~0.8 m that is based on the scaling law presented by [54]. Scaling relationships also relate the ruptured surface, the length, and the width of the plane to the moment magnitude. The theoretical values for the Samos earthquake are much higher than those of the best-fit model. The use of a larger fault plane was precluded by lack of model resolution (as shown by the checkerboard test), which may account for the difference.

The resultant geodetic seismic moment of $3.67 \cdot \times 10^{19}$ N-m was higher than the seismic one (Table 1), but within the range calculated by others (USGS United States Geological Survey [56]; GFZ—GeoForschungsZentrum [57]; GCMT: Global Centroid Moment Tensor [58]). The latter value indicates a slightly larger moment magnitude (Mw7.01) from the one publicized here for the earthquake. Note that the predominant slip occurs in the lower zone of the fault plane (i.e., >6 km depth), which coincides with the main shock zone. Although this zone lacks good resolution (as shown by the checkerboard test), the slip vectors have a strong dip-slip component and a moderate lateral component. The latter is consistent with the focal mechanism for the dip-slip but contradicts the almost pure dip-slip character of the source. The upper eastern part of the plane exhibits a small slip amplitude (<0.5 m), which is in the area where the plane is running offshore and away from the modelled points. In the western end of the fault plane, the misfit deteriorates but shows intense slip motion (>2 m), which may be attributed to the algorithm's effort to fit the observed deformations at the limits of the data. A quite distinctive feature that emerges from the fault-plane solution is the decreased slip motion on the eastern upper part of the plane. This decrease may indicate a possible deepening of the fault's upper edge there to depths greater than 1 km.

For direct comparison to InSAR results, the 3D modelled vector was computed on a grid of 500 m × 500 m that covered Samos. This vector was projected on the LOS direction for the ascending and descending acquired pairs by using the geometric characteristics for each track at time of acquisition (Figure 10).

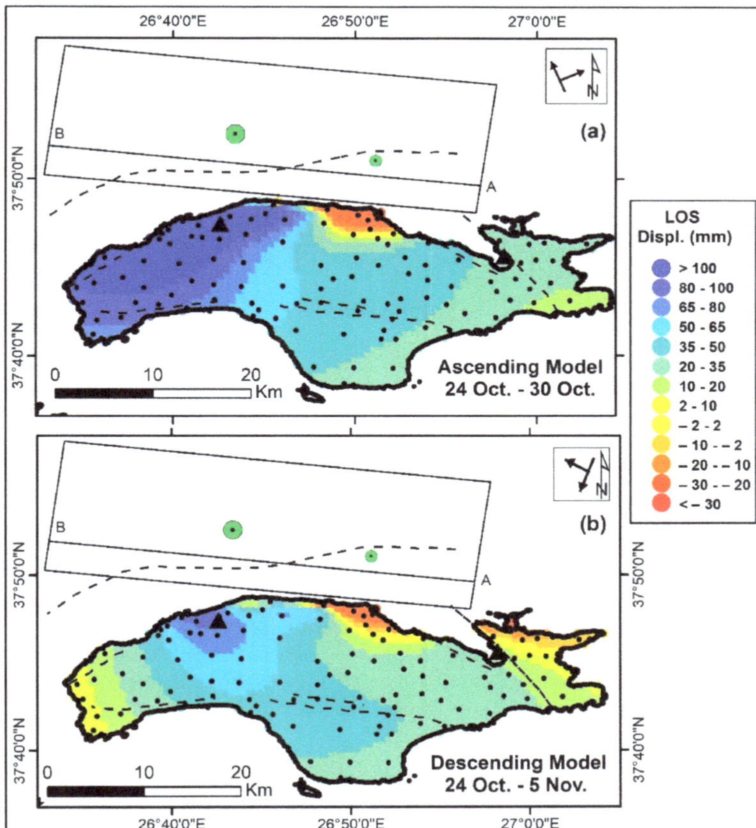


**Figure 10.** Modelled LOS displacement maps after projecting the calculated 3D vector on the LOS direction for the (**a**) ascending and (**b**) descending orbital geometry. Black dots show the InSAR data points used in the inversion process. The rectangle shows the projection of the fault plane on to the surface. Line AB marks the change of the dip angle from 60° to 40° on upper and lower segments, respectively.

The RMS misfit is 0.012 m for all of the data points (GNSS and LOS vectors) that were used in the inversion process for the preferred slip model. The mean scatter between observed and calculated LOS values is 3.61 mm and 3.26 mm for the ascending and descending orbits, respectively (Figure 11a,b). The residuals between the observed and modelled data for both orbital geometries showed larger deviations in the western and southern parts of Samos, as well as for points/targets of low coherence level. In the same areas, the east–west and vertical components also showed the largest discrepancies between observed and calculated values (Figure 11c,d). The mean scatter for the east–west component was −8.1 mm; modeled values were overestimated mainly in the western part. For the vertical component, the mean scatter was 7.2 mm; the larger differences were located in western and southern Samos, as well as in the narrow subsiding zone in the north.

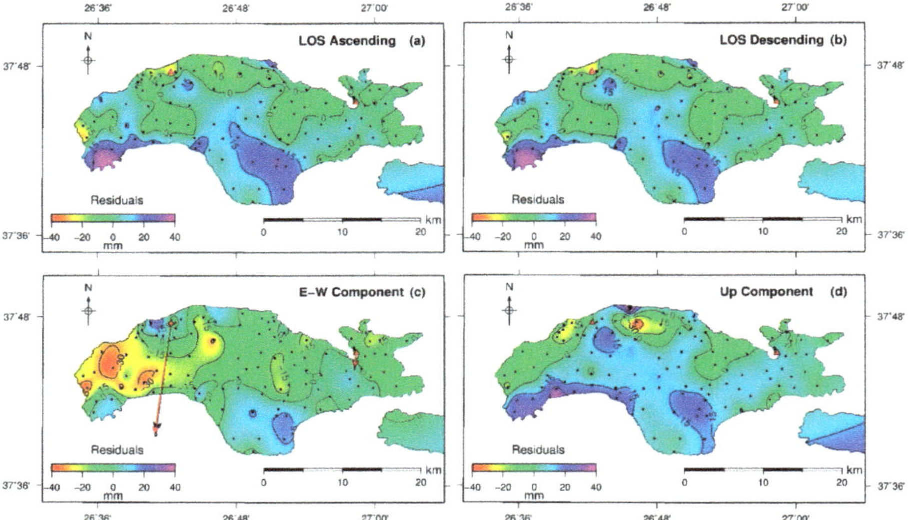

**Figure 11.** Residual maps showing the difference between observed and calculated data for the LOS modelled displacement vectors for (**a**) ascending and (**b**) descending orbits. (**c,d**) Residual maps of E–W and vertical components for the 110-modelled points/targets (black circles). The fit of the GNSS horizontal displacement vectors are included in (**c**) where black arrows are observed, and red arrows are modelled.

The modelled values for the N–S displacement component for the two GNSS stations SAMO and SAMU were underestimated by 4 mm and 8 mm, respectively. The fits for the two Ikarian stations were also underestimated but significantly worse. For the GNSS stations that are located to the north of the epicenter (i.e., CHIO, CHIU, and IZMI), the matching was better with only overestimation (~7 mm) at station IZMI.

## 6. Discussion

The present ground-deformation study is based on joint interpretation of GNSS observations and InSAR data, both before and after the earthquake. In the following, the results of modelling the seismogenic source are discussed in terms of the overall tectonic status of the study area (Figure 12).

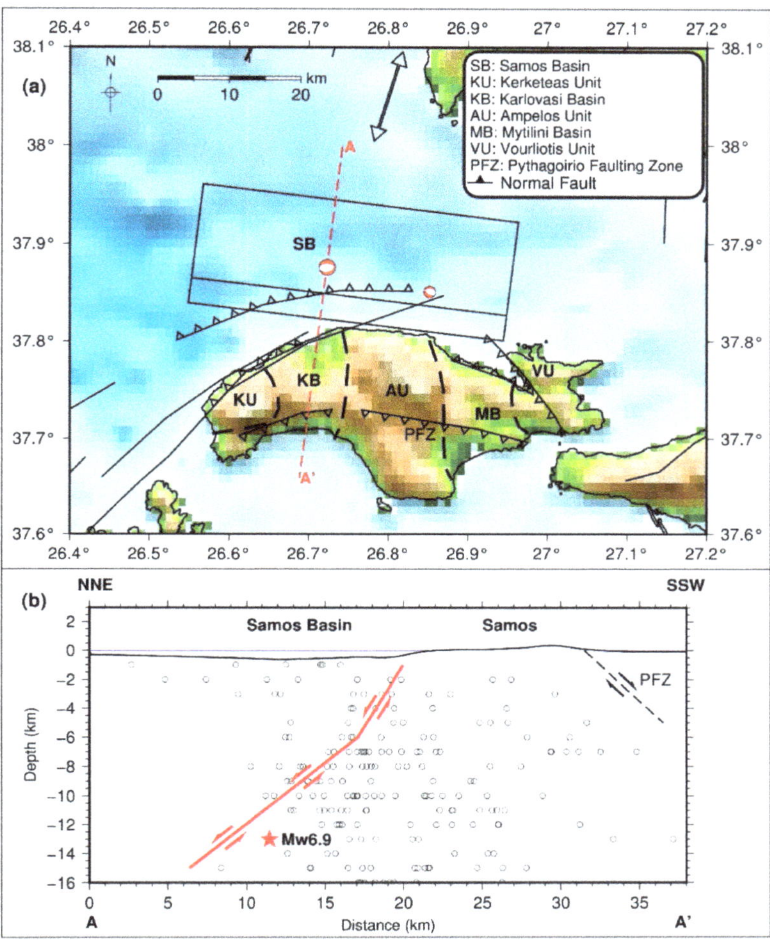



**Figure 12.** (**a**) Map of Samos Island showing the modelled fault plane that has been projected to the surface (black rectangle). Dashed lines delineate the main geotectonic units. The double headed arrow shows the direction of the pre-seismic, regional, extensional strain tensor. (**b**) Cross section along the AA' line showing the modelled plane of the listric fault, which dips to the north (red line). The black circles represent the aftershocks (M ≥ 3) and their primary association with the lower fault segment.

## 6.1. Regional Deformational Characteristics

The pre-seismic kinematic study that was based on continuous GNSS observations over the broad area of Samos reveals a predominant extensional regime, in agreement with previous works [1,2,6,8,10]. The strain field tensor is parallel to the IBTZ, which comprises the main tectonic unit of the Izmir region of western Turkey [7]. This ophiolitic unit is a NE–SW trending zone that is bounded by dextral strike-slip faults that terminate to the northeast of Samos. It lies between the STU to the north and the continental European foreland to the south [10]. The GNSS-derived velocity vectors reveal differential motion along stations CHIO, SAMO, and IKAR to the west and station IZMI to the east, which are compatible with the tectonic regime. Note that Chios is located in the STU, while Samos and Ikaria are in the European foreland [10]. The latter explains the increasing distance between the islands of Ikaria and Chios as described by the baseline changes, while the shortening between Ikaria and Samos is attributable to strike-slip faults in the southern margin of the NIB [17]. The overall extensional condition in the area has resulted in the formation of E–W trending faults, the reactivation of older structures [15], and subsequent triggering of the Mw6.9 earthquake.

The modelling of the earthquake shows that it took place on an almost E–W trending normal fault that dips to the north and lies adjacent to the northern coastal zone of Samos. The extent of the observed deformations from Lesvos to Kalymnos (Figure S1) indicates that the main event was associated with regional tectonic conditions. This conclusion is supported by the magnitude and the regional tectonic characteristics of the area [10], which features Samos at a key position for the dynamic evolution of the Aegean Sea [12]. However, the dimensions of the modelled fault plane are smaller than expected for the given magnitude [54,59].

The displacements that were recorded at the GNSS stations described a strong NNE–SSW extension due to an E–W trending normal fault (Figure 5). The displacement vector from station IZMI reveals a NNE co-seismic motion, while station CHIO reveals a NNW motion, i.e., an opening in-between the two sites, which is opposite to the shortening that occurred during the pre-seismic period. The latter describes the compressional regime in the inter-seismic periods, whereas extensional forces are at play during seismic events. The displacement vectors observed at Ikarian stations are almost parallel to the ones at Samos. This similarity imposes a uniform spatial southward motion for the broader region to the south of the epicentral area, even though the tectonic characteristics and the pre-seismic motion presumed a more distinctive co-seismic differentiation between Samos and Ikaria.

*6.2. Local Deformational Characteristics*

The strongest co-seismic displacement occurred at the western GNSS station SAMO (Table 2), which is close to the epicenter. Its displacement was almost an order of magnitude larger than at the eastern station SAMU. It may be argued that the latter is more distant from the epicenter; however, it had nearly the same horizontal displacement as the station IKAU, which is located at twice the distance from the epicenter. Similarly, the InSAR analysis shows intense deformation in the western part of Samos on both ascending and descending orbits but which substantially attenuates towards the eastern and southern parts (Figure 7). These observations suggest that the activated faulting is less extended towards the east than indicated by post-seismic activity. There are formations in the eastern part of Samos that may inhibit ground motion. Recent studies [14] have revealed a hummocky volcanic relief that is located offshore of the eastern part of Samos; these formations may act as a kinematic discontinuity that effectively reduces ground deformation as exhibited by the GNSS data.

The displacement field from the InSAR analysis highlights the main tectonic units on Samos that have been described in previous works [11,12,14]. Alternating LOS motions between the central (positive values) and the eastern (negative values) parts of the island clearly delineate the area that forms the Vourliotes nappe to the east of Mytilini Basin [12]. The area is marked by a NW–SE normal fault zone (Figure 12). The intense deformation in the central and western parts of Samos coincides with the Ampelos Unit and Karlovasi Basin in the center and to the west, respectively [12]. Another distinct geological unit that shows an alternating LOS pattern is the Kerketeas Unit [12,14]. It is at the westernmost part of Samos and is mapped clearly in the descending interferogram. It is also in NW Samos where both techniques show intense amplitudes on the vertical component (>100 mm), which coincides with intense uplift (100–200 mm) that has been documented along the shoreline [60].

Differential LOS displacement is also apparent in the central southern part of Samos across the Pythagorean fault zone, which is a normal fault that dips to the south at a 45° angle [14,15] (Figure 12). However, the kinematic variation is more apparent in the East–West component map (Figure 8a) and not in the vertical one. The whole area shows a gradual decreasing of uplift towards the south (Figure 8b), which has been verified from shoreline measurements [60].

The intense negative LOS displacement observed on the narrow north central zone may be due to a hanging-wall block of the seismogenic fault, which is bounded by high angle normal faults, as have been mapped by [11]. The modelling reveals that the observed displacement occurs most probably in an area that is north of the surficial extension of the fault plane, which has been modelled to a depth of 1 km depth, such that the subsidence describes the effect of the hanging wall motion to the surface.

When comparing the image of the vertical displacement map (Figure 8b) to field measurements along the coastal zone of Samos [60], an exceptionally good agreement is observed. The discrepancies between the geodetic and the field data are less than 10% for most of the ground-truth points, which verifies the procedure for geodetic correction.

Two segments were modelled along the dipping direction of the modelled fault plane. The dip angle decreased with depth from 60° to 40° from the upper to lower segments, respectively. This change defines the transition from an upper less activated area to a deeper zone where the earthquake and the strongest aftershocks occurred (Figure 12b). The dip-slip motion along the plane is the main component of slip. However, a significant lateral slip is included on the vector, and a strike-slip component does exist for the seismogenic fault, which is consistent with previous studies [17]. The latter does not agree with the focal mechanism solution, which describes a pure dip-slip motion. The checkerboard test shows that the slip distribution along the fault plane is well constrained in the upper segment but less so on the lower one (Appendix C).

Examining the slip distribution on the upper cells in the eastern section of the model fault plane, smaller slip amplitudes (~1 m) are observed compared to the deeper cells (~2.5 m). The latter value may indicate a deepening of the upper edge of the fault to depths larger than the modelled 1 km. This is the area where the hummocky formations were found on SB [14]. The extension of the fault plane to the west terminates close to the NIB and coincides with the decay of seismicity in that direction (Figure 2). It may be considered that the modelled seismogenic fault links together all the main active faults that have been described in the area by [12]. It represents a large fault zone, the activation of which has been triggered by the regional dynamics of the northern Aegean Sea.

Concerning the post-seismic deformations measured by the GNSS data, the period after the earthquake shows that the area is slowly and gradually relaxing. The co-seismic deformation was followed by intense post-seismic relaxation of high deformational rates (~3 mm/day) for the first few days. After the first month, the deformation slowed down and seemed to attain the pre-seismic levels (Figure 5). However, this period may not mark the end of the relaxation period, which is normally expected to last a bit longer in time.

## 7. Conclusions

Pre-seismic geodetic data for the broader region of Samos clearly confirm the NNE–SSW extensional tectonic setting that is expressed by dominant E–W striking normal fault zones. Co-seismic displacements that were recorded by GNSS stations during the Mw6.9 earthquake defined an almost N–S opening of the area, which correlates to the activation of an E–W striking normal fault. The GNSS data from two stations on Samos were used to measure the absolute displacement, and to geodetically correct co-seismic interferograms. Both GNSS and InSAR data exhibited intense displacements in the northern and western parts of the island, but which decay towards its eastern and southern parts. The pattern and amplitude of the InSAR LOS displacement vectors, and the decomposed E–W and vertical components highlighted distinct and differential motions along and across the main geotectonic units.

GNSS data and InSAR deformational maps provided powerful constraints on the fault geometry and slip distribution of the seismogenic plane during the modelling procedure. The best-fit model describes a north dipping listric fault of normal type, which is located to the north and offshore of Samos. The modelled plane is comprised of two segments along the dipping direction, 60° for the upper segment up to 6 km depth, and 40° for the lower one (6–15 km depth). The lower segment is associated with the bulk of aftershocks. The strike direction of the fault (277° CWN) is nearly perpendicular to the pre-seismic extensional field of the region. The length of the plane (~35 km) matches partially with the spatial expansion of the aftershocks. The eastern end of the plane coincides with hummocky relief structures defined in the area, while its western end is situated close to the NIB. The rupture surface is smaller than the estimated one based on scaling laws, while the average slip (1.76 m) is larger. The slip distribution on the modelled plane reveals greater slip on the lower segment, with its upper edge deepening towards the east (i.e., depth > 1 km). The dip-slip component is the dominant one, but a lateral-slip component is also noted. The resultant geodetic seismic moment ($3.67 \cdot 10^{19}$ N-m) indicates an earthquake with a higher magnitude (Mw7.01) than the seismologically calculated one of Mw6.9.

The geodetic observations and the derived deformation field together with the modelling results undoubtedly defined the nodal plane solution of the centroid moment tensor. Areas of strong co-seismic motions were mapped that could possibly be taken into consideration in the civil engineering. Moreover, the slip distribution model could provide vital information to the generation of strong motion synthetic waveforms [61] in case of unavailable near-field strong motion records, contributing to the seismic hazard and risk assessment of Samos Island.

**Supplementary Materials:** The following are available online at https://www.mdpi.com/article/10.3390/rs13091665/s1, Figure S1: Deformational map for the broader area of Samos, where GNSS data have been processed to define co-seismic displacements, including the islands of Lesvos, Leros, and Kalymnos, Figure S2: Time series of the coordinates for the GNSS stations on Kalymnos and Lesvos islands; Figure S3: Time series at 30-s time interval of the coordinates of station SAMU for 30 October 2020.

**Funding:** This research received no external funding.

**Acknowledgments:** The author would like to thank E. Lagios (Department of Geology and Geoenvironment, NKUA, Greece) and B. Damiata (Cotsen Institute of Archaeology, UCLA, USA) for their constructive comments and proof-reading of the manuscript, as well as K. Pavlou and S. Vassilopoulou (Department of Geology and Geoenvironment, NKUA, Greece) for their help with the ArcGIS s/w. Continuous GNSS data were provided by METRICA SA (HexagonSmartNet), and Uranus commercial networks. The seismological analysis was performed by the personnel of the Seismological Laboratory, NKUA (Greece) [25]. Figures were made using the Generic Mapping Tool S/w [62].

**Conflicts of Interest:** The authors declare no conflict of interest.

## Appendix A

BERNESE s/w was used to process the GNSS data [41]. Data from 25 GNSS stations of the EUropean REference Frame (EUREF) and IGS were processed together with the available daily 30s RINEX local GNSS data. For pre-processing, cycle slips and outliers were detected from the RINEX files and phase measurements were used to smooth the code observations. Next, baselines were formed in to single-difference observation files. Lastly, cycle slips were resolved, outliers removed, and ambiguities added to the phase observation files.

The data were then reduced to a Precise Point Positioning (PPP) mode [63] to obtain a priori coordinate solutions that were subsequently introduced to the double-difference method, as a more precise method for static mode solutions [41]. The absolute antenna phase center corrections were used in the processing. Precise orbital solutions were obtained from the Center for Orbit Determination in Europe (CODE). The FES2004 model (http://holt.oso.chalmers.se/loading; accessed on 23 April 2021) was used for the tide loading corrections [64]. The Vienna Mapping Function (http://ggosatm.hg.tuwien.ac.at; accessed on 23 April 2021) and the Neill mapping function were applied for the tropospheric modelling [65]. Ambiguity resolution was applied using several strategies depending on the baseline length between the stations [41]: code-based wide- and narrow-lane techniques for very long baselines (<6000 km), phase-based wide- and narrow-lane for medium baselines (<200 km), and a quasi-ionosphere-free (QIF) strategy for long baselines (<1000–2000 km). The calculated coordinates were evaluated for the repeatability error on a weekly basis and values excluded in cases of large deviations from the weekly solution.

## Appendix B

The LOS displacement vector is composed of projection vectors $d_E$, $d_N$, and $d_U$ in the direction of radar wave. The LOS displacement vector on the 3D motion components is defined by the incidence angle $\theta$ and the azimuth angle $\varphi$ of the satellite heading. The LOS displacement $D_{LOS}$ can be expressed as a function of the real 3D ground surface motion components ($d_E$, $d_N$, and $d_U$) and the SAR imaging angular parameters ($\theta$ and $\varphi$) as

$$D_{LOS} = (-\sin\theta\cos\varphi \quad \sin\theta\sin\varphi \quad \cos\theta) \begin{pmatrix} d_E \\ d_N \\ d_U \end{pmatrix}$$

$$D_{LOS} = -d_E \sin\theta\cos\varphi + d_N \sin\theta\sin\varphi + d_U \cos\theta \tag{A1}$$

The LOS displacement vector can estimate the motion components at a target by combining ascending and descending SAR images. The decomposition equation, presuming there are available data from three interferometric pairs (1, 2, 3) of ascending and descending tracks (i.e., $D_{LOS1}$, $D_{LOS2}$, and $D_{LOS3}$), can be written as $y = A \cdot x$ and analytically by [48]

$$\begin{bmatrix} D_{LOS1} \\ D_{LOS2} \\ D_{LOS3} \end{bmatrix} = A \begin{bmatrix} d_E \\ d_N \\ d_U \end{bmatrix} \tag{A2}$$

where

$$A = \begin{bmatrix} a_1 & b_1 & c_1 \\ a_2 & b_2 & c_c \\ a_3 & b_3 & c_3 \end{bmatrix}$$

$$a_i = -\sin\theta_i\cos\varphi_i, \quad b_i = \sin\theta_i\sin\varphi_i, \quad c_\iota = \cos\theta_i, \quad (\iota = 1, 2, 3)$$

Since the sensitivity of the method on the N–S component is limited, the N component may be omitted, and the second column of the matrix $A$ can be eliminated [49]. Assuming that LOS data are available from only one ascending and descending pair, the linear Equation (A2) reduces to

$$\begin{bmatrix} D_{LOS\ asc} \\ D_{LOS\ desc} \end{bmatrix} = \begin{bmatrix} -\sin\theta_{asc}\cos\varphi_{asc} & \cos\theta_{asc} \\ -\sin\theta_{desc}\cos\varphi_{desc} & \cos\theta_{desc} \end{bmatrix} \begin{bmatrix} d_E \\ d_U \end{bmatrix} \tag{A3}$$

where "*asc*" and "*desc*" stand for ascending and descending orbital trajectory, respectively. To decompose the LOS displacement vector to the E–W and vertical components, Equation (A3) is solved by inversion ($x = A^{-1}y$).

## Appendix C

A checkerboard test was performed for the fault plane to examine the resolution of the slip distribution that was derived from inversion, and to define the optimal discretization for modelling. In Figure A1, the upper part represents the inputted synthetic slip distribution; the lower part shows output from the best-fit slip distribution that was inverted from the synthetic surface deformation and assuming a smoothing factor $k = 100$. A synthetic fault plane that has the same geometric characteristics as the best-fit model was assumed.

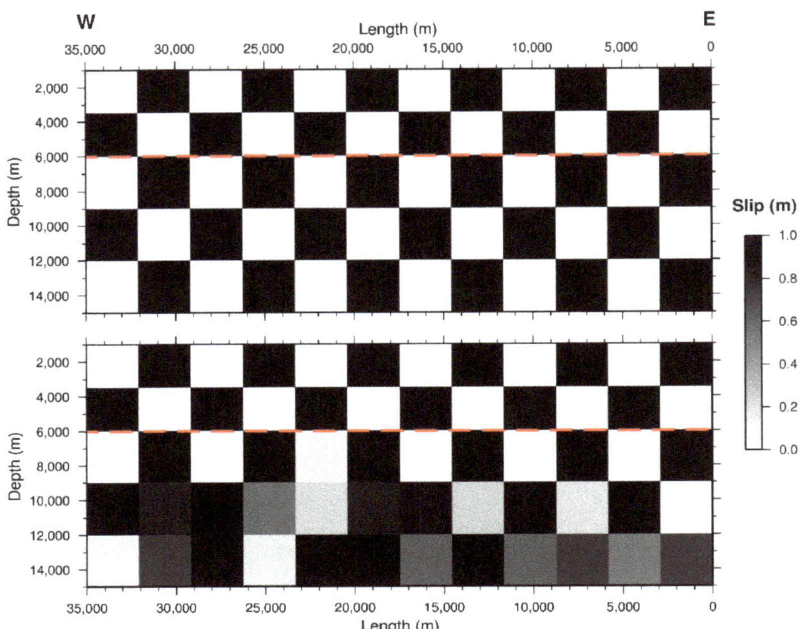

**Figure A1.** Results of checkerboard test (smoothing factor *k* = 100) for the optimal plane discretization. Red dashed line defines the limits of segments along the dip direction. The grey scale gives the magnitude of dip-slip variation.

## References

1. Çirmik, A.; Doğru, F.; Gönenç, T.; Pamukçu, O. The Stress/Strain Analysis of Kinematic Structure at Gülbahçe Fault and Uzunkuyu Intrusive (İzmir, Turkey). *Pure Appl. Geophys.* **2017**, *174*, 1425–1440. [CrossRef]
2. D'Agostino, N.; Métois, M.; Koci, R.; Duni, L.; Kuka, N.; Ganas, A.; Georgiev, I.; Jouanne, F.; Kaludjerovic, N.; Kandić, R. Active crustal deformation and rotations in the southwestern Balkans from continuous GPS measurements. *Earth Planet. Sci. Lett.* **2020**, *539*, 116246. [CrossRef]
3. Reilinger, R.; McClusky, S.; Paradissis, D.; Ergintav, S.; Vernant, P. Geodetic constraints on the tectonic evolution of the Aegean region and strain accumulation along the Hellenic subduction zone. *Tectonophysics* **2010**, *488*, 22–30. [CrossRef]
4. Le Pichon, X.; Kreemer, C. The miocene-to-present kinematic evolution of the eastern mediterranean and middle east and its implications for dynamics. *Annu. Rev. Earth Planet. Sci.* **2010**, *38*, 323–351. [CrossRef]
5. Roumelioti, Z.; Kiratzi, A.; Melis, N. Relocation of the 26 July 2001 Skyros Island (Greece) earthquake sequence using the double-difference technique. *Phys. Earth Planet. Inter.* **2003**, *138*, 231–239. [CrossRef]
6. Kreemer, C.; Chamot-Rooke, N.; Le Pichon, X. Constraints on the evolution and vertical coherency of deformation in the Northern Aegean from a comparison of geodetic, geologic and seismologic data. *Earth Planet. Sci. Lett.* **2004**, *225*, 329–346. [CrossRef]
7. Kreemer, C.; Holt, W.E.; Haines, A.J. An integrated global model of present-day plate motions and plate boundary deformation. *Geophys. J. Int.* **2003**, *154*, 8–34. [CrossRef]
8. Müller, M.; Geiger, A.; Kahle, H.-G.; Veis, G.; Billiris, H.; Paradissis, D.; Felekis, S. Velocity and deformation fields in the North Aegean domain, Greece, and implications for fault kinematics, derived from GPS data 1993–2009. *Tectonophysics* **2013**, *597–598*, 34–49. [CrossRef]
9. Coskun, S.; Dondurur, D.; Cifci, G.; Aydemir, A.; Gungor, T.; Drahor, M.G. Investigation on the tectonic significance of Izmir, Uzunada Fault Zones and other tectonic elements in the Gulf of Izmir, western Turkey, using high resolution seismic data. *Mar. Pet. Geol.* **2017**, *83*, 73–83. [CrossRef]
10. Schmid, S.M.; Fügenschuh, B.; Kounov, A.; Maţenco, L.; Nievergelt, P.; Oberhänsli, R.; Pleuger, J.; Schefer, S.; Schuster, R.; Tomljenović, B.; et al. Tectonic units of the Alpine collision zone between Eastern Alps and western Turkey. *Gondwana Res.* **2020**, *78*, 308–374. [CrossRef]
11. Ring, U.; Okrusch, M.; Will, T. Samos Island, Part I: Metamorphosed and non-metamorphosed nappes, and sedimentary basins. *J. Virtual Explor.* **2007**, *27*, 5. [CrossRef]

12. Roche, V.; Jolivet, L.; Papanikolaou, D.; Bozkurt, E.; Menant, A.; Rimmelé, G. Slab fragmentation beneath the Aegean/Anatolia transition zone: Insights from the tectonic and metamorphic evolution of the Eastern Aegean region. *Tectonophysics* **2019**, *754*, 101–129. [CrossRef]

13. Ring, U.; Laws, S.; Bernet, M. Structural analysis of a complex nappe sequence and late-orogenic basins from the Aegean Island of Samos, Greece. *J. Struct. Geol.* **1999**, *21*, 1575–1601. [CrossRef]

14. Nomikou, P.; Evangelidis, D.; Papanikolaou, D.; Lampridou, D.; Litsas, D.; Tsaparas, Y.; Koliopanos, I. Morphotectonic Analysis along the Northern Margin of Samos Island, Related to the Seismic Activity of October 2020, Aegean Sea, Greece. *Geosciences* **2021**, *11*, 102. [CrossRef]

15. Chatzipetros, A.; Kiratzi, A.; Sboras, S.; Zouros, N.; Pavlides, S. Active faulting in the north-eastern Aegean Sea Islands. *Tectonophysics* **2013**, *597–598*, 106–122. [CrossRef]

16. Kokkalas, S.; Aydin, A. Is there a link between faulting and magmatism in the south-central Aegean Sea? *Geol. Mag.* **2013**, *150*, 193–224. [CrossRef]

17. Sakellariou, D.; Tsampouraki-Kraounaki, K. Plio-quaternary extension and strike-slip tectonics in the Aegean. *Transform. Plate Boundaries Fract. Zones* **2018**, *339*, 374.

18. Kassaras, I.; Kapetanidis, V.; Ganas, A.; Tzanis, A.; Kosma, C.; Karakonstantis, A.; Valkaniotis, S.; Chailas, S.; Kouskouna, V.; Papadimitriou, P. The New Seismotectonic Atlas of Greece (v1.0) and Its Implementation. *Geosciences* **2020**, *10*, 447. [CrossRef]

19. Papadimitriou, P.; Kassaras, I.; Kaviris, G.; Tselentis, G.-A.; Voulgaris, N.; Lekkas, E.; Chouliaras, G.; Evangelidis, C.; Pavlou, K.; Kapetanidis, V.; et al. The 12th June 2017 Mw=6.3 Lesvos earthquake from detailed seismological observations. *J. Geodyn.* **2018**, *115*, 23–42. [CrossRef]

20. Ganas, A.; Elias, P.; Kapetanidis, V.; Valkaniotis, S.; Briole, P.; Kassaras, I.; Argyrakis, P.; Barberopoulou, A.; Moshou, A. The July 20, 2017 M6.6 Kos Earthquake: Seismic and Geodetic Evidence for an Active North-Dipping Normal Fault at the Western End of the Gulf of Gökova (SE Aegean Sea). *Pure Appl. Geophys.* **2019**, *176*, 4177–4211. [CrossRef]

21. Sakkas, V.; Lagios, E. Fault modelling of the early-2014 ~M6 Earthquakes in Cephalonia Island (W. Greece) based on GPS measurements. *Tectonophysics* **2015**, *644*, 184–196. [CrossRef]

22. Lagios, E.; Sakkas, V.; Novali, F.; Bellotti, F.; Ferretti, A.; Vlachou, K.; Dietrich, V. SqueeSAR™ and GPS ground deformation monitoring of Santorini Volcano (1992–2012): Tectonic implications. *Tectonophysics* **2013**, *594*, 38–59. [CrossRef]

23. Lagios, E.; Papadimitriou, P.; Novali, F.; Sakkas, V.; Fumagalli, A.; Vlachou, K.; Del Conte, S. Combined Seismicity Pattern Analysis, DGPS and PSInSAR studies in the broader area of Cephalonia (Greece). *Tectonophysics* **2012**, *524–525*, 43–58. [CrossRef]

24. Papadimitriou, P.; Kapetanidis, V.; Karakonstantis, A.; Spingos, I.; Kassaras, I.; Sakkas, V.; Kouskouna, V.; Karatzetzou, A.; Pavlou, K.; Kaviris, G.; et al. First Results on the Mw = 6.9 Samos Earthquake of 30 October 2020. *Bull. Geol. Soc. Greece* **2020**, *56*, 251–279. [CrossRef]

25. Greece Earthquake Catalogue Search. Available online: http://www.geophysics.geol.uoa.gr (accessed on 23 April 2021).

26. Geodynamics Institute of the National Observatory of Athens (GI-NOA). Available online: www.gein.noa.gr (accessed on 23 April 2021).

27. Turkish Disaster and Emergency Management Presidency AFAD. Available online: https://deprem.afad.gov.tr (accessed on 23 April 2021).

28. Klein, F.W. User's guide to HYPOINVERSE-2000: A Fortran program to solve for earthquake locations and magnitudes. *U.S. Geol. Surv. Prof. Pap* **2002**, 1–123. [CrossRef]

29. Karakonstantis, A. 3-D simulation of Crust and Upper Mantle Structure in the Broader Hellenic Area through Seismic Tomography. Ph.D. Thesis, National and Kapodistrian University of Athens, Athens, Greece, 2017.

30. Waldhauser, F. HypoDD-A Program to Compute Double-Difference Hypocenter Locations. *U.S. Geol. Surv. Open File Rep.* **2001**, 25. [CrossRef]

31. Caporali, A.; Floris, M.; Chen, X.; Nurce, B.; Bertocco, M.; Zurutuza, J. The November 2019 seismic sequence in Albania: Geodetic constraints and fault interaction. *Remote Sens.* **2020**, *12*, 846. [CrossRef]

32. He, Z.; Chen, T.; Wang, M.; Li, Y. Multi-segment rupture model of the 2016 Kumamoto earthquake revealed by InSAR and GPS data. *Remote Sens.* **2020**, *12*, 3721. [CrossRef]

33. Tzanis, A.; Chailas, S.; Sakkas, V.; Lagios, E. Tectonic Deformation in the Santorini Volcanic Complex (Greece) as inferred by joint analysis of Gravity, Magnetotelluric and DGPS Observations. *Geophys. J. Int.* **2020**, *220*, 461–489. [CrossRef]

34. Boixart, G.; Cruz, L.F.; Cruz, R.M.; Euillades, P.A.; Euillades, L.D.; Battaglia, M. Source model for Sabancaya Volcano constrained by DInSAR and GNSS surface deformation observation. *Remote Sens.* **2020**, *12*, 1852. [CrossRef]

35. Gatsios, T.; Cigna, F.; Tapete, D.; Sakkas, V.; Pavlou, K.; Parcharidis, I. Copernicus sentinel-1 MT-InSAR, GNSS and seismic monitoring of deformation patterns and trends at the Methana Volcano, Greece. *Appl. Sci.* **2020**, *10*, 6445. [CrossRef]

36. HexagonSmartNet METRICA S.A. Available online: https://hxgnsmartnet.com (accessed on 23 April 2021).

37. URANUS Net. Available online: http://www.uranus.gr (accessed on 23 April 2021).

38. Onder Cetin, K.; Mylonakis, G.; Sextos, A.; Stewart, J.P. *Seismological and Engineering Effects of the M 7.0 Samos Island (Aegean Sea) Earthquake*; Hellenic Association of Earthquake Engineering: Athens, Greece, 2020.

39. Lekkas, E.; Mavroulis, S.; Gogou, M.; Papadopoulos, G.A.; Triantafyllou, I.; Katsetsiadou, K.-N.; Kranis, H.; Skourtsos, E.; Carydis, P.; Voulgaris, N.; et al. The October 30, 2020, Mw 6.9 Samos (Greece) earthquake. *Newsl. Environ. Disaster Cris. Manag. Strateg.* **2020**, *21*, 1–156.

40. EUREF Permanent GNSS Network–Station Log. Available online: https://www.epncb.oma.be/ftp/station/log_9char/izmi00tur_20201111.log (accessed on 23 April 2021).

41. Dach, R.; Lutz, S.; Walser, P.; Fridez, P. *Bernese GNSS Software Version 5.2*; User Manual; Astronomical Institute, University of Bern, Bern Open Publishing: Bern, Switzerland, 2015.

42. Nocquet, J.-M. Present-day kinematics of the Mediterranean: A comprehensive overview of GPS results. *Tectonophysics* **2012**, *579*, 220–242. [CrossRef]

43. Pesci, A.; Teza, G. Strain rate analysis over the central Apennines from GPS velocities: The development of a new free software. *Boll. Di Geod. E Sci. Affin.* **2007**, *56*, 69–88.

44. Hanssen, R.F. *Radar Interferometry: Data Interpretation and Error Analysis*; Kluwer Academic Publishers: Dordrecht, The Netherlands, 2001.

45. Comerci, V.; Vittori, E. The Need for a Standardized Methodology for Quantitative Assessment of Natural and Anthropogenic Land Subsidence: The Agosta (Italy) Gas Field Case. *Remote Sens.* **2019**, *11*, 1178. [CrossRef]

46. Farolfi, G.; Del Soldato, M.; Bianchini, S.; Casagli, N. A procedure to use GNSS data to calibrate satellite PSI data for the study of subsidence: An example from the north-western Adriatic coast (Italy). *Eur. J. Remote Sens.* **2019**, *52*, 54–63. [CrossRef]

47. Farolfi, G.; Bianchini, S.; Casagli, N. Integration of GNSS and Satellite InSAR Data: Derivation of Fine-Scale Vertical Surface Motion Maps of Po Plain, Northern Apennines, and Southern Alps, Italy. *IEEE Trans. Geosci. Remote. Sens.* **2019**, *57*, 319–328. [CrossRef]

48. Dai, K.; Liu, G.; Li, Z.; Li, T.; Yu, B.; Wang, X.; Singleton, A. Extracting vertical displacement rates in Shanghai (China) with multi-platform SAR images. *Remote Sens.* **2015**, *7*, 9542–9562. [CrossRef]

49. Fuhrmann, T.; Garthwaite, M.C. Resolving three-dimensional surface motion with InSAR: Constraints from multi-geometry data fusion. *Remote Sens.* **2019**, *11*, 241. [CrossRef]

50. Chen, T.; Newman, A.V.; Feng, L.; Fritz, H.M. Slip distribution from the 1 April 2007 Solomon Islands earthquake: A unique image of near-trench rupture. *Geophys. Res. Lett.* **2009**, *36*, L16307. [CrossRef]

51. Feng, L.; Newman, A.V.; Protti, M.; González, V.; Jiang, Y.; Dixon, T.H. Active deformation near the Nicoya Peninsula, northwestern Costa Rica, between 1996 and 2010: Interseismic megathrust coupling. *J. Geophys. Res.* **2012**, *117*, B06407. [CrossRef]

52. GTdef. Available online: https://avnewman.github.io/GTDef/ (accessed on 8 March 2021).

53. Valkaniotis, S.; Briole, P.; Ganas, A.; Elias, P.; Kapetanidis, V.; Tsironi, V.; Fokaefs, A.; Partheniou, H.; Paschos, P. The Mw = 5.6 Kanallaki earthquake of 21 March 2020 in west Epirus, Greece: Reverse fault model from InSAR data and seismotectonic implications for Apulia-Eurasia collision. *Geosciences* **2020**, *10*, 454. [CrossRef]

54. Thingbaijam, K.K.S.; Mai, P.M.; Goda, K. New empirical earthquake source-scaling laws. *Bull. Seismol. Soc. Am.* **2017**, *107*, 2225–2246. [CrossRef]

55. Triantafyllou, I.; Gogou, M.; Mavroulis, S.; Lekkas, E.; Papadopoulos, G.A.; Thravalos, M. The Tsunami Caused by the 30 October 2020 Samos (Aegean Sea) Mw7.0 Earthquake: Hydrodynamic Features, Source Properties and Impact Assessment from Post-Event Field Survey and Video Records. *J. Mar. Sci. Eng.* **2021**, *9*, 68. [CrossRef]

56. USGS Earthquakes Hazards Program. Available online: https://earthquake.usgs.gov/earthquakes/ (accessed on 23 April 2021).

57. GFZ GEOFON. Available online: http://geofon.gfz-potsdam.de/eqinfo/ (accessed on 23 April 2021).

58. Global CMT Web Page. Available online: https://www.globalcmt.org (accessed on 23 April 2021).

59. Wells, D.I.; Coppersmith, K.J. New empirical relationships among magnitude, rupture length, rupture area and surface displacement. *Bull. Seismol. Soc. Am.* **1994**, *84*, 974–1002.

60. Evelpidou, N.; Karkani, A.; Kampolis, I. Relative Sea Level Changes and Morphotectonic Implications Triggered by the Samos Earthquake of 30th October 2020. *J. Mar. Sci. Eng.* **2021**, *9*, 40. [CrossRef]

61. Pischiutta, M.; Akinci, A.; Tinti, H.A. Broad-band ground-motion simulation of 2016 Amatrice earthquake, Central Italy. *Geophys. J. Int.* **2021**, *224*, 1753–1779. [CrossRef]

62. Wessel, P.; Smith, W.H.F.; Scharroo, R.; Luis, J.; Wobbe, F. Generic Mapping Tools: Improved Version Released. *Eos Trans. Agu* **2013**, *94*, 409–410. [CrossRef]

63. Zumberge, J.; Heflin, M.; Jefferson, D.; Watkins, M.; Webb, F.H. Precise point positioning for the efficient and robust analysis of GPS data from large networks. *J. Geophys. Res.* **1997**, *102*, 5005–5017. [CrossRef]

64. Métois, M.; D'Agostino, N.; Avallone, A.; Chamot-Rooke, N.; Rabaute, A.; Duni, L.; Kuka, N.; Koci, R.; Georgiev, I. Insights on continental collisional processes from GPS data: Dynamics of the peri-Adriatic belts. *J. Geophys. Res. Solid Earth* **2015**, *120*, 8701–8719. [CrossRef]

65. Nykiel, G.; Wolak, P.; Figurski, M. Atmospheric opacity estimation based on IWV derived from GNSS observations for VLBI applications. *Gps Solut.* **2018**, *22*, 9. [CrossRef]

*Article*

# Multi-Segment Rupture Model of the 2016 Kumamoto Earthquake Revealed by InSAR and GPS Data

Zhongqiu He [1], Ting Chen [1,2,*], Mingce Wang [1] and Yanchong Li [1]

[1]  School of Geodesy and Geomatics, Wuhan University, Wuhan 430079, China;
zhongqiuhe@whu.edu.cn (Z.H.); mcwang@whu.edu.cn (M.W.); YC_Lee@whu.edu.cn (Y.L.)
[2]  Key Laboratory of Geospace Environment and Geodesy, Ministry of Education, Wuhan University,
Wuhan 430079, China
*  Correspondence: tchen@sgg.whu.edu.cn; Tel.: +86-27-6877-1723

Received: 5 October 2020; Accepted: 9 November 2020; Published: 12 November 2020

**Abstract:** The 2016 Kumamoto earthquake, including two large (Mw ≥6.0) foreshocks and an Mw 7.0 mainshock, occurred in the Hinagu and Futagawa fault zones in the middle of Kyushu island, Japan. Here, we obtain the complex coseismic deformation field associated with this earthquake from Advanced Land Observation Satellite-2 (ALOS-2) and Sentinel-1A Interferometric Synthetic Aperture Radar (InSAR) data. These InSAR data, in combination with available Global Positioning System (GPS) data, are then used to determine an optimal four-segment fault geometry with the jRi method, which considers both data misfit and the perturbation error from data noise. Our preferred slip distribution model indicates that the rupture is dominated by right-lateral strike-slip, with a significant normal slip component. The largest asperity is located on the northern segment of the Futagawa fault, with a maximum slip of 5.6 m at a 5–6 km depth. The estimated shallow slips along the Futagawa fault and northern Hinagu fault are consistent with the displacements of surface ruptures from the field investigation, suggesting a shallow slip deficit. The total geodetic moment release is estimated to be $4.89 \times 10^{19}$ Nm (Mw 7.09), which is slightly larger than seismological estimates. The calculated static Coulomb stress changes induced by the preferred slip distribution model cannot completely explain the spatial distribution of aftershocks. Sensitivity analysis of Coulomb stress change implies that aftershocks in the stress shadow area may be driven by aseismic creep or triggered by dynamic stress transfer, requiring further investigation.

**Keywords:** 2016 Kumamoto earthquake; InSAR; GPS; slip distribution; shallow slip deficit; Coulomb stress change

---

## 1. Introduction

In April 2016, a series of shallow earthquakes struck Kumamoto prefecture of Kyushu island, southwest Japan, causing more than 100 fatalities, more than 1000 injuries, and severe destruction to houses and infrastructure. The sequence started with an Mw 6.2 foreshock on 14 April 2016 (UTC), and the Mw 7.0 mainshock, which is thought to be the largest event in central Kyushu island recorded since 1990, occurred approximately 28 h later. The two events generated strong ground motions that reached the maximum seismic intensity of 7 (the largest value on the Japan Meteorological Agency (JMA) scale). Previous studies related to the kinematic rupture process have shown that the mainshock rupture propagated unilaterally to the northeast direction and terminated near the southwest edge of the Aso volcano [1–3]. The moment tensor solutions from seismic catalogs (e.g., Global Centroid Moment Tensor (GCMT) [4], United States Geological Survey (USGS) [5], and National Research Institute for Earth Science and Disaster Resilience (NIED) F-net [6]) indicate that the mainshock was dominated by right-lateral motion with significant non-double-couple components, suggesting that rather complex mechanisms were involved in this sequence (Figure 1).

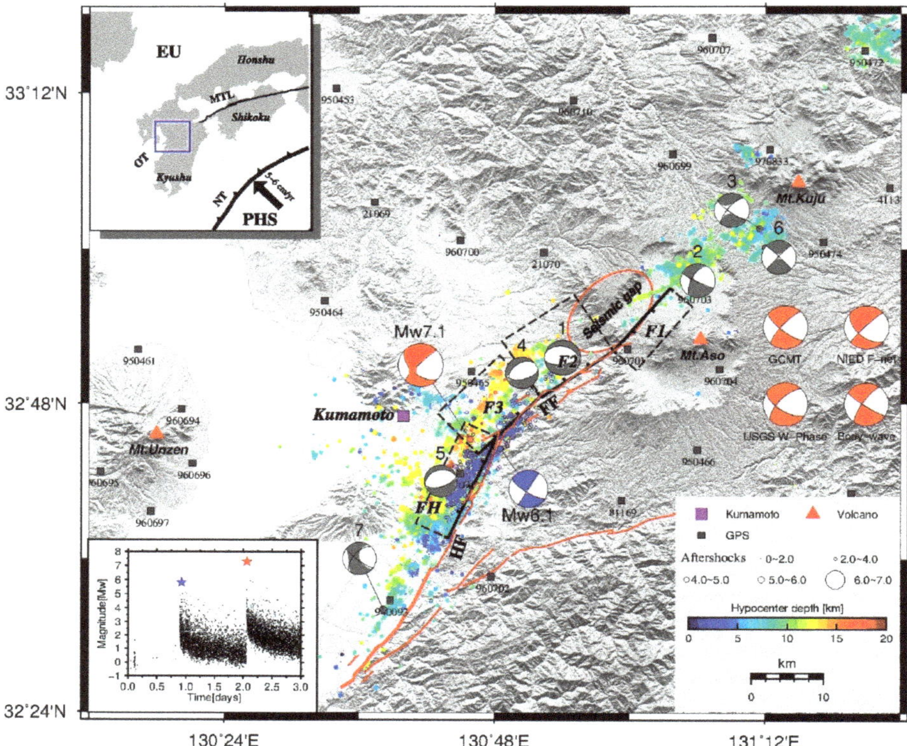

**Figure 1.** Tectonic setting of the 2016 Kumamoto earthquake, Kyushu island, in southwest Japan. The upper-left inset shows the seismogenic background of southwest Japan, while the blue rectangle delineates the study area. The lower-left inset illustrates the magnitude time evolution of the first three-day earthquake sequence, and blue and red stars denote the largest foreshock and mainshock, respectively. Blue and red beach balls indicate the focal mechanism solutions of the foreshock and mainshock determined by NIED F-net [6], respectively, while gray beach balls show aftershocks with Mw >5.0. Moreover, the mainshock focal mechanisms from other earthquake catalogs (e.g., GCMT [4], USGS [5] W-phase, and body wave) are also shown for comparison. Color-filled circles denote the relocated foreshocks (blue) and aftershocks (the filled color varies with hypocentral depth) until 30 April and the circles' size is proportional to the magnitude [3]. The red ellipse illustrates a ~10 km-long seismic gap of aftershocks near the Aso volcano. Active faults are shown with red lines. The gray squares denote the GPS stations. The red triangles show the active volcanoes and the purple square is Kumamoto city. Surface projections of the assumed fault planes (marked as F1, F2, F3, and FH) are represented by black dashed rectangles (the top edge marked by thick lines). PHS, Philippine Sea Plate; NT, Nankai Trough; OT, Okinawa Trough; MTL, Median Tectonic Line; EU, Eurasian plate; FF, Futagawa fault; HF, Hinagu fault.

This sequence produced approximately 34 km-long surface ruptures, mainly along the Futagawa fault zone and its eastern extension to the caldera of the Aso volcano, and the northernmost part of the Hinagu fault zone [7]. The Futagawa and Hinagu fault zones belong to part of the western extension of the Median Tectonic Line (MTL), known as the longest right-lateral strike-slip active fault in Japan. These fault zones are located within the Beppu–Shimabara graben, which is dominantly controlled by north–south extension and some northeast–southwest shear modifications from MTL [8]. The oblique subduction of the Philippine Sea Plate (PHS) and the northern extension of the opening Okinawa Trough (OT) are considered to be the sources that drove the formation of the graben structure [9].

The spatial distribution of seismicity and surface ruptures from the field investigation suggest that multi-segment faults ruptured during the earthquake. To better understand the complicated rupture and stress interaction, it is crucial to clarify the source characteristics of this earthquake. Several previous studies [1–3,10–15] have reported the source models from the inversion of teleseismic, strong-motion, and geodetic datasets, either independently or jointly. However, the fault geometry and maximum slip of various models differ significantly. Typically, Kubo et al. [10] estimated a maximum slip of 3.8 m from strong-motion waveforms with a curved fault model. Their results fitted the waveforms relatively well, yet the maximum slip seemed to be underestimated when compared with the peak slip of 5.7 m from the joint inversion of strong-motion, teleseismic body, and surface waveforms [11]. Fukahata and Hashimoto [12] determined the fault dip angles and slip distribution on two-segment fault planes by inverting the ALOS-2 InSAR data. Their model was well-constrained, but a large misfit (up to 30 cm) occurred near the caldera of Aso volcano, which suggested that their fault model needed to be improved. Additionally, Yue et al. [3] performed the joint inversion of GPS, strong-motion, Synthetic Aperture Radar (SAR) images, and surface offset data with a curved multi-segment fault model. Their results showed the maximum slip exceeding 10 m, which is rare for an Mw 7.0 intraplate event. These studies only assumed northwest-dipping faults, while Ozawa et al. [13], Yoshida et al. [14], and Zhang et al. [15] indicated that a southeast dipping fault within the Aso caldera region might also have ruptured. Therefore, a further study of the fault geometry and slip distribution is required to better understand the source mechanism of this complex earthquake.

In this study, the complex deformation field associated with the 2016 Kumamoto earthquake is firstly derived from both ALOS-2 and Sentinel-1A SAR satellite images. Subsequently, a combination of InSAR and GPS data is utilized to determine the optimal multi-segment fault geometry with the jRi method and to invert the slip distribution on the fault planes. Finally, the Coulomb stress change calculation and its sensitivity analysis are carried out to investigate the possible triggering mechanisms.

## 2. Coseismic Surface Displacements from InSAR and GPS Data

### 2.1. InSAR Data

During the 2016 Kumamoto earthquake, both ALOS-2 and Sentinel-1 SAR satellites captured the coseismic surface deformation field covering the entire epicenter area. The ALOS-2 satellite, equipped with the Phased Array type L-band Synthetic Aperture Radar-2 (PALSAR-2) and operated by the Japan Aerospace Exploration Agency (JAXA), provides excellent surface displacement observations. The L-band ALOS-2 satellite can observe the ground surface from the right or left direction and penetrate the heavy vegetation to obtain a high coherent interferogram. The Sentinel-1 satellite is composed of a constellation of two satellites (A and B) maintained by the European Space Agency (ESA) and provides C-band SAR images to detect crustal deformation signals. This constellation operates the novel Terrain Observation with Progressive Scans (TOPS) mode for its standard acquisition, which has a broad swath of up to 400 km wide, with a spatial resolution down to 5 m and a minimum 12-day repeat cycle [16]. In this study, we collected one pair of images from the ALOS-2/PALSAR-2 descending track (P023D) and two pairs of images from both the descending and ascending track (T163D and T156A) of Sentinel-1A, in order to map the coseismic surface deformation. Detailed information on the SAR data used in this study is given in Table 1.

GAMMA software [17] was utilized to process all SAR images. The interferograms were multi-looked with ratios between the range and azimuth direction of 10:2 and 8:8 for the Sentinel-1A and ALOS-2/PALSAR-2 images, respectively. The 90-m-resolution Digital Elevation Model (DEM) from the Shuttle Radar Topography Mission (SRTM) [18] was used to remove the topographic effects. The interferograms were then filtered with an adaptive-filter method [19] and unwrapped using statistical-cost network-flow approaches [20]. Finally, all interferograms were geocoded to the World Geodetic System (WGS-84) geographic coordinates with a 90 m resolution.

**Table 1.** Synthetic aperture radar (SAR) interferometric pairs used in the study.

| Track | Sensor | Orbit | Mode | Primary Image (yyyy/mm/dd) | Secondary Image (yyyy/mm/dd) | $B_\perp$ (m) | $B_T$ (days) | $\sigma$ (cm) | Np |
|---|---|---|---|---|---|---|---|---|---|
| P023D | ALOS-2 | Descending | Strip-map | 2016/03/07 | 2016/04/18 | 81.5 | 42 | 2.3 | 657 |
| T163D | Sentinel-1A | Descending | IW | 2016/03/27 | 2016/04/20 | 1.9 | 24 | 2.5 | 549 |
| T156A | Sentinel-1A | Ascending | IW | 2016/04/08 | 2016/04/20 | 65.4 | 12 | 3.2 | 568 |

$B_\perp$ represents the perpendicular baseline between the orbits; $B_T$ is the interval between the master and slave images acquisition dates; Interferometric Wide (IW) swath mode is the main acquisition mode over land; $\sigma$, denoting measured error from a combination of various error sources (e.g., ionospheric and atmospheric effects, topographic correction residuals, and orbital and unwrapping errors), was estimated from the far-field area [21]; Np is the number of down-sampling points.

The interferograms obtained from SAR images show significant deformation signals along the mapped active faults associated with the earthquake (Figure 2). Compared with the L-band ALOS-2 interferogram, two C-band Sentinel-1A interferograms can hardly acquire the deformation near the faults due to the worse decorrelation caused by the large displacement gradient, surface ruptures, and landslides. Intensive fringes appear on both the Futagawa fault and the northern part of the Hinagu fault, along which displacement discontinuities can be clearly identified. The unwrapped interferograms of two descending tracks (P023D and T163D) both show three major asymmetric lobes of deformation and detect the line-of-sight subsidence signals at the Aso caldera. Two lobes are located along the mapped Futagawa and Hinagu fault zones, while another is located within the Aso caldera. The observed line-of-sight (LOS) displacements vary from −46 to 18 cm and from −44 to 67 cm for the P023D and T163D tracks, respectively. The slight discrepancy between the two descending tracks arises from the different incidence angles of ALOS-2 (~36°) and Sentinel-1A (~38°) satellite SAR sensors [22]. In the ascending track (T156A), the LOS displacement varies from −97 to 106 cm and the range is much larger than that of the two descending tracks. The LOS displacements on both sides of the Futagawa-Hinagu fault zone display opposite directions, consistent with a strike-slip event. Large displacements of up to 50 cm in the LOS direction are also detected within the Aso caldera.

### 2.2. GPS Data

In addition to InSAR data, coseismic GPS displacements near the epicentral region from the GPS Earth Observation Network system (GEONET) were also collected. The distribution of GPS stations is shown in Figure 3. The horizontal displacements are consistent with the predominant right-lateral rupture, while the vertical displacements are much more complicated. The largest ~1 m horizontal displacement to the southwest and 0.25 m uplift can be observed at the 960701 station, which is located within kilometers of the mapped fault trace near the Aso caldera. However, this station was not included in the inversion because it is located too close to the fault trace, and the displacement could not be properly modeled by Green's functions [3]. Moreover, the small displacements with large errors were also not used in the inversion. Consequently, 45 horizontal and six vertical displacements were used to constrain the fault geometry and slip distribution. It is widely known that InSAR measures large-scale line-of-sight displacements with a high spatial resolution, whereas GPS can provide 3-D continuous displacements with a sub-centimeter accuracy. Therefore, a combination of InSAR and GPS datasets can better constrain the fault geometry and slip distribution associated with earthquakes.

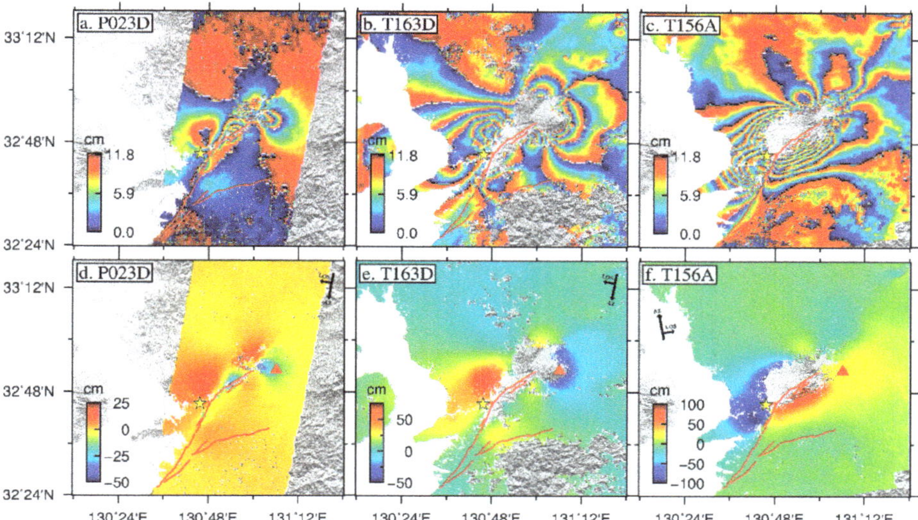

**Figure 2.** Wrapped and unwrapped InSAR maps. (**a–c**) Coseismic interferograms derived from the ALOS-2 and Sentinel-1A images. Each color cycle represents 11.8 cm of line-of-sight (LOS) deformation. (**d–f**) Unwrapped LOS deformation corresponding to (**a–c**). The active faults are shown as a red line. Yellow star and red triangle denote the mainshock epicenter and Aso volcano, respectively. Positive value indicates LOS displacement moving toward the satellite.

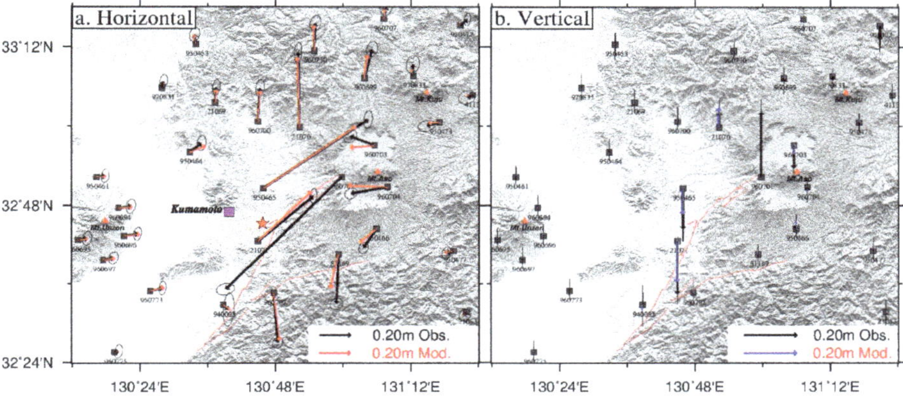

**Figure 3.** Observed co-seismic GPS displacements compared with modeled displacements from the preferred model. (**a**) Horizontal displacements with exaggerated 95% confidence ellipses and (**b**) vertical displacements. Red star and red triangle denote the mainshock epicenter and Aso volcano, respectively.

## 3. Model

The surface displacements derived from InSAR and GPS data were modeled using the rectangular dislocations embedded in a homogeneous, isotropic, and elastic half-space [23] and assuming a Poisson's ratio of 0.25. Millions of highly correlated data points in the unwrapped interferograms were down-sampled with a quadtree sampling method [24] to improve the computational feasibility and efficiency of the model. The InSAR and GPS datasets were weighted equally in the joint inversion, which could fit both datasets quite well, as shown by the relative weight ratio tests (Figure S1).

### 3.1. Determination of the Fault Geometry with the jRi Method

The surface ruptures, active fault traces, InSAR deformation maps, and seismicity indicate that multi-segment fault planes ruptured during the earthquake. Therefore, four-segment fault planes (hereafter termed F1, F2, F3, and FH) are considered to represent the rupture planes, among which segment F1 is set as a southeast dipping fault plane in the eastward extension of the Futagawa fault zone under the Aso caldera, and the other three northwest dipping segments are located along the Futagawa and northern Hinagu fault zones (Figure 1).

However, determining the fault geometry parameters of multiple segments and the slip distribution on the fault planes simultaneously is computationally expensive [25,26]. In this study, the location and extent of the four-segment fault planes could be identified from geologically-mapped fault traces, surface ruptures, and the seismicity distribution, but the fault dip angles of those segments were difficult to determine, especially for the F1 segment located at a seismic gap of the aftershock. Therefore, we considered fixing the location and extent of the four-segment faults and then searching for the optimal fault dip angles that minimized the data misfit between observed and predicted surface displacements. Regularization errors, such as weight root mean square errors (WRMSE) with the form of $\|\mathbf{C}_d^{-1/2}(\mathbf{d} - \mathbf{G}(\boldsymbol{\delta})\mathbf{s})\|_2$ or root mean square errors (RMSE) with the form of $\|(\mathbf{d} - \mathbf{G}(\boldsymbol{\delta})\mathbf{s})\|_2$, are the most common measures of misfit [25,27,28], and the jRi value [29], including both regularization error and perturbation error from data noise, has also been proposed to find the best-fitting model [30]. The jRi value with respect to fault dip angles can be expressed as:

$$jRi(\boldsymbol{\delta}) = \|\mathbf{C}_d^{-1/2}(\mathbf{d} - \mathbf{G}(\boldsymbol{\delta})\mathbf{s})\|_2 - \frac{1}{k}diag(\mathbf{MI}_2\mathbf{M}^T) + \frac{1}{k}diag(\mathbf{MM}^T), \tag{1}$$

where $\mathbf{C}_d^{-1/2}$ is the weight matrix from Cholesky decomposition of the inverse of the data covariance matrix $\mathbf{C}_d$; $\mathbf{G}(\boldsymbol{\delta})$ is the assembled matrix of Green's functions for a vector of dip angle $\boldsymbol{\delta}$ (with $\boldsymbol{\delta} = [\delta_1, \ldots, \delta_n]$, n = 4), relating the slip $\mathbf{s}$ on a dislocation to the observed data $\mathbf{d}$ (i.e., InSAR and GPS datasets); k is the amount of data used in the inversion; $\mathbf{I}_2$ is a matrix consisting of $2 \times 2$ $\mathbf{I}$ matrices, where $\mathbf{I}$ represents a k-by-k identity matrix; and $\mathbf{M} = [\mathbf{I}, -\mathbf{N}]$, with $\mathbf{N} = \mathbf{G}(\boldsymbol{\delta})\mathbf{G}^{-g}$. $\mathbf{G}^{-g}$ is the generalized inverse with the following form:

$$\mathbf{G}^{-g} = \left( \begin{bmatrix} \mathbf{C}_d^{-1/2}\mathbf{G}(\boldsymbol{\delta}) \\ \kappa\mathbf{L} \end{bmatrix}^T \begin{bmatrix} \mathbf{C}_d^{-1/2}\mathbf{G}(\boldsymbol{\delta}) \\ \kappa\mathbf{L} \end{bmatrix} \right)^{-1} \left( \mathbf{C}_d^{-1/2}\mathbf{G}(\boldsymbol{\delta}) \right)^T, \tag{2}$$

where $\mathbf{L}$ is the Laplacian smoothing matrix and $\kappa$ is a regularization factor that determines the slip roughness $\|\mathbf{Ls}\|_2$. The optimal $\kappa = 2$ can be chosen from the trade-off curve [25] between the data misfit and the average slip roughness or the jRi curve [29], as shown in Figure S2.

It has been demonstrated that the determination of the fault dip angle via the jRi method with geodetic data is more robust than the performance of WRMSE or RMSE in simple one-segment fault scenarios, such as the 2015 Mw 7.9 Gorkha Nepal earthquake [30], but further verification was required for the complex multi-segment fault scenario in this study. Therefore, we extended the jRi method to determine the multi-segment fault dip angles and compared the results with the RMSE method to illustrate the effectiveness of the jRi method. More specifically, the optimal fault dip angles of four segments were iteratively searched, while the slip distributions on the four-segment faults were inverted simultaneously, until the jRi or RMSE value exhibited an insignificant difference. In order to avoid the effect of the imprecise estimation of the slip distribution on the fault planes during searching for optimal dip angles, each fault segment was further discretized into 2 km $\times$ 2 km sub-faults.

Figure 4 shows the result of estimating the fault dip angles with the jRi and RMSE methods from InSAR and GPS data. The optimal fault dip angles determined with the jRi method were estimated to be 77°, 57°, 63°, and 74° for the F1, F2, F3, and FH segments, respectively. The fault dip angles of F2, F3, and FH segments were also well-determined with the RMSE method, and only slight differences can be observed when compared with the result from the jRi method. However, the fault dip angle of F1 was poorly determined with the RMSE method, resulting in a shallower dip of 68° compared with the result obtained from the jRi method. The detailed fault parameters of the four-segment fault model used in this study are summarized in Table 2.

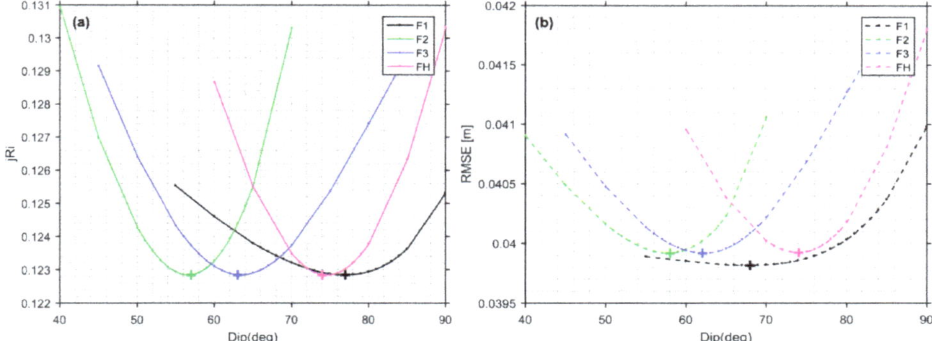

**Figure 4.** Comparison of the estimations of fault dip angles with the jRi and root mean square error (RMSE) methods. (**a**) Plot of the jRi values as a function of dip angles for four segments. (**b**) Plot of the RMSE values. Optimal dip angles corresponding to minimal jRi and RMSE values are marked with cross symbols.

**Table 2.** Fault parameters of the four-segment fault model used in this study.

| Segment | Latitude (°) | Longitude (°) | Depth (km) | Length (km) | Width (km) | Strike (°) | Dip (°) | Max Slip (m) | Moment (Nm) | Mw |
|---------|-----------|-----------|-------|--------|-------|-----------|--------|----------|-----------|-----|
| F1 | 131.0178 | 32.9088 | 0.2 | 12 | 18 | 40 | 77 | 2.3 | | |
| F2 | 130.9222 | 32.8386 | 0.2 | 12 | 18 | 236 | 57 | 5.6 | $4.89 \times 10^{19}$ | 7.09 |
| F3 | 130.8216 | 32.7722 | 0.2 | 12 | 18 | 226 | 63 | 3.3 | | |
| FH | 130.7672 | 32.6918 | 0.2 | 16 | 18 | 205 | 74 | 3.0 | | |

The position (latitude, longitude, and depth) indicates the top-center of each fault segment. The rigidity was assumed to be 32 GPa. The fault dip angles were determined with the jRi method.

## 3.2. Distributed Slip Model

After the fault geometry parameters of the four segments were determined, a bounded variable least squares method [31] was adopted to invert the distributed slip on each segment. Regarding the limitation on slip, all segments were allowed to include both right-lateral strike-slip and free dip-slip components with a magnitude of less than 10 m, which is a reasonable constraint based on the focal mechanism of the mainshock and previous studies [1,2,6]. Moreover, the Laplacian smoothing constraint was also adopted to avoid non-physical oscillation of the slip distribution.

The preferred slip distribution model reveals a single asperity on each of the four segments, with a maximum slip of 5.6 m on the F2 segment at a 5–6 km depth (Figure 5). The predominant slip occurs at a depth of 0–13 km, which confirms the shallow hypocenters of this earthquake sequence. The relocated aftershocks are mostly located at a depth of between 6 and 15 km, where the slip decays quickly. The slip on the F2 segment is mainly right-lateral strike-slip, but with a significant normal slip of ~2.7 m. The shallowest slip on the F2 segment is characterized by 1–2 m strike-slip, together with 0–1.5 m normal slip, which is consistent with the lateral and vertical displacement components of surface ruptures from the field investigation [7]. On the F3 and FH segments, the slip is almost pure right-lateral, but the maximum slip on the F3 segment (3.3 m) is slightly larger than that on the FH segment (3.0 m). The slip on the southeast dipping F1 segment exhibits a dominant normal slip, but smaller magnitude than the other three segments, and the maximum slip is less than 2.5 m. The total geodetic moment release was estimated to be $4.89 \times 10^{19}$ Nm, which is equivalent to a Mw 7.09 event, assuming a shear modulus of 32 GPa. The estimated moment release is in general agreement with that of Zhang et al. [15], but slightly larger than the seismic moment derived from teleseismic data [4–6] or regional strong-motion data [2]. Note that InSAR data cover the largest foreshock (Mw 6.0) and four days of aftershocks, which may explain the slightly higher moment release.

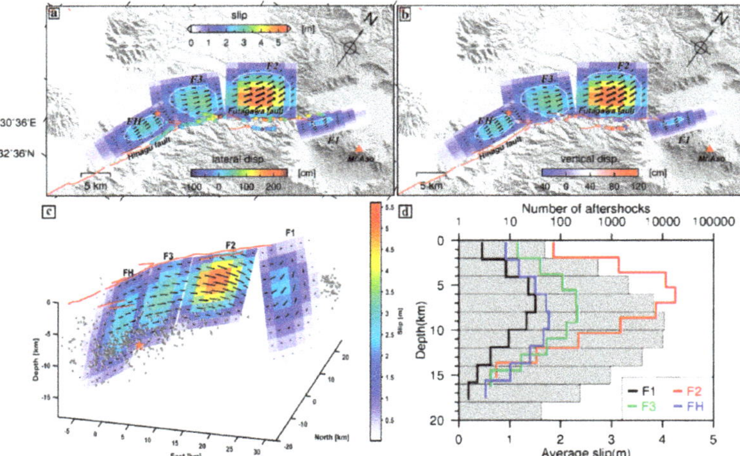

**Figure 5.** Slip distribution of the preferred model inverted from InSAR and GPS data, and comparison with surface ruptures and aftershocks. (**a–b**) Surface projection of the slip distribution. The color-filled circles in (**a**) show the lateral displacements of surface ruptures from the field investigation (positive values for right-lateral displacements), while those in (**b**) show the vertical displacements (positive values for south-side uplift displacements). The red star denotes the epicenter, and red lines show the active faults. The white contours highlight the asperities with a slip magnitude of over 2 m. Four segments are marked with F1, F2, F3, and FH from northeast to southwest. (**c**) 3-D slip distribution viewing from SSE. The black arrows show the slip directions of each sub-fault. (**d**) The slip averaged at different depths on each segment and number of aftershocks versus depth. Slips on four segments are plotted with curves. The gray histogram illustrates the depth distribution of aftershocks.

Figure 6 shows a comparison of observed and predicted InSAR data from our best-fitting slip model, as well as the corresponding residuals. The general deformation patterns from both ALOS-2 and Sentinel-1A observations can be interpreted well by the preferred model. However, there are some near-fault residuals for ALOS-2 descending track P023D with maximum misfit of up to 10 cm, which may be attributed to a combination of complicated near-field deformation and possible nontectonic influences of liquefaction, landslides, and inelastic deformation [32]. The histograms of the residual distribution of each InSAR track are illustrated in Figure 7. The RMSEs were estimated to be 5.64, 2,75, and 2.74 cm for P023D, T163D, and T156A tracks, respectively. The preferred model provides a good fit to the GPS horizontal and vertical displacements shown in Figure 3. A detailed comparison of three components of the observed and predicted GPS displacements is shown in Figure 7. The fits to horizontal displacements are relatively better than those to vertical displacements, with RMSEs of 1.70, 2.35, and 4.38 cm for east, north, and vertical components, respectively.

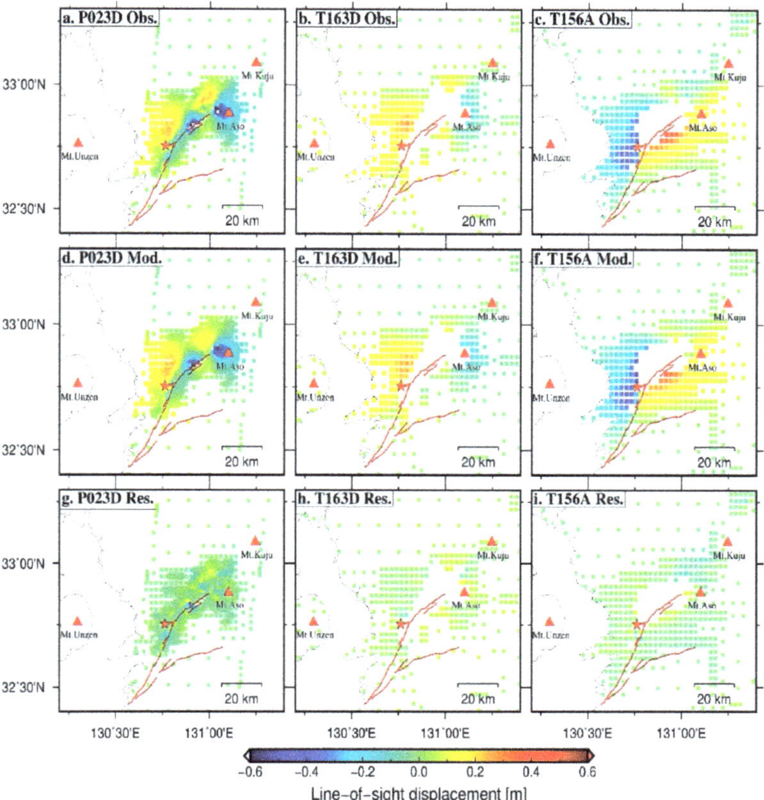

**Figure 6.** Observed and model-predicted InSAR data and the residuals. (**a–c**) Subsampled InSAR line-of-sight displacement data used in the inversion for ALOS-2 descending track P023D, Sentinel-1A descending track T163D, and ascending track T156A, respectively. (**d–f**) Model-predicted InSAR line-of-sight displacements corresponding to (**a–c**) from our preferred model. (**g–i**) Residuals after subtracting the model predictions from the observations. Positive value indicates the LOS displacement moving toward the satellite.

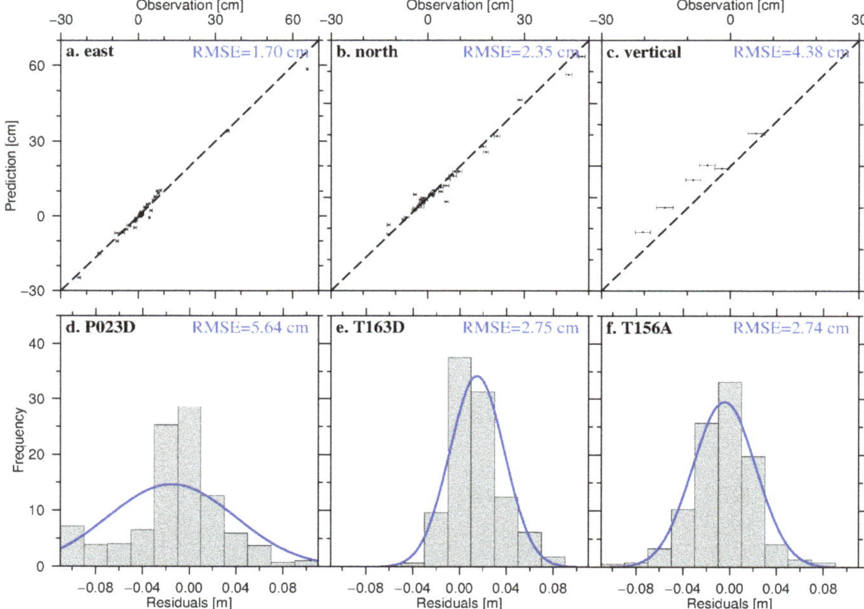

**Figure 7.** Results of GPS and InSAR data fittings. (**a–c**) Comparison of the observations of GPS data and the predictions from the preferred model. (**d–f**) Histograms of residuals after subtracting the model predictions from the observations of InSAR data.

## 4. Coulomb Stress Change Analysis

Numerous previous studies have shown that the Coulomb stress changes caused by an earthquake may trigger subsequent events, change the seismicity rate, and affect the state of nearby faults through stress transfer [33–35]. Based on the Coulomb failure criterion, the change in Coulomb failure stress ($\Delta$CFS) due to an earthquake on the receiver faults can be calculated as follows:

$$\Delta\mathrm{CFS} = \Delta\tau_s + \mu'\Delta\sigma_n, \tag{3}$$

where $\Delta\tau_s$ is the shear stress change parallel to the slip direction; $\mu'$ is the effective friction coefficient, varying between 0 and 0.8, with a moderate value of 0.4 used as default; and $\Delta\sigma_n$ is the normal stress change (positive for tension). Positive values of $\Delta$CFS on the receiver faults with a typical triggering threshold of ~0.1 bar (0.01 MPa) are more likely to promote failure and increase regional seismicity, while negative ones may inhabit failure and decrease seismicity [36].

In order to investigate the possible triggering mechanism of the sequence, the static stress changes were calculated based on the distributed slip model in this study. The calculations were performed using Coulomb 3.4 software [37], with an effective friction coefficient of 0.4 and a shear modulus of 32 GPa. Receiver fault parameters (i.e., strike, dip, and rake) were specified as 220°, 68°, and −165°, respectively, and were derived from the equivalent focal mechanism of our preferred model, which shows a slight discrepancy with focal mechanisms of GCMT [4], USGS [5], and NIED [6] F-net (Table S1). Static stress changes at various depths between 0 and 20 km were compared with the locations of Mw > 3.0 aftershocks that occurred before 30 April 2016. Only large and early aftershocks were selected in our analysis to reduce possible location errors and multiple interactions between aftershocks.

Figure 8 shows the static Coulomb stress, shear stress, and normal stress changes at different depths with an interval of 5 km, compared with the spatial distribution of Mw >3.0 aftershocks.

The localized pattern of Coulomb stress changes is significantly observed in the vicinity of fault zones and dominated by shear stress changes. At a depth of 2.5 km, the calculated Coulomb stress increases larger than 10 bars are mainly located near the most northeastern and southwestern ends of the fault zone, where aftershocks primarily occurred. The pattern of stress changes at a 7.5 km depth is similar to that at a 2.5 km depth, but is more complicated at a 12.5 km depth. The area with positive Coulomb stress changes at a 12.5 km depth is larger than that at a shallower depth, but the majority of aftershocks do not concentrate on the Coulomb stress increase regions. As the depth increases to 17.5 km, the positive $\Delta$CFS area has only a few aftershocks and may have a potential seismic risk in the future. Additionally, if focal mechanisms are available, a more stringent calculation that resolves Coulomb stress changes on the nodal planes associated with the aftershocks can be conducted to assess whether or not the nodal planes are brought closer to failure. The focal mechanisms of aftershocks with Mw >5.0 from NIED F-net were selected to calculate Coulomb stress changes on the two conjugate planes at each hypocenter. The result highlights that six out of the seven aftershocks are positively affected by a Coulomb stress increase on at least one conjugate plane (Table 3). Intriguingly, the largest aftershock (event 1) occurred just 20 min after the mainshock, but the Coulomb stress changes resolved on both conjugate planes show negative values. In general, static Coulomb stress triggering can explain the majority of Mw >5.0 aftershocks, but there are also some Mw >3.0 aftershocks within the stress shadow zone that may be triggered by other factors, which will be discussed in the following section.

**Table 3.** Coulomb stress changes at the hypocenter resolved on the two conjugate nodal planes of the focal mechanisms of aftershocks with Mw >5.0.

| ID | Origin Time (UTC) | | Hypocenter | | | Mw | Nodal Plane 1 | | Nodal Plane 2 | |
|---|---|---|---|---|---|---|---|---|---|---|
| | yyyy/mm/dd | dd:mm | Latitude (°) | Longitude (°) | Depth (km) | | Strike/Dip/Rake | $\Delta$CFS[1] (bar) | Strike/ Dip/Rake | $\Delta$CFS[2] (bar) |
| 1 | 2016/04/15 | 16:45 | 32.8632 | 130.8990 | 10.55 | 5.7 | 286/35/−70 | −18.15 | 81/57/−104 | −20.11 |
| 2 | 2016/04/15 | 18:03 | 32.9638 | 131.0868 | 6.89 | 5.5 | 209/60/−174 | 7.35 | 116/85/−30 | 7.79 |
| 3 | 2016/04/15 | 18:55 | 33.0265 | 131.1910 | 10.89 | 5.5 | 220/72/−167 | 1.09 | 126/78/−19 | 2.02 |
| 4 | 2016/04/16 | 00:48 | 32.8470 | 130.8350 | 15.91 | 5.2 | 230/38/−112 | −4.43 | 77/55/−73 | 6.95 |
| 5 | 2016/04/16 | 07:02 | 32.6992 | 130.7200 | 12.30 | 5.1 | 255/30/−88 | 3.24 | 72/60/−91 | −2.78 |
| 6 | 2016/04/18 | 11:41 | 33.0020 | 131.1998 | 8.64 | 5.4 | 314/86/3 | 2.38 | 224/87/176 | 1.07 |
| 7 | 2016/04/19 | 08:52 | 32.5352 | 130.6353 | 9.96 | 5.3 | 221/60/−169 | 2.45 | 126/81/−30 | 1.51 |

Coulomb stress changes were evaluated using our preferred slip distribution model and an efficient coefficient of friction of 0.4. The focal mechanisms originated from the F-net catalog of the National Research Institute for Earth Science and Disaster Resilience (NIED), Japan.

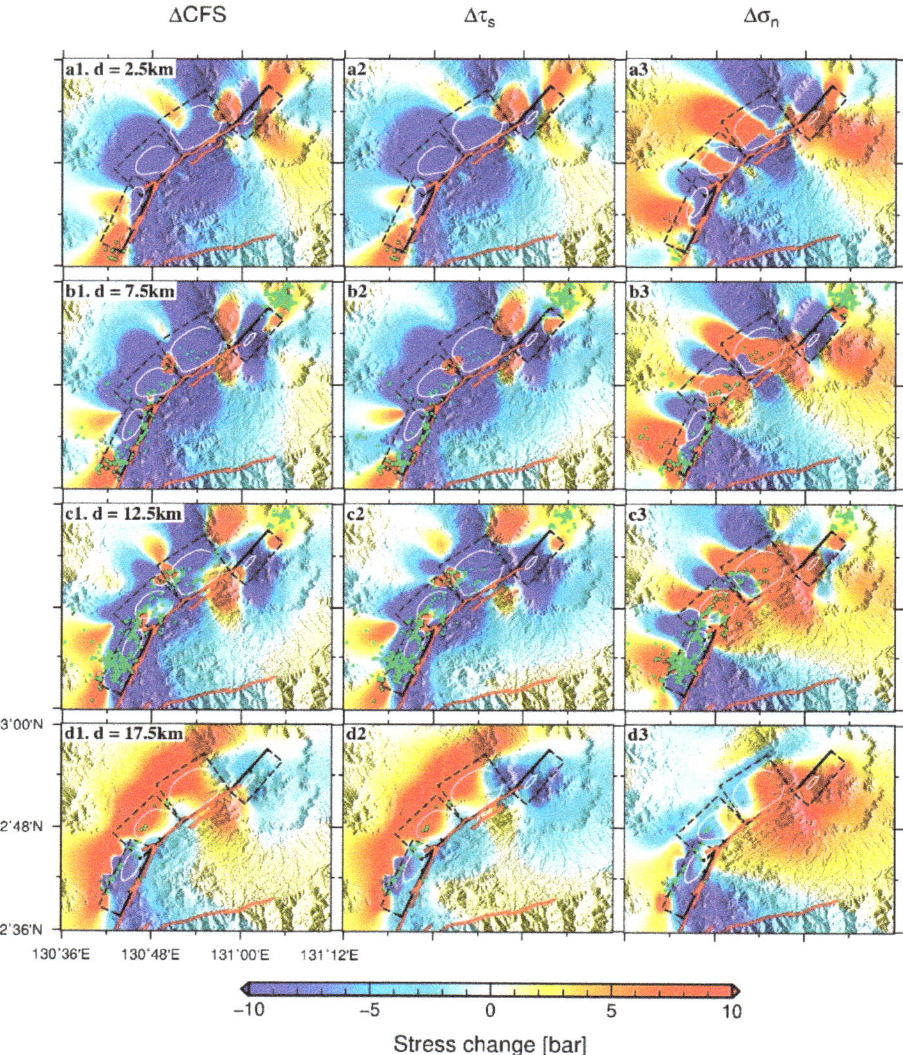


**Figure 8.** Static stress changes caused by the mainshock at different depths, calculated from the preferred model with $\mu' = 0.4$. (**a1–d1**) show the static Coulomb stress changes (positive $\Delta$CFS may promote failure); (**a2–d2**) show the shear stress changes (positive in the direction of slip); (**a3–d3**) show the normal stress changes (positive when the fault is unclamped). The receiver fault is approximately parallel to the seismogenic fault of the mainshock, with strike, dip, and rake of (220°, 68°, and −165°, respectively. Aftershocks with Mw >3.0 are denoted by green dots in each panel, with hypocenter depth ranges of d ± 2.5 km. The white contours highlight the asperities with a slip amplitude of over 2 m. The red lines show the active faults. The black dashed rectangles are surface projections of fault planes.

## 5. Discussion

### 5.1. Seismogenic Fault Geometry

The fault model proposed in this study is composed of four-segment fault planes, including three northwest dipping planes (F2, F3, and FH) along the Futagawa and northern Hinagu faults and one southeast dipping plane (F1) along the northeastward extension of the Futagawa fault under the Aso caldera. The FH segment along the Hinagu fault zone has a strike of N205° and an estimated dip angle of 75°, in good agreement with other previous models [2,12]. The F3 segment with a strike of N226° connects the Futagawa and Hinagu faults, while the F2 segment has a strike of N236° along the Futagawa fault. The dip angles were estimated to be 63° and 57° for the F3 segment and F2 segment, respectively. The F1 segment with a strike of N40° and an optimal dip angle of 77° is located at the northeastward extension of the Futagawa fault, where dextral displacements exceeding 100 cm were observed in the field investigation [7]. Our preferred fault model is generally consistent with previous studies based on surface displacements and/or strong-motion waveforms [12,15]. For example, Fukahata and Hashimoto [12] obtained the dip angles of two faults, attaining values of 61° ± 6° for the Futagawa fault and 74° ± 12° for the Hinagu fault by inverting InSAR data. Zhang et al. [15] combined InSAR, GPS, and strong-motion data to construct a three-segment model, in which the Futagawa fault segment dips 60° to the northwest, the Hinagu fault segment dips 70° to the northwest, and a southeast dipping fault segment near the Aso volcano has a dip angle of 80°.

However, there are several studies proposing northwest dipping fault models in the Aso caldera region based on teleseismic and/or strong-motion data [10,38]. Is a southeast dipping fault model more reasonable than a northwest dipping one? We carried out additional tests to compare the RMSE and jRi values for the southeast dipping fault F1 and the northwest dipping one. The results (Figure S3) indicate that a southeast dipping fault F1 can better fit the observed data than a northwest dipping one. Ozawa et al. [13] suggested that a southeast dipping fault plane near the Aso caldera region should be considered to explain the complex deformation pattern obtained from SAR images. Yoshida et al. [14] also confirmed that the model with a southeast dipping fault predicted the observed subsidence around the central cones of Aso volcano better than a northwest dipping one. The vertical displacements calculated from our preferred model were utilized to confirm the predictive performance by comparing the data with those of previous models. The result (Figure S4) indicates that our four-segment model with a southeast dipping F1 segment can better explain the subsidence around the Aso volcano from the decomposition of InSAR data [3,39].

### 5.2. Comparison with Previous Slip Distribution Models

The checkerboard and jackknife tests (Appendix A) show that the slip distribution model on each segment is well-constrained by the combination of InSAR and GPS data. Our preferred model is in general agreement with previous models based on teleseismic waveforms [1], strong-motion data [2,10], geodetic data [39,40], or combinations of them [3,11,15]. Consensus has been reached on at least three things, which are: (1) multi-segment faults with varying fault geometries ruptured during this earthquake; (2) the ruptures initiated at the Hinagu segment and propagated unilaterally to the northeastern extension of the Futagawa fault under the Aso caldera; and (3) the slip on the Hinagu fault (FH) was almost purely right-lateral strike-slip, and the Futagawa fault (F2 and F3) was dominated by right-lateral strike-slip, along with a significant normal slip component. However, there are some inconsistencies in the maximum slip and slip pattern. Our preferred slip distribution model shows that each segment has a concentrated slip area (asperity) and the largest asperity is located on the F2 segment, with a peak slip of 5.6 m at a 5–6 km depth. The maximum slip (5.6 m) is similar to the results of [1,2,11,12], but smaller than those of [3] (~10 m), [15] (7.4 m), and [41] (6.9 m). The slip pattern discrepancy may be attributed to the different data sets, model regularizations, and parameterizations employed. For example, Yue et al. [3] combined GPS, strong-motion, InSAR data from both ALOS-2 and Sentinel-1A images, and surface offsets to invert the kinematic rupture process on the curved multi-segment fault planes. They selected a smaller smoothing factor based on the consistency between the calculated Coulomb stress and aftershock distribution, which might overestimate the peak slip. Using Sentinel-1A imagery, GPS, and strong motion data, Zhang et al. [15] also inverted the slip distribution on the three-segment fault planes. Compared with our model, Zhang et al. [15] constructed the fault model with a similar southeast-dipping fault near the Aso volcano and generated a rougher slip pattern, choosing a smaller smoothing factor with the L-curve method. In our model, the Futagawa fault is parameterized into two segments (F2 and F3) and has two different asperities, which are consistent with the third and fourth asperity locations in Yue et al.'s model [3]. In summary, based on adequate constraints of InSAR (both ALOS-2 and Sentinel-1A images) and GPS datasets, our preferred model selects an optimal smoothing factor with the jRi method and presents a more plausible slip pattern.

### 5.3. Shallow Slip Deficit

The estimated slips on the shallowest sub-faults along the Futagawa and northern Hinagu faults are consistent with the displacements of surface ruptures from the field investigation [7] (Figure 5). On the F2 segment, the estimated slip has a peak of 5.6 m at a 5–6 km depth and diminishes to an average of 2 m near the surface, which implies a shallow slip deficit (SSD) of ~65% (Figure 5). A similar SSD also occurred on the other three segments. Previous studies have demonstrated that SSD might be caused by velocity-strengthening friction at a shallow depth that can be recovered through shallow afterslip [42,43] or distributed coseismic brittle damage in the uppermost crust layer [44,45]. Moreover, some studies have shown that SSD can also arise from artifacts due to a lack of near-fault data [46] or from model uncertainties [47]. However, the SSD in this study is real because the slip near the surface was estimated well in the inversion, and a similar SSD was also obtained using near-fault data in several previous studies [3,41]. Interestingly, Yue at al. [3] and Moore et al. [48] both observed significant post-seismic deformation close to the faults, which might arise from afterslip at a shallow depth. Therefore, velocity-strengthening friction at shallow depths is more likely to be the mechanism that produces significant SSD, but the "distributed brittle failure" explanation cannot be ruled out from the currently available observations.

### 5.4. Sensitivity Analysis of Coulomb Stress Change

Although the majority of Mw > 5.0 aftershocks with available focal mechanisms support static Coulomb stress triggering, the entire aftershock sequence cannot be explained well by the calculated Coulomb stress changes on the receiver fault that is parallel to the seismogenic fault of the mainshock. A few aftershocks occurred in the stress shadow regions characterized by negative stress changes. Given that the calculation of Coulomb stress change involves three variables ($\Delta\tau_s$, $\Delta\sigma_n$, and $\mu'$ in Equation (3)), the sensitivity of Coulomb stress calculations to the effective coefficient of friction, receiver fault, and source model also needs to be investigated. The effective coefficient of friction varies from 0.2 to 0.8 for different types of fault. The mainshock focal mechanisms from NIED, GCMT, USGS W-phase, and this study were adopted to define receiver faults in four different scenarios, respectively. Furthermore, six available finite fault slip models inverted from different datasets shown in Table S1 were collected to address the sensitivity of the source model. The result of the sensitivity tests (Figures S5–S7) indicates that the source model plays a more important role in the Coulomb stress calculations, especially in the case of a complex rupture event, while the receiver fault geometry and effective friction coefficient have a slight effect on the size of the Coulomb stress change area. A comparison of Coulomb stress changes calculated with different source models shows that the differences are apparent near the faults, and source models (i.e., Yue et al. [3], Zhang et al. [15], and our study) inverted from various datasets jointly can generate a more detailed stress change pattern which can better explain the spatial distribution of seismicity. In all scenarios, there are some common features in that the static Coulomb stress increase significantly triggers aftershocks located in the northern part of the Aso volcano, and a few aftershocks in a stress shadow area may be driven by aseismic creep [3,49] or triggered by dynamic stress transfer [50], which requires further investigation.

## 6. Conclusions

The complex coseismic deformation field associated with the 2016 Kumamoto earthquake sequence is derived from Sentinel-1A and ALOS-2 SAR images. Together with GPS data, InSAR data provide powerful constraints on the fault geometry and slip distribution of the seismogenic fault. Based on surface ruptures, active fault traces, InSAR deformation maps, and seismicity, we constructed a four-segment fault model with one southeast-dipping fault plane and three northwest-dipping fault planes. The fault dip angles of the four segments were determined well with the jRi method, which exhibited a better performance than the conventional RMSE method. Our preferred slip distribution model reveals a single asperity on each segment with a maximum slip of 5.6 m on the F2 segment and suggests a significant shallow slip deficit that may be caused by velocity-strengthening friction at shallow depths. Areas with large slips are spatially complementary to the distribution of aftershocks. The total geodetic moment release was estimated to be $4.89 \times 10^{19}$ Nm, equivalent to an Mw 7.09 event. The Coulomb stress calculation and its sensitivity analysis indicate that static Coulomb stress triggering can only explain part of the aftershock sequence, while aftershocks that occurred in a stress shadow area may be triggered by other factors, such as aseismic creep or dynamic stress transfer.

**Supplementary Materials:** The following are available online at http://www.mdpi.com/2072-4292/12/22/3721/s1: Figure S1: The choice of the relative weight ratio of GPS relative to InSAR datasets; Figure S2: The smoothing factor κ determination; Figure S3. Comparison of the RMSE and jRi values between the southeast dipping fault F1 and the northwest dipping fault F1; Figure S4: Predicted vertical displacements from our preferred model and previous source models; Figure S5: Sensitivity test of Coulomb stress changes to effective coefficients of friction ($\mu' = 0.2$, 0.4, 0.6, and 0.8) at different depths; Figure S6: Sensitivity test of Coulomb stress changes to receiver faults at different depths, calculated from the preferred model with $\mu' = 0.4$; Figure S7: Sensitivity test of Coulomb stress changes to source models at different depths, resolved on the receiver fault of $220°/68°/-165°$ with $\mu' = 0.4$; Figure S8: Checkerboard test for resolution of the joint or individual inversion of InSAR and GPS dat;. Figure S9: Jackknife test results of a. the mean slip, b. the standard derivation (Std.), and c. the coefficient of variation (CV) of slip; Table S1: Source parameters of the 2016 Kumamoto earthquake obtained from the inversion of various datasets.

**Author Contributions:** T.C. and Z.H. conceived the work. Z.H. performed the main experiments and wrote the manuscript. T.C., M.W. and Y.L. analyzed the results. All authors contributed to the reviewing and editing of the paper. All authors have read and agreed to the published version of the manuscript.

**Funding:** This work was supported by the National Key Research and Development Program of China (Grant no. 2018YFC1503605 and 2016YFC1401506).

**Acknowledgments:** We would like to thank the Editor and three anonymous reviewers for their helpful suggestions and comments, which greatly improved the manuscript. We thank Yangmao Wen for helping with InSAR data processing and Yukitoshi Fukahata for providing GPS data. This work benefitted from discussions with Yuqing He and Chendong Zhang. The Sentinel-1A SAR data were provided by the European Space Agency (ESA) through Sentinels Scientific Data Hub, and the ALOS-2/ PALSAR-2 data were provided by JAXA through the RA6 project (ID: 3048). Most figures in this paper were prepared by the Generic Mapping Tools (GMT) software [51].

**Conflicts of Interest:** The authors declare no conflict of interest.

## Appendix A

### Checkerboard and Jackknife Tests

A set of checkerboard tests were performed to investigate the spatial resolution of the slip distribution inverted from InSAR and GPS data, both individually and jointly. The synthetic slip model, whose sub-fault size increases from 2 to 4 km along the downdip direction, has an alternating slip of 0 or 2 m. We first generated simulated InSAR and GPS data, without adding any noise, as we focused on the contribution of each dataset in the inversion. Then, the simulated data were inverted to see how well the synthetic slip model was recovered. The result (Figure S8) shows that InSAR data have a great importance in recovering the slip distribution on each fault segment. GPS data have a poor model resolution due to the sparse distribution and a small number of stations, but still place valuable constraints on the peak slip. The joint inversion presents an improved spatial resolution of the slip distribution compared with individual InSAR or GPS inversion. The slip distribution from joint inversion is well-recovered for the shallow sub-faults (<10 km along the dip direction), but smeared at a deeper depth, especially on the FH segment.

In order to better quantify the robustness of the slip inversion, a jackknife test was conducted by randomly removing 20% of the data and running the inversion with the same model parameters [52]. This process was repeated 200 times, and the mean slip model, the standard deviation of slip, and the coefficient of variation (CV) were then calculated. The CV given by dividing the mean by the standard deviation is a good measure of the slip variability on each sub-fault. Sub-faults with a low CV have a more reliable slip than those with a high CV. The result (Figure S9) shows that the main slip area on each segment is stable due to their CV being lower than 0.2, while a high CV occurs at the edge of each segment.

## References


1. Yagi, Y.; Okuwaki, R.; Enescu, B.; Kasahara, A.; Miyakawa, A.; Otsubo, M. Rupture process of the 2016 Kumamoto earthquake in relation to the thermal structure around Aso volcano. *Earth Planets Space* **2016**, *68*, 118. [CrossRef]
2. Asano, K.; Iwata, T. Source rupture processes of the foreshock and mainshock in the 2016 Kumamoto earthquake sequence estimated from the kinematic waveform inversion of strong motion data. *Earth Planets Space* **2016**, *68*, 147. [CrossRef]
3. Yue, H.; Ross, Z.E.; Liang, C.; Michel, S.; Fattahi, H.; Fielding, E.; Moore, A.; Liu, Z.; Jia, B. The 2016 Kumamoto Mw = 7.0 Earthquake: A significant event in a fault–volcano system. *J. Geophys. Res. Solid Earth* **2017**, *122*, 9166–9183. [CrossRef]
4. Global Centroid Moment Tensor Catalogue (GCMT). Available online: http://www.globalcmt.org/CMTsearch.html (accessed on 30 June 2020).
5. United States Geological Survey (USGS). Available online: https://earthquake.usgs.gov/earthquakes/eventpage/us20005iis/moment-tensor (accessed on 30 June 2020).
6. National Research Institute for Earth Science and Disaster Resilience (NIED). Available online: https://www.fnet.bosai.go.jp/event/tdmt.php?_id=20160415162400&LANG=en (accessed on 30 June 2020).
7. Shirahama, Y.; Yoshimi, M.; Awata, Y.; Maruyama, T.; Azuma, T.; Miyashita, Y.; Mori, H.; Imanishi, K.; Takeda, N.; Ochi, T.; et al. Characteristics of the surface ruptures associated with the 2016 Kumamoto earthquake sequence, central Kyushu, Japan. *Earth Planets Space* **2016**, *68*, 191. [CrossRef]
8. Matsumoto, S.; Nakao, S.; Ohkura, T.; Miyazaki, M.; Shimizu, H.; Abe, Y.; Inoue, H.; Nakamoto, M.; Yoshikawa, S.; Yamashita, Y.; et al. Spatial heterogeneities in tectonic stress in Kyushu, Japan and their relation to a major shear zone Seismology. *Earth Planets Space* **2015**, *67*, 172. [CrossRef]
9. Kato, A.; Nakamura, K.; Hiyama, Y. The 2016 Kumamoto earthquake sequence. *Proc. Jpn. Acad. Ser. B Phys. Biol. Sci.* **2016**, *92*, 358–371. [CrossRef]
10. Kubo, H.; Suzuki, W.; Aoi, S.; Sekiguchi, H. Source rupture processes of the 2016 Kumamoto, Japan, earthquakes estimated from strong-motion waveforms. *Earth Planets Space* **2016**, *68*, 161. [CrossRef]
11. Hao, J.; Ji, C.; Yao, Z. Slip history of the 2016 Mw 7.0 Kumamoto earthquake: Intraplate rupture in complex tectonic environment. *Geophys. Res. Lett.* **2017**, *44*, 743–750. [CrossRef]
12. Fukahata, Y.; Hashimoto, M. Simultaneous estimation of the dip angles and slip distribution on the faults of the 2016 Kumamoto earthquake through a weak nonlinear inversion of InSAR data. *Earth Planets Space* **2016**, *68*, 204. [CrossRef]
13. Ozawa, T.; Fujita, E.; Ueda, H. Crustal deformation associated with the 2016 Kumamoto Earthquake and its effect on the magma system of Aso volcano. *Earth Planets Space* **2016**, *68*, 186. [CrossRef]
14. Yoshida, K.; Miyakoshi, K.; Somei, K.; Irikura, K. Source process of the 2016 Kumamoto earthquake (Mj7.3) inferred from kinematic inversion of strong-motion records. *Earth Planets Space* **2017**, *69*, 64. [CrossRef]

15. Zhang, Y.; Shan, X.; Zhang, G.; Gong, W.; Liu, X.; Yin, H.; Zhao, D.; Wen, S.; Qu, C. Source model of the 2016 Kumamoto, Japan, earthquake constrained by InSAR, GPS, and strong-motion data: Fault slip under extensional stress. *Bull. Seismol. Soc. Am.* **2018**, *108*, 2675–2686. [CrossRef]

16. Yague-Martinez, N.; Prats-Iraola, P.; Gonzalez, F.R.; Brcic, R.; Shau, R.; Geudtner, D.; Eineder, M.; Bamler, R. Interferometric Processing of Sentinel-1 TOPS Data. *IEEE Trans. Geosci. Remote Sens.* **2016**, *54*, 2220–2234. [CrossRef]

17. Werner, C.; Wegmüller, U.; Strozzi, T.; Wiesmann, A. GAMMA SAR and interferometric processing software. In Proceedings of the ERS ENVISAT Symposium, Gothenburg, Sweden, 16–20 October 2000.

18. Farr, T.G.; Rosen, P.A.; Caro, E.; Crippen, R.; Duren, R.; Hensley, S.; Kobrick, M.; Paller, M.; Rodriguez, E.; Roth, L.; et al. The shuttle radar topography mission. *Rev. Geophys.* **2007**, *45*. [CrossRef]

19. Goldstein, R.M.; Werner, C.L. Radar interferogram filtering for geophysical applications. *Geophys. Res. Lett.* **1998**, *25*, 4035–4038. [CrossRef]

20. Chen, C.W.; Zebker, H.A. Two-dimensional phase unwrapping with use of statistical models for cost functions in nonlinear optimization. *J. Opt. Soc. Am. A* **2001**, *18*, 338. [CrossRef]

21. He, P.; Wen, Y.; Xu, C.; Chen, Y. Complete three-dimensional near-field surface displacements from imaging geodesy techniques applied to the 2016 Kumamoto earthquake. *Remote Sens. Environ.* **2019**, *232*, 111321. [CrossRef]

22. Wen, Y.; Xu, C.; Liu, Y.; Jiang, G. Deformation and source parameters of the 2015 Mw 6.5 earthquake in Pishan, Western China, from Sentinel-1A and ALOS-2 data. *Remote Sens.* **2016**, *8*, 134. [CrossRef]

23. Okada, Y. Internal deformation due to shear and tensile faults in a half-space. *Bull. Seismol. Soc. Am.* **1992**, *82*, 1018–1040.

24. Jónsson, S.; Zebker, H.; Segall, P.; Amelung, F. Fault slip distribution of the 1999 Mw 7.1 Hector Mine, California, earthquake, estimated from satellite radar and GPS measurements. *Bull. Seismol. Soc. Am.* **2002**, *92*, 1377–1389. [CrossRef]

25. Burgmann, R.; Ayhan, M.E.; Fielding, E.J.; Wright, T.J.; McClusky, S.; Aktug, B.; Demir, C.; Lenk, O.; Turkezer, A. Deformation during the 12 November 1999 Duzce, Turkey, earthquake, from GPS and InSAR data. *Bull. Seismol. Soc. Am.* **2002**, *92*, 161–171. [CrossRef]

26. Fukahata, Y.; Wright, T.J. A non-linear geodetic data inversion using ABIC for slip distribution on a fault with an unknown dip angle. *Geophys. J. Int.* **2008**, *173*, 353–364. [CrossRef]

27. Chen, T.; Newman, A.V.; Feng, L.; Fritz, H.M. Slip distribution from the 1 April 2007 Solomon Islands earthquake: A unique image of near-trench rupture. *Geophys. Res. Lett.* **2009**, *36*, 6–11. [CrossRef]

28. Xu, C.; Liu, Y.; Wen, Y.; Wang, R. Coseismic slip distribution of the 2008 Mw 7.9 Wenchuan earthquake from joint inversion of GPS and InSAR data. *Bull. Seismol. Soc. Am.* **2010**, *100*, 2736–2749. [CrossRef]

29. Barnhart, W.D.; Lohman, R.B. Automated fault model discretization for inversions for coseismic slip distributions. *J. Geophys. Res. Solid Earth* **2010**, *115*, 1–17. [CrossRef]

30. Luo, H.; Chen, T.; Xu, C.; Sha, H. Fault dip angle determination with the jRi criterion and coulomb stress changes associated with the 2015 Mw 7.9 Gorkha Nepal earthquake revealed by InSAR and GPS data. *Tectonophysics* **2017**, *714*, 55–61. [CrossRef]

31. Stark, P.; Parker, R. Bounded-variable least-squares: An algorithm and applications. *Comput. Stat.* **1995**, 1–13.

32. Fujiwara, S.; Morishita, Y.; Nakano, T.; Kobayashi, T.; Yarai, H. Non-tectonic liquefaction-induced large surface displacements in the Aso Valley, Japan, caused by the 2016 Kumamoto earthquake, revealed by ALOS-2 SAR. *Earth Planet. Sci. Lett.* **2017**, *474*, 457–465. [CrossRef]

33. Reasenberg, P.A.; Simpson, R.W. Response of regional seismicity to the static stress change produced by the Loma Prieta earthquake. *Science* **1992**, *255*, 1687–1690. [CrossRef]

34. King, G.C.P.; Stein, R.S.; Lin, J. Static stress changes and the triggering of earthquakes. *Bull. Seismol. Soc. Am.* **1994**, *84*, 935–953.

35. Freed, A.M. Earthquake triggering by static, dynamic, and postseismic stress transfer. *Annu. Rev. Earth Planet. Sci.* **2005**, *33*, 335–367. [CrossRef]

36. Toda, S.; Stein, R.S.; Reasenberg, P.A.; Dieterich, J.H.; Yoshida, A. Stress transferred by the 1995 Mw = 6.9 Kobe, Japan, shock: Effect on aftershocks and future earthquake probabilities. *J. Geophys. Res. Solid Earth* **1998**, *103*, 24543–24565. [CrossRef]

37. Toda, S.; Stein, R.S.; Sevilgen, V.; Lin, J. *Coulomb 3.3 Graphic-Rich Deformation and Stress-Change Software for Earthquake, Tectonic, and Volcano Research and Teaching—User Guide*; U.S. Geological Survey Open-File Report 2011–1060; U.S. Geological Survey: Reston, VA, USA, 2011; 63p.

38.  Kobayashi, H.; Koketsu, K.; Miyake, H. Rupture processes of the 2016 Kumamoto earthquake sequence: Causes for extreme ground motions. *Geophys. Res. Lett.* **2017**, *44*, 6002–6010. [CrossRef]
39.  Himematsu, Y.; Furuya, M. Fault source model for the 2016 Kumamoto earthquake sequence based on ALOS-2/PALSAR-2 pixel-offset data: Evidence for dynamic slip partitioning. *Earth Planets Space* **2016**, *68*, 169. [CrossRef]
40.  Jiang, H.; Feng, G.; Wang, T.; Bürgmann, R. Toward full exploitation of coherent and incoherent information in Sentinel-1 TOPS data for retrieving surface displacement: Application to the 2016 Kumamoto (Japan) earthquake. *Geophys. Res. Lett.* **2017**, *44*, 1758–1767. [CrossRef]
41.  Scott, C.; Champenois, J.; Klinger, Y.; Nissen, E.; Maruyama, T.; Chiba, T.; Arrowsmith, R. The 2016 M7 Kumamoto, Japan, earthquake slip field derived from a joint inversion of differential lidar topography, optical correlation, and InSAR surface displacements. *Geophys. Res. Lett.* **2019**, *46*, 6341–6351. [CrossRef]
42.  Tse, S.T.; Rice, J.R. Crustal earthquake instability in relation to the depth variation of frictional slip properties. *J. Geophys. Res.* **1986**, *91*, 9452. [CrossRef]
43.  Marone, C. Laboratory-derived friction laws and their application to seismic faulting. *Annu. Rev. Earth Planet. Sci.* **1998**, *26*, 643–696. [CrossRef]
44.  Fialko, Y.; Sandwell, D.; Simons, M.; Rosen, P. Three-dimensional deformation caused by the Bam, Iran, earthquake and the origin of shallow slip deficit. *Nature* **2005**, *435*, 295–299. [CrossRef]
45.  Kaneko, Y.; Fialko, Y. Shallow slip deficit due to large strike-slip earthquakes in dynamic rupture simulations with elasto-plastic off-fault response. *Geophys. J. Int.* **2011**, *186*, 1389–1403. [CrossRef]
46.  Xu, X.; Tong, X.; Sandwell, D.T.; Milliner, C.W.D.; Dolan, J.F.; Hollingsworth, J.; Leprince, S.; Ayoub, F. Refining the shallow slip deficit. *Geophys. J. Int.* **2016**, *204*, 1867–1886. [CrossRef]
47.  Huang, M.H.; Fielding, E.J.; Dickinson, H.; Sun, J.; Gonzalez-Ortega, J.A.; Freed, A.M.; Bürgmann, R. Fault geometry inversion and slip distribution of the 2010 Mw 7.2 El Mayor-Cucapah earthquake from geodetic data. *J. Geophys. Res. Solid Earth* **2017**, *122*, 607–621. [CrossRef]
48.  Moore, J.D.P.; Yu, H.; Tang, C.H.; Wang, T.; Barbot, S.; Peng, D.; Masuti, S.; Dauwels, J.; Hsu, Y.J.; Lambert, V.; et al. Imaging the distribution of transient viscosity after the 2016 Mw 7.1 Kumamoto earthquake. *Science* **2017**, *356*, 163–167. [CrossRef] [PubMed]
49.  Perfettini, H.; Avouac, J.P. Postseismic relaxation driven by brittle creep: A possible mechanism to reconcile geodetic measurements and the decay rate of aftershocks, application to the Chi-Chi earthquake, Taiwan. *J. Geophys. Res. Solid Earth* **2004**, *109*. [CrossRef]
50.  Nissen, E.; Elliott, J.R.; Sloan, R.A.; Craig, T.J.; Funning, G.J.; Hutko, A.; Parsons, B.E.; Wright, T.J. Limitations of rupture forecasting exposed by instantaneously triggered earthquake doublet. *Nat. Geosci.* **2016**, *9*, 330–336. [CrossRef]
51.  Wessel, P.; Smith, W.H.F.; Scharroo, R.; Luis, J.; Wobbe, F. Generic mapping tools: Improved version released. *Eos Trans. Am. Geophys. Union* **2013**, *94*, 409–410. [CrossRef]
52.  Melgar, D.; Ganas, A.; Geng, J.; Liang, C.; Fielding, E.J.; Kassaras, I. Source characteristics of the 2015 Mw 6.5 Lefkada, Greece, strike-slip earthquake. *J. Geophys. Res. Solid Earth* **2017**, *122*, 2260–2273.

**Publisher's Note:** MDPI stays neutral with regard to jurisdictional claims in published maps and institutional affiliations.

*Article*

# Seismogenic Source Model of the 2019, $M_w$ 5.9, East-Azerbaijan Earthquake (NW Iran) through the Inversion of Sentinel-1 DInSAR Measurements

**Emanuela Valerio** [1], **Mariarosaria Manzo** [1], **Francesco Casu** [1], **Vincenzo Convertito** [2], **Claudio De Luca** [1], **Michele Manunta** [1], **Fernando Monterroso** [1], **Riccardo Lanari** [1,*] and **Vincenzo De Novellis** [1]

[1]  Istituto per il Rilevamento Elettromagnetico dell'Ambiente, IREA-CNR, 80124 Napoli, Italy; valerio.e@irea.cnr.it (E.V.); manzo.mr@irea.cnr.it (M.M.); casu.f@irea.cnr.it (F.C.); deluca.c@irea.cnr.it (C.D.L.); manunta.m@irea.cnr.it (M.M.); monterroso.f@irea.cnr.it (F.M.); denovellis.v@irea.cnr.it (V.D.N.)
[2]  Istituto Nazionale di Geofisica e Vulcanologia, Osservatorio Vesuviano, 80124 Napoli, Italy; vincenzo.convertito@ingv.it
*  Correspondence: lanari.r@irea.cnr.it

Received: 13 March 2020; Accepted: 21 April 2020; Published: 24 April 2020

**Abstract:** In this work, we investigate the $M_w$ 5.9 earthquake occurred on 7 November 2019 in the East-Azerbaijan region, in northwestern Iran, which is inserted in the tectonic framework of the East-Azerbaijan Plateau, a complex mountain belt that contains internal major fold-and-thrust belts. We first analyze the Differential Synthetic Aperture Radar Interferometry (DInSAR) measurements obtained by processing the data collected by the Sentinel-1 constellation along ascending and descending orbits; then, we invert the achieved results through analytical modelling, in order to better constrain the geometry and characteristics of the seismogenic source. The retrieved fault model shows a rather shallow seismic structure, with a center depth at about 3 km, approximately NE–SW-striking and southeast-dipping, characterized by a left-lateral strike-slip fault mechanism (strike = 29.17°, dip = 79.29°, rake = −4.94°) and by a maximum slip of 0.80 m. By comparing the inferred fault with the already published geological structures, the retrieved solution reveals a minor fault not reported in the geological maps available in the open literature, whose kinematics is compatible with that of the surrounding structures, with the local and regional stress states and with the performed field observations. Moreover, by taking into account the surrounding geological structures reported in literature, we also use the retrieved fault model to calculate the Coulomb Failure Function at the nearby receiver faults. We show that this event may have encouraged, with a positive loading, the activation of the considered receiver faults. This is also confirmed by the distribution of the aftershocks that occurred near the considered surrounding structures. The analysis of the seismic events nucleated along the left-lateral strike-slip minor faults of the East-Azerbaijan Plateau, such as the one analyzed in this work, is essential to improve our knowledge on the seismic hazard estimation in northwestern Iran.

**Keywords:** the 2019 East-Azerbaijan earthquake; strike-slip fault; Sentinel-1 DInSAR measurements; analytical modelling; Coulomb Failure Function

---

## 1. Introduction

On 7 November 2019 (22:47 UTC), a $M_w$ 5.9 earthquake took place in the East-Azerbaijan (hereinafter referred to as E-Azerbaijan earthquake), in northwestern Iran, about 100 km east of Tabriz, the fourth largest city of Iran, with over two million citizens (Figure 1a). The event was clearly felt in most parts of the E-Azerbaijan province, killing at least five people, injuring hundreds, and causing widespread damage to the surrounding villages. During the next days, the mainshock was followed by more than 170 events with $2.5 \leq M_w \leq 4.8$, recorded by the Broadband Iranian Network (BIN) [1]. The mainshock was recorded by different institutions: the Iranian Seismological Center (IRSC) [2], the United States Geological Survey (USGS) [3], the GFZ-Potsdam (GEOFON) [4] and the Harvard Global CMT catalog [5,6], that furnished four different hypocentral locations and focal mechanisms parameters. Despite such different solutions, the analysis of the available focal mechanisms and hypocenters suggests that the E-Azerbaijan earthquake nucleated along a strike-slip fault, located within the complex structural setting of the Bozgush Range [7,8] (Figure 1b,c). Although no medium-high intensity earthquakes have affected this seismogenic zone during the last years, an extensive active deformation and the consequent seismicity have interested the considered Iranian region and have represented a serious risk to the local cities and villages since ancient times. The historical seismic activity rate of this region is high, as testified by the many documented instrumental or historical earthquakes with magnitudes above $M_w$ 6.0 [9]. In particular, one of the strongest historical seismic events took place on 22 March 1879, with $M_w$ 6.7 along the South Bozgush Fault (black star in Figure 1b), causing 2000 casualties and heavy damage to the surrounding villages.

The seismicity of the E-Azerbaijan Plateau can be related to the oblique Arabia-Eurasia convergence, which is accommodated in northwestern Iran, partly as seismic activity along strike-slip and thrust faults [7,10–17] (Figure 1a). Accordingly, the E-Azerbaijan Plateau represents a complex mountain belt, which contains three internal major fold-and-thrust belts and, in particular, the Bozgush Range, bounded by two major faults, the North- and South-Bozgush Faults [7,8] (Figure 1b).

A wealth of data exists from historical seismicity and the related focal mechanisms [9,10,18–20], stress state reconstruction [7], field observations [21] and geodetic measurements around the involved seismogenic zone [14,16]. Thanks to the integration of these multidisciplinary data, it is possible to provide a detailed picture of the kinematics and the main geological structures of the region. Moreover, the fault geometry and slip distribution analysis for medium-large earthquakes are very significant for improving our knowledge of both the active faults distribution and the associated seismic hazard of a given region.

In this work, we first exploit the Differential Synthetic Aperture Radar Interferometry (DInSAR) measurements obtained by processing the data collected by the Sentinel-1 (S1) satellite of the Copernicus European Program along ascending and descending orbits. Then, we apply an analytical modelling approach to the computed coseismic DInSAR displacements, with the aim of better constraining the kinematics of the main seismic source. In addition, starting from the retrieved fault model characteristics and by considering the already known surrounding geological structures, we perform an analysis of the Coulomb stress transfer on the nearby faults, in order to investigate possible fault interaction processes.

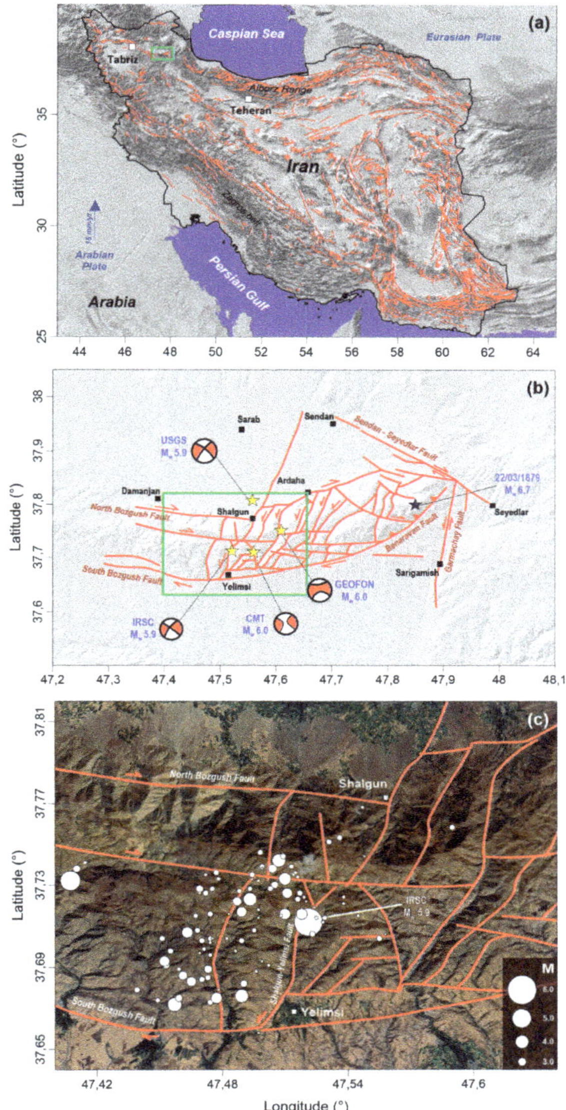



**Figure 1.** Geodynamic scenario. (**a**) Tectonic Map of Iran (WGS84), where the main tectonic structures (derived from [22]) are reported with red lines. The green rectangle identifies the zone considered in the following panel and the reported plate velocity is derived from [16]. (**b**) Detailed structural map of the considered seismogenic area, in which the main geological lineaments are highlighted by red lines (derived from [8]). The different proposed epicentral locations and focal mechanisms are also shown, as well as the strongest historical event occurred in the considered area (black star). The green rectangle identifies the zone considered in the following panel. (**c**) Distribution of the seismicity recorded from 7 to 9 November 2019 (white dots), shown as a function of magnitude (the higher the magnitude, the bigger the circles). The main local structures are also indicated with red lines (derived from [8]). In all the panels, the reported data are superimposed on the 1 arcsec Shuttle Radar Topography Mission (SRTM) Digital Elevation Model (DEM) of the zone.

## 2. Tectonic Setting

E-Azerbaijan, in northwestern Iran, is characterized by a complex active tectonics, linked to the main geodynamic regime controlled by the oblique convergence between the Arabian and Eurasian plates, with a present-day deformation rate of about 13–15 mm/year [7,10,11,14,16] (Figure 1a). In particular, this region represents a segment of the Alpine-Himalayan orogenic belt, whose origin is ascribable to several intense geodynamic processes (i.e., multiple accretion and continental collision consequent to the subduction and closure of the Tethys Ocean), which occurred between the late Eocene and the Oligocene epochs [7,18]. The local expression of this large-scale tectonic framework is represented by intense deformations and seismic activity, mainly located along strike-slip faults and thrusts, and analyzed through GPS measurements and earthquakes focal mechanisms (GCMT) catalog [12,16,23,24], respectively. Indeed, the recorded historical [25] and instrumental seismicity [19] in Iran suggests an intracontinental deformation concentrated in several mountain belts surrounding relatively aseismic blocks (i.e., central Iran, Lut and South Caspian blocks [14]).

Our study area is inserted in the tectonic context of the E-Azerbaijan Plateau, an articulated mountain belt containing three internal major fold-and-thrust belts (the Arasbaran, Ghoshe Dagh and Bozgush Ranges) [7], and in particular of the Bozgush Range, bounded by two major faults, the North- and South-Bozgush Faults (Figure 1b). Several interpretations about their kinematics have been proposed in literature: Zamani and Masson [7] suggest that these faults can be identified as major thrusts associated with the original paleogeographic configuration and build upon inherited, pre-existing structures in the late Cenozoic; on the other hand, Faridi et al. [8] propose a right-lateral component according to the stream deflections and the present day kinematic measurements [23]. However, all the authors suggest that the Bozgush Range can be interpreted as a positive flower structure, also known as a pop-up zone [26]. Moreover, recent structural investigations in the mountain discovered local range-parallel dextral faults and range-transverse conjugate sinistral faults, with a strong normal component in Neogene and Quaternary sediments. The eastern Bozgush Range terminates on the roughly N–S-striking Garmachay sinistral fault that, with the Shalgun–Yelimsi Fault, defines the eastern Bozgush Range as left-stepping step-over breached and fragmented by multiple faults. These many minor faults accommodate about a 30° counterclockwise rotation of the East Bozgush Mountain with respect to the West Bozgush Range and accommodated uplift of the Precambrian and Paleozoic rocks on the southeastern flanks of the range. In addition to the sub-parallel NW–SE right-lateral strike-slip major faults, which attracted much attention because they are responsible for historical and destructive earthquakes, the study area also includes sinistral NNE–SSW-striking faults, which, in spite of their recent activity, contribute to the accommodation of the Holocene tectonic regime [8] (Figure 1b,c).

## 3. DInSAR Measurements

In order to investigate the ground displacements associated with the considered seismic event, we exploited the Differential SAR Interferometry (DInSAR) technique [27], which allows the analysis of surface displacement phenomena, by providing a measurement of the ground deformation projection along the radar line-of-sight (LOS). The SAR data considered to retrieve the ground deformation associated with the occurred seismic event were acquired by the Sentinel-1 (S1) constellation of the Copernicus European Program. Benefiting from the short revisit time and the small spatial baseline separation of the S1 constellation, we generated several coseismic differential interferograms (Figure 2), with a spatial resolution of about 80 m. Among them we selected, for the seismic source modelling discussed in the following, those less affected by undesired phase artifacts (atmospheric phase delays, decorrelation noise, etc.), thus preserving good spatial coverage and interferometric coherence, which is very relevant in order to correctly retrieve the deformation pattern through the exploited phase unwrapping procedures [28,29]. In particular, the employed S1 data were acquired on 15 October and 20 November 2019, (Figure 2c,g), and on 16 October and 9 November 2019 (Figure 2e,h) along the ascending (ASC) and the descending (DESC) orbits, respectively (Table 1). On both interferograms, several fringes located near the epicentral area are clearly visible (Figure 2g,h); note that each fringe corresponds to a LOS-displacement of about 2.8 cm (i.e., half of the employed S1 C-band wavelength $\lambda = 5.56$ cm). Subsequently, starting from these interferograms, we generated their corresponding LOS displacement maps (Figure 3a and b) through an appropriate phase unwrapping operation [29]. In particular, the ascending track (Figure 3a) presents a deformation pattern characterized by negative values down to about −4 cm and positive values up to about 7 cm, indicating a sensor-to-target distance increase and decrease, respectively; moreover, the descending track (Figure 3b) shows a clearer deformation pattern, characterized by both negative LOS displacement values down to about −5 cm and positive LOS displacement values up to about 5 cm.

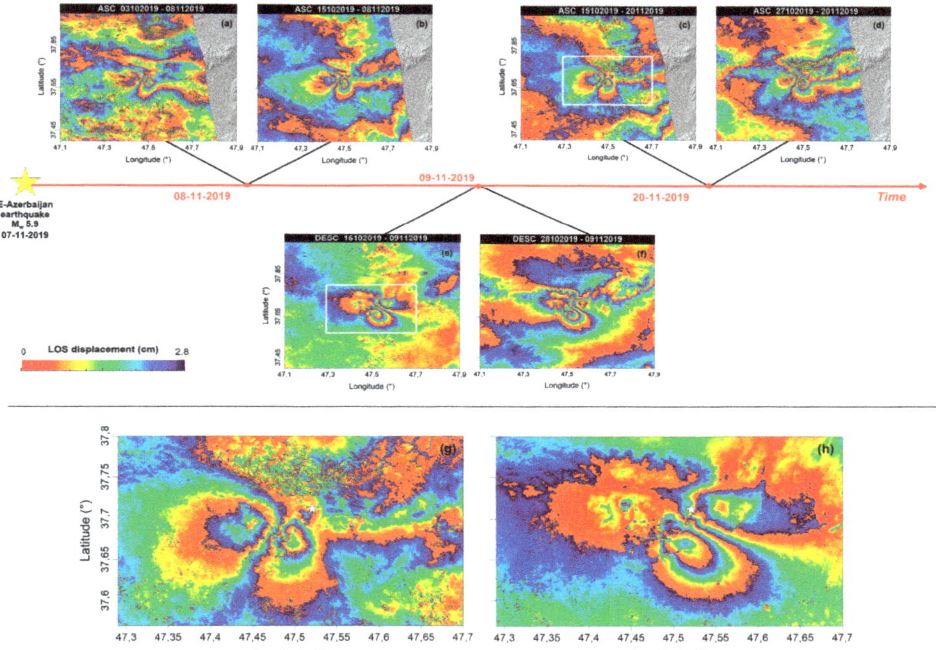

**Figure 2.** Exploited DInSAR measurements. Interferograms (wrapped) generated from Sentinel-1 data pairs acquired along ascending (ASC) orbits on (**a**) 3 October and 8 November 2019, (**b**) 15 October and 8 November 2019, (**c**) 15 October and 20 November 2019, (**d**) 27 October and 20 November 2019, and along descending (DESC) orbits on (**e**) 16 October and 9 November 2019 and (**f**) 28 October and 9 November 2019. The yellow star and the white rectangle represent the $M_w$ 5.9 E- Azerbaijan mainshock and the area considered in panel (**g**) and (**h**), respectively. The zoom of the interferograms considered to perform the subsequently discussed source modelling (panels (**c**) and (**e**)) is reported in panel (**g**) and (**h**). The white star identifies the $M_w$ 5.9 E- Azerbaijan mainshock.

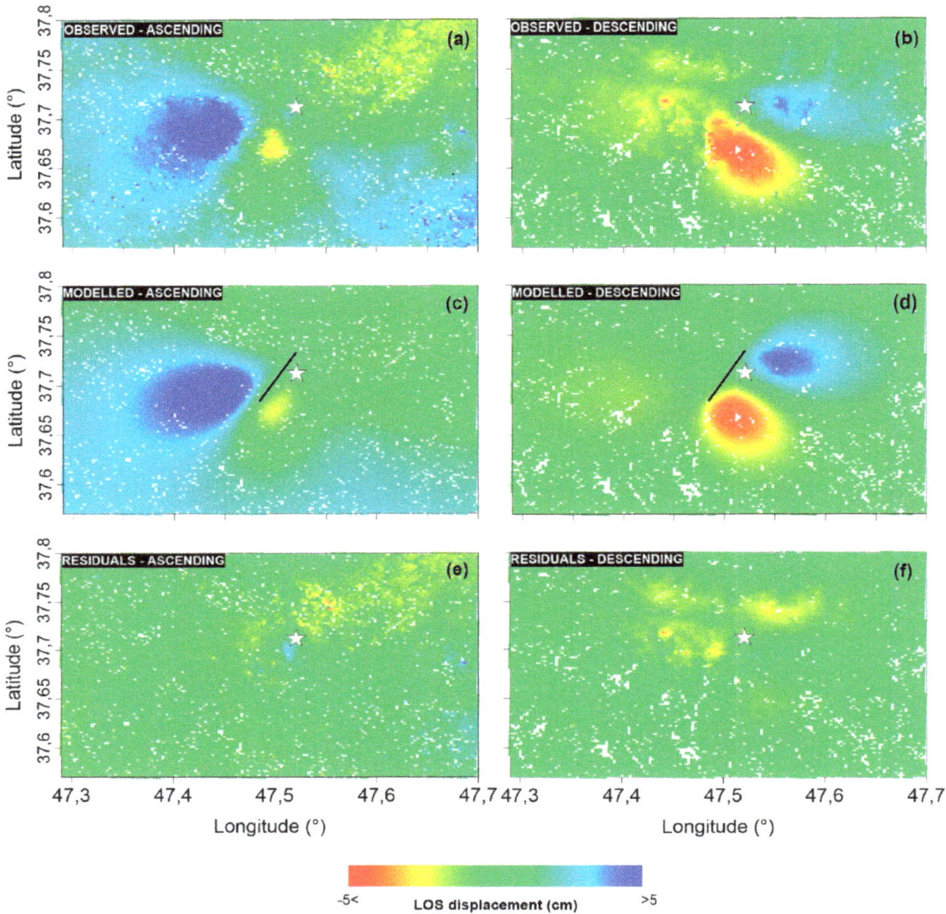



**Figure 3.** Source modelling results. Line-of-sight (LOS) projected displacement maps for S1 ascending (**a**) and descending (**b**) orbits interferograms. LOS projected displacement maps computed from the retrieved analytical model for the S1 ascending (**c**) and descending (**d**) orbits interferograms reported in panels (**a**) and (**b**). Their corresponding residual maps are shown in (**e**) and (**f**), respectively. The white star and the black line indicate the $M_w$ 5.9 E-Azerbaijan mainshock and the retrieved fault plane solution, respectively.

**Table 1.** Main characteristics of the interferometric pairs considered for the seismic source modelling of the E-Azerbaijan earthquake.

| Sensor | Dinsar Pair | Orbit | Perpendicular Baseline (m) | Track |
|--------|-------------|-------|----------------------------|-------|
| Sentinel-1 | 15102019–20112019 | ASC | 149 | 174 |
| Sentinel-1 | 16102019–09112019 | DESC | −12.19 | 6 |

## 4. Modelling

### 4.1. Analytical Modelling

In order to retrieve the fault parameters, we jointly inverted the S1 DInSAR displacements acquired from ascending and descending orbits (see previous section), by performing a consolidated two-step approach that consists of a non-linear optimization to constrain the fault geometry assuming a uniform slip, followed by a linear inversion to retrieve the slip distribution on the fault plane (details about the adopted algorithms can be found in [30]).

The LOS displacements retrieved from the DInSAR interferograms were modelled with a finite dislocation fault in an elastic and homogeneous half-space [31], also applying a compensation for the topography (in order to take into account the real depth from the ground surface) and assessing possible offsets and ramps affecting the DInSAR measurements. Moreover, they were preliminarily sampled over a regular grid (240 m in all the considered area) to reduce the computation load.

Starting from a non-linear inversion algorithm based on the Levenberg–Marquardt (LM) least-squares approach [32], we searched for the source parameters and, thanks to multiple random restarts implemented within the LM approach, it was possible to catch the global minimum during the optimization process.

The results of the non-linear inversion are reported in Figure 3, where we show the LOS-projected original (Figure 3a,b) and modelled (Figure 3c,d) DInSAR measurements, as well as their residuals (Figure 3e,f), obtained as the difference between the original and the modelled data [30]. The retrieved results reveal that the modelled seismic source allowed to effectively replicate the DInSAR measurements and the best-fit solution consists of a left-lateral strike-slip fault. In particular, the retrieved fault plane is located approximately at 3000 ($\pm$ 633) m depth b.s.l. (i.e., the depth of the center of the fault plane) and presents a strike of 29.17° ($\pm$ 5), a dip of 79.29° ($\pm$ 7), a rake of $-4.94$° ($\pm$ 7), and a uniform slip of about 0.73 m ($\pm$ 0.23), distributed over a source of about 6220 ($\pm$ 1540) m $\times$ 5630 ($\pm$ 2530) m (see Table 2 and Figure S2).

The second step of our modeling is represented by the linear inversion process with the computation of the non-uniform slip distribution, in order to have a more accurate estimate of the slip along the fault plane. In particular, the linear inversion was performed by using as starting model that obtained from the previously realized non-linear inversion (Table 2) and inverting the following system:

$$\begin{bmatrix} \mathbf{d}_{\text{DInSAR}} \\ 0 \end{bmatrix} = \begin{bmatrix} \mathbf{G} \\ \mathbf{k}\cdot\nabla^2 \end{bmatrix}\cdot\mathbf{m}$$

where $\mathbf{d}_{\text{DInSAR}}$ is the DInSAR displacements vector, $\mathbf{G}$ is the Green's matrix with the point-source functions, $\mathbf{m}$ is the vector of slip values, and $\nabla^2$ is a smoothing Laplacian operator weighted by an empirical coefficient $\mathbf{k}$ to guarantee a reliable slip varying across the fault; further constraints were introduced by imposing non-negative slip values obtained via non-linear inversion [33]. The fault length and width were extended to reduce the border effects as much as possible; the fault plane was subdivided into patches of $0.4 \times 0.4$ km.

The final slip distribution over the modelled fault plane is shown in Figure 4a and reaches a maximum value of 0.8 m. The retrieved seismic source has a geodetic moment of $8.63 \times 10^{17}$ Nm, corresponding to a moment magnitude of 5.92, coinciding with the value estimated by the IRSC ($M_w$ 5.9) (Figure 4). Aside from a slight correlation between slip and width and slip and rake, the variance analysis indicates that the implemented inversion technique is able to properly solve all the parameters. The best model parameters (best-fit) only slightly differ from the mean values and this is evident from the 1D distributions, which resemble Gaussian distributions, and from the computed standard deviations (see Figure S2).

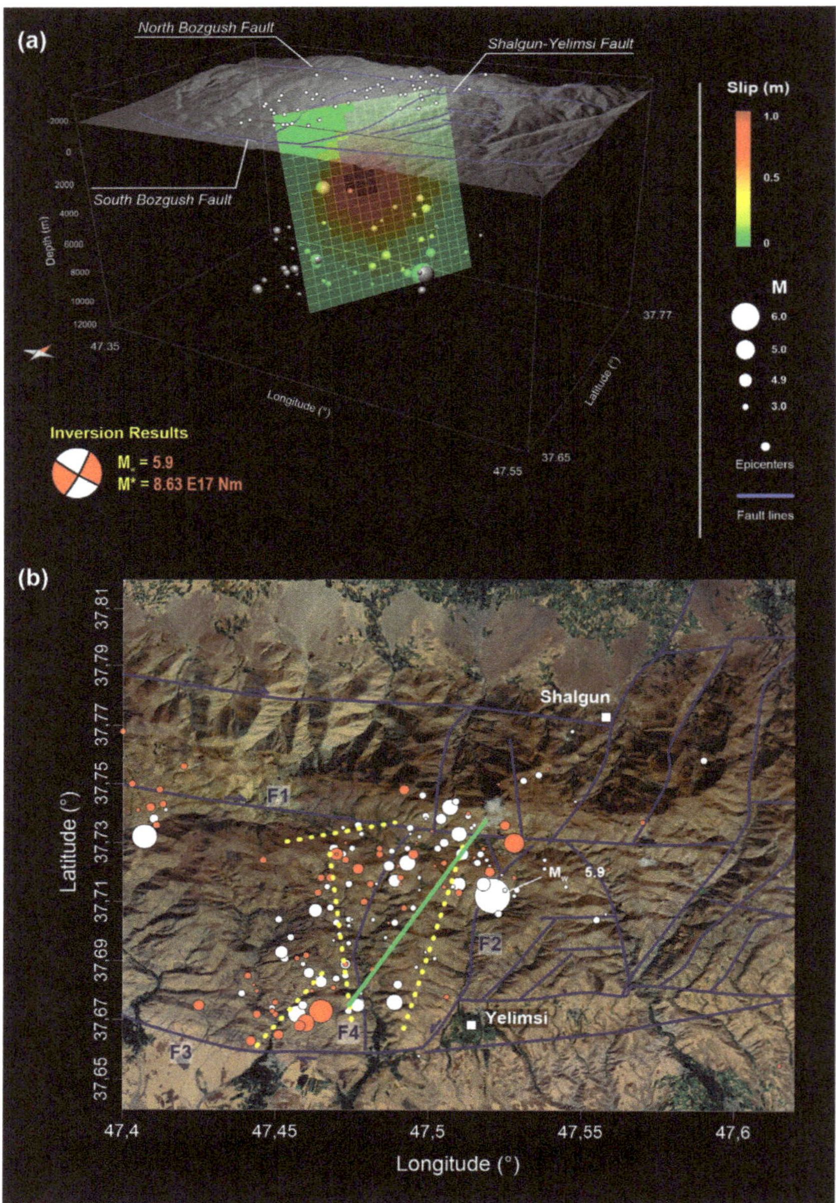



**Figure 4.** Linear inversion results. (**a**) Distributed slip (each patch extends for about $0.4 \times 0.4\,\text{km}^2$) and seismicity distribution (white circles and dots), displayed in 3D view. The computed focal mechanism is also reported. (**b**) Seismicity distribution recorded from 7 to 9 November 2019 (white circles) and from 10 November to 24 December 2019 (red circles) is shown as a function of magnitude (the higher the magnitude, the bigger the circles). The retrieved fault and the distribution of the fractures [21] generated by the seismic event are reported with the green line and the yellow dashed lines, respectively. The main local structures (derived from [8]) are also indicated with blue lines (F1, F2 = Shalgun-Yelimsi Fault, F3 = South Bozgush Fault and F4) and are superimposed on the 1 arcsec SRTM DEM of the zone.

**Table 2.** Best-fit parameters of the seismic source retrieved from non-linear inversion modelling. The 1-σ uncertainty is also reported.

| Parameter | Best-fit |
|---|---|
| Length (m) | 6219.80 (±1540) |
| Width (m) | 5629.90 (±2530) |
| Center Depth (m) | 3096.60 (±633) |
| Dip (deg) | 79.29 (±7) |
| Strike (deg) | 29.17 (±5) |
| Rake (deg) | −4.94 (±7) |
| East (m) | 720,969.80 (±667) |
| North (m) | 4,176,230.30 (±737) |
| Slip (m) | 0.73 (±0.23) |

*4.2. Coulomb Failure Function*

Static stress changes play an important role in the occurrence of earthquakes, which tend to nucleate on faults that have experienced an increase in Coulomb stress by previous events, and typically reach values included in a range of 0.1–10 bars [34–36]. Since the early 1990s, several studies have shown that the space-time earthquake clustering can be explained by faults interaction [37,38]. Indeed, when an earthquake occurs, the state of the stress in the surrounding volume is altered as a consequence of the seismic dislocation and, if the parameters describing the seismogenic fault and the receiver faults are known, the change in the stress field can be computed through the standard elasticity theory. In particular, under the Coulomb failure hypothesis, it is possible to verify if failure on a receiver fault is promoted or not by the slip on the seismogenic source fault by using the Coulomb failure function (CFF) [38,39], which is defined as:

$$\text{CFF} = \Delta\tau + \mu'\Delta\sigma_n$$

where $\Delta\tau$ is the shear stress change, $\Delta\sigma_n$ is the normal stress change and $\mu'$ is the effective fault friction coefficient on the receive fault.

In this work, we use the fault model retrieved from the analytical modelling of the DInSAR measurements, discussed in Section 3, as source fault, and assume a uniform slip distribution to compute the Coulomb stress changes at a reference depth of 5 km that corresponds to the mean depth value of the aftershocks recorded by the BIN during the 7 November–24 December 2019 time interval. For this purpose, we use the software Coulomb 3.3 [40,41].

Starting from the geological structures reported in Faridi et al. [8] and according to the Coulomb 3.3 software convention [40,41], we set the receiver fault mechanism to strike = 95°, dip = 80° and rake 180° for the Fault 1 (F1); 190°, 80°, 0° for the Fault 2 (F2), known as the Shalgun–Yelimsi Fault; 275°, 80°, 180° for the Fault 3 (F3), known as the South Bozgush Fault; and 190°, 80°, 0° for the Fault 4 (F4). For each selected receiver fault mechanism, we tested two effective friction coefficients (i.e., 0.4 and 0.6) (Figure 5). We verified that this latter coefficient (i.e., $\mu'$ = 0.6) only slightly modifies the Coulomb stress change map. Given the mechanism of the source fault (strike = 29°, dip = 80°, rake = −4.9°), the receiver faults having the larger portion falling in a positive CFF area are F1 and F3 (Figure 5a,b,e,f and Figure S3). However, this does not preclude the activation of the other two faults F2 and F4 (Figure 5c,d,g,h and Figure S3). Overall, our results indicate that the main event may have encouraged (i.e., positively stressed), with a positive loading, the activation of all the considered receiver faults. This is confirmed also by the distribution of the aftershocks (black dots in Figure 5) that occurred in proximity or exactly on the considered faults.

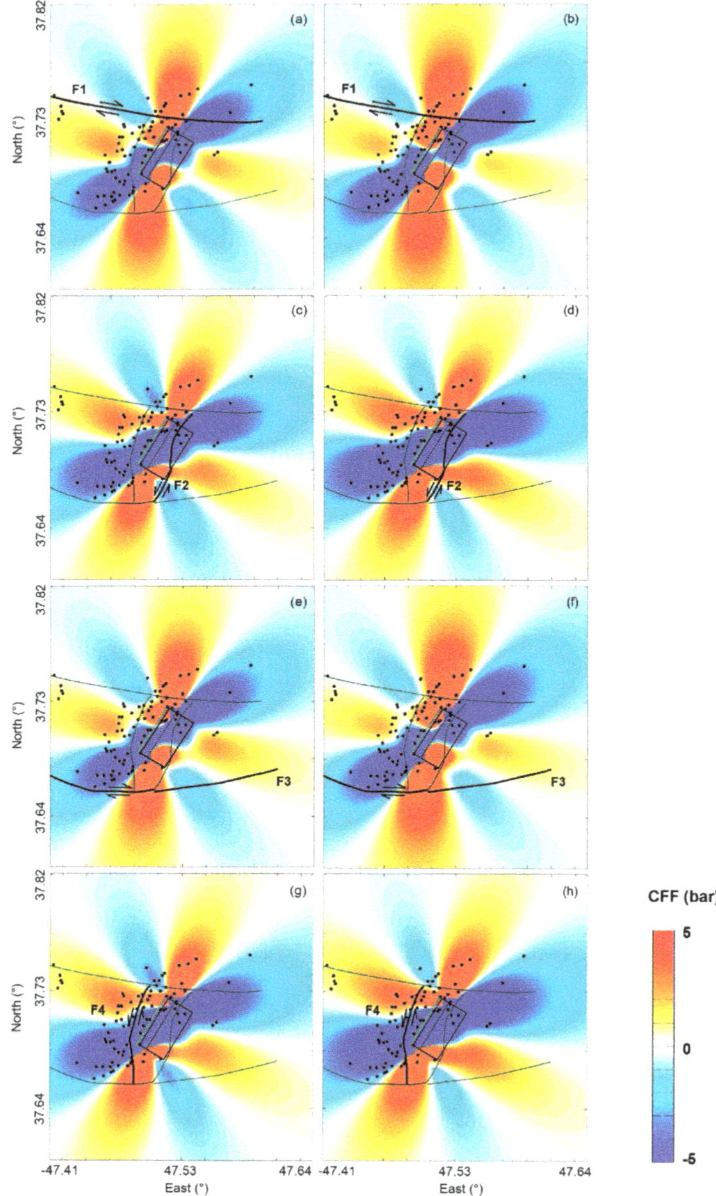

**Figure 5.** Coulomb Failure Function. Coulomb stress change maps computed at a reference depth of 5 km. For each selected receiver fault (F1, F2 = Shalgun–Yelimsi Fault, F3 = South Bozgush Fault and F4) mechanism, we tested two effective friction coefficients: $\mu' = 0.4$ reported in panels (**a,c,e,g**), and $\mu' = 0.6$ in panels (**b,d,f,h**). The aftershocks recorded during the period 7 November–24 December 2019 are reported with black dots. Solid lines represent the receiver faults on which the Coulomb stress change is calculated, whereas lines are not involved in that specific computation. All the reported geological structures are derived from [8].

## 5. Discussion

Starting from the retrieved modelling results, we suggest that the seismic source responsible for the 7 November 2019, $M_w$ 5.9, E-Azerbaijan earthquake consists of a predominantly left-lateral strike-slip fault (strike = 29.17°, dip = 79.29°, rake = −4.94°), with a center depth at about 3 km, located within the complex structural setting of the Bozgush Range [7,8]. By comparing our solution with the different focal mechanisms and hypocentral locations provided by the IRSC, the USGS, the GEOFON and the CMT, we remark that the geometric parameters of the seismogenic source presented in this study mostly agree with the fault plane solutions proposed by the IRSC. Our model indicates a slip distribution (retrieved maximum slip of 0.80 m), mainly located near the ground surface, which can explain the displacement field retrieved by the S1 DInSAR measurements; therefore, we can suggest that our solution can be considered compatible with a shallow source of the earthquake occurring in the complex tectonic setting of the Bozgush Range [7,8]. Furthermore, by analyzing the hypocentral distribution of the earthquakes nucleated after the mainshock and until 24 December 2019, we remark that the seismicity shows an evident southeast-dipping high angle alignment, which is in good agreement with our modelled fault plane (Figure 4). Moreover, the aftershocks distribution shows a good relationship with the CFF values computed along the surrounding structures [35] (Figure 5); indeed, also without accounting for the specific fault mechanism, about 35% of the aftershocks are located in positive CFF areas and, therefore, we can suggest that their origin can be attributable to the static stress transfer.

According to the retrieved results, and by comparing our modelled fault with the detailed structural framework reported in Faridi et al. [8], we suggest that our solution reveals a minor fault located west of the Shalgun–Yelimsi Fault and not mapped in the geological maps available in the open literature [7,8,21,26], whose kinematics is compatible with that of the surrounding structures and with the local and regional stress states (Figure 4). Moreover, Zamani and Masson [7] and Faridi et al. [8] have furnished a detailed reconstruction of the subsurface geology of the considered seismogenic area and have produced some geological sections of the examined region. Starting from these sections, we can suggest that the geometry and characteristics of the retrieved source are in good agreement with those of the represented adjacent faults.

The occurrence of the analyzed seismic event along a left-lateral structure is also confirmed by the fieldwork, the photogrammetry and the drone study, performed by a team of the Northwest branch and the Seismotectonics and Seismology Department of Geological Survey of Iran [21]. In addition, as presented in [21], the reported failures caused by the mainshock nucleation are located in proximity of our modelled structure and correspond mostly to some dynamic surface and slope instabilities (e.g., centimetric cracks, rockfalls, landslides, stone jumping features, and change in spring water-colors), which have been especially concentrated within the hangingwall of the causative fault zone. Moreover, we remark that the fractures generated by this event, as reported in [21], are secondary failures compatible with the kinematics of our modelled fault (Figure 4b). We further remark that, if we should consider as seismogenic source a structure consistent with the reported location and orientation of the Shalgun–Yelimsi Fault [8], the geodetic inversion of the exploited DInSAR measurements would result in a best-fit solution whose residuals are significantly worse than those achieved for our model (Figure S4). This is an additional confirmation of the validity of our findings.

The retrieved fault solution can be related with the reconstruction of the stress states performed by Zamani and Masson [7]; in particular, the complex active tectonics of the studied E-Azerbaijan region is characterized by several compressive structures, such as the Arasbaran, the Ghoshe Dagh and the Bozgush fold-and-thrust-belts, which have been generated by two distinct compressional stress systems, with NE–SW and NW–SE directions. The evolution of these compressive structures have also been associated with considerable Neogene to Quaternary volcanic activity and NE–SW and NW–SE strike-slip faults. The origin and kinematics of the local left-lateral strike-slip faults, such as the causative fault of the considered E-Azerbaijan earthquake, and the related seismicity, can be linked to the accommodation of the nearly N–S shortening between the Arabian and Eurasian plates, with the subsequent eastward extrusion of regional crustal scale blocks [42,43].

## 6. Conclusions

In this work, we have investigated the $M_w$ 5.9 E-Azerbaijan (NW Iran) earthquake, in order to retrieve the causative fault of this seismic event. To this aim, we have exploited the available DInSAR measurements obtained by processing the SAR data collected by the Sentinel-1 constellation along ascending and descending orbits. Moreover, we have applied an analytical modelling two-step approach to the computed coseismic DInSAR displacements, in order to better constrain the kinematics of the main source.

Our main findings can be summarized as follows:

- The source model reveals a rather shallow seismic structure approximately NE–SW-striking and characterized by a left-lateral strike-slip, southeast-dipping faulting mechanism. The retrieved source reveals a minor fault not mapped in the geological maps available in the open literature, but it is characterized by a kinematics compatible with that of the surrounding structures, the local and regional stress states and with some of the field observations.
- Starting from the retrieved fault model characteristics and by considering the known surrounding geological structures, we have performed an analysis of the Coulomb stress transfer on the nearby faults, in order to investigate possible fault interaction processes. Our results indicate that the considered receiver faults may have been positively stressed by the main event and this is confirmed by the aftershocks distribution.
- The analysis of the seismic events nucleated along the left-lateral strike-slip minor faults of the East-Azerbaijan Plateau, such as the one analyzed in this work, is essential to improve our knowledge of the seismic hazard estimation in northwestern Iran.

**Supplementary Materials:** The following are available online at http://www.mdpi.com/2072-4292/12/8/1346/s1, Figure S1: Sentinel-1 DInSAR measurements, Figure S2: The performance analysis relevant to the source model, Figure S3: Coulomb stress change sections, Figure S4: Shalgun-Yelimsi fault modelling results.

**Author Contributions:** Conceptualization, E.V., V.D.N.; methodology, E.V., F.C., V.C., C.D.L., V.D.N., R.L., M.M., M.M.; validation, V.C., V.D.N., R.L.; data curation, E.V., C.D.L., F.M.; writing—original draft preparation, E.V., V.D.N., R.L., M.M.; visualization, E.V., V.D.N.; supervision, V.D.N., R.L., M.M. All authors have read and agreed to the published version of the manuscript.

**Funding:** This research received no external funding.

**Acknowledgments:** This work has been supported by the 2019–2021 IREA-CNR and Italian Civil Protection Department agreement, H2020 EPOS-SP (GA 871121), ENVRI-FAIR (GA 824068) projects, the I-AMICA project (Infrastructure of High Technology for Environmental and Climate Monitoring-PONa3_00363). The Sentinel-1 data have been furnished through the Copernicus Program of the European Union. The DEM of the investigated zone was acquired through the SRTM archive. We thank Simone Atzori for his valuable suggestions.

**Conflicts of Interest:** The authors declare no conflict of interest.

## References

1. Broadband Iranian Network. Available online: http://www.iiees.ac.ir/en/iranian-national-broadband-seismic-network/ (accessed on 15 January 2020).
2. Iranian Seismological Center. Available online: http://irsc.ut.ac.ir/ (accessed on 15 January 2020).
3. U.S. Geological Survey. Earthquake Facts and Statistics. 2017. Available online: https://earthquake.usgs.gov/earthquakes/browse/stats.php (accessed on 22 March 2018).
4. GEOFON Data Centre. GEOFON Seismic Network. Deutsches GeoForschungsZentrum GFZ. *Other/Seism. Netw.* **1993**. [CrossRef]
5. Dziewonski, A.M.; Chou, T.A.; Woodhouse, J.H. Determination of earthquake source parameters from waveform data for studies of global and regional seismicity. *J. Geophys. Res.* **1981**, *86*, 2825–2852. [CrossRef]

6.	Ekström, G.; Nettles, M.; Dziewonski, A.M. The global CMT project 2004–2010: Centroid-moment tensors for 13,017 earthquakes. *Phys. Earth Planet. Inter.* **2012**, *200–201*, 1–9. [CrossRef]

7.	Zamani, B.; Masson, F. Recent tectonics of East (Iranian) Azerbaijan from stress state reconstructions. *Tectonophysics* **2014**, *611*, 61–82. [CrossRef]

8.	Faridi, M.; Burg, J.P.; Nazari, H.; Talebian, M.; Ghorashi, M. Active faults pattern and interplay in the Azerbaijan region (NW Iran). *Geotectonics* **2017**, *51*, 428–437. [CrossRef]

9.	Berberian, M.; Yeats, R.S. Patterns of historical earthquake rupture in the Iranian Plateau. *Bull. Seism. Soc. Am.* **1999**, *89*, 120–139.

10.	Jackson, J.; McKenzie, D. Active tectonics of the Alpine-Himalayan Belt between western Turkey and Pakistan. *Geophys. J. Int.* **1984**, *77*, 185–264. [CrossRef]

11.	Jackson, J.; McKenzie, D. The relationship between plate motions and seismic moment tensors, and the rates of active deformation in the Mediterranean and Middle East. *Geophys. J. Int.* **1988**, *93*, 45–73. [CrossRef]

12.	Jackson, J. Partitioning of strike-slip and convergent motion between Eurasia and Arabia in eastern Turkey and Caucasus. *J. Geophys. Res.* **1992**, *97*, 12471–12479. [CrossRef]

13.	McClusky, S.; Balassanian, S.; Barka, A.; Demir, C.; Ergintav, S.; Georgiev, I.; Gurkan, O.; Hamburger, M.; Hurst, K.; Kahle, H.; et al. Global Positioning System constraints on plate kinematics and dynamics in the eastern Mediterranean and Caucasus. *J. Geophys. Res. Solid Earth* **2000**, *105*, 5695–5719. [CrossRef]

14.	Vernant, P.; Nilforoushan, F.; Hatzfeld, D.; Abbassi, M.R.; Vigny, C.; Masson, F.; Nankali, H.; Martinod, J.; Ashtiani, A.; Bayer, R.; et al. Present-day crustal deformation and plate kinematics in the Middle East constrained by GPS measurements in Iran and northern Oman. *Geophys. J. Int.* **2004**, *157*, 381–398. [CrossRef]

15.	Copley, A.; Jackson, J. Active tectonics of the Turkish–Iranian Plateau. *Tectonics* **2006**, *25*. [CrossRef]

16.	Reilinger, R.; McClusky, S.; Vernant, P.; Lawrence, S.; Ergintav, S.; Cakmak, R.; Ozener, H.; Kadirov, F.; Guliev, I.; Stepanyan, R.; et al. GPS constraints on continental deformation in the Africa–Arabia–Eurasia continental collision zone and implications for the dynamics of plate interactions. *J. Geophys. Res. Solid Earth* **2006**, *111*. [CrossRef]

17.	Copley, A.; Faridi, M.; Ghorashi, M.; Hollingsworth, J.; Jackson, J.; Nazari, H.; Oveisi, B.; Talebian, M. The 2012 August 11 Ahar earthquakes: Consequences for tectonics and earthquake hazard in the Turkish–Iranian Plateau. *Geophys. J. Int.* **2013**, *196*, 15–21. [CrossRef]

18.	Jackson, J.; Haines, J.; Holt, W. The accommodation of Arabia-Eurasia plate convergence in Iran. *J. Geophys. Res. Solid Earth* **1995**, *100*, 15205–15219. [CrossRef]

19.	Engdahl, E.R.; van der Hilst, R.; Buland, R. Global teleseismic earthquake relocation with improved travel times and procedures for depth determination. *Bull. Seism. Soc. Am.* **1998**, *88*, 722–743.

20.	Engdahl, E.R.; Jackson, J.A.; Myers, S.C.; Bergman, E.A.; Priestley, K. Relocation and assessment of seismicity in the Iran region. *Geophys. J. Int.* **2006**, *167*, 761–778. [CrossRef]

21.	Solaymani Azad, S.; Esmaeili, C.; Roustai, M.; Vajedian, S.; Sartipi, A.; Khosh Zare, T.; Rajab Zadeh, H.R. The Geological features of the Torkmanchai NW Iran Earthquake on November 8, 2019 (Mw=5.9). *Geol. Surv. Miner. Exploit. Iran* **2019**. Available online: https://www.gsi.ir/en/news/24204/the-geological-features-of-the-torkmanchai-nw-iran-earthquake-on-november-8-2019-mw-5.9- (accessed on 4 March 2020).

22.	Major Faults in Iran (flt2cg), U.S. Government's Open Data. Available online: https://catalog.data.gov/dataset/major-faults-in-iran-flt2cg-73f76 (accessed on 1 April 2020).

23.	Djamour, Y.; Vernant, P.; Nankali, H.; Tavakoli, F. NW Iran–eastern Turkey present-day kinematics: Results from the Iranian permanent GPS network. *Earth Planet. Sci. Lett.* **2011**, *307*, 27–34. [CrossRef]

24.	Berberian, M.; Arshadi, S. On the evidence of the youngest activity of the North Tabriz Fault and the seismicity of Tabriz city. *Geol. Surv. Iran Rep.* **1976**, *39*, 397–418.

25.	Ambraseys, N.N.; Melville, C.P. *A History of Persian Earthquakes*; Cambridge University Press: Cambridge, UK, 2005.

26.	Solaymani Azad, S.; Philip, H.; Dominguez, S.; Hessami, K.; Shahpasandzadeh, M.; Foroutan, M.; Tabassi, H.; Lamothe, M. Paleoseismological and morphological evidence of slip rate variations along the North Tabriz fault (NW Iran). *Tectonophysics* **2015**, *640–641*, 20–38. [CrossRef]

27.	Massonnet, D.; Rossi, M.; Carmona, C.; Adragna, F.; Peltzer, G.; Feigl, K.; Rabaute, T. The displacement field of the Landers earthquake mapped by radar interferometry. *Nature* **1993**, *364*, 138. [CrossRef]

28.	Fornaro, G.; Franceschetti, G.; Lanari, R.; Rossi, D.; Tesauro, M. Interferometric SAR phase unwrapping using the finite element method. *IEE Proc. Radar Sonar Navig.* **1997**, *144*, 266–274. [CrossRef]

29. Costantini, M. A novel phase unwrapping method based on network programming. *IEEE Trans. Geosci. Remote Sens.* **1998**, *36*, 813–821. [CrossRef]

30. Atzori, S.; Hunstad, I.; Chini, M.; Salvi, S.; Tolomei, C.; Bignami, C.; Stramondo, S.; Trasatti, E.; Antonioli, A.; Boschi, E. Finite fault inversion of DInSAR coseismic displacement of the 2009 L'Aquila earthquake (Central Italy). *Geophys. Res. Lett.* **2009**, *36*, L15305. [CrossRef]

31. Okada, Y. Surface deformation due to shear and tensile faults in a half-space. *Bull. Seism. Soc. Am.* **1985**, *75*, 1135–1154.

32. Marquardt, D.W. An algorithm for least-squares estimation of nonlinear parameters. *J. Soc. Ind. Appl. Math.* **1963**, *11*, 431–441. [CrossRef]

33. Menke, W. *Geophysical Data Analysis: Discrete Inverse Theory*, 1st ed.; Academic Press Inc.: New York, NY, USA, 1984; ISBN 9780080507323.

34. Harris, R.A. Introduction to Special Section: Stress Triggers, Stress Shadows, and Implications for Seismic Hazard. *J. Geophys. Res. Solid Earth* **1998**, *103*, 24347–24358. [CrossRef]

35. Stein, R.S. The role of stress transfer in earthquake occurrence. *Nature* **1999**, *402*, 605–609. [CrossRef]

36. Hill, D.P. Dynamic stresses, Coulomb failure, and remote triggering. *Bull. Seism. Soc. Am.* **2008**, *98*, 66–92. [CrossRef]

37. Stein, R.; King, G.C.P.; Lin, J. Change in failure stress on the southern San Andreas fault system caused by the 1992 magnitude 7.4 Landers earthquake. *Science* **1992**, *258*, 1328–1332. [CrossRef] [PubMed]

38. King, G.C.P.; Stein, R.S.; Lin, J. Static stress changes and the triggering of earthquakes. *Bull. Seismol. Soc. Amer.* **1994**, *84*, 935–953.

39. Reasenberg, P.; Simpson, R. Response of regional seismicity to the static stress change produced by the Loma Prieta Earthquake. *Science* **1992**, *255*, 1687–1690. [CrossRef] [PubMed]

40. Lin, J.; Stein, R.S. Stress triggering in thrust and subduction earthquakes, and stress interaction between the southern San Andreas and nearby thrust and strike-slip faults. *J. Geophys. Res.* **2004**, *109*, B02303. [CrossRef]

41. Toda, S.; Stein, R.S.; Richards-Dinger, K.; Bozkurt, S. Forecasting the evolution of seismicity in southern California: Animations built on earthquake stress transfer. *J. Geophys. Res.* **2005**, *110*. [CrossRef]

42. Su, Z.; Yang, Y.; Li, Y.; Xu, X.; Zhang, J.; Zhou, X.; Ren, J.; Wang, E.; Hu, J.-C.; Zhang, S.; et al. Coseismic displacement of the 5 April 2017 Mashhad earthquake (Mw 6.1) in NE Iran through Sentinel-1A TOPS data: New implications for the strain partitioning in the southern Binalud Mountains. *J. Asian Earth Sci.* **2019**, *169*, 244–256. [CrossRef]

43. Ghods, A.; Shabanian, E.; Bergman, E.A.; Faridi, M.; Donner, S.; Mortezanejad, G.; Zanjani, A.A. The Varzaghan–Ahar, Iran, Earthquake Doublet (Mw 6.4, 6.2): Implications for the geodynamics of northwest Iran. *Geophys. J. Int.* **2015**, *203*, 522–540. [CrossRef]

*Article*

# New Insights into Long-Term Aseismic Deformation and Regional Strain Rates from GNSS Data Inversion: The Case of the Pollino and Castrovillari Faults

Gabriele Cambiotti [1,*], Mimmo Palano [2], Barbara Orecchio [3,†], Anna Maria Marotta [1,†], Riccardo Barzaghi [4,†], Giancarlo Neri [3,†] and Roberto Sabadini [1,†]

[1] Department of Earth Sciences, Università degli Studi di Milano, 20133 Milan, Italy; anna.maria.marotta@unimi.it (A.M.M.); roberto.sabadini@unimi.it (R.S.)
[2] Istituto Nazionale di Geofisica e Vulcanologia, Osservatorio Etneo, 95125 Catania, Italy; mimmo.palano@ingv.it
[3] Department of Physics and Earth Sciences, Messina University, 98166 Messina, Italy; orecchio@unime.it (B.O.); geoforum@unime.it (G.N.)
[4] Politecnico di Milano, DICA, 20133 Milan, Italy; riccardo.barzaghi@polimi.it
* Correspondence: gabriele.cambiotti@unimi.it
† These authors contributed equally to this work.

Received: 30 June 2020; Accepted: 6 September 2020; Published: 9 September 2020

**Abstract:** We present a novel inverse method for discriminating regional deformation and long-term fault creep by inversion of GNSS velocities observed at the spatial scale of intraplate faults by exploiting the different spatial signatures of these two mechanisms. In doing so our method provides a refined estimate of the upper bound of the strain accumulation process. As case study, we apply this method to a six year GNSS campaign (2003–2008) set up in the southern portion of the Pollino Range over the Castrovillari and Pollino faults. We show that regional deformation alone cannot explain the observed deformation pattern and implies high geodetic strain rate, with a WSW-ENE extension of $86 \pm 41 \times 10^{-9}$/yr. Allowing for the possibility of fault creep, the modelling of GNSS velocities is consistent with their uncertainties and they are mainly explained by a shallow creep over the Pollino fault, with a normal/strike-slip mechanism up to 5 mm/yr. The regional strain rate decrease by about 70 percent and is characterized by WNW-ESE extension of $24 \pm 28 \times 10^{-9}$/yr. The large uncertainties affecting our estimate of regional strain rate do not allow infering whether the tectonic regime of the area is extensional or strike-slip, although the latter is slightly more likely.

**Keywords:** regional deformation; fault creep; GNSS velocities; inverse theory

## 1. Introduction

Geodetic strain rates from GNSS observations, once compared with past and ongoing seismicity, provide a first estimate of the amount of deformation that is accommodated by earthquakes [1–3]. This comparison, nevertheless, still leaves open the issue of whether the difference between seismic and GNSS deformation is accommodated by aseismic phenomena or by elastic accumulation preparing the next earthquake [4]. In this respect, geodetic strain rates and earthquake catalogs can only provide an upper bound for the building of the elastic stress.

The main seismic and aseismic phenomena occur at the plate boundary and interseismic deformation from GNSS network are usually modelled by the back-slip approach [5,6] and within the framework of elastic or viscoelastic Earth models [7,8]. These models allow estimating the spatial distribution of the interseismic locking of the interplate faults and, therefore, quantify their elastic loading during the seismic cycle. Considering the large spatial scales of the interseismic deformation at plate boundaries, even sparse GNSS networks can provide valuable information, but difficulties arise in discriminating transient and (long-term) inter seismic deformations due to the limited time intervals spanned by the GNSS data time series [7].

Conversely, interseismic deformation within the plates are the results of elastic deformation mainly caused by the relative motion between plates and by inelastic deformations such as brittle fractures and plastic and ductile flows [9]. The treatment of the lithosphere as a viscous fluid is widely adopted in models of long timescale geological processes. In particular, over geological times and regional spatial scales, this complexity can be taken into account by means of an effective viscosity representing the vertical averaged strength of the lithosphere, as in Splendore and Marotta [10]. This effective viscosity is a key parameter within a thin sheet modelling [11], which allows estimating the surface velocities and the geodetic strain rates due to active tectonics and to provide a first understanding of the average intraplate stress e.g., [12–15]. These insights are appropriate for the modelling of steady-state tectonic deformation and can be hardly translated into the understanding of how strain is accumulated and released at the time scales of the seismic cycle and at the spatial scales of individual faults. Indeed, by definition, viscoelastoplastic Earth models filter out any short-term and small-scale fluctuations of the elastic stress field.

Due to the these difficulties in dealing with intraplate deformations at small time and spatial scales, GNSS observations are usually used to investigate coseismic and postseismic deformations e.g., [16,17] or transient deformations due to temporary changes in the fault creep [18,19], at least when these phenomena yield deformations high enough to be discriminated from long-term trends. In light of this, by fitting linear trends to GNSS data time series, these approaches always avoid the physical modelling of long-term interseismic deformation and, so, they lose the possibility of understanding how much of this deformation is accommodated by long-term tectonics or by steady-state fault creep or, at least, fault-creep lasting more than the time interval spanned by GNSS data. In other words, we are left with the possibility that the whole amount of long-term strain rate inferred from the linear trends at different GNSS stations is building elastic stress and, then, preparing the next earthquakes, when instead some fraction of it could be caused by inelastic processes that we are actually able to model.

To remedy to this lack of the previous approaches and further refine the upper bound for the elastic loading inferred from geodetic strain rates, we develop a novel approach for modelling GNSS velocities that allow estimating long-term creep over intraplate faults by taking advantages of the different spatial signatures of the main physical mechanism responsible for crustal displacements. As case study, we consider the area of the Pollino Range, which represents the most prominent seismic gap within the southern Apennines [20] and, so, likely affected by aseismic processes able to counteract the intraplate elastic loading. In particular, we focus our attention on the southern area of the Pollino Range and part of the Sybari plain. As shown in Figure 1, this area is crossed by two main faults: the Pollino master fault along the southern slope of Mount Pollino massif and the Castrovillari second-order cross fault [21]. While the whole area of the Pollino Range has been affected only by moderate seismicity in the range of magnitude 5.5∼6 in the last centuries [22], its southern portion is characterized by an even lower seismic activity (see Figure 2). Moreover, also the recent 2010–2014 seismic swarm (main shock $M_W$ 5.1) accompanied by a significant transient creep (equivalent to $M_W$ 5.5) has affected the Pollino Range a little further north, activating the Rotonda-Campotenese normal fault system and the Morano Calabro-Piano di Ruggio fault [18], but not its most southern area. The distinctive behaviour of this area is also supported by seismic tomography [23] which has identified it as a low P- and S-wave velocity zone, compared to the northern part of the Pollino Range and the further south Mount Sila massif (see Figure 2), and by the correlation between the high seismicity in high P-wave velocity zones and low seismicity in low P- and S-wave velocity zones. Based on the regional pattern of P-wave receiver functions, seismicity and SKS anisotropy, Chiarabba et al. [24] have suggested that these faults decouple the deformation across the region, from the one driven by the delamination of the southern Apennines to the one related to the retreat mechanism in the Calabrian Subduction System fore arc.

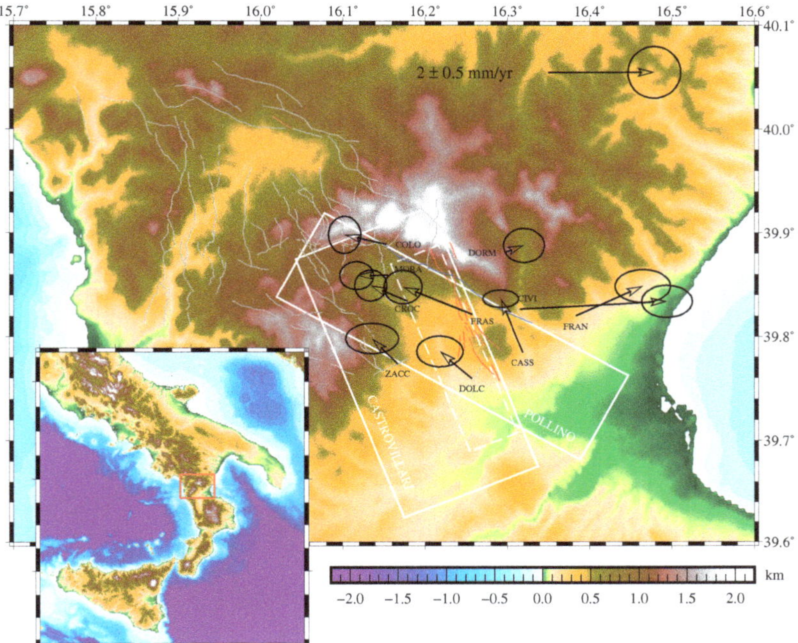


**Figure 1.** Surface velocities from the GNSS campaigns (in Apulia fixed reference frame, black arrows) from Sabadini et al. [25], the fault traces (gray, red and blue lines) identified by Brozzetti et al. [26] and the Castrovillari and Pollino faults used in this study (white solid rectangles). Red and blue lines are the fault traces associated with the Castrovillari and Pollino faults, respectively, and the white dashed rectangle indicates the Castrovillari fault from the Database of Individual Seismogenic Sources [27]. The top edge of the DISS fault is slightly shifted southwestward because is at 1 km depth rather than at the Earth surface. The topography of the area is shown in the background and the inset shows the south Italy and the location of the plotted region (red rectangle).

Geomorphic and trenching investigations identified at least six surface-faulting earthquakes of magnitude 6.5~7 since late Pleistocene age on the Castrovillari and Pollino faults, with the most recent on the Pollino fault and dated in between the XIII and XV century A.D., although no evidence has been found in the historical records Figure 2 [20,21]. Therefore both faults proved to be capable of generating earthquakes of magnitude up to the typical maximum magnitude of Calabria and of the southern Apennines, as large as 7. It is thus important to determine whether aseismic processes are playing a role in inhibiting the seismic potential of these faults or rather to increase it.

The possibility of fault creep on the Castrovillari fault has been already proposed by Sabadini et al. [25] on the basis of surface horizontal velocities estimated at 10 sites by a six year campaign GNSS network from 2003 up to 2008 (see Figure 1). Starting from this previous study, we hereinafter make a step ahead in the modelling of GNSS velocities (i) allowing for the possibility of creep even on the nearby Pollino fault, which could have affected the observed surface velocity as well, and (ii) developing a new inverse method able to discriminate the local contribution due to these creeping faults from the regional one due to large scale intraplate deformations. By jointly estimating the fault creep and the regional deformation from GNSS data inversion, we then aim at defining a more realistic upper bound for the elastic deformation of the southern part of the Pollino Range than that directly provided by geodetic strain rates.

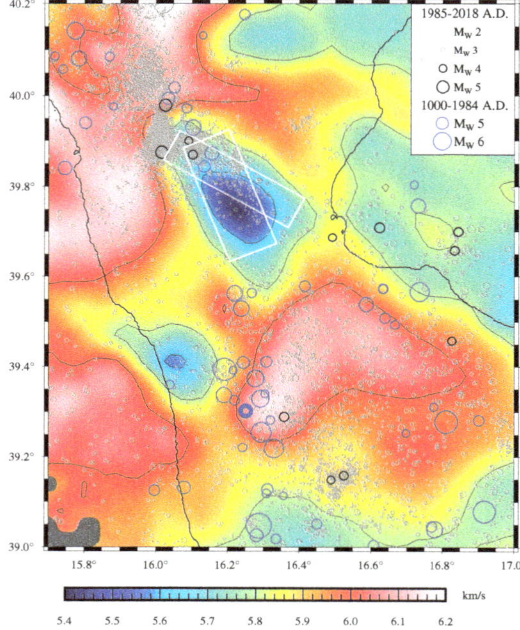

Figure 2. Historical (from 1000 to 1984, blue circles) and recent (from 1895 to present, gray and black circles) earthquakes from the CPTI11 and ISIDe catalogs [28,29]. The velocities of the seismic P-waves at 9 km depth from Totaro et al. [23] are shown in the background.

## 2. Forward Model

We assume that the investigated region is elastic but for the possibility of inelastic deformation localized on the Castrovillari and Pollino faults, and decompose the horizontal velocities **v** at the Earth surface as follows

$$\mathbf{v}(\phi,\theta) = \begin{pmatrix} v_\phi(\phi,\theta) \\ v_\theta(\phi,\theta) \end{pmatrix} = \mathbf{v}^{\mathrm{reg}}(\phi,\theta) + \mathbf{v}^{\mathrm{creep}}(\phi,\theta) \tag{1}$$

where $\phi$ and $\theta$ are the longitude and latitude at the observation point, $v_\phi$ and $v_\theta$ are the east and north components of the velocity field, and $\mathbf{v}^{\mathrm{reg}}$ and $\mathbf{v}^{\mathrm{creep}}$ are the velocities due to regional deformation and fault creep, respectively. For the sake of simplicity, we do not discuss the up component of the velocity because it is not used in the following inversion of the GNSS data by Sabadini et al. [25].

We note that not all the regional deformation must be elastic simply because aseismic processes other than those here considered could contribute to it. In this respect and as already argued in Section 1, the above decomposition shall be regarded as a conservative choice aiming at refining the upper bound for elastic deformation. If aseismic processes other than those on the faults here considered are modelled, this upper bound will be further refined.

### 2.1. Regional Deformation

Following Savage et al. [30], within the assumption that the regional deformation varies on spatial scales larger than those of the investigated region and that the topography elevation is small compared to the horizontal distances involved, the horizontal velocity due to regional deformation can be expanded in Taylor series at a reference point $(\phi_0, \theta_0)$ as follows

$$\mathbf{v}^{\text{reg}}(\phi,\theta) = \mathbf{W}(\phi,\theta)\,\mathbf{w} + \mathbf{E}(\phi,\theta)\,\mathbf{e} \tag{2}$$

where $\mathbf{W}$, $\mathbf{E}$, $\mathbf{w}$ and $\mathbf{e}$ are the following $2 \times 3$–dimensional matrices and 3–dimensional arrays

$$\mathbf{W}(\phi,\theta) = \begin{pmatrix} 1 & \sin\theta_0\,d\phi & -a\,d\theta \\ -\sin\theta_0\,d\phi & 1 & a\cos\theta_0\,d\phi \end{pmatrix}, \qquad \mathbf{E}(\phi,\theta) = a \begin{pmatrix} \cos_0\theta_0\,d\phi & 0 & d\theta \\ 0 & d\theta & \cos\theta_0\,d\phi \end{pmatrix},$$

$$\mathbf{w} = \begin{pmatrix} v_\phi^{\text{reg}}(\phi_0,\theta_0) \\ v_\theta^{\text{reg}}(\phi_0,\theta_0) \\ \omega \end{pmatrix}, \qquad \mathbf{e} = \begin{pmatrix} \epsilon_{\phi\phi} \\ \epsilon_{\theta\theta} \\ \epsilon_{\theta\phi} \end{pmatrix} \tag{3}$$

Here $a$ is the Earth radius, $\epsilon_{\alpha\beta}$ (with $\alpha,\beta = \phi,\theta$) are the horizontal components of the regional strain rate in the geographic reference system, $\omega$ is the regional rotation rate and

$$d\phi = \phi - \phi_0, \qquad d\theta = \theta - \theta_0 \tag{4}$$

In light of this, the regional horizontal velocity field depends on six unknowns, $\mathbf{w}$ and $\mathbf{e}$. We note that the parameters $\mathbf{w}$ describe the rigid-body motion and are equivalent to the three rotation parameters defining the Euler vector [30,31].

In view of the assumptions of elastic body and traction free at Earth surface used to obtain Equation (2), we note also that the vertical components of the regional strain rates read

$$\epsilon_{r\phi} = 0, \qquad \epsilon_{r\theta} = 0, \qquad \epsilon_{rr} = -\frac{\lambda}{\lambda + 2\mu}\left(\epsilon_{\phi\phi} + \epsilon_{\theta\theta}\right) \tag{5}$$

with $\mu$ and $\lambda$ being the two Lamé parameters. In the perspective of linking the regional strain rates to the tectonic regimes, it is convenient to express them in terms of the deviatoric components

$$\epsilon'_{\alpha\beta} = \epsilon_{\alpha\beta} - \frac{1}{3}\delta_{\alpha\beta}\left(\epsilon_{\phi\phi} + \epsilon_{\theta\theta} + \epsilon_{rr}\right) \qquad \forall\,\alpha,\beta = \phi,\theta,r \tag{6}$$

with $\delta_{\alpha\beta}$ being the Kronecker delta. After some straightforward algebra, we write

$$\mathbf{e} = \mathbf{D}\,\mathbf{q} \tag{7}$$

where

$$\mathbf{D} = \begin{pmatrix} 1-\gamma & -\gamma & 0 \\ -\gamma & 1-\gamma & 0 \\ 0 & 0 & 1 \end{pmatrix}, \qquad \mathbf{q} = \begin{pmatrix} \epsilon'_{\phi\phi} \\ \epsilon'_{\theta\theta} \\ \epsilon'_{\theta\phi} \end{pmatrix} \tag{8}$$

with $\gamma = 2/3\,(1 - \alpha)$ and $\alpha = (\lambda + \mu)/(\lambda + 2\,\mu)$. In light of this, Equation (2) can be recast in the following form

$$\mathbf{v}^{\text{reg}}(\phi,\theta) = \mathbf{W}(\phi,\theta)\,\mathbf{w} + \mathbf{Q}(\phi,\theta)\,\mathbf{q} \tag{9}$$

with

$$\mathbf{Q}(\phi,\theta) = \mathbf{E}(\phi,\theta)\,\mathbf{D} \tag{10}$$

For the GNSS data inversion, we will assume that the elastic medium is Poissonian, that is $\lambda = \mu$.

### 2.2. Fault Creep

As it concerns the horizontal velocity due to fault creep, it can be computed from the elastostatic Green functions for tangential dislocations e.g., [32]

$$
\mathbf{v}^{\mathrm{creep}}(\phi,\theta) = \sum_{p=1}^{2} \sum_{i=1}^{2} \int_{\mathcal{F}_p} \mathbf{g}_i(\phi,\theta;\mathbf{x})\, s_i^{(p)}(\mathbf{x})\, dS(\mathbf{x})
\tag{11}
$$

where $\mathcal{F}_p$ and $s_i^{(p)}$ are the surfaces and the strike-slip ($i=1$) and up-dip ($i=2$) aseismic slip rates for the Castrovillari ($p=1$) and Pollino ($p=2$) faults, $\mathbf{x}$ and $dS$ are the vector position identifying the points of the fault surfaces and the infinitesimal surface area, and $\mathbf{g}_i$ is the elastostatic Green function for the strike-slip ($i=1$) and up-dip ($i=2$) point-like unit slip.

As depicted in Figure 1 and listed in Table 1, we assume that the fault surfaces are planar and that the fault creep extend, at most, over rectangular areas of lengths $L_p$ and widths $W_p$, with $p=1$ and 2 for the Castrovillari and Pollino faults, respectively. We establish their geometry and position on the basis of the identified fault traces in the Pollino Range [26] and geomorphic and trenching investigations [20,21]. In particular, the rectangular areas extend up to the Earth surface and their top edges (i.e., the lines of strike) are chosen to approximate the fault traces identified by Brozzetti et al. [26]. Compared to previous published values [27,33], the length and, particularly, the width of the two faults have been enlarged in view of the fact that we will constrain the fault creep distribution to be zero at the internal edges of the faults and in the perspective of estimating its extension by the data inversion rather than prescribing it with an a priori choice of small fault surfaces.

**Table 1.** Fault parameter used in this study.

|  | Castrovillari | Pollino |
|---|---|---|
| dip | 60° | 70° |
| strike | 158° | 119° |
| length | $L_1 = 30\,\mathrm{km}$ | $L_2 = 36\,\mathrm{km}$ |
| width | $W_1 = 30\,\mathrm{km}$ | $W_2 = 30\,\mathrm{km}$ |
| top depth | 0 km | 0 km |

Following Yabuki and Matsu'ura [34], we parametrize the creep distributions over the fault surfaces using bicubic splines defined over $2 \times 2\,\mathrm{km}^2$ regular grids. In particular, we use bicubic splines modified in such a way that, at the left, right and bottom edges, the fault creep and its first-order spatial derivatives are zero while, at the top edge, only the first order directional derivative along dip is zero. The latter constraint allows the slip to reach the Earth surface and, at the same time, avoids anomalous results at this edge. We provide details about the regular grid and the bicubic splines in Appendix A.

Within this framework, the total number of modified bicubic splines for each fault is $N_p = (N_L^p - 1)\,N_W^p$, with $N_L^p$ and $N_W^p$ being the number of intervals of the regular grid along the strike, $l$, and dip, $w$, coordinates, and the fault creep distribution reads

$$
s_i^{(p)}(\mathbf{x}(l,w)) = \sum_{j=1}^{N_L^p - 1} \sum_{k=0}^{N_W^p - 1} s_{ijk}^{(p)}\, \widehat{\Phi}_{j,k}(l,w)
\tag{12}
$$

for $(l,w) \in [0,L_p] \times [0,W_p]$ and zero otherwise. Here, $\widehat{\Phi}_{j,k}(l,w)$ are the modified bicubic splines as defined in Appendix A, $s_{ijk}^{(p)}$ are the coefficients of the modified bicubic splines that, for each fault, we collect into the $S_p$–dimensional array $\mathbf{s}_p$, with $S_p = 2\,N_p$.

After substitution of Equation (12) into Equation (11), we thus obtain the linear relation between the velocities at the Earth surface and the fault creep coefficients

$$\mathbf{v}^{\text{creep}}(\phi,\theta) = \sum_{p=1}^{2} \sum_{i=1}^{2} \sum_{j=1}^{N_L^p-1} \sum_{k=-1}^{N_W^p-1} \mathbf{k}_{ijk}^{(p)}(\phi,\theta)\, s_{ijk}^{(p)} \tag{13}$$

with

$$\mathbf{k}_{ijk}^{(p)}(\phi,\theta) = \int_0^{L_p} \int_0^{W_p} \mathbf{g}_i(\phi,\theta;\mathbf{x}(l,w))\, \widehat{\Psi}_{j,k}(l,w)\, dl\, dw \tag{14}$$

In the following we shall also quantify the roughness of the fault creep distribution that we define as

$$X_S^p = \sum_{i=1}^{2} \int_0^{L_p} \int_0^{W_p} \left[ \left( \frac{\partial^2 s_i^{(p)}(\mathbf{x}(l,w))}{\partial l^2} \right)^2 + \left( \frac{\partial^2 s_i^{(p)}(\mathbf{x}(l,w))}{\partial w^2} \right)^2 + 2\left( \frac{\partial^2 s_i^{(p)}(\mathbf{x}(l,w))}{\partial l\, \partial w} \right)^2 \right] dl dw \tag{15}$$

and, as detailed in Yabuki and Matsu'ura [34] and Cambiotti et al. [35], we recast in the following bilinear form

$$X_S^p = \mathbf{s}_p^{\mathrm{T}}\, \mathbf{S}_p\, \mathbf{s}_p \tag{16}$$

where $\mathbf{S}_p$ are $S_p \times S_p$–dimensional matrices, the elements of which can be obtained by substituting Equation (12) into Equation (15).

## 3. Inverse Problem

Within the framework outlined in Section 2, Equations (1), (2), (7) and (13), the relation between data and model is linear and reads

$$\widetilde{\mathbf{y}} = \mathbf{y}(\mathbf{m}) + \mathbf{z} \tag{17}$$

where $\widetilde{\mathbf{y}}$ and $\mathbf{z}$ is the $Y$–dimensional arrays of the observed surface velocities and errors (with $Y$ being the number of GNSS data), and $\mathbf{y}$ is the $Y$–dimensional array of modelled surface velocities

$$\mathbf{y}(\mathbf{m}) = \mathbf{K}\,\mathbf{m} \tag{18}$$

with $\mathbf{m}$ being the $M$–dimensional array collecting all the model parameters (with $M = 6 + S_1 + S_2$)

$$\mathbf{m} = \begin{pmatrix} \mathbf{w} \\ \mathbf{q} \\ \mathbf{s}_1 \\ \mathbf{s}_2 \end{pmatrix} \tag{19}$$

and $\mathbf{K}$ being the $Y \times M$–dimensional data kernel matrix

$$\mathbf{K} = \begin{pmatrix} \mathbf{W} & \mathbf{Q} & \mathbf{K}_1 & \mathbf{K}_2 \end{pmatrix} \tag{20}$$

Here, $\mathbf{W}$, $\mathbf{Q}$ are $Y \times 3$–dimensional matrices and $\mathbf{K}_p$ are $Y \times S_p$–dimensional matrices ($p = 1, 2$), the row of which are obtained evaluating Equations (3), (10) and (14) at the geographic coordinates of the GNSS sites.

The model parameter space $\mathcal{M}$ is given by the Cartesian products of the parameter spaces of each subset of model parameters is $\mathcal{M} = \mathcal{W} \times \mathcal{Q} \times \mathcal{S}_1 \times \mathcal{S}_2 = \mathbb{R}^M$, with $\mathcal{W}, \mathcal{Q} = \mathbb{R}^3$ for $\mathbf{w}$ and $\mathbf{q}$, and $\mathcal{S}_p = \mathbb{R}^{S_p}$ for $\mathbf{s}_p$.

### 3.1. Information from Observations and a Priori Constraints

In addition to the model parameters, we shall also consider four hyper-parameters that we collect in the following 4-dimensional array

$$
\mathbf{h} = \begin{pmatrix} \alpha \\ \beta_1 \\ \beta_2 \\ \rho \end{pmatrix}
\tag{21}
$$

where $\alpha > 0$ and $\beta_p > 0$ weight information from observation and from a priori constriants on the roughness of $p$-th fault creep distribution [34,35]. We note that each fault is controlled by a different hyper-parameter because its roughness must not be necessarily connected with the roughness of other faults. The fourth hyper-parameter, $\rho > 0$, is introduced in this study and weights a priori constraints on the regional strain rates, which will be discussed in a while. The hyper-parameter space is simply $\mathcal{H} = \mathbb{R}^4_+$.

Following Yabuki and Matsu'ura [34] and Cambiotti et al. [35], the probability density function (PDF) for the data given the model parameters and hyper-parameters and the a priori PDF for the $p$-th fault creep distribution given the hyper-parameters can be written as follows

$$
f(\tilde{\mathbf{y}} \,|\, \mathbf{m}, \alpha) = \frac{1}{\sqrt{(2\,\pi\,\alpha)^M \, \|\mathbf{Y}\|}} \exp\left( -\frac{(\tilde{\mathbf{y}} - \mathbf{K}\,\mathbf{m})^{\mathrm{T}} \, \mathbf{Y}^{-1} \, (\tilde{\mathbf{y}} - \mathbf{K}\,\mathbf{m})}{2\,\alpha} \right)
\tag{22}
$$

$$
f(\mathbf{s}_p \,|\, \beta_p) = \sqrt{\frac{\|\mathbf{S}_p\|}{(2\,\pi\,\beta_p)^{S_p}}} \exp\left( -\frac{\mathbf{s}_p^{\mathrm{T}} \, \mathbf{S}_p \, \mathbf{s}_p}{2\,\beta_p} \right)
\tag{23}
$$

where $\mathbf{Y}$ is the covariance matrix of the observational errors. We recall that the inclusion of the hyper-parameter $\alpha$ allows a simple representation of observational and modelling errors and assume that they obey a Gaussian distribution with zero mean and covariance matrix $\mathbf{C} = \alpha\,\mathbf{Y}$, where $\alpha$ must be estimated as part of the inversion as an assessment of a posteriori uncertainties [35,36]. In particular, the latter can be interpreted as follows: there are modelling errors when $\alpha > 1$, they are negligible when $\alpha = 1$ and the physical model is over-parametrized (i.e., it fits the observations too well) when $\alpha < 1$.

We introduce also the a priori PDF for the deviatoric regional strain rates given the hyper-parameters, which we define as follows

$$
f(\mathbf{q} \,|\, \rho) = \sqrt{\frac{\|\mathbf{I}_2\|}{(2\,\pi\,\rho)^3}} \exp\left( -\frac{\mathbf{q}^{\mathrm{T}} \, \mathbf{I}_2 \, \mathbf{q}}{2\,\rho} \right)
\tag{24}
$$

where $\mathbf{I}_2$ is the $3 \times 3$–dimensional matrix such that the bilinear form $\mathbf{q}^{\mathrm{T}} \, \mathbf{I}_2 \, \mathbf{q}$ yields the second invariant of the deviatoric strain rate tensor, $I_2 = \epsilon'_{ij}\,\epsilon'_{ij}$,

$$
\mathbf{I}_2 = \begin{pmatrix} 2 & 1 & 0 \\ 1 & 2 & 0 \\ 0 & 0 & 2 \end{pmatrix}
\tag{25}
$$

This additional prior information is introduced in order to make fair the joint estimate of regional strain rate and fault creep. Indeed, the a priori constraints of zero fault creep and of first-order derivatives at the fault edges (see Section 2.2 and Appendix A) and small slip roughness (weighted by the hyper-parameter $\beta_p$) favour, within the present inverse method, small fault creep; the additional prior information above does the same for the regional strain rate.

In the end, we do not consider any prior information on the parameters describing the rigid-body motion and the hyper-parameters themselves, meaning that their PDF is the uniform PDF over $\mathcal{W}$ and

$\mathcal{H}$. The a priori PDF for the model parameters and hyper-parameters is simply proportional to the PDF of the model parameters given the hyper-parameters

$$f(\mathbf{m}, \mathbf{h}) \propto f(\mathbf{s}_1 \,|\, \beta_1)\, f(\mathbf{s}_2, \,|\, \beta_2)\, f(\mathbf{q} \,|\, \rho) \tag{26}$$

We avoid any a priori constraints on the rigid-body motion becouse they would make the inverse method dependent on the specific reference system on which the GNSS velocities are expressed.

### 3.2. Posterior Probability Density Function

According to the Bayes' theorem [34,37], the PDF for the data and the a priori PDF, Equations (22) and (26), can be combined in order to obtain the posterior PDF for all the model parameters and hyperparameters given the data

$$p(\mathbf{m}, \mathbf{h}) = c\, f(\widetilde{\mathbf{y}} \,|\, \mathbf{m}, \alpha)\, f(\mathbf{m}, \mathbf{h}) \tag{27}$$

where $c$ is a constant of normalization.

As discussed in Fukuda and Johnson [36], the approach proposed by Yabuki and Matsu'ura [34] is based on the Akaike Bayesian information criterion (ABIC) and does not account for the uncertainties on the hyperparameters, which is appropriate only if the marginal PDF for the hyperparameters has a dominant and sharp peak at its maximum. In this respect, we prefer to consider the fully Bayesian approach by Fukuda and Johnson [36] and estimate the mean model and hyper-parameters and standard deviations by averaging over the whole model and-hyper-parameter spaces. In particular, in the following, we will make use of the expectation, $E(\cdot)$, and covariance, $C(\cdot, \cdot)$, operators defined as follows

$$E(\mathbf{f}) = \int_{\mathcal{H}} \int_{\mathcal{M}} \mathbf{f}(\mathbf{m}, \mathbf{h})\, p(\mathbf{m}, \mathbf{h})\, d\mathbf{m}\, d\mathbf{h}, \qquad C(\mathbf{f}) = E(\mathbf{f}\, \mathbf{f}^{\mathrm{T}}) - E(\mathbf{f})\, E(\mathbf{f})^{\mathrm{T}} \tag{28}$$

with $\mathbf{f}$ being an arbitrary function or function array of the model paramaters and hyper-parameters.

### 3.3. Prior Information on the Tectonic Regimes

After some minor modifications, the present inverse method can be adapted to include prior information about the tectonic regime and its orientation. For the case of the regional deformation, two principal axes of the deviatoric strain rates, say $\hat{x}_1$ and $\hat{x}_2$, are parallel to the Earth surface and the third one, say $\hat{x}_3$, is perpendicular to it. Denoting with $\epsilon_i'$ the principal value associated with the $i$-th principal axis and ordering the first two directions in such a way that $\epsilon_1' > \epsilon_2'$, we can discriminate different tectonic regimes according to the following inequalities

$$\begin{array}{ll} \epsilon_3' < \epsilon_2' < \epsilon_1' & \text{extensional} \\ \epsilon_2' < \epsilon_1' < \epsilon_3' & \text{compressional} \\ \epsilon_2' < \epsilon_3' < \epsilon_1' & \text{strike-slip} \end{array} \tag{29}$$

that, using $\epsilon_3' = -\epsilon_1' - \epsilon_2'$, become

$$\begin{array}{ll} -\frac{1}{2}\epsilon_1' < \epsilon_2' < \epsilon_1' & \text{extensional} \\ \epsilon_2' < -2\,\epsilon_1' & \text{compressional} \\ -2\,\epsilon_1' < \epsilon_2' < -\frac{1}{2}\,\epsilon_1' & \text{strike-slip} \end{array} \tag{30}$$

We note that extensional and strike-slip regimes imply $\epsilon_1' > 0$, while compressional regimes do not, and that the intermediate principal values for extensional, compressional and strike-slip regimes are $\epsilon_2'$, $\epsilon_1'$ and $\epsilon_3'$, respectively. Furthermore, the direction of maximum extension is $\hat{x}_1$ for extensional and strike-slip regimes, while the direction of maximum compression is $\hat{x}_2$ for compressional regimes.

Within this framework, instead of the deviatoric regional strain rates, Equation (8), we can choose the following set of three parameters which make straightforward to recognise the tectonic regimes

$$\mathbf{q}' = \begin{pmatrix} \epsilon_1' \\ \zeta \\ \xi \end{pmatrix} \tag{31}$$

where $\xi$ is the azimuth of the first principal direction, $\hat{\mathbf{x}}_1$, and $\zeta = \epsilon_2'/\epsilon_1'$ is the ratio between the two horizontal principal values. After some straigtforward algebra, we can obtain the following relation between the old and new model parameters

$$\mathbf{q}(\mathbf{q}') = \hat{\mathbf{q}}(\zeta, \xi)\, \epsilon_1' \tag{32}$$

with

$$\hat{\mathbf{q}}(\xi, \zeta) = \begin{pmatrix} \cos^2 \xi + \zeta \sin^2 \xi \\ \sin^2 \xi + \zeta \cos^2 \xi \\ (1 - \zeta) \sin \xi \cos \xi \end{pmatrix} \tag{33}$$

The model space for $\mathbf{q}'$ is given by

$$\mathcal{Q}' = \mathcal{Q}'_E \cup \mathcal{Q}'_C \cup \mathcal{Q}'_S \subset \mathbb{R}^3 \tag{34}$$

where $\mathcal{Q}'_E$, $\mathcal{Q}'_C$ and $\mathcal{Q}'_S$ are the following subsets

$$\mathcal{Q}'_E = (0, \infty) \times (-1/2, 1] \times (0, \pi) \subset \mathbb{R}^3,$$

$$\mathcal{Q}'_C = (0, \infty) \times (-\infty, -2] \times (0, \pi) \cup (-\infty, 0) \times [2, \infty) \times (0, \pi) \subset \mathbb{R}^3,$$

$$\mathcal{Q}'_S = (0, \infty) \times (-2, -1/2) \times (0, \pi) \subset \mathbb{R}^3 \tag{35}$$

which correspond to all the possible model parameters $\mathbf{q}'$ describing extensional, compressional and strike-slip tectonic regimes, respectively.

Due to the non linearity of Equation (32), the a posteriori PDF modifies according to the Jacobian rule [37]

$$p'(\mathbf{w}, \mathbf{q}', \mathbf{s}_1, \mathbf{s}_2, \mathbf{h}) = p(\mathbf{w}, \mathbf{q}(\mathbf{q}'), \mathbf{s}_1, \mathbf{s}_2, \mathbf{h})\, J(\mathbf{q}') \tag{36}$$

with

$$J(\mathbf{q}') = \left| (1 - \zeta)\, \epsilon_1'^2 \right| \tag{37}$$

This form of the a posteriori PDF can be used both to investigate the probabilities of different tectonic regimes, depending directly on the new parameters which make straightforward to recognise the tectonic regimes to which they refer (particularly $\xi$ and $\zeta$), and to include prior information on the tectonic regime by multiplying it with a priori PDF as function of these new parameters or, more simply, investigating the a posteriori PDF over a specific subset of model space $\mathcal{Q}'$ given in Equation (34).

## 4. Results

To disclose the behaviour of the long-lasting quiescent Castrovillari normal fault, a campaign GNSS network composed by 10 sites over an area of about $30\,\text{km} \times 17\,\text{km}$ centred on the normal-faulting area has thus been set up [25]. Implementation of the campaign sites was carefully performed to ensure a submillimetre forced centring and antenna height measurements. The GNSS measurements started in 2003 and five campaigns have been performed up to 2008. Annual occupations coincided with the September-October (dry) period in an effort to minimize seasonal effects. To avoid displacement artifacts that can be introduced via differing receiver/antenna, all measurements throughout the six-year experiment were performed by using the same instrumentation set-up. During each campaign, data were collected over a period of four consecutive days, with daily sessions of at least 8 h; some network sites have been observed continuously during each period.

Table 2 reports the GNSS velocities by Sabadini et al. [25] after transformation from IGS05 to IGS08 and rotation into an Apulia fixed reference frame [33] in order to better highlight the peculiar deformation pattern of the area (Figure 1). Despite the fact that the local GNSS campaigns were carried out for investigating the Castrovillari fault, it guarantees some coverage of the nearby Pollino fault and, so, we extend the present analysis also to the latter. The CIVI, DORM and FRAN stations, indeed, fall within the foot-walls of both faults, while the other stations are above them, mainly in the hanging-walls.

**Table 2.** Site id, longitude, latitude, east and north component of estimated GNSS velocities and relative uncertainties (in mm/yr) used in this study. The original GNSS velocities from Sabadini et al. [25] have been transformed from IGS05 to IGS08 and, then, rotated into an Apulia fixed reference frame [33].

| GNSS Data | | | | | | |
|---|---|---|---|---|---|---|
| CASS | 16.31969 | 39.78493 | 1.33 | 4.95 | 0.34 | 0.18 |
| COLO | 16.15402 | 39.88887 | 0.99 | 4.14 | 0.31 | 0.37 |
| CROC | 16.17953 | 39.83458 | 1.07 | 4.24 | 0.32 | 0.31 |
| DOLC | 16.25764 | 39.75962 | 1.14 | 4.45 | 0.44 | 0.31 |
| DORM | 16.29699 | 39.88124 | 2.16 | 4.05 | 0.39 | 0.34 |
| FRAN | 16.38424 | 39.82068 | 3.03 | 4.46 | 0.52 | 0.33 |
| FRAS | 16.25836 | 39.82188 | 0.45 | 4.45 | 0.36 | 0.33 |
| MORA | 16.15572 | 39.85936 | 1.18 | 3.94 | 0.32 | 0.28 |
| ZACC | 16.16781 | 39.77370 | 1.26 | 4.44 | 0.50 | 0.31 |
| CIVI | 16.31628 | 39.82693 | 4.54 | 4.05 | 0.48 | 0.32 |

In the following discussion, observed and modelled velocities will be compared in the local fixed reference frame, i.e., after the removal of the rigid body motion estimated by the GNSS data inversions themselves

$$\delta\mathbf{v}(\phi,\theta) = \mathbf{v}(\phi,\theta) - \mathbf{Q}(\phi,\theta)\,\mathbf{w} \tag{38}$$

The reference point with respect to which we perform the Taylor expansion of the regional velocity field is $(\phi_0, \theta_0) = (16.23°, 39.81°)$, about in the middle of the GNSS network.

### 4.1. Regional Deformations without Fault Creep

To provide a first upper bound for the regional deformation, let us first implement the inversion scheme presented in Section 3 for estimating only the regional deformation, i.e., assuming no fault creep. In other words, we investigate the a posteriori PDF given zero fault creep, $p(\mathbf{w}, \mathbf{q}, \alpha, \rho) = p(\mathbf{m}, \mathbf{h} \,|\, \mathbf{s}_1 = \mathbf{s}_2 = \mathbf{0})$.

Table 3 (scheme R) lists the mean and standard deviation for the model parameters of the regional deformation, and Figure 3A shows the comparison between observed and modelled surface velocities. The estimated strain rate is primarily consistent with a strike-slip tectonic regime, with WSW-ENE extension of $86.5 \pm 41.2 \times 10^{-9}$/yr and a slightly smaller compression of $-62.3 \pm 36.3 \times 10^{-9}$/yr along the perpendicular direction. The large uncertainties of this estimate, nevertheless, does not rule out the possibility of an extensional tectonic regime. The direction of maximum extension is consistent with the large scale deformation pattern estimated by Devoti et al. [38] for this area and, at smaller spatial scales, a little further north in the Pollino Range, by Cheloni et al. [18]. On the other hand, our estimate is more than twice greater than those of these authors (a few tens of nanostrain per year and $34 \pm 7 \times 10^{-9}$, respectively). This discrepancy can be explained by considering that our larger estimate has been obtained using a local GNSS network, with GNSS stations close to fault traces and both in the hanging and foot walls of the Castrovillari and Pollino faults, and, so, it is more sensitive to near-field deformations caused by the fault creep rather than just to the regional deformation. In other words, this preliminary result, together with the low seismic activity of the Castrovillari and Pollino faults in the past millenium and the low P- and S-wave velocities of the surrounding area Figure 2; [23,28], indicate that most of this high geodetic strain rate must be accommodated by long-lasting fault creep, at least from 2003 up to 2008 and ever earlier in time, since 1995, as it results by the campaign GNSS network and DInSAR data [25].

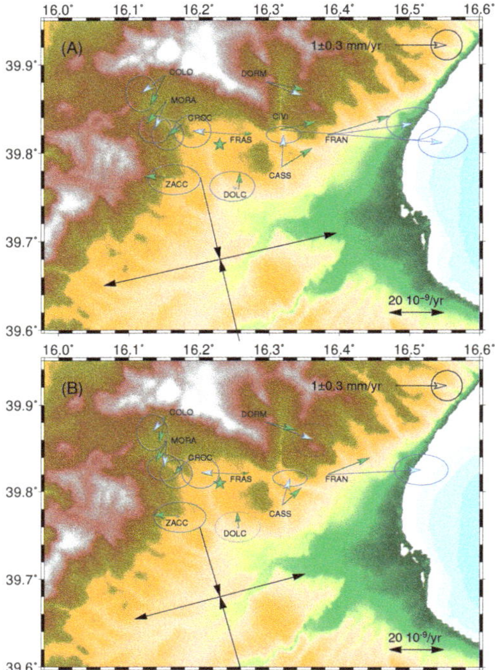


**Figure 3.** (**A**) Comparison between observed and modelled surface velocities (blue and green arrows, respectively) according to the inversion scheme R (see Table 3), after the removal of the estimated rigid-body motion. The ellipses represents the one-sigma errors from GNSS data analysis (observational errors) and should be rescaled by the square root of the estimated hyper-parameter $\alpha$ in order to account for the modelling errors ($\sqrt{\alpha} = 2.18$ for this inversion). The regional strain rate is also drawn, further south with respect to the reference point ($\phi_0, \theta_0$) of the Taylor series expansion (green star). (**B**) As panel A, but for the inversion scheme R* (see Table 3, $\sqrt{\alpha} = 1.45$), where the CIVI station has not been used.

An additional indication that these two faults are creeping comes from the fact that the regional deformation alone hardly explain the surface velocities observed by the FRAS and CIVI stations. In particular, even after the removal of the rigid-body motion, the observed velocity for CIVI station is still high, almost 3 mm/yr, while the modelled one is less than 1 mm/yr. These large discrepancies between observed and modelled ones make large also the estimate of the hyper-parameter $\alpha = 4.75$, meaning that modelling errors are still large and, then, motivating the possibility of fault creep in order to improve the modelling of the observed GNSS velocities.

To understand how much the CIVI station affects our results, we performed the same GNSS data inversion without using this station (see Table 3, scheme R*, and Figure 3B). Even in this case the regional deformation alone hardly explain the surface velocities observed at two GNSS stations (now FRAS and FRAN) and the estimate of the hyper-parameter $\alpha = 2.11$, although smaller, still indicates the presence of modelling errors. The direction of maximum extension is almost unchanged ($\xi = 74.6$)

and the maximum strain rate is $63.9 \pm 29.0 \times 10^{-9}$/yr, smaller than the previous one by only the 25 per cent. This result, in addition to confirm the necessity of allowing fault creep, enlightens the role of the CIVI station. Further checks on the impact of this station on our conclusions will be considered in the following discussion.

**Table 3.** The mean estimates and standard deviations of the east, $v_\phi$, and north, $v_\theta$, components of the regional velocity (in mm/yr) at the the reference point $(16.23°, 39.81°)$, of the rotation rate, $\omega$ (in $10^{-9}$/yr) and of the mean components of the deviatoric strain rates $\epsilon'_{11}$, $\epsilon'_{22}$ and $\epsilon'_{12}$ (in $10^{-9}$/yr) with respect to the two horizontal principal axes, $\epsilon'_{ij} = \hat{x}_i \cdot \epsilon' \cdot \hat{x}_j$ $(i,j = 1,2)$. The results are presented for the different case of GNSS data inversion considered in the main text, with and without fault creep and with prior information about the tectonic regime and the azimuth of the direction of maximum extension, $\zeta$. For the inversion schemes R* and FR* the CIVI station is not used.

| | | | | Inversion Scheme | | | |
|---|---|---|---|---|---|---|---|
| | Nickname | R | FR | FR-E$_{77}$ | FR-E$_{115}$ | R* | FR* |
| **Features** | Creep | no | yes | yes | yes | no | yes |
| | Regime | all | all | Extensional | Extensional | all | all |
| | Azimuth | all | all | 77.1° | 115.0° | all | all |
| **Estimates** | $\alpha$ | 4.75 | 1.10 | 1.32 | 1.05 | 2.11 | 0.46 |
| | $v_\phi$ | $-0.43 \pm 0.30$ | $-0.13 \pm 0.28$ | $-0.18 \pm 0.26$ | $-0.08 \pm 0.24$ | $-0.57 \pm 0.20$ | $-0.28 \pm 0.23$ |
| | $v_\theta$ | $0.49 \pm 0.21$ | $0.15 \pm 0.27$ | $0.25 \pm 0.26$ | $0.19 \pm 0.24$ | $0.51 \pm 0.14$ | $0.26 \pm 0.22$ |
| | $\omega$ | $-10.8 \pm 34.6$ | $-0.2 \pm 32.5$ | $-9.7 \pm 29.9$ | $2.3 \pm 29.0$ | $-3.0 \pm 23.5$ | $-0.9 \pm 24.0$ |
| | $\epsilon'_1$ | $86.5 \pm 41.2$ | $24.2 \pm 27.6$ | $47.7 \pm 25.8$ | $40.2 \pm 18.9$ | $63.9 \pm 29.0$ | $29.1 \pm 22.2$ |
| | $\epsilon'_2$ | $-62.3 \pm 36.3$ | $-13.9 \pm 31.0$ | $-11.0 \pm 14.5$ | $-9.9 \pm 11.5$ | $-54.1 \pm 24.8$ | $-18.9 \pm 23.5$ |
| | $\epsilon'_{12}$ | $0 \pm 29.9$ | $0 \pm 26.8$ | $0$ | $0$ | $0 \pm 20.2$ | $0 \pm 20.8$ |
| | $\zeta$ | 77.1° | 115.0° | 77.1° | 115.0° | 74.6° | 98.4° |
| | | Figure 3A | Figure 4 | Figure 8 | Figure 9 | Figure 3B | Figure 5 |

## 4.2. Regional Deformations and Fault Creep

Let us now consider the full inversion scheme, where observed surface velocities are explained by both regional deformations and fault creep over the Castrovillari and Pollino faults. Table 3 (scheme FR) lists the mean and standard deviation for the model parameters of the regional deformation, and Figure 4 shows the comparison between observed and modelled surface velocities (blue and orange arrows, respectively), as well as the fault creep distribution and its standard deviation. The contribution to the modelled velocities due to only the regional strain is also shown (green arrows). The estimated strain rates, together with their uncertainties, are consistent with both extensional and strike-slip tectonic regimes, with WNW-ESE extension of $24.2 \pm 27.6 \times 10^{-9}$/yr and a compression of $-13.9 \pm 31.0 \times 10^{-9}$/yr along the perpendicular direction (that is about an half of the extension rate in amplitude). The direction of maximum extension is now 115°, rotated of about 38° clockwise with respect to that estimated by the inversion scheme R. This refined estimate of the regional strain rate, based on the FR inversion scheme rather than the R one, corroborates the transitional character of the investigated area enlightened by the analysis of the focal mechanism from waveform inversion, separating the WSW-ENE extensional domain of the southern Apennines from the NW-SE extensional one of the Calabrian Arc [39,40].

The fault creep distributions for the Castrovillari and Pollino faults indicate a normal/right-lateral strike-slip mechanism with average rake angle of $-126°$ and $-143°$ and maxima of 2.1 mm/yr and 5.2 mm/yr close to the fault traces, in correspondence with the FRAS and CIVI stations, respectively. The standard deviation for both the fault creep distribution is about 2∼3 mm/yr and goes to zero at the internal edges due to the a priori constraints that the fault creep distribution is zero outside the adopted rectangular areas. In light of this, we can assess that the fault creep on the Pollino fault is statistically significant (i.e., it differs from zero by about two-sigma errors), while that on the Castrovillari fault is not (it is about one-sigma error).

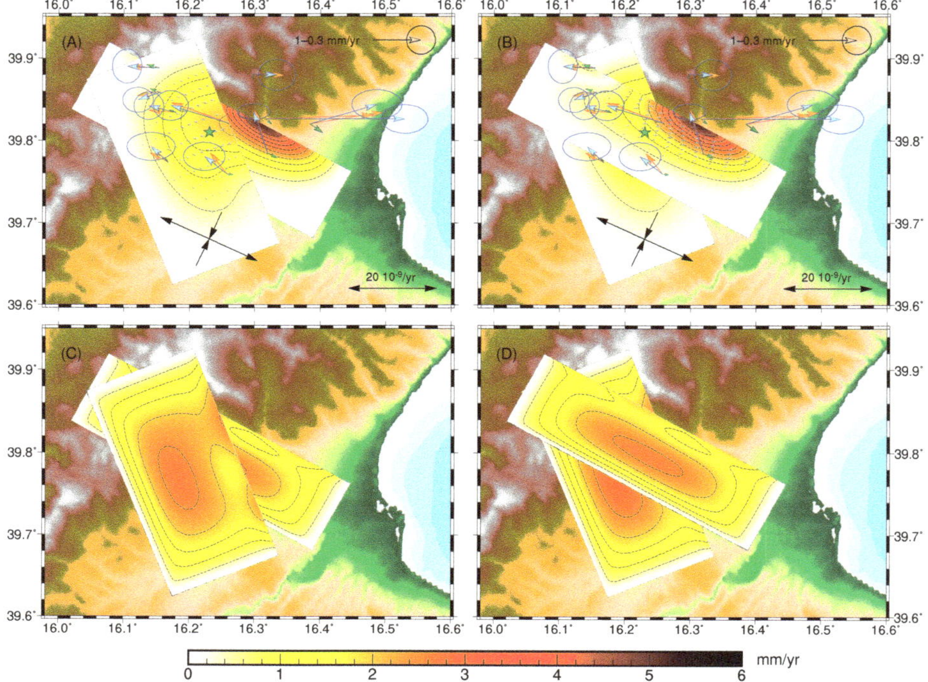

**Figure 4.** (**A,B**) Mean fault creep distributions and (**C,D**) standard deviation over the Castrovillari and Pollino faults (the left and right columns show the Castrovillari fault over the Pollino one and viceversa), estimated according to the inversion scheme FR (see Table 3). The gray arrows represent the direction of the fault creep and the contour lines are given every 0.5 mm/yr. In panels A and B, the blue and red arrows are the observed and modelled surface velocities, after the removal of the estimated rigid-body motion. The ellipses represents the one-sigma errors from GNSS data analysis (observational error) and should be rescaled by the square root of the estimated hyper-parameter $\alpha$ in order to account for the modelling errors ($\sqrt{\alpha} = 1.05$ for this inversion). The estimated regional strain rate is also drawn, further south with respect to the reference point ($\phi_0, \theta_0$) of the Taylor series expansion (green star), and the green arrows represent the velocities due to this strain rate.

Comparing Figures 3 and 4, we note that the observed and modelled velocities are in better agreement for the inversion scheme FR than they are for the inversion scheme R. This results also from the estimate of the hyper-parameter $\alpha$, that is now $\alpha = 1.10$ instead of $\alpha = 4.75$ (see Table 3), meaning that the discrepancies between observed and modelled velocities are consistent with the observational errors, while the modelling errors are negligible. This reduction of the hyper-parameter $\alpha$ confirms the fact that fault creep is the necessary physical mechanism in order to explain the velocity changes on the small spatial scale of the campaign GNSS network. On the other hand, different from Sabadini et al. [25], our updated results indicate that the creep is mainly occurring on the Pollino fault rather than the Castrovillari fault, which was the only fault considered in this previous study. Furthermore, we note that the observed velocities are mainly explained by the fault creep rather than by the regional deformation. The contribution to surface velocities due to the regional strain rate (green arrows), indeed, is only a small fraction of the modelled velocities which also include the contribution from the fault creep (orange arrows) and, in amplitude, is smaller than the one-sigma errors affecting the observed surface velocities. This fact thus explains why the estimated regional strain rate is consistent with the case of no regional strain (the maximum extension rate of $24.2 \times 10^{-9}$/yr is indeed smaller than its one-sigma error of $27.6 \times 10^{-9}$/yr) and points out how the large variations of the observed surface velocities from one GNSS station to another of the local GNSS campaign network are mainly due to fault creep, at least on the Pollino fault.

Our estimates, both of the regional strain rates and of the fault creep distributions, are affected by large uncertainties due to the small number of GNSS data used in the inversion, as well as due to the smallness of the geographical area that they cover. Increasing the number of GNSS station would certainly decrease these uncertainties and, particularly, improve the resolving power on the fault creep distribution. Enlarging the covered geographic area, instead, would affect mainly the estimate of the regional deformation because larger baseline lengths between GNSS stations better constrain large spatial scale strain rates. On the other hand, they could be affected by deformation caused by other creeping faults other than those here considered. Furthermore, regional strain rates could vary on larger spatial scales, making more inaccurate the first-order expansion of the regional velocity field, Equation (2). Therefore, in this first attempt to discriminate intraplate regional deformation from fault creep, we have preferred to consider only the campaign GNSS network by Sabadini et al. [25], without adding complexities such as including additional GNSS stations from other GNSS networks far away from the near-field of the Pollino and Castrovillari faults.

An even more stringent issue that prevent us from including additional data in the present analysis is that all the permanent GNSS stations in the proximity of the Pollino and Castrovillari faults were set up only in recent years. In this respect, these additional observed crustal velocities would refer to time periods different (and without overlapping) from that of the GNSS campaign used here and they thus reflect averaged fault creep which can be different from that occurring during the six year GNSS campaign (2003–2008). Significant variations in the fault creeps over the years have to be expected due to transient aseismic processes as, for instance, the one observed by Cheloni et al. [18] from 2010 to 2014, a little further north to the Pollino and Castrovillari faults. Because our inverse method aims at estimating the long-term trend due to aseismic processes in a specific time windows and it does not take into account their evolution in time, these variations from one time periods to another would introduce modelling errors that biases the estimate of a hardly quantifiable amount. Further improvements in the understanding of complex spatial pattern and temporal evolution of displacement time series at intraplate regions can be achieved in the future by combining our method, based on the discrimination of the spatial patterns of the different physical processes, with the classical ones, mainly based on the alone temporal evolution.

As already done in Section 4.1 about the effects of the CIVI station on the estimate of the only regional deformation, we performed the same GNSS data inversion for estimating both regional deformation and fault creep without using this station (see Table 3, scheme FR*, and Figure 5). The estimate of the regional strain rate is slightly larger than that obtained within the inversion scheme FR (the maximum extension and compression are $29.1 \pm 22.2 \times 10^{-9}$/yr and $-18.9 \pm 23.5 \times 10^{-9}$/yr instead of $24.2 \pm 27.6 \times 10^{-9}$/yr and $-13.9 \pm 31.0 \times 10^{-9}$/yr) and the direction of maximum extension is slightly rotated anti-clockwise by about 17° ($\xi = 98.4°$ instead of 115.0°). Furthermore, the fault creep distribution over the Castrovillari fault is twice smaller, with the shallow maximum of 1.0 mm/yr instead of 2.1 mm/yr, while that over the Pollino fault decreases by about the 30 per cent, with the shallow maximum of 3.5 mm/yr instead of 5.2 mm/yr. These differences in the estimates, when compared with their uncertainties, are minor and the same main considerations based on the inversion scheme FR still hold even if the CIVI station is not used in the inversion. In particular, the normal/right-lateral strike-slip mechanism of the Pollino fault is confirmed also by the inversion scheme FR* and the regional strain rate are consistent with both strike-slip and normal tectonic regimes, with a maximum extension of about a few tens of nanostrain per year.

In contrast to the inversion scheme FR where the agreement between observed and modelled GNSS velocities is consistent with observational errors from GNSS data processing, the inversion scheme FR* yields a smaller estimate of the hyper-parameter $\alpha$, that is now $\alpha = 0.46$ instead of $\alpha = 1.10$, meaning that the model fits too well the GNSS velocities. This fact can be seen as an indication that observational errors have been overestimated during the GNSS data processing or that the model is overparameterized compared to the few data at our disposal. As already mentioned above and as reminder for next local GNSS campaigns, a denser GNSS network will certainly improve our results, making them less sensitive to single GNSS stations.

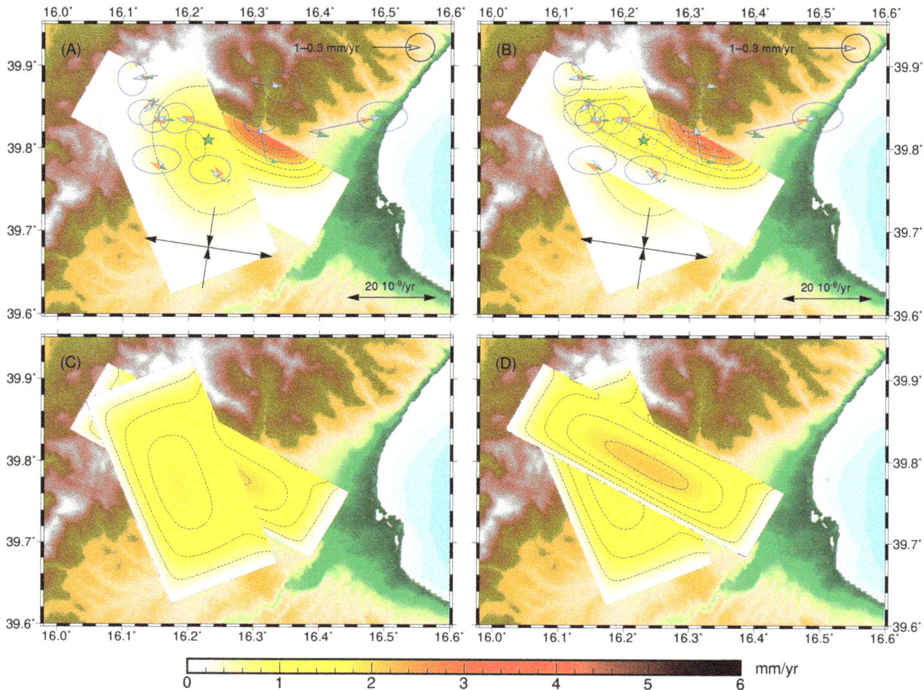

**Figure 5.** As Figure 4 but for the inversion scheme FR* (see Table 3), where the CIVI station has not been used. In this case $\sqrt{\alpha} = 0.68$, meaning that the fault creep model fits too well the observations or that observational errors from GNSS data analysis have been overestimated.

### 4.3. Regional Deformations and Fault Creep and Prior Information about the Tectonic Regimes

To better understand the results obtained from the inversion scheme FR and, at the same time, to enlighten the full potential of our novel inverse method, let us now consider the inclusion of prior information about the tectonic regimes as described in Section 3.3.

Figure 6A shows the marginal PDF for extensional and strike-slip tectonic regimes, varying both the azimuth of the maximum extension direction, $\xi$, and the ratio between the two horizontal principal values, $\zeta$,

$$p'(\xi, \zeta) = \int_0^\infty \left( \int_{\mathcal{W} \times \mathcal{S}_1 \times \mathcal{S}_2} p'(\mathbf{w}, \mathbf{q}', \mathbf{s}_1, \mathbf{s}_2, \mathbf{h}) \, d\mathbf{w} \, d\mathbf{s}_1 \, d\mathbf{s}_2 \, d\mathbf{h} \right) d\epsilon_1' \tag{39}$$

with $p'(\mathbf{w}, \mathbf{q}', \mathbf{s}_1, \mathbf{s}_2, \mathbf{h})$ being given by Equation (36). We have not considered compressional tectonic regimes because both the inversion schemes R and FR have already shown that it is the less likely one. In Figure 6B, instead, we report the maximum extension rate estimated varying the tectonic regimes. The marginal PDFs for the only ratio between the two horizontal principal values, $\zeta$,

$$p'(\zeta) = \int_0^\pi p'(\xi, \zeta) \, d\xi \tag{40}$$

and the only azimuth of the maximum extension direction, $\xi$,

$$p'(\xi) = \int_{-2}^{1} p'(\xi,\zeta)\,\mathrm{d}\zeta \tag{41}$$

are shown in Figure 7.

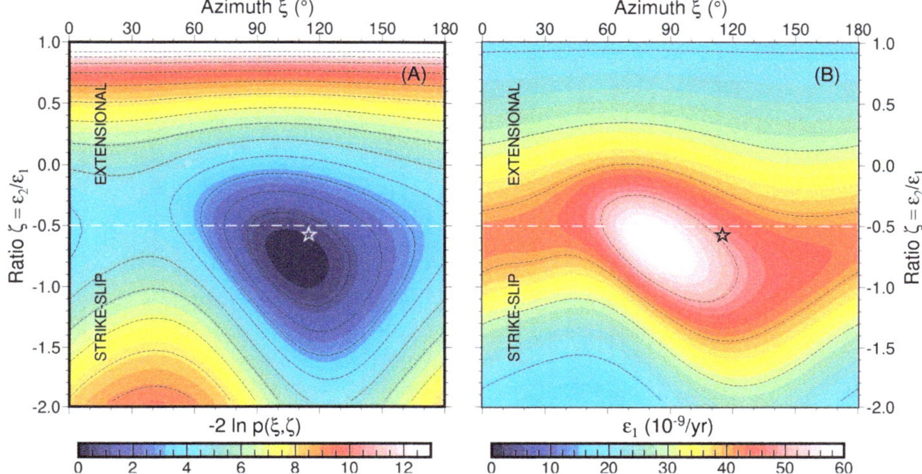

**Figure 6.** (**A**) Marginal PDF $p'(\xi,\zeta)$ for the azimuth of maximum horizontal extension, $\zeta$, and the ratio between horizontal principal strain rates, $\zeta = \epsilon_2/\epsilon_1$, of extensional and strike-slip tectonic regimes for the inversion scheme FR (see Table 3). (**B**) Mean of the maximum horizontal strain rates $E(\epsilon_1 \mid \xi,\zeta)$ at fixed tectonic regimes as function of the azimuth $\xi$ and the ratio $\zeta$. The (white and black) stars indicate the estimated tectonic regime for the inversion scheme FR and the contour lines are given every 1 unit in panel A and $10^{-8}/\mathrm{yr}$ in panel B.

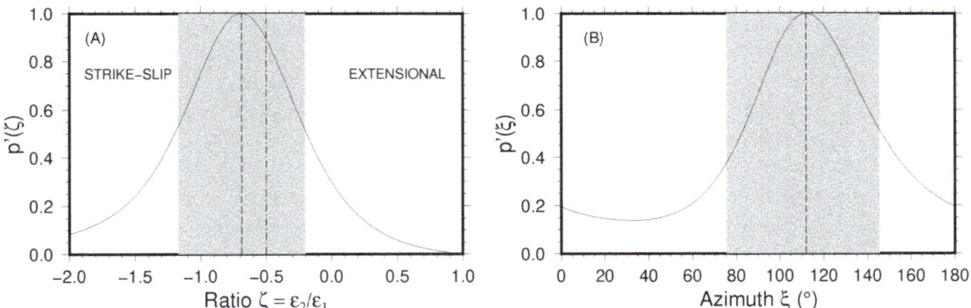

**Figure 7.** Marginal PDFs for (**A**) the ratio between horizontal principal strain rates, $\zeta = \epsilon_2/\epsilon_1$, of extensional and strike-slip tectonic regimes and (**B**) the azimuth of maximum horizontal extension, $\xi$. The vertical dashed line indicates the maximum probability (set to 1) and the grey area denotes the 68 per cent confidence interval. The vertical dashed-dot line in panel A indicate the transition between strike-slip and extensional tectonic regimes.

The marginal PDF has its maximum close to the tectonic regime estimated within the inversion scheme FR ($\xi = 115°$ and $\zeta = -0.57$) and it is bell shaped, but for some asymmetries across the transition from strike-slip to normal tectonic regimes. We note also that the strike-slip tectonic regimes are more likely than the normal ones, although the latter cannot be excluded. The different location of the maximum of the marginal PDF and the tectonic regime estimated by the inversion scheme FR is attributable to the different meaning of the two approaches, with and without prior information about the tectonic regimes. Indeed, within the inversion scheme FR, the tectonic regime is inferred by the mean strain rate tensor (which can be seen as it has been obtained by averaging the strain rate tensors for every tectonic regimes weighted by their probabilities), while the marginal PDF is obtained by considering each tectonic regime separately. This difference also explains why the maximum extension rate of $24.2 \times 10^{-9}/\mathrm{yr}$ estimated within the inversion scheme FR is twice smaller than the maximum extension rate estimated using the same tectonic regime inferred from the inversion scheme FR as prior information (see the value in correspondence of the black star in Figure 6B that is about $50 \times 10^{-9}/\mathrm{yr}$). However, as shown in Figure 6B, it is noteworthy to notice that the maximum extension rate for the normal and strike-slip tectonic regimes is always smaller than $60 \times 10^{-9}/\mathrm{yr}$ and therefore even smaller than the estimate of $86.5 \times 10^{-9}/\mathrm{yr}$ obtained within the inversion scheme R, when the fault creep is not allowed.

As shown in Figure 7, from the marginal PDFs for the only ratio between the two horizontal principal values, $\zeta$, and the only azimuth of the maximum extension direction, $\xi$, we can quantify the 68 per cent confidence intervals (grey regions), which are $\zeta \in [-1.167, -0.202]$ and $\xi \in [75.4, 145.5]$ or, equivalently, $\zeta = -0.687^{+0.485}_{-0.480}$ and $\xi = 112.0°^{+33.5}_{-36.5}$, with $\zeta = -0.687$ and $\xi = 112.0°$ being the parameters at which the PDFs have their maxima (vertical dashed lines).

To gain insight on the impact of prior information about the tectonic regimes on the results, we now consider the two representative inversion schemes FR-E$_{77}$ and FR-E$_{115}$ where we fix a priori the direction of maximum extension at $\xi = 77.1°$ and $115.0°$, respectively (these are the same azimuths obtained from the inversion schemes R and FR), and we require that the tectonic regime is normal, as commonly expected for the Apennines chain (that is we consider all the ratios between the horizontal principal values from $-1/2$ to $1$, $\zeta \in (-1/2, 1)$). As before, Table 3 (schemes FR-E$_{77}$ and FR-E$_{115}$) lists the mean and standard deviation for the model parameters of the regional deformation, and Figures 8 and 9 show the comparisons between observed and modelled velocities (blue and orange arrows, respectively), as well as the fault creep distributions and their standard deviation.

We note that the fault creep distributions are not affected by the inclusion of the prior information about the tectonic regime, with the exception of some minor differences which are smaller than the uncertainties as, for instance, the smaller creep obtained over the Castrovillari fault, with a deeper maximum of about $1.6\,\mathrm{mm/yr}$. The contribution from the regional strain rate to the modelled surface velocities (green arrows), instead, is greater than that obtained within the inversion scheme FR and, indeed, the estimates of the maximum extension rate (46.8 and $40.0 \times 10^{-9}/\mathrm{yr}$ for the inversion schemes FR-E$_{77}$ and FR-E$_{115}$, respectively) is about twice larger than that for the inversion scheme FR without prior information. Nevertheless, as noticed above, this increase in the estimated extension rates must be ascribed to the fact that the same estimate with the inversion scheme without prior information results from a weighted average of the strain rate tensors of every tectonic regimes which, differing in the direction of maximum extension, compensate each other to some extent.

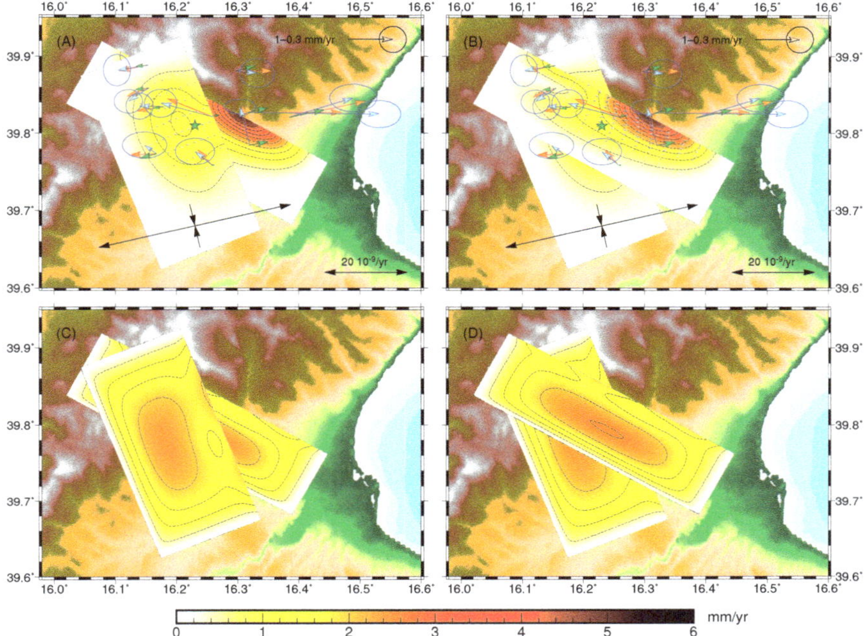

**Figure 8.** As Figure 4 but for the case FR-E$_{77}$ of GNSS data inversion (see Table 3, $\sqrt{\alpha} = 1.15$).

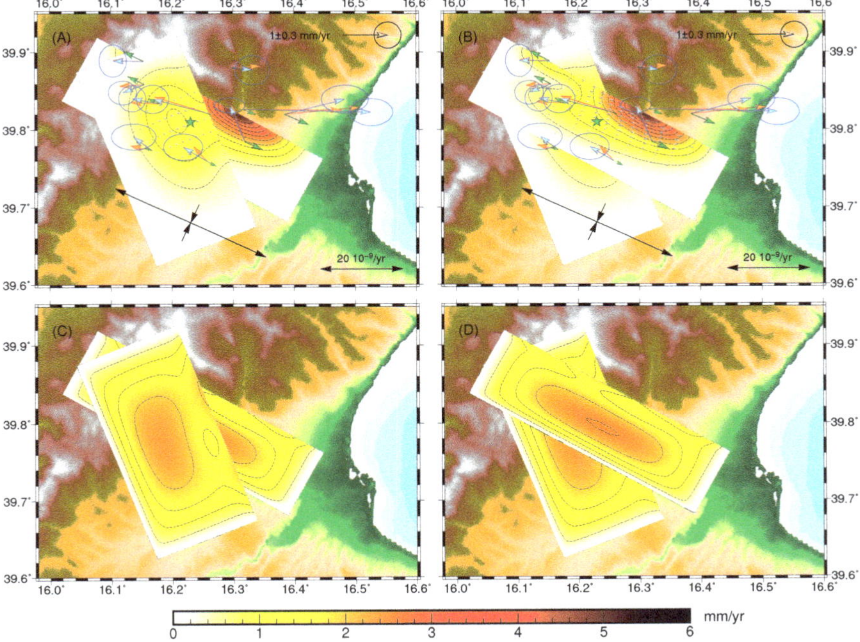

**Figure 9.** As Figure 4 but for the case FR-E$_{115}$ of GNSS data inversion (see Table 3, $\sqrt{\alpha} = 1.02$).

## 5. Discussion

The inverse method presented in Section 3 have been adopted to explain the surface horizontal velocities obtained from a six year GNSS campaign [25] over the Castrovillari and Pollino faults, cutting the southern portion of the Pollino seismic gap [20]. Achieved results have provided valuable information on the regional deformation of the area and on the aseismic processes occurring on both faults. More in detail, the model where only the regional deformation is considered (i.e., the scheme R) yields large residuals between modelled and observed velocities, larger by more than two times the observational uncertainties (see Figure 3 and the estimate of the hyper-parameter $\alpha$ reported in Table 3). Different, when the fault creep is allowed (i.e., the scheme FR), the residuals become consistent with the observational uncertainties (see Figure 4 and again Table 3), indicating that a more sophisticated model is necessary for explaining the observations. In particular, this second inversion clearly indicates, above the estimated uncertainties, that the Pollino fault is creeping, with a maximum slip rate of 5.2 mm/yr one. The six year GNSS campaign, indeed, points out that the largest velocities occur at those GNSS sites close to the Pollino fault trace (marked by blue line in Figure 1) and that such a spatial pattern can be hardly explained only on the basis of a regional deformation. Moreover, such a creeping behavior of the Pollino fault, while confirm the results achieved in Sabadini et al. [25], are well supported by the seismic observations which identified it as a low P- and S-wave velocity crustal zone (see Figure 1), therefore suggesting an anomalous thermal state of the crust. This anomalous thermal state of the crust would inhibit frictional fault sliding in favour of creeping deformation, as testified by the lack of large earthquakes (M > 6) in the last centuries.

The limitations due to the number and the geographical distribution of the adopted GNSS sites, nevertheless, does not allow us to constrain the regional deformation with high accuracy. In particular, it remains an ambiguity between extensional or strike-slip tectonic regimes, although the latter is slightly more likely as shown in Figure 7A. The inversion scheme FR, indeed, yields a maximum extension of $24.2 \pm 27.6$/yr with an azimuth of $115°$ and maximum compression of $-13.9 \pm 31$/yr along the perpendicular direction. In light of this, although the estimated regional deformation is characterized by high uncertainties of the order of a few tens of nanostrain per year (comparable with the expected value of strain rate in southern Italy), the data from the GNSS measurements, jointly with the novel inverse method, support the hypothesis of fault creeping, at least as it concerns the Pollino fault. Improvements in the estimate of the model parameters will be achieved in the future by exploiting more data and planning their spatial distribution on the basis of synthetic tests performed within the framework of the present inverse method. Furthermore, the planning of additional GNSS campaign measurements of the existing GNSS episodic network will benefit of the increasing number of continuous GNSS stations installed in the area since the past decade. As already argued in Section 4.2, it will be important to guarantee the overlapping of the time intervals spanned by continuous and campaign GNSS stations. Indeed, the present inverse method, being based on the crustal velocities over a specific time interval, can only provide an estimate of the average fault creep over the same time interval.

To verify the sensitivity of our results to specific GNSS stations, we have performed a similar inversion (i.e., the scheme FR*) removing the CIVI station, which presents the largest velocity in the Apulia fixed reference frame. The results are similar to that obtained using all the data (i.e, the scheme FR) but for the fact that the estimated regional strain rate is slightly larger (the maximum extension and compression are $29.1 \pm 22.2 \times 10^{-9}$/yr and $-18.9 \pm 23.5 \times 10^{-9}$/yr instead of $24.2 \pm 27.6 \times 10^{-9}$/yr and $-13.9 \pm 31.0 \times 10^{-9}$/yr) and the Pollino fault creep decreases by about the 30 per cent, with the shallow maximum of 3.5 mm/yr instead of 5.2 mm/yr. Furthermore, in order to consider the possibility of including prior information about the tectonic regime, we have discussed the estimates obtained requiring that a normal tectonic regime and fixing the direction of maximum extension with the azimuth obtained by the inversion scheme R and FR.

## 6. Conclusions

We have developed a novel inverse method for discriminating regional deformation and fault creep from surface velocities observed by a GNSS network at the spatial scale of intraplate faults which allows including, or not, prior information on the tectonic regime. Surface velocities caused by the fault creep are modelled according to the dislocation theory e.g., [32] while the rigid-body motion and regional strain rates describing the regional deformation are assumed to be homogeneous in space and dealt with as additional model parameters [30]. Then, we have applied this inverse method for explaining the surface velocities obtained from the GNSS network over the Castrovillari and Pollino faults [25].

In contrast to other studies which only focus on transient aseismic processes [18], this method allows us to estimate the whole fault creep over the time window spanned by the GNSS time series within intraplate areas. This is a crucial advancement because it allows us to deal even with cases of long-lasting creep of intraplate faults, the effects of which are linear in time and consequently they cannot be easily discriminated by the secular trends due to regional strain rates on the basis of time series analyses. They are herein rather discriminated by exploiting the different spatial signatures of the regional deformation (homogeneous in space when the investigated area is small and does not exceed the typical spatial scales at which tectonic processes change, generally fifty to hundred kilometres) and of the fault creep (localised in the near-field of the creeping faults). In light of this, the present inverse method is effective in recognizing creeping faults by jointly estimating the fault creep distribution and regional strain rates, as well as the rigid-body motion of the investigated area. In doing so, our estimated regional strain rate constitutes a refinement of the upper bound of the strain accumulation process presently based on the geodetic strain rate, including the main aseismic processes affecting the crust sampled by the GNSS network.

In this respect, taking advantages of the different spatial signatures of the main physical mechanism responsible for crustal displacements, the present method complements the already available methods which aim to detect departures from linear trends in the GNSS time series and interpret them in terms of effects of transient fault creeps. Further improvements in the understanding of the complex spatial pattern and temporal evolution of displacement time series in intraplate regions, where both transient and long-term aseismic processes can be relevant, will be obtained by combining this method with the previous ones mainly based on the alone temporal evolution.

For the case study here considered, the southern portion of the Pollino seismic gap, the geodetic strain rate estimated by the observed GNSS velocities is characterized by a maximum extension rate of about $86.5 \times 10^{-9}$/yr, as it results from the inversion scheme R which does not account for any aseismic processes. Allowing for the possibilities of fault creep over both the Castrovillari and Pollino faults in the inversion scheme R, this estimate is reduced by more than 70 per cent, to about $24.2 \times 10^{-9}$/yr and the agreement between observed and modelled surface velocities greatly improves mainly due to a normal/right-lateral strike-slip creep of the Pollino fault, with maximum slip rates of about 5 mm/yr. This inversion also predicts a smaller creep over the Castrovillari fault, with maximum of 2 mm/yr. Nevertheless, despite the Pollino fault, the latter estimate is smaller than its uncertainties and, then, we cannot conclude that also the Castrovillari fault is creeping.

As expected, taking into account the main sources of aseismic deformation, the estimate of the regional strain rate becomes much smaller than the geodetic strain rate obtained within the inversion scheme R. Furthermore, as indicated by the estimate of the hyper-parameter $\alpha$, the modelling of the observed surface velocities greatly improves and the discrepancies between observed and modelled surface velocities are consistent with observational errors, without the need for invoking additional modelling errors. The estimated regional strain rate, nevertheless, is affected by large uncertainties and this makes difficult to assess whether the southern area of the Pollino Range is characterised by a normal or strike-slip tectonic regime, as well as to define accurately its orientation.

The inclusion of prior information about the tectonic regime does not alter significantly the fault creep distributions of the Pollino and Castrovillari faults, but it can double the estimates of maximum extension due to regional deformation, as shown by the two representative inversion schemes FR-E$_{77}$ and FR-E$_{115}$. The estimated maximum extensions, however, are still (about twice) smaller than the geodetic strain rate obtained by the inversion scheme R.

Due to the complex tectonic setting of the investigated area, we cannot argue in favour of specific prior information to be used. On the other hand, these results show that it is important to jointly consider regional deformation and aseismic processes in order to correctly interpret crustal motion and that we can obtain novel insights on the regional strain rates, although affected by large uncertainties.

**Author Contributions:** Conceptualization, All; methodology, G.C.; software, G.C. and M.P.; validation, G.C.; formal analysis, G.C. and M.P.; investigation, All; resources, All; data curation, R.B. and M.P.; writing–original draft preparation, G.C. and R.S.; writing–review and editing, All; visualization, G.C.; supervision R.S. and G.N. All authors have read and agreed to the published version of the manuscript.

**Funding:** This research received no external funding.

**Conflicts of Interest:** The authors declare no conflict of interest.

## Appendix A. Modified Bicubic Splines

We define a regular grid over the rectangular rupture area by taking the nodes at steps of lengths $\Delta L$ along strike and $\Delta W$ down dip

$$\Delta L = \frac{L}{N_L}, \qquad \Delta W = \frac{W}{N_W} \tag{A1}$$

where $N_L$ and $N_W$ are the numbers of intervals along strike and down dip, respectively

Following Yabuki and Matsu'ura [34], the bicubic spline centred at the $(j,k)$-th node, that is $(x_1, x_2) = (j\,\Delta L, k\,\Delta W)$, reads

$$\Psi_{j,k}(x_1, x_2) = \Phi_j(x_1/\Delta L)\,\Phi_k(x_2/\Delta W) \tag{A2}$$

with

$$\Phi_i(x) = \Phi(x - i), \qquad \Phi(x) = \frac{1}{6}\begin{cases} (2+x)^3 & -1 \geq x \geq -2 \\ 4 - 3(x+2)\,x^2 & 0 \geq x > -1 \\ 4 + 3(x-2)\,x^2 & 1 \geq x > 0 \\ (2-x)^3 & 2 \geq x > 1 \end{cases} \tag{A3}$$

With this definition, for $j = -1, N_L + 1$ and $k = -1, N_W + 1$, the bicubic spline $\Phi_{j,k}$ differs from zero within the rectangular area where we allow the fault creep, i.e. $(x_1, x_2) \in [0, L] \times [0, W]$, for a total of $N = (N_L + 3)(N_W + 3)$ basis functions. We note that also bicubic splines centred outside the rectangle, but adjacent to the edges, differ from zero inside the rectangle and that, in general, they do not go to zero approaching the edges embedded in the solid Earth, as well as their derivatives.

To satisfy the requirement that the fault creep distribution goes to zero approaching the internal (left, right and bottom) edges of the fault and that its first-order derivatives at all the edges (including the top edge at the Earth's surface as well), we can combine the bicubic splines centred along the edges and outside the rectangle with those adjacent to the edges and centred inside the rectangle. In this respect, we shall consider only modified bicubic splines for $j = 1, N_L - 1$ and $k = 0, N_W - 1$, for a total of $N = (N_L - 1) N_W$ basis function.

In this perspective, we first define the following two kinds of modified cubic splines (distinguished by the tilde and the hat symbols)

$$\widehat{\Phi}_i^{\pm}(x) = \Phi_i(x) - \frac{1}{2}\Phi_{i\pm1}(x) + \Phi_{i\pm2}(x), \qquad \widetilde{\Phi}_i^{\pm}(x) = \Phi_i(x) + \Phi_{i\pm2}(x) \tag{A4}$$

in such a way that

$$\widehat{\Phi}_i^{\pm}(x)\Big|_{x=i\pm1} = 0, \qquad \frac{\widehat{\Phi}_i^{\pm}(x)}{\partial x}\Big|_{x=i\pm1} = \frac{\widetilde{\Phi}_i^{\pm}(x)}{\partial x}\Big|_{x=i\pm1} = 0 \tag{A5}$$

Then, we define the modified bicubic splines $\widehat{\Psi}_{j,k}$ at the bottom (left and right) corners

$$\widehat{\Psi}_{1,N_W-1}(x_1,x_2) = \widehat{\Phi}_1^-(x_1/\Delta L)\,\widehat{\Phi}_{N_W-1}^+(x_2/\Delta W)$$

$$\widehat{\Psi}_{N_L-1,N_W-1}(x_1,x_2) = \widehat{\Phi}_{N_L-1}^+(x_1/\Delta L)\,\widehat{\Phi}_{N_W-1}^+(x_2/\Delta W) \tag{A6}$$

at the top (left and right) corners

$$\widehat{\Psi}_{1,1}(x_1,x_2) = \widehat{\Phi}_1^-(x_1/\Delta L)\,\widetilde{\Phi}_1^-(x_2/\Delta W), \qquad \widehat{\Psi}_{N_L-1,1}(x_1,x_2) = \widehat{\Phi}_{N_L-1}^+(x_1/\Delta L)\,\widetilde{\Phi}_1^-(x_2/\Delta W) \tag{A7}$$

at the bottom edge

$$\widehat{\Psi}_{j,NW-1}(x_1,x_2) = \Phi_j(x_1/\Delta L)\,\widehat{\Phi}_{N_W-1}^+(x_2/\Delta W) \tag{A8}$$

at the top edge

$$\widehat{\Psi}_{j,1}(x_1,x_2) = \Phi_j(x_1/\Delta L)\,\widetilde{\Phi}_1^-(x_2/\Delta W) \tag{A9}$$

and at the left and right edges

$$\widehat{\Psi}_{1,k}(x_1,x_2) = \widehat{\Phi}_1^-(x_1/\Delta L)\,\Phi_k(x_2/\Delta W), \qquad \widehat{\Psi}_{N_L-1,k}(x_1,x_2) = \widehat{\Phi}_{N_L-1}^+(x_1/\Delta L)\,\Phi_k(x_2/\Delta W) \tag{A10}$$

for $j = 2, \cdots, N_L - 2$ and $k = 0, N_W - 2$. All the other bicubic splines, instead, already satisfy the requirements and do not need any modification

$$\widehat{\Psi}_{j,k}(x_1,x_2) = \Psi_{j,k}(x_1,x_2) \tag{A11}$$

for $j = 2, \cdots, N_L - 2$ and $k = 0$ and $k = 2, \cdots, N_W - 2$.

With these modification we have

$$\widehat{\Psi}_{j,k}(x_1,x_2)\Big|_{x_2=W} = 0, \qquad \frac{\partial\widehat{\Psi}_{j,k}(x_1,x_2)}{\partial x_2}\Big|_{x_2=W} = \frac{\partial\widehat{\Psi}_{j,k}(x_1,x_2)}{\partial x_2}\Big|_{x_2=0} = 0 \tag{A12}$$

at the bottom and top edges, $\forall\, x_1 \in [0,L]$, and

$$\widehat{\Psi}_{j,k}(x_1,x_2)\Big|_{x_1=0} = \widehat{\Psi}_{j,k}(x_1,x_2)\Big|_{x_1=L} = 0, \qquad \frac{\partial\widehat{\Psi}_{j,k}(x_1,x_2)}{\partial x_1}\Big|_{x_1=0} = \frac{\partial\widehat{\Psi}_{j,k}(x_1,x_2)}{\partial x_1}\Big|_{x_1=L} = 0 \tag{A13}$$

at the left and right edges, $\forall\, x_2 \in [0,W]$.

The above expressions holds for $N_L \geq 3$ and $N_W \geq 1$. For the sake of completeness, we note that it is possible to consider also the case $N_L = 2$ for which we have only $N = (N_W + 1)$ modified bicubic splines. Let us first introduce the following modified cubic spline

$$\widehat{\Phi}_i(x) = \Phi_i(x) - \frac{\Phi_{i+1}(x) + \Phi_{i-1}(x)}{2} + \Phi_{i-2}(x) + \Phi_{i+2}(x) \tag{A14}$$

and, then, define the modified bicubic splines at the bottom edge (that is also the bottom left and right corners)

$$\widehat{\Psi}_{1,N_W-1}(x_1, x_2) = \widehat{\Phi}_1(x_1)\, \widehat{\Phi}^+_{N_W-1}(x_2) \tag{A15}$$

and the remaining ones (along the left and right edges)

$$\widehat{\Psi}_{1,k}(x_1, x_2) = \widehat{\Phi}_1(x_1)\, \Phi_k(x_2) \tag{A16}$$

for $k = -1, N_W - 2$.

## References


1. Clarke, P.J.; Davies, R.R.; England, P.C.; Parsons, B.; Billiris, H.; Paradissis, D.; Veis, G.; Cross, P.A.; Denys, P.H.; Ashkenazi, V.; et al. Crustal strain in central Greece from repeated GPS measurements in the interval 1989–1997. *Geophys. J. Int.* **1998**, *135*, 195–214. [CrossRef]
2. Mazzotti, S.; Leonard, L.; Cassidy, J.; Rogers, G.; Halchuk, S. Seismic hazard in western Canada from GPS strain rates versus earthquake catalog. *J. Geophys. Res.* **2011**, *116*, B12310. [CrossRef]
3. Palano, M.; Imprescia, P.; Agnon, A.; Gresta, S. An improved evaluation of the seismic/geodetic deformation-rate ratio for the Zagros Fold-and-Thrust collisional belt. *Geophys. J. Int.* **2018**, *213*, 194–209. [CrossRef]
4. Masson, F.; Chery, J.; Hatzfeld, D.; Martinod, J.; Vernant, P.; Tavakoli, F.; Ghafory-Ashtiani, M. Seismic versus aseismic deformation in Iran inferred from earthquakes and geodetic data. *Geophys. J. Int.* **2005**, *160*, 217–226. [CrossRef]
5. Savage, J.C.; Burford, R.O. Geodetic determination of relative plate motion in Central California. *J. Geophys. Res.* **1973**, *78*, 832–845. [CrossRef]
6. Savage, J.C. A dislocation mode of strain accumulation and release at a subduction zone. *J. Geophys. Res.* **1983**, *88*, 4984–4996. [CrossRef]
7. McCaffrey, R. Interseismic locking on the Hikurangi subduction zone: Uncertainties from slow-slip events. *J. Geophys. Res.* **2014**, *119*, 7874–7888. [CrossRef]
8. Li, S.; Moreno, M.; Bedford, J.; Rosenau, M.; Oncken, O. Revisiting viscoelastic effects on interseismic deformation and locking degree: A case study of the Peru-North Chile subduction zone. *J. Geophys. Res.* **2015**, *120*, 4522–4538. [CrossRef]
9. Ranalli, G. *Rheology of the Earth*, 2nd ed.; Chapman and Hall: London, UK, 1995.
10. Splendore, R.; Marotta, A. Crust-mantle mechanical structure in the Central Mediterranean region. *Tectonophysics* **2013**, *603*, 89–103. [CrossRef]
11. England, P.; McKenzie, D. A thin viscous sheet model for continental deformation. *R. Astron. Soc. Geophys. J.* **1982**, *70*, 295–321; Erratum in **1983**, *73*, 523–532. [CrossRef]
12. Medvedev, S.; Podladchikov, Y. New extended thin-sheet approximation for geodynamic applications–I. Model formulation. *Geophys. J. Int.* **1999**, *136*, 567–585. [CrossRef]
13. Marotta, A.; Sabadini, R. The signatures of tectonics and glacial isostatic adjustment revealed by the strain rate in Europe. *Geophys. J. Int.* **2004**, *157*, 865–870. [CrossRef]
14. Marotta, A.; Mitrovica, J.; Sabadini, R.; Milne, G. Combined effects of tectonics and glacial isostatic adjustment on intraplate deformation in central and northern Europe: Applications to geodetic baseline analyses. *J. Geophyis. Res.* **2004**, *109*, B01413. [CrossRef]
15. Jiménez-Munt, I.; Garcia-Castellanos, D.; Negredo, A.; Platt, J. Gravitational and tectonic forces controlling postcollisional deformation and the present-day stress field of the Alps: Constraints from numerical modeling. *Tectonics* **2005**, *24*, TC5009. [CrossRef]
16. Hearn, E. What can GPS data tell us about the dynamics of post-seismic deformation? *Geophys. J. Int.* **2003**, *155*, 753–777. [CrossRef]
17. Hutton, W.; DeMets, C.; Sanchez, O.; Suarez, G.; Stock, J. Slip kinematics and dynamics during and after the 1995 October 9 $M_w$ = 8.0 Colima-Jalisco earthquake, Mexico, from GPS geodetic constraints. *Geophys. J. Int.* **2001**, *146*, 637–658. [CrossRef]

18. Cheloni, D.; D'Agostino, N.; Selvaggi, G.; Avallone, A.; Fornaro, G.; Giuliani, R.; Reale, D.; Sansosti, E.; Tizzani, P. Aseismic transient during the 2010–2014 seismic swarm: Evidence for longer recurrence of M $\geq$ 6.5 earthquakes in the Pollino gap (Southern Italy)? *Sci. Rep.* **2017**, *7*, 576. [CrossRef]
19. Murray, J.; Minson, S.; Svarc, J. Slip rates and spatially variable creep on faults of the northern San Andreas system inferred through Bayesian inversion of Global Positioning System data. *J. Geophys. Res.* **2014**, *119*, 6023–6047. [CrossRef]
20. Cinti, F.; Moro, M.; Pantosti, D.; Cucci, L.; D'Addezio, G. New constraints on the seismic history of the Castrovillari fault in the Pollino gap (Calabria, southern Italy). *J. Seismol.* **2002**, *6*, 199–217. [CrossRef]
21. Michetti, A.; Ferreli, L.; Serva, L.; Vittori, E. Geological evidence for strong historical earthquakes in an "aseismic" region: The Pollino case. *J. Geodyn.* **1997**, *24*, 67–86. [CrossRef]
22. Totaro, C.; Presti, D.; Billi, A.; Gervasi, A.; Orecchio, B.; Guerra, I.; Neri, G. The ongoing seismic sequence at the Pollino Mountains. *Italy Seismol. Res. Lett.* **2013**, *84*, 955–962. doi:10.1785/0220120194. [CrossRef]
23. Totaro, C.; Koulakov, I.; Orecchio, B.; Presti, D. Detailed crustal structure in the area of the southern Apennines–Calabrian Arc border from local earthquake tomography. *J. Geodyn.* **2014**, *82*, 87–97. [CrossRef]
24. Chiarabba, C.; Agostinetti, N.; Bianchi, I. Lithospheric fault and kinematic decoupling of the Apennines system across the Pollino range. *Geophys. Res. Lett.* **2016**, *43*, 3201–3207. [CrossRef]
25. Sabadini, R.; Aoudia, A.; Barzaghi, R.; Crippa, B.; Marotta, A.; Borghi, A.; Cannizzaro, L.; Calcagni, L.; Dalla Via, G.; Rossi, G.; et al. First evidences of fast creeping on a long-lasting quiescent earthquake normal-fault in the Mediterranean. *Geophys. J. Int.* **2009**, *179*, 720–732. [CrossRef]
26. Brozzetti, F.; Cirillo, D.; de Nardis, R.; Cardinali, M.; Lavecchia, G.; Orecchio, B.; Presti, D.; Totaro, C. Newly identified active faults in the Pollino seismic gap, southern Italy, and their seismotectonic significance. *J. Struct. Geol.* **2017**, *94*, 13–31. [CrossRef]
27. DISS Working Group. Database of Individual Seismogenic Sources (DISS), Version 3.0.4: A Compilation of Potential Sources for Earthquakes Larger than M 5.5 in Italy and Surrounding Areas. 2007. Available online: http://www.ingv.it/DISS/ (accessed on 15 July 2017).
28. Rovida, A.; Camassi, R.; Gasperini P.; Stucchi, M. CPTI11, la versione 2011 del Catalogo Parametrico dei Terremoti Italiani, Milano, Bologna. 2011. Available online: http://emidius.mi.ingv.it/CPTI (accessed on 15 July 2017).
29. ISIDe Working Group. Italian Seismological Instrumental and Parametric Databases. 2016. Available online: http://cnt.rm.ingv.it/iside (accessed on 15 July 2017).
30. Savage, J.; Gan, W.; Svarc, J. Strain accumulation and rotation in the Eastern California Shear Zone. *J. Geophys. Res.* **2001**, *106*, 21995–22007. [CrossRef]
31. Shen, Z.K.; Jackson, D.; Ge, B. Crustal deformation across and beyond the Los Angeles basin from geodetic measurements. *J. Geophys. Res.* **1996**, *101*, 27957–27980. [CrossRef]
32. Okada, Y. Internal deformation due to shear and tensile faults in a half-space. *Seismol. Soc. Am.* **1992**, *82*, 1018–1040.
33. Ferranti, L.; Palano, M.; Cannavò, F.; Mazzella, M.; Oldow, J.; Gueguen, E.; Mattia, M.; Monaco, C. Rates of geodetic deformation across active faults in southern Italy. *Tectonophysics* **2014**, *621*, 101–122. [CrossRef]
34. Yabuki, T.; Matsu'ura, M. Geodetic data inversion using a Bayesian information criterion for spatial distribution of fault slip. *Geophys. J. Int.* **1992**, *109*, 363–375. [CrossRef]
35. Cambiotti, G.; Zhou, X.; Sparacino, F.; Sabadini, R.; Sun, W. Joint estimate of the rupture area and slip distribution of the 2009 L'Aquila earthquake by a Bayesian inversion of GPS data. *Geophys. J. Int.* **2017**, *209*, 992–1003. [CrossRef]
36. Fukuda, J.; Johnson, K. A fully Bayesian inversion for spatial distribution of fault slip with objective smoothing. *Bull. Seism. Soc. Am.* **2008**, *98*, 1128–1146. [CrossRef]
37. Tarantola, A. *Inverse Problem Theory and Methods for Model Parameter Estimation*; SIAM: Philadelphia, PA, USA, 2005.
38. Devoti, R.; Esposito, A.; Pietrantonio, G.; Pisani, A.; Riguzzi, F. Evidence of large scale deformation patterns from GPS data in the Italian subduction boundary. *EPSL* **2011**, *311*, 230–241. [CrossRef]

39. Orecchio, B.; Prestia, D.; Totaro, C.; D'Amico, S.; Neri, G. Investigating slab edge kinematics through seismological data: Thenorthern boundary of the Ionian subduction system (south Italy). *J. Geodyn.* **2015**, *88*, 23–35. [CrossRef]
40. Totaro, C.; Orecchio, B.; Presti, D.; Scolaro, S.; Neri, G. Seismogenic stress field estimation in the Calabrian Arc region (south Italy) from a Bayesian approach. *Geophys. Res. Lett.* **2016**, *43*, 8960–8969. [CrossRef]

 *remote sensing*

*Article*

# Geodetic Deformation versus Seismic Crustal Moment-Rates: Insights from the Ibero-Maghrebian Region

**Federica Sparacino [1], Mimmo Palano [1,*], José Antonio Peláez [2] and José Fernández [3]**

[1]   Istituto Nazionale di Geofisica e Vulcanologia, Osservatorio Etneo - Sezione di Catania, 95125 Catania, Italy; federica.sparacino@ingv.it

[2]   Department of Physics, University of Jaén, 23071 Jaén, Spain; japelaez@ujaen.es

[3]   Institute of Geosciences (CSIC, UCM), C/ Doctor Severo Ochoa, 7, Ciudad Universitaria, 28040 Madrid, Spain; jft@mat.ucm.es

*   Correspondence: mimmo.palano@ingv.it; Tel.: +39-095-7165800

Received: 27 January 2020; Accepted: 12 March 2020; Published: 16 March 2020

**Abstract:** Seismic and geodetic moment-rate comparisons can reveal regions with unexpected potential seismic hazards. We performed such a comparison for the Southeastern Iberia—Maghreb region. Located at the western Mediterranean border along the Eurasia–Nubia plate convergence, the region has been subject to a number of large earthquakes (M ≥ 6.5) in the last millennium. To this end, on the basis of available geological, tectonic, and seismological data, we divided the study area into twenty-five seismogenic source zones. Many of these seismogenic source zones, comprising the Western Betics, the Western Rif mountains, and the High, Middle, and Saharan Atlas, are characterized by seismic/geodetic ratio values lower than 23%, evidencing their prevailing aseismic behavior. Intermediate seismic/geodetic ratio values (between 35% and 60%) have been observed for some zones belonging to the Eastern Betics, the central Rif, and the Middle Atlas, indicating how crustal seismicity accounts only for a moderate fraction of the total deformation-rate budget. High seismic/geodetic ratio values (> 95%) have been observed along the Tell Atlas, highlighting a fully seismic deformation.

**Keywords:** earthquake catalogs; GNSS; seismic/aseismic behavior; earthquake hazards; Eurasia-Nubia plate

## 1. Introduction

Seismic and geodetic moment-rate comparisons provide relevant insights into the seismic hazard of regions subjected to significant crustal deformation. Valuable examples come from numerous tectonic regions worldwide, such as Iran [1,2], western Canada [3], western USA [4,5], Greece [6,7], southern Italy [8–10], Himalayas [11–13], and the East African Rift [14]. Geodetic moment-rate represents a measure of crustal deformation and might include both anelastic and elastic components, whereas seismic moment-rates measure the deformation accommodated by faulting. Although seismic and geodetic moment-rate estimations are affected by several physical uncertainties, their comparison has enabled identifying regions where the total deformation-rate budget is entirely released by crustal seismicity [3,4,10], as well as regions where the excess deformation-rate can be released either in aseismic slip across faults or through large future earthquakes [1,2,5,13,14]. Regarding this last point, a common practice is to estimate the number of large earthquakes (usually called "missing earthquakes") needed so that the moment released by seismicity balances the deficit of moment derived from geodetic measurements [3,4,13].

The most common factors affecting moment-rate estimations are related to the completeness and temporal length of available seismic catalogs, which may be little adequate when it comes to capturing the seismic cycle of a given region; therefore, providing a lower estimation of the seismic moment-rate. Moreover, factors related to the reliability of the geodetic data, for instance, an insufficient time span to sample both the seismic and the aseismic spectrum, velocity uncertainties, stations density, and network geometry, long-term deformation transient, etc., may strongly affect the estimations. Furthermore, this comparison could be challenging in regions such as the Southeastern Iberia–Maghreb (western Mediterranean border; Figure 1), characterized by low-rate tectonic deformation and long seismic cycles with unusually high-magnitude (M > 7) earthquakes and seismicity patterns that are highly clustered in time.

Here, based on a combination of original Global Navigation Satellite System (GNSS) observations integrated with solutions coming from literature, scalar geodetic moment-rates have been compared with those derived from earthquake catalogs in order to provide a statistical estimation of the deformation-rate budget for the Southeastern Iberia–Maghreb region. In order to mitigate possible misreadings arising from the above mentioned physical uncertainties, we adopted different approaches in computing seismic moment-rates. In particular, the seismic moment-rates have been computed by adopting the moment summation [15] and the truncated cumulative Gutenberg–Richter distribution [16] approaches. Uncertainties in moment-rate estimations have been computed by adopting a logic tree approach [17].

## 2. Background Setting

The tectonic setting of the Southeastern Iberia–Maghreb region results from the Cenozoic orogenic evolution within the complex framework of the Eurasia–Nubia plate convergence (e.g., [18,19]). The present-day plate tectonic setting was reached in the earliest Miocene, with the welding of the Iberian Peninsula to the Eurasia plate. At that time, the main plate boundary between Africa and Eurasia in the Western Mediterranean region developed along the transpressive fault zone connecting the Açores triple junction to the Rif and Tell Atlas through the Gibraltar Orogenic Arc (e.g., [20,21]). In Southern Iberia, the Betics show a mainly North-North-West (NNW) direction of tectonic transport, turning around the Gibraltar Strait to a prevailing South-West (SW)-directed thrusting along the Rif while an overall South-South-East (SSE)-directed tectonic transport dominated the Tell Atlas fold-thrust belt. The inner side of the Gibraltar Orogenic Arc is occupied by the Alboran basin (Figure 1), a back-arc basin formed mainly during the early Miocene (e.g., [22]).

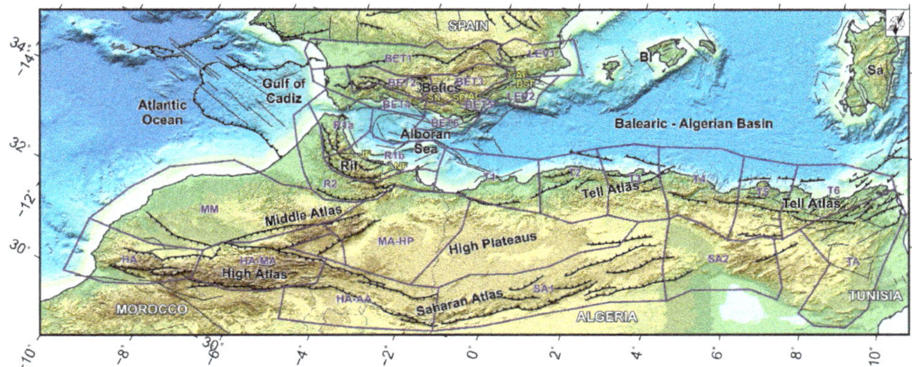


**Figure 1.** Simplified tectonic map of Northwestern Africa and Southeastern Iberia. Mapped faults are re-drawn from [23] and references therein. Seismic source zones (Betics seismic source zones—BET1, BET2, BET3, BET4, BET5, BET6; Atlas seismogenic sources—HA, HA-AA, HA-MA, SA1, SA2, TA represent High, Middle, Saharan, and Tunisian Atlas, as well as the northeastern part of the Anti-Atlas; Levante seismogenic regions—LEV1, LEV2; Moroccan Meseta-High Plateaus seismic source zones—MA-HP, MM; Rif seismogenic sources—R1a, R1b, R2; Tell seismogenic source zones—T1, T2, T3, T4, T5, T6) analyzed in this study are also reported as blue solid polygons [24,25]. The map is plotted in an oblique Mercator projection.

A number of on- and off-shore active fault systems dissect the study area and are responsible for a large amount of the current seismic release of the region. Among them, the Trans Alboran shear zone (a major NE-SW trending left-lateral strike-slip system) cuts the Alboran basin, and branches into the Eastern Betics shear zone and into the Nekor/Al-Hoceima seismic zone, northward and southward, respectively (e.g., [26–28]). In addition, the Trans Alboran shear zone is westward connected to southern Spanish off-shore by the right-lateral Yusuf fault [29]. In western Algeria, active faulting occurs within the Tell Atlas, along NE-SW-trending, right-stepping en-echelon reverse faults [30]. In eastern Algeria, active reverse faulting occurs in a broader area [31–33] and is coupled with right-lateral strike slip faulting on EW-trending faults [34]. Active faulting is recognized along the Aurès Mountains and the southern Atlas region [35].

Southern Iberia and Maghreb regions have been subject by the occurrence of large earthquakes (with estimated magnitude M ≥ 6.5) in the last centuries [36–38]. Global Navigation Satellite System (GNSS)-based measurements show that active deformation involves both Southern Iberia and Maghreb regions with rates up to 6 mm/yr (e.g., [23,39]). Such a GNSS-based active deformation has been interpreted in terms of elastic block modeling (e.g., [39,40]. However, the detection of a significant aseismic deformation component on the Betic–Rif system (~75%; see [41] for details) highlights that the comparison of seismic and geodetic deformation-rates would not be balanced across the region. Results achieved in this study support the inference on the aseismic deformation behavior for most of the Betics and Rif regions, revealing a fully seismic deformation along most of the Tell Atlas.

## 3. Seismological Background and Seismic Moment-Rates

### 3.1. Historical Earthquakes

Available historical seismic catalogs for southern Iberia and Maghreb regions highlight the occurrence of large (M ≥ 6.5) earthquakes in Tunisia since AD 856 [36–38,42], including not well-documented damaging earthquakes since several centuries before Christ (BC). However, in spite of efforts made in recent updates [25] focused on seismic hazard studies, the accuracy of these catalogs is not uniform due both to the off-shore occurrence of numerous earthquakes and to the sparsely populated area. A number of studies focusing on this region have evidenced that many small and moderate earthquakes have most likely been overlooked, and that available seismic catalogs can be considered complete for M ≥ 6 earthquakes since approximately 1500, 1700, and 1850, for southern Iberia, northern Morocco and northern Algeria areas, respectively (e.g., [25,37,38] and references therein). Historical earthquakes are largely concentrated along the Azores–Gibraltar fault zone, along the Betic–Rif system, and along the Tell Atlas fold-thrust belt (Figure 2a). In detail, the strongest historical earthquakes (M ≥ 7.0) are mainly concentrated on the Azores–Gibraltar fault system (AD 1309, 1356, 1761, and 1755 events) and along the Tell Atlas fold-thrust belt (AD 856, 1365, 1716, 1722, 1790, 1832, 1867, and 1891 events).

*3.2. Instrumental Seismicity*

The distribution of instrumental seismicity (earthquakes with M ≥ 2.5) as well as the main seismotectonic features (focal plane solutions for crustal events with M ≥ 3.0) of the study area are reported in Figure 2b,c, respectively. The bulk of instrumental seismicity, corresponding to low and moderate magnitude earthquakes, is spread across the Nubia–Eurasia plate boundary. It shows clear epicenter alignments between Azores and Gibraltar, and from Algeria to Tunisia, providing evidence of active faulting along the plate boundary belt. Along the Azores-Gibraltar fault zone and the Gulf of Cádiz, earthquakes have depths up to 40 km, being characterized by fault plane solutions with both reverse and strike-slip features (Figure 2b,c). Along North Algeria and Tunisia regions, earthquakes usually have hypocentral depths shallower than 30 km and are characterized by fault plane solutions with prevailing reverse features, coupled with subordinate strike-slip faulting features (Figure 2b,c). Crustal seismicity spreads over a ~300 km wide band between Northern Morocco and Southern Spain, where a number of epicenter alignments suggests the presence of individual active seismogenic faults; therefore, depicting a complex network of NE-SW and WNW-ESE faults (with lengths from tens to some hundreds of kilometers) that accommodate the Nubia–Eurasia convergence (Figure 2b). Seismicity at intermediate depths concentrates in the 50–100 km depth interval, in a 150-km-long band that extends from Southern Spain to the Alboran Sea and abruptly disappears toward east. In addition, the occurrence of deep earthquakes down to more than 600 km depth below the Granada Basin has been documented since 1954 (see [43,44] and references therein for additional details). These deep earthquakes are likely related to a sunk lithospheric slab whose relation with the current surface deformation is still debated (e.g., [45]).

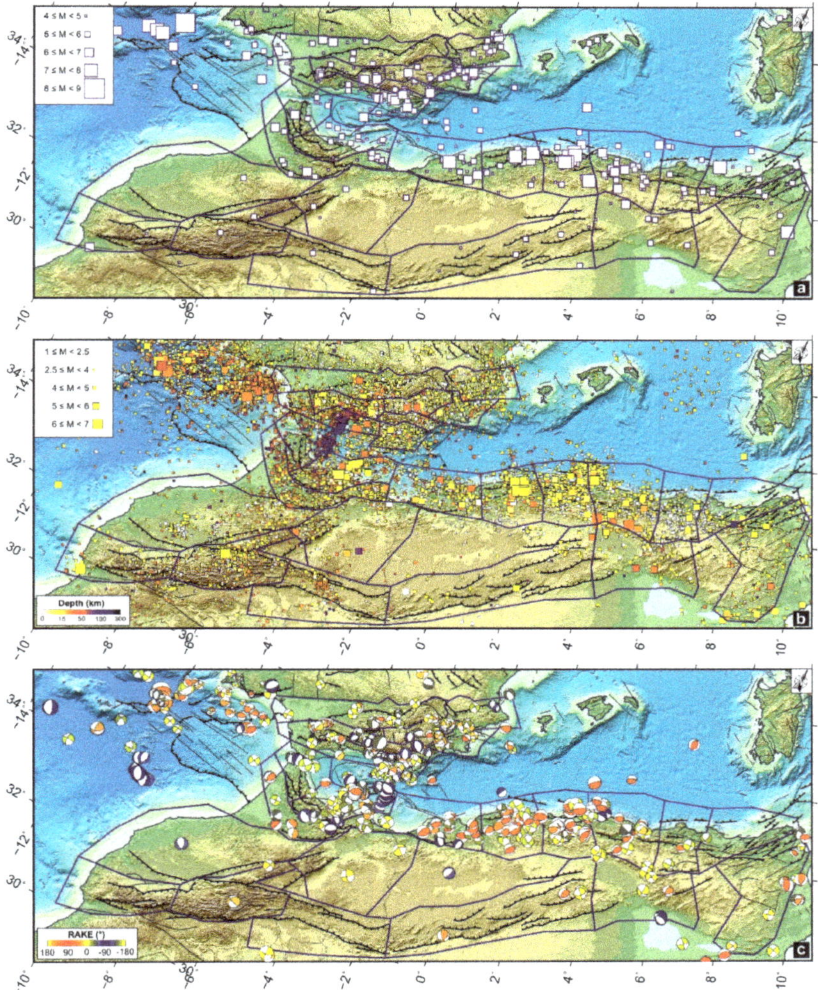

**Figure 2.** (**a**) Historical earthquakes (blue squares; M ≥ 4.0) occurring during AD 856–1950 [36–38]. (**b**) Instrumental crustal seismicity (M ≥ 2.5) occurring in the investigated area since 1910 [46]. Symbols are colored according to focal depth. (**c**) Lower hemisphere, equal area projection for fault plane solutions (with M ≥ 3) compiled from the investigated area [23,47–49]; fault plane solutions are colored according to rake: red indicates pure thrust faulting, blue is pure normal faulting, and yellow is strike-slip faulting.

### 3.3. Seismogenic Source Zones

The study area has been divided into twenty-five crustal seismogenic source zones (Figure 1) by taking into account the seismic zonation available in recent literature [24,25,50], which have been defined on the basis of the latest tectonic, geological, and seismological considerations. Some of these zones are well-defined on the basis of constraints coming from the analysis of the earthquake catalog (stationarity of the completeness periods, evaluation of the mean activity rate) and from a set of geological and seismotectonic considerations, such as style, geometry, and distribution of fault systems (with direct evidence of Holocene activity) and their relation to the local stress and deformation regimes.

Other zones, although defined on similar seismotectonic or geological information, are lacking a clear correspondence between the contemporary seismic activity and the Holocene tectonic activity.

The seismic sources adopted for the Moroccan–Algeria–Tunisian region come from [24] and slightly updated in [50]. For this region, the general criterion adopted in [24,50] was to delimit areas with similar tectonic features and a more or less homogeneous stress pattern and seismicity parameters. Individual faults were not specifically considered because of the very scarce availability of kinematic, seismic and paleoseismic observations in this region. They were only considered ensuring that known faults or fault segments were not divided and partially included in different adjacent sources.

The seismic sources adopted for southeastern Spain belong to the last official upgrade of the Spanish seismic hazard maps [25], known as the 'ByA12' zonation (Zonificación de Bernal y asociados). This zonation is prevailing based on the available seismotectonic and geological information and was aimed at seismic hazard computations [25]. Such a zonation is characterized by some small zones sampled by zero or just one GNSS site; therefore, we joined some of them into large zones following more generalized seismic and tectonic criteria (Figure 1). We performed such an additional step because the computed strain-rates cannot be taken as an accurate estimate for small zones, due to the lack of spatial constraints from geodetic data (i.e., the strain-rates information would be derived from the velocity interpolation only).

A short description of the twenty-five seismogenic source zones is given below (see also Table A1 in Appendix A for additional details).

- Tell seismogenic source zones (T1, T2, T3, T4, T5, and T6). Tell region is the one with the largest density of earthquakes, including some of the most destructive events in the area, such as the 10 October, 1980 Chlef ($M_W$ 7.3) and the more recent 21 May, 2003 Boumerdes ($M_W$ 6.9) earthquakes. The differentiation among these six zones is mainly related to the different seismic features, clearly revealed when computing seismicity parameters. These zones are based on the previous works by [51–53]. T1 to T4 sources display a dominant compression striking NNW-SSE, whereas the T5 source exhibits a N-S-striking compressional strike-slip regime (e.g., [24,49,54]). Few focal mechanisms are available for the T6 source and therefore the mean stress regime is poorly constrained (Figure 2c).
- Rif seismogenic sources (R1a, R1b, and R2). The Rif region has been divided into three source zones on the basis of well-known active faults (e.g., the Jebah and Nekor faults) and seismic activity distribution in the Rif and in the southwestern part of the Alboran Sea. Large earthquakes, such as the recent 25 January, 2016 ($M_w$ 6.3) Al Hoceima event or the historical 1624 May, 11 ($M_S$ 6.7) Fez event, occurred along some of these active faults. R1b and R2 sources show a horizontal compression, striking NW-SE to NNW-SSE, as well as a perpendicular horizontal extension [50]. There are few mechanisms for R1a source, but they show a pure WSW-ENE horizontal extension [54].
- Atlas seismogenic sources (HA, HA-MA, HA-AA, SA1, SA2, and TA). Overall, they include the High, Middle, Saharan, and Tunisian Atlas, as well as the northeastern part of the Anti-Atlas. These zones have been struck by moderate to large magnitude earthquakes, such as the 856 3, December ($M_S$ 7.0) Tunisia event, the 1731 ($M_W$ 6.4) and the 29 February, 1960 ($M_W$ 5.7) Agadir events. Some of these zones are also based on the previous works by [51–53]. HA-MA, SA1, SA2, and TA sources exhibit a near pure compressional stress regime, with a wide range of attitudes (NNW-SSE, NW-SE, N-S, and NW-SE, respectively; [24,55,56]. The stress field for the other seismogenic source zones is poorly constrained because of the limited number of available focal mechanisms (Figure 2c).
- Moroccan Meseta-High Plateaus seismic source zones (MM and MA-HP). They include a region with scarce seismicity and characterized by the occurrence of only low and low to moderate earthquakes, such as the 1979 January, 17 ($M_S$ 4.6) Khenifra and the 29 November, 1980 ($M_S$ 4.9) Jerada events. The MA-HP source is based on a previous one already described in [52,53]. The MM source is characterized by the presence of NNE-SSW and NE-SW oriented faults, as well as faults

- parallel to the Atlantic coast [57]. The stress field for both seismogenic source zones is poorly constrained because of the limited number of available focal mechanisms (Figure 2c).
- Betic seismic source zones (BET1, BET2, BET3, BET4, BET5, and BET6). These seismogenic sources correspond to the seismogenic region called Betic Cordilleras, also embracing the Subbetic, western part of the Prebetic units, and the Guadalquivir Basin, in its northern part. With the exception of the 29 March, 1954 ($M_w$ 7.8) deep seismic event (depth of ~625 km), both instrumental and historical earthquakes did not exceed M 6.3 [24]. As mentioned above, initial sources delimited in [25] have been grouped into larger seismogenic sources. BET1 source zone embraces a complex network of faults striking WSW-ENE on the Guadalquivir Basin and NE-SW to NNE-SSW on the Sierra de Segura area, which hosted some historical destructive earthquakes, for instance the 1357 ($M_w$ 6.1) Andalucia and the 1504 ($M_w$ 6.1) Carmona events. BET2 source zone is characterized by faults striking N140/160E and N70E to E-W, as well as faults striking NW-SE which hosted destructive earthquakes, such as the 1431 ($M_w$ 6.11) Granada event and the 1884 ($M_w$ 6.30) Arenas del Rey event. BET3 source zone is characterized by low to moderate instrumental seismicity with a $M_w$ 6.17 destructive event occurring in 1531 close to the town of Baza. BET4 source zone is characterized by low to moderate instrumental seismicity with a Mw 6.10 destructive event occurring in 1910 close to the small coastal town of Adra. BET5 source zone includes, from west to east, the Sierra Nevada, the Sierra de los Filabres, and the Almanzora Corridor areas. This zone is characterized by NW-SE to NNW-SSE oriented faults which are considered the seismogenic sources for seismicity striking the region. BET6 source zone includes the Adra-Sierra Alhamilla and Almería areas, incorporating to some extent the off-shore platform. In this source, existing fault systems mostly strike NW-SE and NNE-SSW, hosting historical destructive earthquakes, such as the 1522 ($M_w$ 6.07) Almería event. Regarding the stress regime, BET2 and BET4 source zones are characterized by a primary NNW-SSE sub-horizontal compression, while BET1, BET3, BET5, and BET6 zones are characterized by a complex pattern resulting from the interaction of different sources of stress (see [54,55] for additional details).
- LEV1 and LEV2 seismogenic source zones. This area, named as Levante seismogenic region [25], covers the easternmost part of the Prebetic and Internal Betics and is dominated by a prevailing strike slip stress field with a near E-W extension and a N-S compression [54]. Both seismogenic source zones are characterized by the occurrence of several low to moderate magnitude earthquakes and some damaging historical events, rarely exceeding M > 6 such as the 1396 ($M_W$ 6.00) Tavernes de Valldigna and the 1645 ($M_S$ 6.25) Muro de Alcoy for LEV1 and the 1829 ($M_W$ 6.25) Torrevieja event for LEV2 [36]. Among the main tectonic features included in this source, the Cádiz-Alicante and the Caudete-Elda-Elche fault systems are worth noting. Regarding the LEV2 seismogenic source zone, different fault systems host the seismicity, such as the striking N140-160E San Miguel de Salinas and Torrevieja faults and the NW-SE Bajo Segura fault.

### 3.4. Seismic Moment-Rates

For each seismogenic source zone, the seismic moment-rate was calculated by adopting a truncated cumulative Gutenberg–Richter distribution [16]:

$$\dot{M}_{seis} = \varphi \frac{b}{(c-b)} 10^{[(c-b)M_{max}+a+d]} \tag{1}$$

where $M_{max}$ is the magnitude of the largest earthquake that could occur within the investigated region; $\varphi$ is a correction for the stochastic magnitude–moment relation; taking into account an average error of 0.2 on magnitudes, we assumed $\varphi = 1.27$ [16]. $c$ and $d$ are the coefficients of the magnitude ($M$)–scalar moment ($M_{seis}$) relation:

$$\log M_{seis} = cM + d \tag{2}$$

According to [58] we set c = 1.5 and d = 9.05. *a* and *b* are coefficients (seismicity rate and slope, respectively) of the Gutenberg–Richter recurrence relation:

$$\log N_M = a - bM \tag{3}$$

being $N_M$ the cumulative number of earthquakes with magnitude *M* and larger.

**Table 1.** Earthquake catalog statistic and estimated seismic moment-rates ($\dot{M}_{seis}$) with associated errors for each seismic source zone (SSZ). Moment-rates have been estimated by using both a (#) truncated cumulative Gutenberg–Richter distribution [16] and the (*) the moment summation [6] approach. Seismicity rate (a) and slope (b) of the Gutenberg–Richter recurrence relation with its associated error are reported, as well as the values of the maximum magnitude observed ($M_{obs}$) and estimated ($M_{max}$). Abbreviations are as follows: P, [37]; H, [38], S, [36]. Because of the large uncertainties of MA-HP and SA1, we also reported the recomputed *a* and *b* values (see the main text for details); in curved brackets the original values.

| SSZ | $M_{obs}$ | $a$ | $b$ | $M_{max}$ | $\dot{M}_{seis}$ # ($10^{16}$ Nm/yr) | $\dot{M}_{seis}$ * ($10^{16}$ Nm/yr) |
|---|---|---|---|---|---|---|
| BET1 | 6.10 (20 October 1883)[P] | 4.03 | 1.13 ± 0.01 | 6.7 ± 0.4 | $1.41^{+0.33}_{-0.27}$ | $0.68^{+0.47}_{-0.24}$ |
| BET2 | 7.80 (29 March 1954)[P] | 4.43 | 1.13 ± 0.01 | 6.7 ± 0.4 | $3.53^{+0.82}_{-0.67}$ | $1.31^{+0.91}_{-0.46}$ |
| BET3 | 6.17 (30 September 1531)[S] | 3.87 | 1.13 ± 0.01 | 6.6 ± 0.4 | $0.89^{+0.21}_{-0.17}$ | $0.60^{+0.42}_{-0.21}$ |
| BET4 | 6.10 (16 June 1910)[P] | 3.84 | 1.13 ± 0.01 | 6.7 ± 0.4 | $0.91^{+0.21}_{-0.17}$ | $0.62^{+0.43}_{-0.21}$ |
| BET5 | 6.10 (9 November 1518)[H] | 3.78 | 1.13 ± 0.01 | 6.6 ± 0.4 | $0.73^{+0.17}_{-0.14}$ | $0.45^{+0.31}_{-0.16}$ |
| BET6 | 6.50 (22 September 1522) [H] | 4.24 | 1.13 ± 0.01 | 6.7 ± 0.3 | $2.28^{+0.41}_{-0.35}$ | $0.16^{+0.11}_{-0.06}$ |
| HA | 6.40 (1731)[P] | 2.06 | 0.69 ± 0.06 | 6.6 ± 0.4 | $3.09^{+2.62}_{-1.42}$ | $3.23^{+2.25}_{-1.13}$ |
| HA-AA | 5.80 (22 June 1941)[P] | 1.69 | 0.55 ± 0.05 | 6.3 ± 0.3 | $3.90^{+2.58}_{-1.56}$ | $1.36^{+0.95}_{-0.48}$ |
| HA-MA | 5.30 (11 August 2007) | 5.81 | 1.35 ± 0.01 | 5.5 ± 0.3 | $5.53^{+0.39}_{-0.36}$ | $2.23^{+1.55}_{-0.78}$ |
| LEV1 | 6.00 (18 December 1396)[S] | 3.21 | 0.92 ± 0.02 | 6.9 ± 0.4 | $3.68^{+1.52}_{-1.08}$ | $0.54^{+0.37}_{-0.19}$ |
| LEV2 | 6.25 (21 March 1829)[S] | 3.38 | 0.92 ± 0.02 | 6.7 ± 0.4 | $4.17^{+1.71}_{-1.21}$ | $1.83^{+1.27}_{-0.64}$ |
| MA-HP | 5.30 (29 November 1980)[H] | 2.71 (4.08) | 0.76 ± 0.07 (1.04 ± 0.41) | 5.6 ± 0.3 | $1.05^{+0.7}_{-0.4}$ $(1.46^{+13.81}_{-1.27})$ | $0.71^{+0.49}_{-0.25}$ |
| MM | 5.10 (17 January 1979)[P] | 7.20 | 1.66 ± 0.15 | 5.6 ± 0.4 | $2.98^{+19.96}_{-2.37}$ | $0.35^{+0.24}_{-0.12}$ |
| R1a | 5.50 (29 September 1822)[P] | 2.62 | 0.83 ± 0.14 | 5.7 ± 0.4 | $0.49^{+0.73}_{-0.29}$ | $2.45^{+1.70}_{-0.85}$ |
| R1b | 6.40 (24 February 2004)[P] | 2.58 | 0.67 ± 0.06 | 6.8 ± 0.3 | $19.27^{+14.49}_{-8.28}$ | $3.69^{+2.57}_{-1.29}$ |
| R2 | 6.70 (11 May 1624)[P] | 4.96 | 1.19 ± 0.10 | 6.7 ± 0.4 | $5.96^{+5.96}_{-2.88}$ | $3.66^{+2.55}_{-1.28}$ |
| SA1 | 5.80 (4 April 1924)[H] | 1.77 (5.50) | 0.60 ± 0.06 (1.27 ± 0.26) | 6.3 ± 0.4 | $2.62^{+2.29}_{-1.23}$ $(7.00^{+27.13}_{-3.88})$ | $1.84^{+1.28}_{-0.64}$ |
| SA2 | 6.00 (12 February 1946)[H] | 4.67 | 1.07 ± 0.08 | 6.1 ± 0.4 | $6.96^{+5.18}_{-2.93}$ | $7.13^{+4.97}_{-2.49}$ |
| T1 | 7.00 (9 October 1790)[H] | 3.25 | 0.79 ± 0.04 | 7.4 ± 0.4 | $50.60^{+33.32}_{-20.09}$ | $24.85^{+17.31}_{-8.68}$ |
| T2 | 7.30 (10 October 1980)[H] | 2.81 | 0.65 ± 0.05 | 7.8 ± 0.3 | $300.12^{+224.74}_{-128.66}$ | $155.60^{+108.40}_{-54.33}$ |
| T3 | 7.30 (2 January 1867)[H] | 3.50 | 0.81 ± 0.07 | 7.5 ± 0.4 | $79.15^{+80.02}_{-39.76}$ | $30.39^{+21.17}_{-10.61}$ |
| T4 | 6.90 (21 May 2003)[H] | 3.86 | 0.86 ± 0.05 | 7.3 ± 0.3 | $65.18^{+40.60}_{-24.98}$ | $19.31^{+13.45}_{-6.74}$ |
| T5 | 5.80 (27 October 1985)[H] | 4.46 | 1.08 ± 0.01 | 5.9 ± 0.3 | $3.18^{+0.62}_{-0.52}$ | $1.36^{+0.95}_{-0.47}$ |
| T6 | 7.00 (27 November 1722)[H] | 2.85 | 0.73 ± 0.03 | 7.0 ± 0.4 | $23.48^{+14.22}_{-8.86}$ | $12.67^{+8.83}_{-4.43}$ |
| TA | 7.00 (3 December 856)[H] | 2.50 | 0.65 ± 0.17 | 7.0 ± 0.4 | $30.71^{+92.35}_{-23.12}$ | $3.37^{+2.35}_{-1.18}$ |

Adopted parameters are detailed in Table 1; as a first step, we used the values already reported in literature [24,25]. As described in [24,25], *a* and *b* parameters have been directly computed from available catalogs by adopting the maximum likelihood method [59]. The $M_{max}$ values and associated uncertainties have been estimated by adopting different approaches. Regarding the Spanish seismogenic source zones, $M_{max}$ values have been estimated by adopting empirical laws [60] linking magnitude values to the dimensions of mapped active faults. For the zones characterized by a lack tectonic data, $M_{max}$ values have been estimated directly from the seismic catalog by adopting a cumulative exponential-truncated Gutenberg–Richter recurrence relation. Regarding the seismic source zones located along the African margin, $M_{max}$ values have been estimated from the seismic catalog by adopting the Kijko–Sellevoll approach [61], a method usually applied on regions where only a limited number of large earthquakes is available. Because the seismogenic source zones MA-HP and SA1 are characterized by $b$-values affected by large uncertainties (0.41 and 0.26, respectively), we performed new estimations by using as input event records reported in the International Seismological Centre (ISC) bulletin (see Figure A2 in Appendix B for details). Finally, estimated seismic moment-rates with an associated 67% confidence interval are reported in Table 1.

The scalar seismic moment-rate can be estimated also by adopting the moment summation approach [15]:

$$\dot{M}_{seis} = \frac{1}{\Delta T} \sum_{n=1}^{N} M_{seis}^{(n)} \tag{4}$$

where $N$ is the number of events occurring during a given time interval $\Delta T$ in the volume $A \cdot H_s$ (with $A$, the surface area and $H_s$, the seismogenic thickness), $M_{seis}$ is the scalar seismic moment of the $n$-$th$ earthquake from the N total earthquakes. In the following, we focused mainly on the results obtained from Equation (1); however, in order to test sensitivity, we considered results achieved by adopting Equation (4) (see Table 1).

## 4. GNNS Data and Geodetic Moment-Rates

### 4.1. GNSS Data Processing

Here we analyse an extensive GNSS dataset which, covering 20 years of observations (from 1999.00 up to 2019.00), includes more than 300 continuous GNSS sites available at EUREF Permanent GNSS Network [62], at Crustal Dynamics Data Information System [63], at UNAVCO [64] and from local and regional networks [23]. Time series of this GNSS dataset cover different time spans, ranging from 3.5 to 20 years with an average duration of 8.7 years. We also included 25 episodic GNSS sites located in Morocco with measurements carried out during the 1999.80–2006.71-time interval [40]. The GNNS phase observations were processed by using the GAMIT/GLOBK 10.7 software [65] following the strategy described in [66]. As a final processing step, a consistent set of positions and velocities in the International Terrestrial Reference Frame ITRF2014 reference frame [67] has been computed.

In order to improve the spatial density of the geodetic velocity field over the studied area, we integrated our solutions with those reported in [39,68–70]. In particular, we aligned published solutions to our ITRF2014 one by solving for Helmert transformation parameters that minimize the root mean square (RMS) of differences between velocities at common sites. The resulting velocity field, aligned to the ITRF2014, is reported in Table S1. Such a velocity field is characterized by an average site spatial density of $3.5 \times 10^{-4}$ site km$^2$ (i.e., ca. 11 sites within a radius of 100 km), with differences between the regions (Figure 3). The Betics have a high density of sites ($8.3 \times 10^{-4}$ site km$^2$), the Rif region has a density of sites of $3.0 \times 10^{-4}$ site km$^2$, while the Tell and the Saharan regions have a density of sites of $0.9 \times 10^{-5}$ site km$^2$. In order to highlight regional and crustal deformation patterns over the investigated area, we align our ITRF2014 velocities to both a Eurasian and Nubian reference frames (see Table S1). The estimated velocity fields are reported in Figure 3.

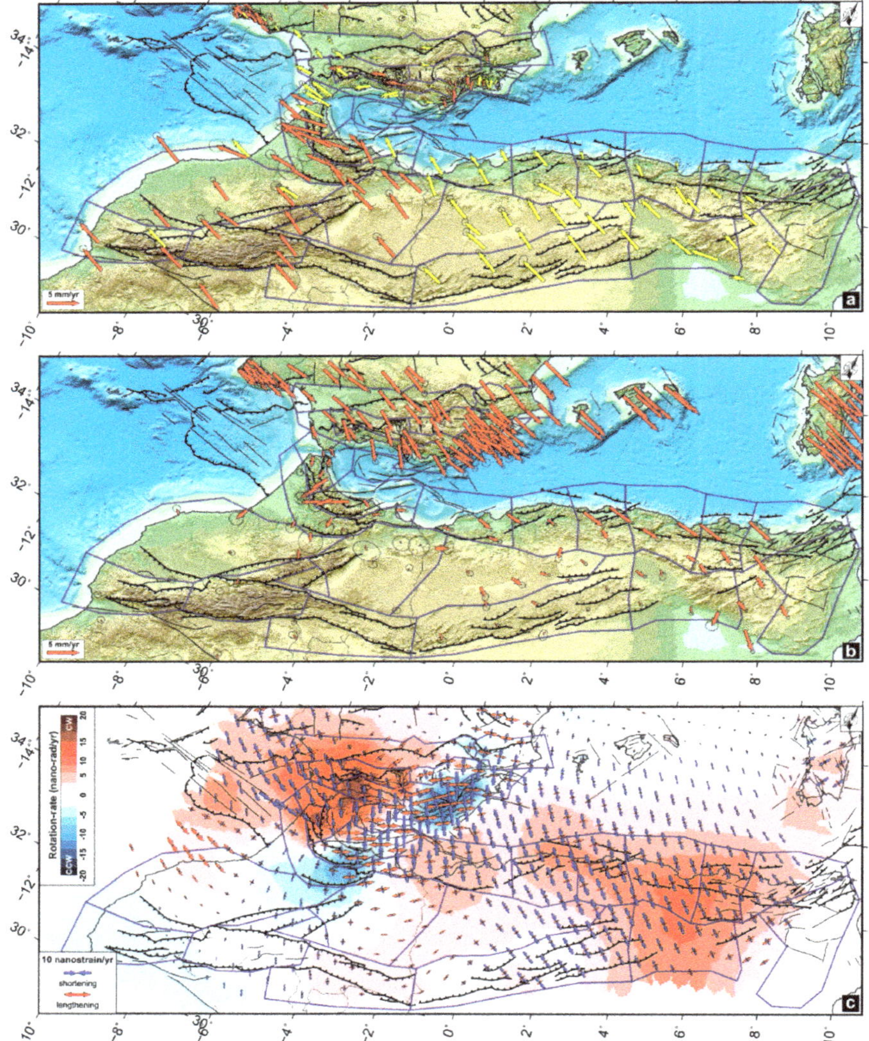

**Figure 3.** Global Navigation Satellite System (GNSS) velocities and 95% confidence ellipses in a fixed (**a**) Eurasian and (**b**) Nubia reference frame (see Table S1 in the Supplementary Material section). Continuous and episodic GNSS stations are reported as yellow and red arrows, respectively. (**c**) Geodetic strain-rate field: red and blue arrows represent the greatest extensional ($\varepsilon_{Hmax}$) and contractional ($\varepsilon_{hmin}$) horizontal strain-rates, respectively; the colors in the background show the rotation strain-rate (which describes the rotations with respect to a downward positive vertical axis with clockwise (cw) positive and counterclockwise (ccw) negative, as in normal geological conventions). The map is plotted in an oblique Mercator projection.

Looking at the velocity field referred to a Eurasian reference frame (Figure 3a), the primary pattern is given by a crustal shortening of ~5 mm/yr related to the Nubia–Eurasia oblique convergence. Such a crustal shortening is largely adsorbed on the Alboran and the Balearic–Algerian basins along off-shore structures. Eastern Betics are characterized by geodetic velocities showing a NW-to-NE fan-shaped pattern with rates ranging from ~3 mm/yr near the coast to ~0.8 mm/yr inland (Figure 3). To the east, the Baleares promontory is characterized by a motion of ~0.5 mm/yr toward East. Stations located on the southern sector of the Betics move toward SW while stations located in the eastern Rif move toward NW; therefore, highlighting a NNW-SSE to N-S contraction of the Alboran Basin. Stations located on the central sector of the Gibraltar Arc are characterized by a westward motion, leading to an E-W stretching of the western side of the Alboran Basin. Moreover, the Gibraltar Arc is affected by an E-W contraction of ~0.3 mm/yr, as deduced by differentiating the motion between the stations located externally and internally to the arc (Figure 3a). On northern Algeria along the coastal area, geodetic velocities decrease progressively from west (~3.0 mm/yr) to east (~0.5 mm/yr). Southward, the region encompassing the Middle, High, and Saharan Atlas as well as the western High Plateaus, is characterized by velocities similar to those predicted by the rigid Eurasia–Nubia plate motion [23,55], suggesting that the region belongs to stable Nubia. This last feature is also readily visible in the Nubian reference frame (Figure 3b), which shows small geodetic velocities (~0.3 mm/yr) in the Middle, High, Saharan Atlas and in the western High Plateaus, and velocities with values up to 5 mm/yr along the coastal area of the Tell Atlas and the Aurès Mountains.

*4.2. Strain-Rate Field Estimation*

The horizontal strain-rates have been estimated on a regular 0.5° × 0.5° grid over the investigated area by adopting the method reported in [71]. This method allows introducing different spatial weighting functions of data (e.g., uniform Gaussian or quadratic spatial weighting function), enabling to obtain a finer resolution, especially on regions characterized by sparsely distributed data. Based on some preliminary tests (see the Appendix C for additional details), the horizontal strain-rate field has been estimated by adopting a weighting threshold of 24 and by using a Gaussian function for distance weighting and Voronoi cell for areal weighting, respectively.

The estimated horizontal strain-rates and the rotation-rates are reported in Figure 3b. A large region comprising the eastern part of the Saharan Atlas, the Aurès Mountains and the Balearic–Algerian basin, is characterized by a prevailing shortening of ~10 nanostrain/yr along the NW-SE orientation. On the Algerian–Tunisian side, this shortening is coupled with clockwise rotation-rates up to 15 nanoradian/yr, suggesting the occurrence of transpressive right-lateral faulting between the Aurès Mountains and the contiguous eastern Saharan Atlas/High Plateau block. The High Atlas and the western part of the Saharan Atlas are characterized by an elongation of ~5 nanostrain/yr along a prevailing E-W direction. Westward, the regions comprising the High Atlas and the Middle Atlas are characterized by a shortening of ~5 nanostrain/yr along the N-S orientation, passing to a NW-SE orientation along the westernmost sector of the Rif. All these regions are characterized by small rotation-rates (~2 nanoradian/yr), highlighting their low tectonic activity. The region comprising the eastern Rif, the Alboran Basin and the Betics are characterized by a shortening of ~12 nanostrain/yr along the NW-SE orientation coupled with an elongation up to ~16 nanostrain/yr along the NE-SW orientation. This region is characterized by a complex rotation-rate pattern: clockwise rotation-rates up to 20 nanoradian/yr can be observed along the north-western Rif and the western Betics, while counterclockwise rotation-rates of ~15 nanoradian/yr and ~10 nanoradian/yr can be observed along the eastern Betics and the eastern Rif, respectively.

*4.3. Geodetic Moment-Rates Computation*

The geodetic moment-rate (and its associated uncertainty) was estimated for each seismic source zone with volume $A{\cdot}H_s$ as [72]:

$$\dot{M}_{geod} = 2\mu H_s A[Max(\left|\varepsilon_{H\max}\right|, \left|\varepsilon_{h\min}\right|, \left|\varepsilon_{H\max} + \varepsilon_{h\min}\right|)] \tag{5}$$

where $\mu$ is the shear modulus of the crust, $\varepsilon_{Hmax}$ and $\varepsilon_{hmin}$ are the principal horizontal strain-rates, and *Max* is a function returning the highest value of the arguments. Since the geodetic moment-rate is proportional to the chosen seismogenic thickness $H_s$, in order to choose appropriate values, we analyzed the depth distributions of earthquakes with magnitude M ≥ 1 for each seismic source zone (see Appendix B for details). After some preliminary estimations, the seismic source zones have been merged into 3 main regions, defining approximate seismogenic thicknesses of 15, 18, and 30 km for Betics (BET1, BET2, BET3, BET4, BET5, BET6, LEV1, and LEV2), Rif—western Atlas (R1a, R1b, R2, T1, MM, HA, HA-MA, MA-HP, HA-AA, SA1, and TA) and Tell Atlas (SA2, T2, T3, T4, T5, and T6), respectively. $H_s$ values estimated for the Betics agree well with the ones reported (~15 km) in literature and estimated by using well-located earthquakes (see [73] and references therein). Our estimated $H_s$ value (18 km) for the Tell Atlas region falls within the range of 15–20 km proposed in recent literature (e.g., [31,74,75]). Therefore, in the following computations we adopted the values of 15 and 18 km as estimated from the depth distributions of earthquakes for the Betics and Tell regions, respectively (see Table 2). Regarding the Rif—western Atlas regions, [41] assumed a fixed $H_s$ value of 15 km for a wide area including the Rif region, while [76] observed that ~85% of seismicity located in Central High Atlas has focal depths lesser than 20 km. Based on these considerations, and since the completeness of the catalog will vary considerably throughout the period of time studied and throughout the study region, we cautiously have chosen an $H_s$ value of 20 km for the Rif—western Atlas region.

**Table 2.** Geodetic moment-rate ($\dot{M}_{geod}$) with associated error and adopted parameter values for each seismic source zone (SSZ). *A*, area of the SSZ; $\mu$, shear modulus for crustal rocks; $H_s$, seismogenic thickness with a 10% variation for the lower and upper ranges (see Appendix B for additional details). Regarding the shear modulus $\mu$, a value of $30 \times 10^{10}$ N/m$^2$ has been used, typical of average crustal rocks [77], and allowed for a 5% variation for the lower and upper ranges.

| SSZ | $A$ [km$^2$] | $\mu$ [$10^{10}$ N/m$^2$] | $H_s$ [km] | $\dot{M}_{geod}$ [$10^{16}$ Nm/yr] |
|---|---|---|---|---|
| BET1 | $3.48 \times 10^4$ | $30.0 \pm 1.5$ | $15 \pm 1.5$ | $22.3 \pm 10.2$ |
| BET2 | $1.07 \times 10^4$ | $30.0 \pm 1.5$ | $15 \pm 1.5$ | $6.2 \pm 3.0$ |
| BET3 | $1.02 \times 10^4$ | $30.0 \pm 1.5$ | $15 \pm 1.5$ | $9.7 \pm 3.8$ |
| BET4 | $1.48 \times 10^4$ | $30.0 \pm 1.5$ | $15 \pm 1.5$ | $12.0 \pm 3.9$ |
| BET5 | $5.47 \times 10^3$ | $30.0 \pm 1.5$ | $15 \pm 1.5$ | $5.6 \pm 2.2$ |
| BET6 | $9.77 \times 10^3$ | $30.0 \pm 1.5$ | $15 \pm 1.5$ | $17.1 \pm 6.0$ |
| HA | $3.38 \times 10^4$ | $30.0 \pm 1.5$ | $20 \pm 2.0$ | $4.5 \pm 7.1$ |
| HA-AA | $4.58 \times 10^4$ | $30.0 \pm 1.5$ | $20 \pm 2.0$ | $18.7 \pm 11.0$ |
| HA-MA | $3.59 \times 10^4$ | $30.0 \pm 1.5$ | $20 \pm 2.0$ | $12.9 \pm 5.0$ |
| LEV1 | $1.73 \times 10^4$ | $30.0 \pm 1.5$ | $15 \pm 1.5$ | $9.2 \pm 6.1$ |
| LEV2 | $1.07 \times 10^4$ | $30.0 \pm 1.5$ | $15 \pm 1.5$ | $11.4 \pm 4.7$ |
| MA-HP | $5.52 \times 10^4$ | $30.0 \pm 1.5$ | $20 \pm 2.0$ | $41.7 \pm 18.2$ |
| MM | $8.97 \times 10^4$ | $30.0 \pm 1.5$ | $20 \pm 2.0$ | $73.2 \pm 33.6$ |
| R1a | $1.76 \times 10^4$ | $30.0 \pm 1.5$ | $20 \pm 2.0$ | $17.1 \pm 7.6$ |
| R1b | $2.33 \times 10^4$ | $30.0 \pm 1.5$ | $20 \pm 2.0$ | $34.9 \pm 9.1$ |
| R2 | $2.66 \times 10^4$ | $30.0 \pm 1.5$ | $20 \pm 2.0$ | $39.3 \pm 15.5$ |
| SA1 | $7.68 \times 10^4$ | $30.0 \pm 1.5$ | $20 \pm 2.0$ | $70.0 \pm 21.6$ |
| SA2 | $4.30 \times 10^4$ | $30.0 \pm 1.5$ | $18 \pm 1.8$ | $31.1 \pm 10.7$ |
| T1 | $3.12 \times 10^4$ | $30.0 \pm 1.5$ | $20 \pm 2.0$ | $37.4 \pm 10.9$ |
| T2 | $2.72 \times 10^4$ | $30.0 \pm 1.5$ | $18 \pm 1.8$ | $23.2 \pm 7.3$ |
| T3 | $2.67 \times 10^4$ | $30.0 \pm 1.5$ | $18 \pm 1.8$ | $24.5 \pm 7.1$ |
| T4 | $3.68 \times 10^4$ | $30.0 \pm 1.5$ | $18 \pm 1.8$ | $29.0 \pm 8.7$ |
| T5 | $2.46 \times 10^4$ | $30.0 \pm 1.5$ | $18 \pm 1.8$ | $15.7 \pm 5.5$ |
| T6 | $3.01 \times 10^4$ | $30.0 \pm 1.5$ | $18 \pm 1.8$ | $16.3 \pm 6.7$ |
| TA | $5.22 \times 10^4$ | $30.0 \pm 1.5$ | $20 \pm 2.0$ | $29.4 \pm 12.5$ |

Table 2 lists the estimated geodetic moment-rates for all the identified source zones along with the associated uncertainties given from the ones related to the strain-rates, the seismogenic thicknesses and the shear modulus (for these last two parameters we assumed an uncertainty of 10% and 5% with respect to the chosen value, respectively).

## 5. Results and Discussions

Here we provide a detailed statistical evaluation of the seismic/geodetic moment-rates ratio for the Ibero–Maghrebian region by comparing GNSS-based moment-rates with those derived from seismic catalogs.

### 5.1. Moment-Rates

The estimated seismic moment-rates are reported in Table 1 and Figure 4. As above mentioned, the estimations have been performed by adopting two different approaches: the truncated Gutenberg–Richter distribution ([16]; Figure 4a) and the moment summation ([15]; Figure 4b).

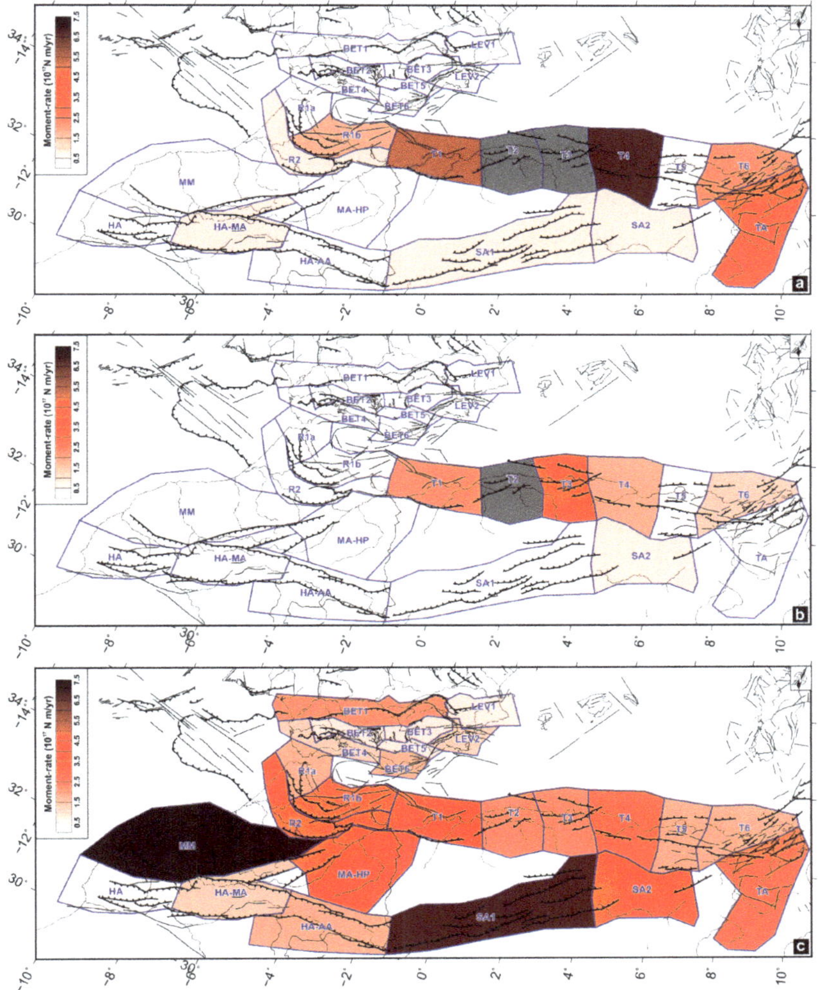

**Figure 4.** Estimated moment-rates from seismicity, by considering the truncated Gutenberg–Richter distribution (**a**) and by considering the cumulative Kostrov summation method (**b**) and from (**c**) geodetic data (see Tables 1 and 2 for details). SSZ characterized by moment-rates larger than 7.5 × 1017 Nm/yr are reported in grey. The maps are plotted in an oblique Mercator projection.

Estimations coming from the former approach, with the exception of very few zones (e.g., HA, SA2, and R1a; see Figure A4 in Appendix D), are generally larger (~40% on average) than the ones estimated from the latter approach; however, both estimations overlap within their associated uncertainties. The truncated Gutenberg–Richter distribution estimates the magnitude distribution of the earthquakes through the computation of the $b$ value (Equation (3)), assuming that earthquake distribution is ergodic over the investigated region and represents an adequate sample of long-term seismicity, therefore allowing to take into account the probable incompleteness of the existing catalog (e.g., [78]). The moment summation method is strongly influenced by the catalog completeness and suffers from the possible lack of large earthquakes with high recurrence rate compared to the catalog duration. Based on these observations, in the following we refer to the results coming from the truncated Gutenberg–Richter distribution approach.

Computed values range in the interval $4.85 \times 10^{15} - 3.0 \times 10^{18}$ Nm/yr. The higher values ($\geq 1.0 \times 10^{17}$ Nm/yr) are observed along the eastern sector of the Rif (R1b) and along the Tell Atlas region (e.g., T1, T2, T3, T4, T6, and TA), clearly highlighting the high seismic release of this belt. Although T5 belongs to the Tell Atlas region, it shows an estimated moment-rate of $\sim 3.2 \times 10^{16}$ Nm/yr, a very low value when compared with surrounding zones. Betics seismogenic source zones (BET1, BET2, BET3, BET4, BET5, and BET6) are characterized by values in the $7.3 \times 10^{15} - 3.5 \times 10^{16}$ Nm/yr range, exhibiting a low to moderate seismic release. The Levante seismogenic region (LEV1 and LEV2) shows a moderate seismic release with an estimated moment-rate of $\sim 4.0 \times 10^{16}$ Nm/yr. A similar pattern can be observed on the Rif regions (R1a, R2), along the Moroccan Meseta - High Plateaus (MM and MA-HP) and along the Atlas seismogenic sources (HA, HA-AA, HA-MA, SA1, and SA2), with estimated seismic moment-rates ranging in the $4.9 \times 10^{15} - 7.0 \times 10^{16}$ Nm/yr interval.

The estimated geodetic moment-rates and associated uncertainties are reported in Table 2; computed values range in the interval $4.46 \times 10^{16} - 7.32 \times 10^{17}$ Nm/yr. Based on appraised values, at least four main interval ranges can be distinguished. In particular, moment-rate values larger than $7.0 \times 10^{17}$ Nm/yr have been estimated for MM and SA1 zones. Moment-rate values spanning the $3.0 \times 10^{17} - 4.2 \times 10^{17}$ Nm/yr interval are observed for the eastern sector of the Rif (R1b and R2), for the westernmost zone of Tell Atlas (T1) and for the High Plateaus region (MA-HP) and for SA2 zone. Moment-rate values spanning the $2.0 \times 10^{17} - 2.9 \times 10^{17}$ Nm/yr interval are observed for the central sector of the Tell Atlas region (T2, T3 and T4), the western sector of Betics (BET1) and for SA2. Moment-rate values spanning the $1.1 \times 10^{17} - 1.9 \times 10^{17}$ Nm/yr interval are observed on the eastern sectors of the Tell Atlas (T5 and T6), on LEV2 zone, on the High Atlas region (HA-MA and HA-AA), on the western Rif (R1a) and on the southern Betics (BET4 and BET6). Moment-rate values lower than $1.0 \times 10^{17}$ Nm/yr are observed on the central and eastern Betics (BET2, BET3 and BET5) and on LEV1 zone. The geodetic moment-rate computed for the TA seismogenic source zone is based on strain-rates information derived from the velocity interpolation only, therefore this estimation cannot be taken as an accurate estimate, due to the lack of spatial constraints from geodetic data (TA is not covered by GNSS stations). In addition, the geodetic moment-rate inferred for zone HA is statistically negligible because of the large uncertainties associated with the estimated value. Therefore, based on these considerations, the TA and HA seismogenic source zones have been excluded from the comparison reported below.

### 5.2. Seismic/Geodetic Moment-Rates Ratio

In the following, we define the seismic/geodetic moment-rates ratio (expressed as a percentage) as the "Seismic coupling coefficient" (hereinafter SCC; [2]), the geodetic moment-rate being a measure of both elastic and anelastic loading rates and the seismic moment-rate a measure of the elastic unloading rate. The theoretical range of SCC is between 0% and 100%. A SCC value close to 100% indicates that a large amount of the deformation budget is released through earthquakes. Conversely, a low ratio indicates an apparent seismic moment deficit, suggesting either a proportion of aseismic deformation (i.e., ongoing unloading by creep and other plastic process) or accumulating strain not released by seismicity (i.e., elastic storage).

Indeed, SCC can exceed the value of 100%, i.e., the seismic moment-rate is larger than the geodetic ones. This is commonly observed when seismic catalogs cover a very short time interval and contain one or more large earthquakes [2]. The temporal length of a seismic catalog as well as its completeness governs the adequacy of catalogs to estimate seismic moment-rates over a given region (e.g., [78]). Roughly speaking, a seismic catalog with a short duration (100–300 years) could be little adequate to capture the seismic cycle of a given region; indeed, seismic moment-rates estimated from seismic catalogs require that the average earthquake recurrence interval should be shorter than the catalog duration (see [2,4] for additional details). This means that the used seismic catalog would be long enough to capture a complete earthquake cycle for an individual seismogenic fault or alternatively, to provide a statistical sample of all phases of the seismic cycle, including of course earthquakes, for a region containing multiple seismogenic faults. Moreover, 50–100 year-long instrumental catalogs are the most common source of data used to derive earthquake statistics worldwide, under the assumption that such a time span is adequate to derive earthquake return periods over timescales of 500–5000 years, typically used in probabilistic seismic hazard analysis [3]. As mentioned above, available historical seismic catalogs for southern Iberia and Maghreb regions highlight the occurrence of large (M $\geq$ 6.5) earthquakes since AD 856 and can be considered complete for M $\geq$ 6 earthquakes since approximately 1500, 1700 and 1850, for the Southern Iberia, northern Morocco and northern Algeria areas, respectively [25,37,38]. Being aware that on the one hand, the earthquake occurrences and statistics may not be steady state over the whole timescales of our catalog and that on the other hand, the truncated Gutenberg–Richter distribution allows taking into account the probable incompleteness of the catalog [78], in the following we carefully discuss and frame our results in comparison with available and independent geological and geophysical observations.

Regarding geodetic moment-rate estimations, the GNSS measurements should sample a large spatial scale, so that they are not affected by local strain accumulation on individual faults, and a long enough time interval, so that measurement uncertainties have a minimal effect on velocity estimations and adequately sample both seismic, aseismic and long-term deformation transients. As noted above, our GNSS dataset shows different spatial density over the investigated area, with high density of sites on the Iberian region and low density along the Maghrebian region. Time series of this GNSS dataset cover different sample periods ranging from 3.5 to 20 years (with an average duration of 8.7 years) and, with the exception of a few sites located in the R1b source zone, generally do not include large earthquakes that could significantly contribute with co-seismic and post-seismic displacements to the estimated geodetic velocities. Therefore, our estimated regional velocity and the related strain rate field are statistically significant in most of the study area.

Estimated SCC values are reported in Table 3 and in Figure 5; based on the achieved SCC values, we considered three principal ranges of values: less than 23%, between 35% and 60%, and more than 95%.

**Table 3.** Comparison of the seismic coupling coefficient (expressed as a percentage), which takes into account the moment-rates estimated from the (#) truncated cumulative Gutenberg–Richter distribution [16] and the (*) the moment summation [15] approach. Confidence intervals (67%) are also reported. SCC = seismic coupling coefficient.

| SSZ | SCC (%) (*) | SCC (%) (#) |
|---|---|---|
| BET1 | $3.05^{+2.55}_{-1.76}$ | $6.31^{+3.26}_{-3.14}$ |
| BET2 | $21.02^{+17.89}_{-12.62}$ | $56.63^{+30.66}_{-29.66}$ |
| BET3 | $6.22^{+4.96}_{-3.25}$ | $9.22^{+4.17}_{-3.98}$ |
| BET4 | $5.12^{+3.94}_{-2.45}$ | $7.55^{+3.04}_{-2.86}$ |
| BET5 | $8.02^{+6.43}_{-4.23}$ | $12.94^{+5.94}_{-5.67}$ |
| BET6 | $0.94^{+0.73}_{-0.46}$ | $13.36^{+5.26}_{-5.11}$ |
| HA-AA | $7.30^{+6.67}_{-5.01}$ | $20.89^{+18.54}_{-14.91}$ |
| HA-MA | $17.24^{+13.71}_{-8.94}$ | $42.84^{+16.70}_{-16.65}$ |
| LEV1 | $5.84^{+5.60}_{-4.36}$ | $40.04^{+31.11}_{-28.84}$ |
| LEV2 | $16.01^{+12.95}_{-8.63}$ | $36.55^{+21.19}_{-18.38}$ |
| MA-HP | $1.70^{+1.40}_{-0.95}$ | $2.50^{+1.99}_{-1.49}$ |
| MM | $0.48^{+0.40}_{-0.27}$ | $4.07^{+27.33}_{-3.74}$ |
| R1a | $14.32^{+11.85}_{-8.11}$ | $2.84^{+4.47}_{-2.12}$ |
| R1b | $10.59^{+7.88}_{-4.62}$ | $55.20^{+43.96}_{-27.78}$ |
| R2 | $9.31^{+7.45}_{-4.90}$ | $15.16^{+16.29}_{-9.45}$ |
| SA1 | $2.62^{+2.00}_{-1.22}$ | $3.74^{+3.47}_{-2.09}$ |
| SA2 | $22.93^{+17.82}_{-11.24}$ | $22.38^{+18.33}_{-12.16}$ |
| T1 | $66.39^{+50.10}_{-30.13}$ | $135.17^{+97.25}_{-66.45}$ |
| T2 | $670.14^{+512.26}_{-314.94}$ | $1292.56^{+1049.84}_{-687.28}$ |
| T3 | $124.05^{+93.67}_{-56.40}$ | $323.03^{+339.88}_{-187.58}$ |
| T4 | $66.61^{+50.54}_{-30.69}$ | $224.82^{+155.50}_{-109.52}$ |
| T5 | $8.64^{+6.75}_{-4.29}$ | $20.23^{+8.18}_{-7.88}$ |
| T6 | $77.96^{+63.02}_{-41.98}$ | $144.41^{+105.60}_{-80.46}$ |

A large sector of the study area, comprising Betics (BET1, BET3, BET4, BET5, BET6), Rif (R1a, R2), High, Middle, and Saharan Atlas (MM, MA-HP, HA-AA, SA1, SA2), and one region in the Tell Atlas (T5), is characterized by SCC values lower than 23%. This result is highly consistent with previous estimations computed for the Betics and Rif areas (see [41] for details) which highlighted how about 24% of crustal deformation is released seismically. A number of indicators such as surface heat flow, seismic tomography and local rheological models provided evidence of a weak crustal rheology; therefore, supporting the inference on the aseismic behavior of some sectors of the region. In detail, surface heat flow measurements available for the area show high values (100–120 mW m$^{-2}$; [79,80]) along the eastern sector of the Alboran basin and moderate values (60–75 mW m$^{-2}$) along the Betics [81] and Rif [82] regions. In the central and southern Gulf of Cadiz, the surface heat flow is characterized by values close to 40–50 mW m$^{-2}$ [74], assuming typical values for stable continental/oceanic lithosphere [83]. Local seismic tomographies have inferred a low-velocity zone at 18-km depth beneath the central Betics, indicating variations in lithology and/or in the rigidity of the lower crust rocks (e.g., [84]). Local rheological models (e.g., [85]) highlighted how the crustal yield strength as well as the inferred depth of the brittle-ductile transition follow the curved shape of the Gibraltar arc with maximum depths of 12–9 km, whereas in the Rif and the Betics, such a transition became shallower eastward (~6–5 km depth; Figure 5). All these geological and geophysical evidences support the inference on the aseismic behavior, at least of the Betics and Rif regions.

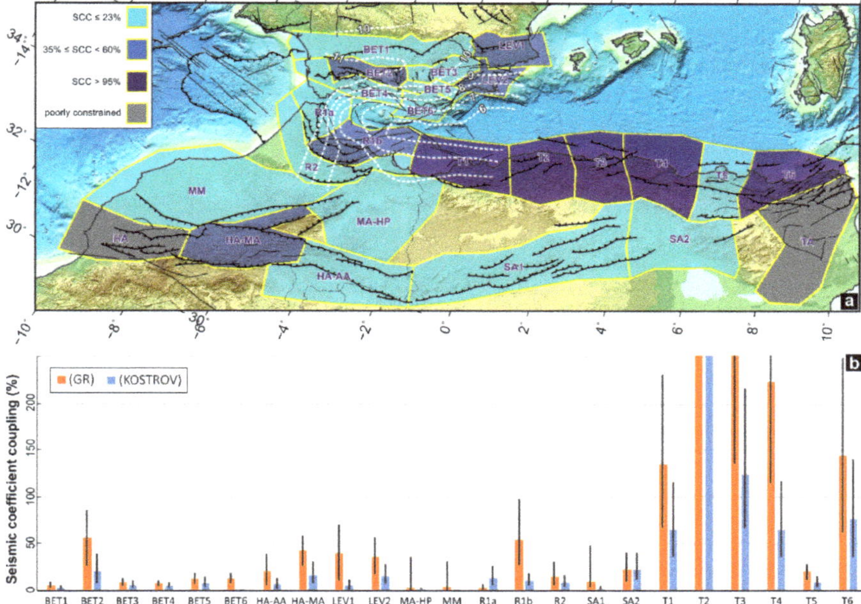

**Figure 5.** (**a**) Seismic coupling coefficient (SCC), expressed as a percentage of the seismic/geodetic moment-rate ratio as computed in this study (see SCC (*) values in Table 3). As discussed in the main text, three principal ranges of SCC values have been identified: low (less than 23%), intermediate (between 35% and 60%) and high (more than 95%). In grey are reported the SSZ excluded from the comparison, i.e., HA and TA (see the text for more details). The white dashed line represents the depth (in km) of the brittle-ductile transition as reported in [85]. (**b**) Comparison of the seismic coupling coefficient (expressed as a percentage) which takes into account the moment-rates estimated from both the approaches discussed in the text. Confidence intervals (67%) are also reported.

Intermediate SCC values (between 35% and 60%) have been observed for HA-MA, R1b, LEV1, LEV2 and BET2 seismogenic source zones; therefore, suggesting how these regions account only for a moderate seismic fraction of the total deformation-rate budget. Some of these regions are characterized by well-known active faults which frequently generate moderate earthquakes such as LEV2 and R1b. On these zones, a significant contribution to the measured crustal deformation can be attributed to aseismic post-seismic mechanisms such as afterslip and/or viscoelastic relaxation, as for instance documented for earthquakes striking the R1b source zone in the last decades [86].

The higher SCC values (>95%) have been observed along the Tell Atlas zones (T1, T2, T3, T4, T6), suggesting how the measured crustal deformation over these zones is mostly released through earthquakes. All these source zones are characterized by SCC values larger than 100%. These values are not surprising, since, along the Tell Atlas zones, the crustal shortening of ~5 mm/yr related to the Nubia–Eurasia oblique convergence is largely adsorbed on off-shore reverse structures bordering the North African margin [23,39], therefore a relevant portion of the long-term elastic strain accumulation is not captured by the on-shore GNSS stations. In addition, a number of moderate to large earthquakes have epicenters concentrated off-shore (not sampled by GNSS stations), such as the 1790 Oran (M~7) [87], the 1856 Djijelli Mw 7.2 [88], and the 2003 Boumerdes Mw 6.8 [89] earthquakes, therefore, their released seismic energy corresponds to strain with no full counterpart on the on-shore surface.

Both these features lead to an underestimation of geodetic strain-rate. As above mentioned, the geodetic moment-rate estimations are affected by the assumed computational parameters; even if increasing the seismogenic thickness value up to 25 km and/or computational parameters (grid size and weighting threshold) for strain-rates estimation, the SCC values for Tell Atlas zones remain confined to values larger than 100%. All these considerations, coupled with the occurrence of large destructive events (M > 7) both in historical and instrumental periods adds realistic constraints on the seismic behavior of the region. In such a context the low SCC value inferred for T5 source zone (~20%) appears puzzling. The seismic moment-rate is based on a maximum magnitude $M_{max}$ of 5.8, about 1 order of magnitude unit lesser than the values estimated for the other Tell Atlas source zones. It must be noted that an increase of 1 magnitude unit leads to an increase by a factor of ~5.4 of the moment-rate (Equation (1)), therefore leading to a SCC value close to 100%. These considerations, while on one hand suggest that the low SCC value inferred for T5 source zone could be poorly representative of its seismic behavior, on the other suggest that the possibility of forthcoming earthquakes in the region may increase. Since in this region the occurrence of a moderate earthquake (1985 Constantine Mw 5.8 event) is documented in the available seismic catalogs, it appears to be a potential seismic gap like other well studied regions of the world (e.g., [90,91].

Although the seismic and geodetic moment-rates comparison implies numerous assumptions and simplifications, it provides significant insights into the potential seismic hazard of tectonically active regions, essentially for time-dependent seismic hazard assessments. As noted above, valuable examples come from numerous tectonic regions worldwide such as Iran [1,2], western Canada [3], western USA [4,5], Greece [6,7], southern Italy [8–10], Himalaya [11–13], and the East African Rift [14]. Despite the different approaches adopted by these authors, all their estimations have enabled achieving valuable results that document cases with a good agreement between seismic and geodetic moment-rates and also cases where geodetic moment-rates are significantly larger than the seismic ones. However, uncertainties related to the physical significance of the deformation-rates mismatch over varying spatial and temporal scales are currently poorly understood. The continuous growth in continuous seismic and GNNS networks is allowing the acquisition of spatially extensive datasets at an increasing number of tectonic areas worldwide, therefore leading to an improved comprehension of such a physical significance.

## 6. Conclusive Remarks

Based on the SCC estimations previously presented and discussed, we identified three main regions with different seismic behavior along the western Mediterranean border.

- A large sector of the study area, comprising the western Betics, the western Rif, and the High, Middle, and Saharan Atlas is characterized by SCC values lower than 23%. Such a result coupled with geological and geophysical evidence supports the inference on the aseismic deformation (aseismic release processes) behavior at least for most of the Betics and Rif regions.
- Intermediate SCC values (between 35% and 60%) have been observed for some regions belonging to the eastern Betics, the central Rif, and the Middle Atlas. On these regions, crustal seismicity accounts only for a moderate fraction of the total deformation-rate budget.
- Higher SCC values (>95%) have been observed along the Tell Atlas, highlighting a fully seismic deformation. By considering the low SCC value inferred for T5 source zone, we speculated on the possibility of this being a seismic gap and hosting impending earthquakes.

**Supplementary Materials:** The following are available online at http://www.mdpi.com/2072-4292/12/6/952/s1, Table S1: Main features for each seismogenic source zone analyzed in this study, Table S2: Site coordinates and velocities referred to ITRF2014, Eurasian and Nubia reference frame for all GNSS sites analyzed in this study.

**Author Contributions:** Conceptualization, M.P. and F.S.; methodology, M.P.; software, M.P., F.S., and J.A.P.; validation, M.P., F.S., J.A.P., and J.F.; formal analysis, M.P., F.S., and J.A.P.; investigation, M.P.; resources, M.P., F.S., J.A.P., and J.F.; data curation, M.P., F.S., J.A.P., and J.F.; supervision, M.P.; project administration, M.P., J.A.P. and J.F.; funding acquisition, J.A.P. and J.F. All authors have read and agreed to the published version of the manuscript.

**Funding:** The research performed in this study was partially supported by the Spanish CGL2015-65602-R, CGL2016-80687-R and RTI2018-093874-B-100 projects, and the Programa Operativo FEDER Andalucía 2014-2020 – call made by the University of Jaén 2018.

**Acknowledgments:** We would like to thank all the people involved in the installation and maintenance of the GNSS stations and for making data available. Data used in this paper are available in the supporting information. The maps in this paper were produced using the public domain Generic Mapping Tools (GMT) software [92]. We would like to thank the Editor Bommie Xiong and the anonymous reviewers for their constructive and valuable suggestions on the earlier drafts of this manuscript. We thank S. Conway for the revision of the English form of this manuscript.

**Conflicts of Interest:** The authors declare no conflict of interest.

## Appendix A

**Table A1.** Main features for each seismogenic source zone (SSZ) analyzed in this study. The last column reports the number of GNSS sites, and site density for km$^2$ (in brackets) for each SSZ.

| SSZ | Stress Regime | Seismological Parameters | Level of Knowledge of Tectonic Structures | References | GNSS Data |
|---|---|---|---|---|---|
| BET1 | NNW-SSE compression and perpendicular horizontal extension | well constrained | High | [25,54,93,94] | 11 (3.2 ·10$^{-4}$) |
| BET2 | Strike-slip to NW-SE compression | well constrained | High | [25,54,93,94] | 15 (14.1 ·10$^{-4}$) |
| BET3 | NE extension to strike-slip | well constrained | High | [25,54,93,94] | 4 (3.9 ·10$^{-4}$) |
| BET4 | Strike-slip to NW-SE compression | well constrained | High | [25,54,93,94] | 5 (3.4 ·10$^{-4}$) |
| BET5 | NE extension to strike-slip | well constrained | High | [25,54,93,94] | 8 (14.6 ·10$^{-4}$) |
| BET6 | NE extension to strike-slip | well constrained | High | [25,54,93,94] | 3 (3.1 ·10$^{-4}$) |
| HA | NNW-SSE horizontal compression to strike-slip | poorly constrained | Medium | [24,50,93] | 2 (0.6 ·10$^{-4}$) |
| HA-AA | NNW-SSE horizontal compression to strike-slip | poorly constrained | Medium | [24,50,93] | 2 (0.4 ·10$^{-4}$) |
| HA-MA | NNW-SSE horizontal compression to strike-slip | poorly constrained | Medium-High | [24,50,93] | 3 (0.8 ·10$^{-4}$) |
| LEV1 | NE extension to strike-slip | well constrained | High | [25,54,93,94] | 6 (3.5 ·10$^{-4}$) |
| LEV2 | NE extension to strike-slip | well constrained | High | [25,54,93,94] | 22 (20.7 ·10$^{-4}$) |
| MA-HP | NNW-SSE compression and perpendicular horizontal extension | poorly constrained | Medium | [24,50,93] | 8 (1.4 ·10$^{-4}$) |
| MM | NNW-SSE pure compressional | poorly constrained | Medium | [24,50,93] | 7 (0.8 ·10$^{-4}$) |
| R1a | NNW-SSE compression and perpendicular horizontal extension | well constrained | High | [24,50,54,93,94] | 9 (5.1 ·10$^{-4}$) |
| R1b | NNW-SSE compression and perpendicular horizontal extension | well constrained | High | [24,50,54,93,94] | 3 (1.3 ·10$^{-4}$) |
| R2 | NW-SE compression and perpendicular horizontal extension | well constrained | Medium-High | [24,50,54,93] | 7 (2.6 ·10$^{-4}$) |
| SA1 | NW-SE horizontal compression to strike-slip | poorly constrained | Medium | [24,50,93] | 6 (0.8 ·10$^{-4}$) |
| SA2 | N-S horizontal compression | poorly constrained | Medium | [24,50,93] | 6 (1.4 ·10$^{-4}$) |
| T1 | NNW-SSE pure compressional to strike-slip | well constrained | High | [24,50,54,93,94] | 4 (1.3 ·10$^{-4}$) |
| T2 | NNW-SSE pure compressional to strike-slip | well constrained | High | [24,50,54,93,94] | 2 (0.7 ·10$^{-4}$) |
| T3 | NNW-SSE pure compressional | well constrained | High | [24,50,54,93] | 3 (1.1 ·10$^{-4}$) |
| T4 | NNW-SSE pure compressional | well constrained | High | [24,50,54,93] | 4 (1.1 ·10$^{-4}$) |
| T5 | N-S compressional strike-slip | well constrained | High | [24,50,54,93] | 3 (1.2 ·10$^{-4}$) |
| T6 | NNW-SSE pure compressional | well constrained | High | [24,50,56,93] | 1 (0.3 ·10$^{-4}$) |
| TA | NNW-SSE pure compressional | poorly constrained | High | [24,50,56,93] | 0 (0.0) |

## Appendix B

To choose the most appropriate seismogenic thickness for the investigated region, the depth distributions of earthquakes from the ISC online catalog [46] were used. To this end, by taking into account all earthquakes with magnitude M ≥ 1, for each seismic zone we analyzed (i) the depth distribution—time plots and (ii) the cumulative frequency plots. Since several source zones share a similar pattern, we therefore merged all the seismic source zones into three main regions (Figure S1a) in order to obtain more robust results. Hence we re-analyzed, for each merged region, the depth distribution—time plots (Figure A1b–d) and the cumulative frequency plots (Figure A1e–g). Cumulative frequency plots of depth distributions of earthquakes show that 90% of events occur at depths less than 15, 18 and 30 km for Betics (BET1, BET2, BET3, BET4, BET5, BET6, LEV1, and LEV2), Tell Atlas (T2, T3, T4, T5, T6, and SA2), and Rif—western Atlas (R1a, R1b, R2, T1, MM, HA, HA-MA, HA-AA, SA1, MA-HP, and TA), respectively.

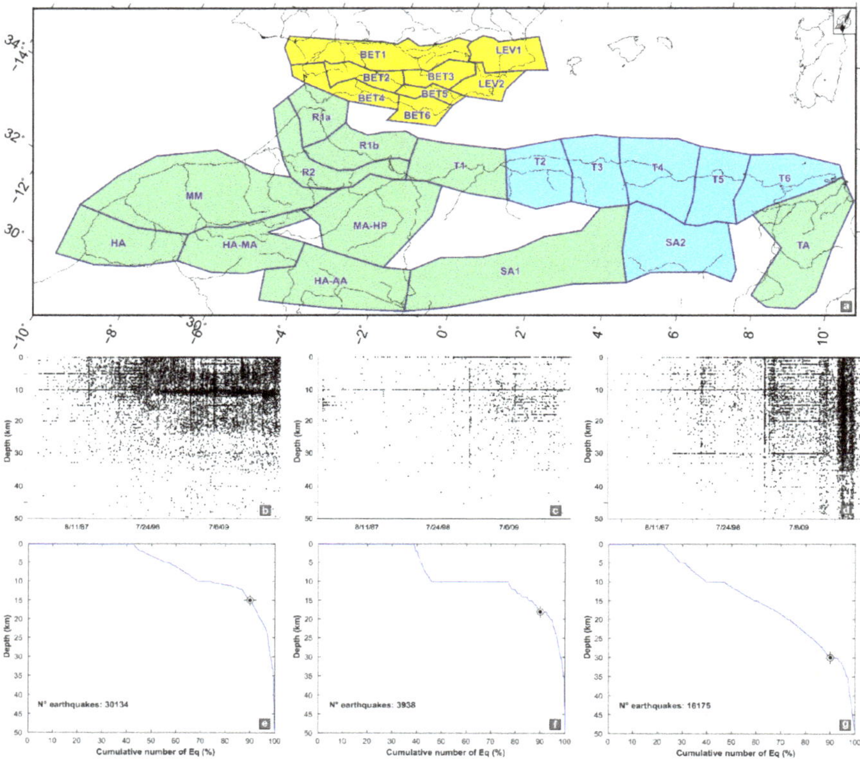


**Figure A1.** (**a**) Seismic source zones subdivision into three main regions. In yellow, the Betics seismogenic sources (BET1, BET2, BET3, BET4, BET5, BET6, LEV1, and LEV2); in green the Rif—western Atlas seismogenic sources (R1a, R1b, R2, T1, MM, HA, HA-MA, HA-AA, SA1, MA-HP, and TA); in light-blue the Tell Atlas seismogenic sources (T2, T3, T4, T5, T6, and SA2). (**b**) Depth distribution—time plot of the Betics seismogenic sources. (**c**) Depth distribution—time plot of the Tell Atlas seismogenic sources. (**d**) Depth distribution—time plot of the Rif—western Atlas seismogenic sources. (**e**) Cumulative frequency plots of depth distribution of earthquakes; 90% of events occur at depths less than 15 km for the Betics main region. (**f**) Cumulative frequency plots of depth distribution of earthquakes; 90% of events occur at depths less than 18 km for the Tell Atlas main region. (**g**) Cumulative frequency plots of depth distribution of earthquakes; 90% of events occur at depths less than 30 km for the Rif - western Atlas main region.

The ISC routinely produces catalogs of earthquake hypocenter locations relying on data contributed by seismological agencies from around the world. A large number of earthquakes have their focal depth assigned to 0, 5, 10, 15, 20, 25, and 30 km, because of difficulties in computing reliable depth estimations.

All events with M ≥ 3.5 are manually reviewed and relocated by an ISC analyst. The location provided by the ISC is based entirely on P-wave travel-time tables derived from a global 1D Earth velocity model (e.g., the ak135 model [95]). Therefore, many ISC hypocenters are poorly constrained in focal depth, especially in regions not extensively covered by seismic stations, as are the Rif and Atlas ones, and must be interpreted with caution.

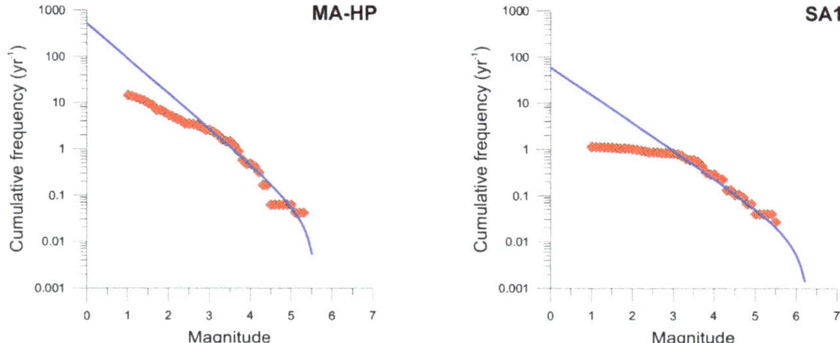


**Figure A2.** Cumulative frequency-magnitude distributions (red diamonds) of earthquakes for MA-HP and SA1 seismogenic source zones. The blue line represents the truncated Gutenberg–Richter function according to (1).

**Appendix C**

A variety of methods, based on different approaches, have been developed to compute strains (see [71] for an overview). Since the spatial distribution of our velocity field data is heterogeneous, in order to verify that the resulting strain-rate field provides a good representation of the regional deformation field of the investigated area, we performed some additional computations by adopting the method described in [71].

Four solutions are computed using a different combination of data weighting functions (12, 18, 24, and 32). The computation was performed on a 0.5° × 0.5° regular grid by adopting a Voronoi cell approach for areal weighting. Achieved results are reported in Figure A3: arrows show the greatest extensional ($\varepsilon_{Hmax}$) and contractional ($\varepsilon_{hmin}$) horizontal strain-rates, while the color in the background shows the second invariant of the strain-rate tensor (defined as $\varepsilon_{2inv} = \sqrt{\varepsilon_E^2 + \varepsilon_N^2 + 2\varepsilon_{EN}^2}$, where $\varepsilon_E$, $\varepsilon_N$ and $\varepsilon_{EN}$ are three strain-rate components in an east and north Cartesian coordinate system).

Considering results reported in Figure A3, the strain-rate pattern suggests that as the weighting threshold decreases from 32 (Figure A3d) to 18 (Figure A3b), the strain-rate model picks up more tectonic strain signals along main tectonic features within the investigated area. However, as the weighting threshold decreases from 18 (Figure A3b) to 12 (Figure A3a), the differential strain-rate pattern starts to highlight very local signals. Some of these local signals are located in regions with dense data population (e.g., south-eastern Spain) while others are located in regions not covered by GNSS data, therefore suggesting a significant deteriorating of the strain-rate pattern at some places.

Balancing the trade-off between the resolution and robustness, we chose the strain-rate pattern resulting from a weighting threshold of 24 as the optimal model for the characterization of the strain-rate field in the Ibero–Maghrebian region.

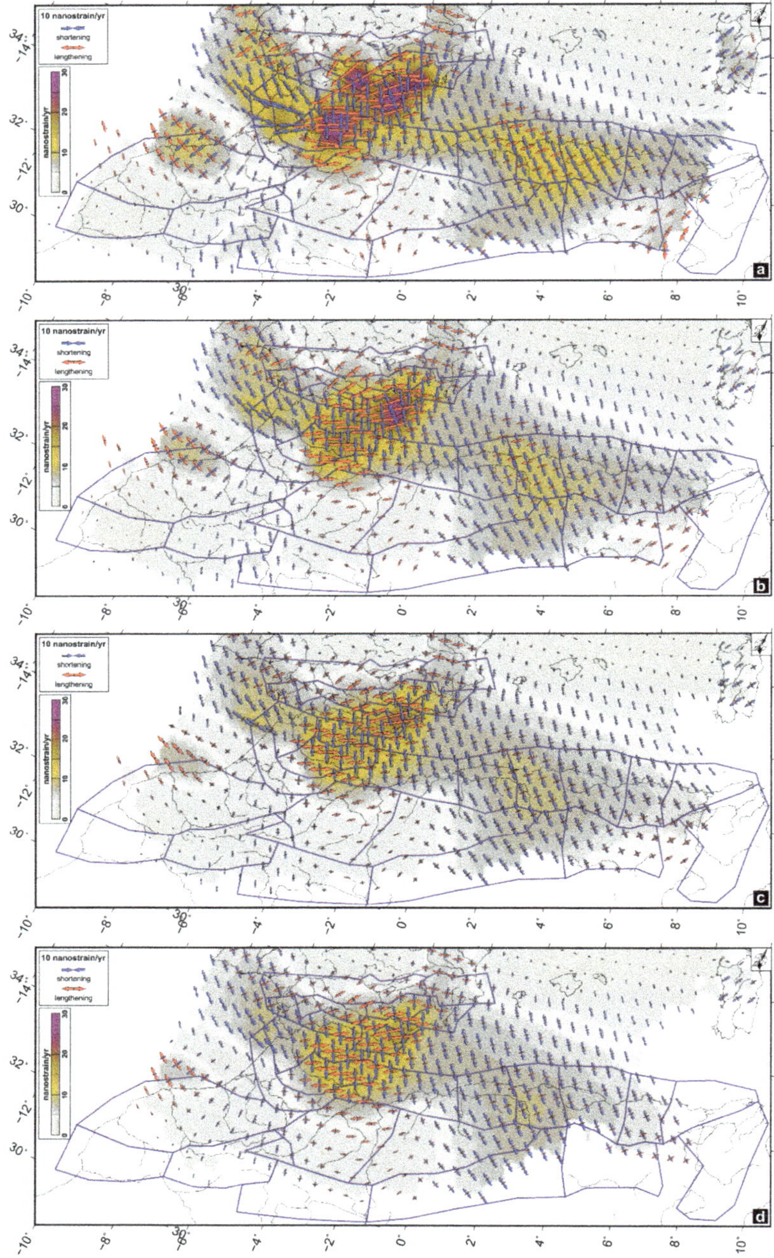

**Figure A3.** Estimated strain-rate field according to a Gaussian function for distance weighting of 12 (panel a), 18 (panel b), 24 (panel c), and 32 (panel d).

## Appendix D

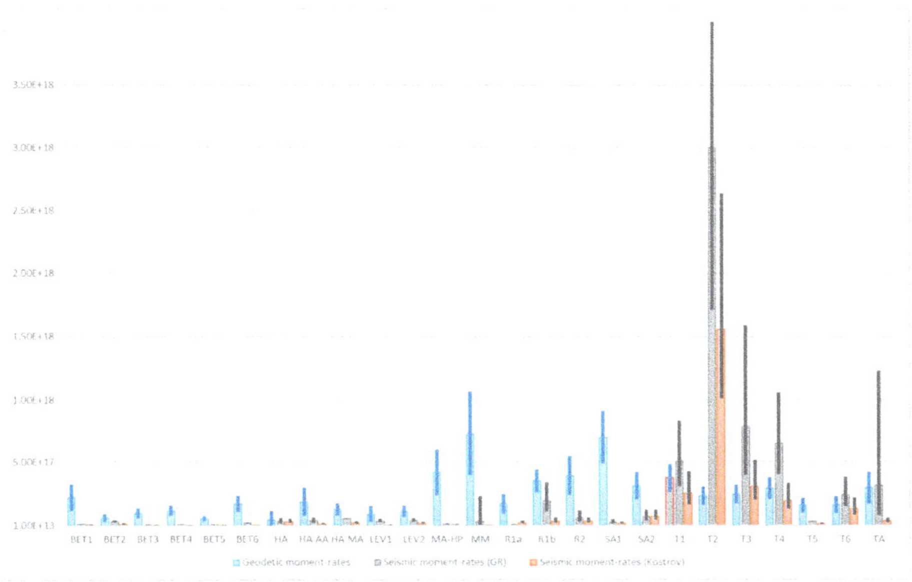

**Figure A4.** Moment-rate (N·m/yr) estimates computed in this study (see Tables 1 and 2 in the main text for details). Uncertainties (67% confidence interval) are also reported.

## References


1. Masson, F.; Chéry, J.; Hatzfeld, D.; Martinod, J.; Vernant, P.; Tavakoli, F.; Ghafory-Ashtiani, M. Seismic versus aseismic deformation in Iran inferred from earthquakes and geodetic data. *Geophys. J. Int.* **2005**, *160*, 217–226. [CrossRef]
2. Palano, M.; Imprescia, P.; Agnon, A.; Gresta, S. An improved evaluation of the seismic/geodetic deformation-rate ratio for the Zagros Fold-and-Thrust collisional belt. *Geophys. J. Int.* **2018**, *213*, 194–209. [CrossRef]
3. Mazzotti, S.; Leonard, L.J.; Cassidy, J.F.; Rogers, G.C.; Halchuk, S. Seismic hazard in western Canada from GPS strain rates versus earthquake catalog. *J. Geophys. Res.* **2011**, *116*, B12310. [CrossRef]
4. Pancha, A.; Anderson, J.G.; Kreemer, C. Comparison of seismic and geodetic scalar moment rates across the Basin and Range Province. *Bull. Seismol. Soc. Am.* **2006**, *96*, 11–32. [CrossRef]
5. Bos, A.G.; Spakman, W. Kinematics of the southwestern U.S. deformation zone inferred from GPS motion data. *J. Geophys. Res.* **2005**, *110*, B08405. [CrossRef]
6. Rontogianni, S. Comparison of geodetic and seismic strain rates in Greece by using a uniform processing approach to campaign GPS measurements over the interval 1994–2000. *J. Geodyn.* **2010**, *50*, 381–399. [CrossRef]
7. Chousianitis, K.; Ganas, A.; Evangelidis, C.P. Strain and rotation rate patterns of mainland Greece from continuous GPS data and comparison between seismic and geodetic moment release. *J. Geophys. Res.* **2015**, *120*, 3909–3931. [CrossRef]
8. Jenny, S.; Goes, S.; Giardini, D.; Kahle, H.G. Seismic potential of southern Italy. *Tectonophysics* **2006**, *415*, 81–101. [CrossRef]
9. Palano, M.; Cannavò, F.; Ferranti, L.; Mattia, M.; Mazzella, M.E. Strain and stress fields in the Southern Apennines (Italy) constrained by geodetic, seismological and borehole data. *Geophys. J. Int.* **2011**, *187*, 1270–1282. [CrossRef]

10. D'Agostino, N. Complete seismic release of tectonic strain and earthquake recurrence in the Apennines (Italy). *Geophys. Res. Lett.* **2014**, *41*, 1155–1162. [CrossRef]

11. Bilham, R.; Ambraseys, N. Apparent Himalayan slip deficit from the summation of seismic moments for Himalayan earthquakes, 1500–2000. *Curr. Sci.* **2005**, *88*, 1658–1663.

12. Bungum, H.; Lindholm, C.D.; Mahajan, A.K. Earthquake recurrence in NW and central Himalaya. *J. Asian Earth Sci.* **2017**, *138*, 25–37. [CrossRef]

13. Stevens, V.L.; Avouac, J.P. Millenary Mw > 9.0 earthquakes required by geodetic strain in the Himalaya. *Geophys. Res. Lett.* **2017**, *43*, 1118–1123. [CrossRef]

14. Déprez, A.; Doubre, C.; Masson, F.; Ulrich, P. Seismic and aseismic deformation along the East African Rift System from a reanalysis of the GPS velocity field of Africa. *Geophys. J. Int.* **2013**, *193*, 1353–1369. [CrossRef]

15. Kostrov, V. Seismic moment and energy of earthquakes, and seismic Row of rock. *Izv. Acad. Sci. USSR Phys. Solid Earth* **1974**, *1*, 23–44.

16. Hyndman, R.D.; Weichert, D.H. Seismicity and rates of relative plate motion on the plate boundaries of western North America. *Geophys. J. R. Astron. Soc.* **1983**, *72*, 59–82. [CrossRef]

17. Mazzotti, S.; Adams, J. Rates and uncertainties on seismic moment and deformation in eastern Canada. *J. Geophys. Res.* **2005**, *110*. [CrossRef]

18. Michard, A.; Chalouan, A.; Feinberg, H.; Goffé, B.; Montigny, R. How does the Alpine belt end between Spain and Morocco? *Bull. Soc. Géol. France* **2002**, *173*, 3–15. [CrossRef]

19. Platt, J.P.; Whitehouse, M.J.; Kelley, S.P.; Carter, A.; Hollick, L. Simultaneous extensional exhumation across the Alboran Basin: Implications for the causes of late orogenic extension. *Geology* **2003**, *31*, 251–254. [CrossRef]

20. McKenzie, D.P. Plate tectonics of the Mediterranean region. *Nature* **1970**, *226*, 239–243. [CrossRef]

21. Andrieux, J.; Fontbote, J.M.; Mattauer, M. Sur un modèle explicatif de l'arc de Gibraltar. *Earth Planet. Sci. Lett.* **1971**, *12*, 191–198. [CrossRef]

22. Comas, M.C.; Platt, J.P.; Soto, J.I.; Watts, A.B. The origin and tectonic history of the Alboran Basin: Insights from Leg 161 results. In *Proceedings of the Ocean Drilling Program*; Scientific Results; Zahn, R., Comas, M.C., Klaus, A., Eds.; Ocean Drilling Program: College Station, TX, USA, 1999; Volume 161, pp. 555–580.

23. Palano, M.; González, P.J.; Fernández, J. The diffuse plate boundary of Nubia and Iberia in the Western Mediterranean: Crustal deformation evidence for viscous coupling and fragmented lithosphere. *Earth Planet. Sci. Lett.* **2015**, *430*, 439–447. [CrossRef]

24. Peláez, J.A.; Henares, J.; Hamdache, M.; Sanz de Galdeano, C. A seismogenic zone model for seismic hazard studies in northwestern Africa in Moment tensor solutions. In *A Useful Tool for Seismotectonics*; D'Amico, S., Ed.; Springer Natural Hazards: Berlin/Heidelberg, Germany, 2018; pp. 643–680.

25. CNIG. *Update of Seismic Hazard Maps of Spain*; Centro Nacional de Información Geográfica: Madrid, Spain, 2013.

26. Bourgois, J.A.; Mauffret, A.; Ammar, N.A.; Demnati, N.A. Multichannel seismic data imaging of inversion tectonics of the Alboran Ridge (western Mediterranean Sea). *Geo Mar. Lett.* **1992**, *12*, 117–122. [CrossRef]

27. Masana, E.; Martínez-Díaz, J.J.; Hernández-Enrile, J.L.; Santanach, P. The Alhama de Murcia fault (SE Spain), a seismogenic fault in a diffuse plate boundary: Seismotectonic implications for the Ibero-Magrebian region. *J. Geophys. Res.* **2004**, *109*. [CrossRef]

28. Gracia, E.; Pallas, R.; Soto, J.I.; Comas, M.C.; Moreno, X.; Masana, E.; Santanach, P.; Diez, S.; Garcia, M.; Dañobeitia, J. Active faulting offshore SE Spain, (Alboran Sea): Implications for earthquake hazard assessment in the southern Iberian Margin. *Earth Planet. Sci. Lett.* **2006**, *241*, 734–749. [CrossRef]

29. Ballesteros, M.; Rivera, J.; Muñoz, A.; Muñoz-Martín, A.; Acosta, J.; Carbó, A.; Uchupi, E. Alboran basin, Southern Spain. Part II: Neogene tectonic implications for the orogenic float model. *Mar. Petrol. Geol.* **2008**, *25*, 75–101. [CrossRef]

30. Meghraoui, M.; Cisternas, A.; Philip, H. Seismotectonics of the lower Cheliff basin: Structural background of the El Asnam (Algeria) earthquake. *Tectonics* **1986**, *5*, 809–836. [CrossRef]

31. Meghraoui, M.; Pondrelli, S. Active faulting and transpression tectonics along the plate boundary in North Africa. *Ann. Geophys.* **2012**, *55*. [CrossRef]

32. Bahrouni, N.; Bouaziz, S.; Soumaya, A.; Ben Ayed, N.; Attafi, K. Active deformation analysis and evaluation of earthquake hazard in Gafsa region (Southern Atlas of Tunisia). *Geophys. Res. Abstr.* **2013**, *15*, EGU2013-1009.

33. Rabaute, A.; Chamot-Rooke, N. *Active Tectonics of the Africa-Eurasia Boundary from Algiers to Calabria. Map at 1:500,000 Scale*; Geosubsight: Paris, France, 2014; ISBN 978-2-9548197-0-9.

34. Maouche, S.; Abtout, A.; Merabet, N.E.; Aïfa, T.; Lamali, A.; Bouyahiaoui, B.; Ayache, M. Tectonic and Hydrothermal Activities in Debagh, Guelma Basin (Algeria). *J. Geophys. Res.* **2013**, 409475. [CrossRef]

35. Ben Hassen, M.; Deffontaines, B.; Turki, M.M. Recent tectonic activity of the Gafsa fault through morphometric analysis: Southern Atlas of Tunisia. *Quat. Int.* **2014**, *338*, 99–112. [CrossRef]

36. The SHARE European Earthquake Catalogue. Available online: www.emidius.eu/SHEEC/ (accessed on 26 January 2020).

37. Peláez, J.A.; Chourak, M.; Tadili, B.A.; Aït Brahim, L.; Hamdache, M.; López Casado, C.; Martínez Solares, J.M. A catalog of main Moroccan earthquakes from 1045 to 2005. *Seismol. Res. Lett.* **2007**, *78*, 614–621. [CrossRef]

38. Hamdache, M.; Peláez, J.A.; Talbi, A.; López Casado, C. A unified catalog of main earthquakes for northern Algeria from a.d. 856 to 2008. *Seismol. Res. Lett.* **2010**, *81*, 732–739. [CrossRef]

39. Bougrine, A.; Yelles-Chaouche, A.K.; Calais, E. Active deformation in Algeria from Continuous GPS measurements. *Geophys. J. Int.* **2019**, *217*, 572–588. [CrossRef]

40. Koulali, A.; Ouazar, D.; Tahayt, A.; King, R.W.; Vernant, P.; Reilinger, R.E.; McClusky, S.; Mourabit, T.; Davila, J.M.; Amraoui, N. New GPS constrains on active deformation along the Africa-Iberia plate boundary. *Earth Planet. Sci. Lett.* **2011**, *308*, 211–217. [CrossRef]

41. Stich, D.; Martín, J.B.; Morales, J. Deformación sísmica y asísmica en la zona Béticas-Rif-Alborán. *Rev. Soc. Geol. Esp.* **2007**, *20*, 311–319.

42. Martínez Solares, J.M.; Mezcua, J. *Seismic Catalog of the Iberian Peninsula (880 BC-1900)*; IGN: Madrid, Spain, 2002. (In Spanish)

43. Buforn, E.; Udías, A.; Madariaga, R. Intermediate and deep earthquakes in Spain. *Pure Appl. Geophys.* **1991**, *136*, 375–393. [CrossRef]

44. Buforn, E.; Pro, C.; Udías, A.; del Fresno, C. The 2010 Granada, Spain, deep earthquake. *Bull. Seismol. Soc. Am.* **2011**, *101*, 2418–2430. [CrossRef]

45. Cunha, T.A.; Matias, L.M.; Terrinha, P.; Negredo, A.M.; Rosas, F.; Fernandes, R.M.S.; Pinheiro, L.M. Neotectonics of the SW Iberia margin, Gulf of Cadiz and Alboran Sea: A reassessment including recent structural, seismic and geodetic data. *Geophys. J. Int.* **2012**, *188*, 850–872. [CrossRef]

46. International Seismological Centre. Available online: www.isc.ac.uk/ (accessed on 26 January 2020).

47. Dziewonski, A.M.; Chou, T.-A.; Woodhouse, J.H. Determination of earthquake source parameters from waveform data for studies of global and regional seismicity. *J. Geophys. Res.* **1981**, *86*, 2825–2852. [CrossRef]

48. Ekström, G.; Nettles, M.; Dziewonski, A.M. The global CMT project 2004–2010: Centroid-moment tensors for 13,017 earthquakes. *Phys. Earth Planet. Inter.* **2012**, *200–201*, 1–9. [CrossRef]

49. Custódio, S.; Lima, V.; Vales, D.; Cesca, S.; Carrilho, F. Imaging active faulting in a region of distributed deformation from the joint clustering of focal mechanisms and hypocentres: Application to the Azores-western Mediterranean region. *Tectonophysics* **2016**, *676*, 70–89. [CrossRef]

50. Peláez, J.A.; Henares, J.; Hamdache, M.; Sanz de Galdeano, C. An updated seismic model for northwestern Africa. In Proceedings of the 16th European Conference on Earthquake Engineering, Thessaloniki, Greece, 18–21 June 2018.

51. Aoudia, A.; Vaccari, F.; Suhadolc, P.; Meghraoui, M. Seismogenic potential and earthquake hazard assessment in the Tell Atlas of Algeria. *J. Seism.* **2000**, *4*, 79–98. [CrossRef]

52. Peláez, J.A.; Hamdache, M.; López Casado, C. Seismic hazard in northern Algeria using spatially smoothed seismicity. Results for peak ground acceleration. *Tectonophysics* **2003**, *372*, 105–119. [CrossRef]

53. Peláez, J.A.; Hamdache, M.; López Casado, C. Updating the probabilistic seismic hazard values of Northern Algeria with the 21 May 2003 M 6.8 Algiers earthquake included. *Pure Appl. Geophys.* **2005**, *162*, 2163–2177. [CrossRef]

54. Henares, J.; López Casado, C.; Sanz de Galdeano, C.; Delgado, J.; Peláez, J.A. Stress fields in the Ibero-Maghrebian region. *J. Seismol.* **2003**, *7*, 65–78. [CrossRef]

55. Palano, M.; González, P.J.; Fernández, J. Strain and stress fields along the Gibraltar Orogenic Arc: Constraints on active geodynamics. *Gondwana Res.* **2013**, *23*, 1071–1088. [CrossRef]

56. Soumaya, A.; Ben Ayed, N.; Delvaux, D.; Ghanmi, M. Spatial variation of present-day stress field and tectonic regime in Tunisia and surroundings from formal inversion of focal mechanisms: Geodynamics implications for central Mediterranean. *Tectonics* **2015**, *34*, 1154–1180. [CrossRef]

57. Aït Brahim, L.; Chotin, P.; Hinaj, S.; Abdelouafi, A.; Nakhcha, C.; Dhont, D.; Sossey Alaoui, F.; Bouaza, A.; Tabyaoui, H.; Chaouni, A. Paleostress evolution in the Moroccan African margin from Triassic to present. *Tectonophysics* **2002**, *357*, 187–205. [CrossRef]

58. Hanks, T.C.; Kanamori, H. A moment magnitude scale. *J. Geophys. Res.* **1979**, *84*, 2348–2350. [CrossRef]

59. Weichert, D.H. Estimation of the earthquake recurrence parameters for unequal observation observation periods for different magnitudes. *Bull. Seismol. Soc. Am.* **1980**, *70*, 1337–1346.

60. Wells, D.L.; Coppersmith, K.J. New empirical relationship among magnitude, rupture length, rupture width, rupture area, and surface displacement. *Bull. Seismol. Soc. Am.* **1994**, *84*, 974–1002.

61. Kijko, A.; Singh, M. Statistical tools for maximum possible earthquake magnitude estimation. *Acta Geophys.* **2011**, *59*, 674–700. [CrossRef]

62. EUREF. Permanent GNSS Network. Available online: www.epncb.oma.be (accessed on 26 January 2020).

63. CDDIS. NASA's Archive of Space Geodesy Data. Available online: https://cddis.nasa.gov/ (accessed on 26 January 2020).

64. UNAVCO. Available online: www.unavco.org (accessed on 26 January 2020).

65. Herring, T.A.; King, R.W.; Floyd, M.A.; McClusky, S.C. *Introduction to GAMIT/GLOBK, Release 10.7*; Massachusetts Institute of Technology: Cambridge, UK, 2017; Available online: www-gpsg.mit.edu (accessed on 26 January 2020).

66. Palano, M. On the present-day crustal stress, strain-rate fields and mantle anisotropy pattern of Italy. *Geophys. J. Int.* **2015**, *200*, 969–985. [CrossRef]

67. Altamimi, Z.; Rebischung, P.; Métivier, L.; Collilieux, X. ITRF2014: A new release of the international terrestrial reference frame modeling nonlinear station motions. *J. Geophys. Res.* **2016**, *121*, 6109–6131. [CrossRef]

68. Echeverría, A.; Khazaradze, G.; Asensio, E.; Gárate, J.; Martín Dávila, J.; Suriñach, E. Crustal deformation in eastern Betics from CuaTeNeo GPS network. *Tectonophysics* **2013**, *608*, 600–612. [CrossRef]

69. Garate, J.; Martin-Davila, J.; Khazaradze, G.; Echeverria, A.; Asensio, E.; Gil, A.J.; de Lacy, M.C.; Armenteros, J.A.; Ruiz, A.M.; Gallastegui, J.; et al. Topo-Iberia project: CGPS crustal velocity field in the Iberian Peninsula and Morocco. *GPS Solut.* **2015**, *19*, 287–295. [CrossRef]

70. Cabral, J.; Mendes, V.B.; Figueiredo, P.; da Silveira, A.B.; Pagarete, J.; Ribeiro, A.; Dias, R.; Ressurreição, R. Active tectonics in Southern Portugal (SW Iberia) inferred from GPS data. Implications on the regional geodynamics. *J. Geodyn.* **2017**, *112*, 1–11. [CrossRef]

71. Shen, Z.-K.; Wang, M.; Zeng, Y.; Wang, F. Optimal interpolation of spatially discretized geodetic data. *Bull. Seismol. Soc. Am.* **2015**, *105*, 2117–2127. [CrossRef]

72. Savage, J.C.; Simpson, R.W. Surface strain accumulation and the seismic moment tensor. *Bull. Seismol. Soc. Am.* **1997**, *87*, 1345–1353.

73. Stich, D.; Martínez-Solares, J.M.; Custódio, S.; Batlló, J.; Martín, R.; Teves-Costa, P.; Morales, J. Seismicity of the Iberian Peninsula. In *The Geology of Iberia: A Geodynamic Approach*; Springer: Cham, Switzerland, 2020; pp. 11–32. [CrossRef]

74. Mahsas, A.; Lammali, K.; Yelles, K.; Calais, E.; Freed, A.; Briole, P. Shallow afterslip following the 2003 May 21, Mw = 6.9 Boumerdes earthquake, Algeria. *Geophys. J. Int.* **2008**, *172*, 155–166. [CrossRef]

75. Lin, J.; Stein, R.; Meghraoui, M.; Toda, S.; Ayadi, A.; Dorbath, C.; Belabbes, S. Stress transfer among en echelon and opposing thrusts and tear faults: Triggering caused by the 2003 Mw = 6.9 Zemmouri, Algeria, earthquake. *J. Geophys. Res.* **2011**, *116*, B03305. [CrossRef]

76. Onana, P.N.E.; Toto, E.A.; Zouhri, L.; Chaabane, A.; El Mouraouah, A.; Brahim, A.I. Recent seismicity of Central High Atlas and Ouarzazate basin (Morocco). *Bull. Eng. Geol. Environ.* **2011**, *70*, 633–641. [CrossRef]

77. Turcotte, D.; Schubert, G. *Geodynamics*; Cambridge University Press: Cambridge, UK, 2002; p. 456.

78. Ward, S.N. On the consistency of earthquake moment rates, geological fault data, and space geodetic strain: The United States. *Geophys. J. Int.* **1998**, *134*, 172–186. [CrossRef]

79. Torné, M.; Fernàndez, M.; Comas, M.C.; Soto, J.I. Lithospheric structure beneath the Alboran Basin: Results from 3D gravity modeling and tectonic relevance. *J. Geophys. Res.* **2000**, *105*, 3209–3228. [CrossRef]

80. Polyak, B.G.; Fernàndez, M.; Khutorskoy, M.D.; Soto, J.I.; Basov, I.A.; Comas, M.C.; Khain, V.Y.; Alonso, B.; Agapova, G.V.; Mazurova, I.S.; et al. Heat flow in the Alboran Sea, western Mediterranean. *Tectonophysics* **1996**, *263*, 191–218. [CrossRef]

81. Fernàndez, M.; Marzán, I.; Correia, A.; Ramalho, E. Heat flow, heat production, and lithospheric thermal regime in the Iberian Peninsula. *Tectonophysics* **1998**, *291*, 29–53. [CrossRef]
82. Rimi, A.; Chalouan, A.; Bahi, L. Heat flow in the westernmost part of the alpine mediterranean system (The Rif, Morocco). *Tectonophysics* **1998**, *285*, 135–146. [CrossRef]
83. Pollack, H.N.; Chapman, D.S. On the regional variation of heat flow, geotherms, and lithospheric thickness. *Tectonophysics* **1977**, *38*, 279–296. [CrossRef]
84. Serrano, I.; Zhao, D.P.; Morales, J. 3-D crustal structure of the extensional Granada Basin in the convergent boundary between the Eurasian and African plates. *Tectonophysics* **2002**, *344*, 61–79. [CrossRef]
85. Fernàndez-Ibáñez, F.; Soto, J.I. Crustal rheology and seismicity in the Gibraltar Arc (western Mediterranean. *Tectonics* **2008**, *27*, TC2007. [CrossRef]
86. González, P.J.; Palano, M.; Fernández, J. Study of the present-day tectonics and seismogenetic sources of the Al-Hoceima region (Morocco) using GPS and MTINSAR. In Proceedings of the Fringe 2009 Workshop, Frascati, Italy, 30 November–4 December 2009; Volume 30.
87. Yelles Chauche, A.K.; Kherroubi, A.; Beldjoudi, H. The large Algerian earthquakes (267 A.D.-2017). *Física de la Tierra* **2017**, *29*, 159–182. [CrossRef]
88. Roger, J.; Hébert, H. The 1856 Djijelli (Algeria) earthquake and tsunami: Source parameters and implications for tsunami hazard in the Balearic Islands. *Nat. Hazards Earth Syst. Sci.* **2008**, *8*, 721–731. [CrossRef]
89. Hamdache, M.; Peláez, J.A.; Yelles Chauche, A.K. The Algiers, Algeria earthquake (Mw 6.8) of 21 May 2003: Preliminary report. *Seismol. Res. Lett.* **2004**, *75*, 360–367. [CrossRef]
90. Rong, Y.; Jackson, D.D.; Kagan, Y.Y. Seismic gaps and earthquakes. *J. Geophys. Res.* **2003**, *108*, 2471. [CrossRef]
91. Gupta, H.; Gahalaut, V.K. Seismotectonics and large earthquake generation in the Himalayan region. *Gondwana Res.* **2014**, *25*, 204–213. [CrossRef]
92. Wessel, P.; Smith, W.H. The generic mapping tools (GMT). *Eos Trans. Am. Geophys. Union* **1995**, *76*, 329. [CrossRef]
93. Heidbach, O.; Rajabi, M.; Cui, X.; Fuchs, K.; Müller, B.; Reinecker, J.; Reiter, K.; Tingay, M.; Wenzel, F.; Xie, F.; et al. The World Stress Map database release 2016: Crustal stress pattern across scales. *Tectonophysics* **2018**, *744*, 484–498. [CrossRef]
94. de Vicente, G.; Cloetingh, S.; Muñoz Martín, A.; Olaiz, A.; Stich, D.; Vegas, R.; Galindo Zaldívar, J.; Fernández Lozano, J. Inversion of moment tensor focal mechanisms for active stresses around the microcontinent Iberia: Tectonic implications. *Tectonics* **2008**, *27*, TC1009. [CrossRef]
95. Kennett, B.L.N.; Engdahl, E.R.; Buland, R. Constraints on seismic velocities in the Earth from travel times. *Geophys. J. Int.* **1995**, *122*, 108–124. [CrossRef]

*remote sensing*



*Article*

# Shrinking of Ischia Island (Italy) from Long-Term Geodetic Data: Implications for the Deflation Mechanisms of Resurgent Calderas and Their Relationships with Seismicity

Alessandro Galvani [1,*], Giuseppe Pezzo [1], Vincenzo Sepe [1] and Guido Ventura [1,2]

1   Istituto Nazionale di Geofisica e Vulcanologia, Via di Vigna Murata 605, 00143 Roma, Italy; giuseppe.pezzo@ingv.it (G.P.); vincenzo.sepe@ingv.it (V.S.); guido.ventura@ingv.it (G.V.)
2   Istituto per lo Studio degli Impatti Antropici e Sostenibilità in Ambiente Marino, Consiglio Nazionale delle Ricerche (CNR), Capo Granitola (TP), 91021 Campobello di Mazara, Italy
*   Correspondence: alessandro.galvani@ingv.it

**Abstract:** The identification of the mechanisms responsible for the deformation of calderas is of primary importance for our understanding of the dynamics of magmatic systems and the evaluation of volcanic hazards. We analyze twenty years (1997–2018) of geodetic measurements on Ischia Island (Italy), which include the Mt. Epomeo resurgent block, and is affected by hydrothermal manifestations and shallow seismicity. The data from the GPS Network and the leveling route show a constant subsidence with values up to $-15 \pm 2.0$ mm/yr and a centripetal displacement rate with the largest deformations on the southern flank of Mt. Epomeo. The joint inversion of GPS and levelling data is consistent with a 4 km deep source deflating by degassing and magma cooling below the southern flank of Mt. Epomeo. The depth of the source is supported by independent geophysical data. The Ischia deformation field is not related to the instability of the resurgent block or extensive gravity or tectonic processes. The seismicity reflects the dynamics of the shallow hydrothermal system being neither temporally nor spatially related to the deflation.

**Keywords:** GNSS; velocity field; resurgent caldera; subsidence; earthquakes; degassing processes; modelling

**Citation:** Galvani, A.; Pezzo, G.; Sepe, V.; Ventura, G. Shrinking of Ischia Island (Italy) from Long-Term Geodetic Data: Implications for the Deflation Mechanisms of Resurgent Calderas and Their Relationships with Seismicity. *Remote Sens.* **2021**, *13*, 4648. https://doi.org/10.3390/rs13224648

Academic Editor: João Catalão Fernandes

Received: 28 October 2021
Accepted: 15 November 2021
Published: 18 November 2021

**Publisher's Note:** MDPI stays neutral with regard to jurisdictional claims in published maps and institutional affiliations.

## 1. Introduction

The deformation of calderas may be associated with different processes, including the magma accumulation, lateral migration or withdrawal, increase or decrease of gas pressure of hydrothermal systems, and variations in the degassing rate of magma chambers [1]. While uplift phases are indicative of resurgence and may be precursors of volcanic eruptions, testifying to a pressurization of the magmatic system due to magma/gas accumulation or upward magma/fluid migration, subsidence episodes are more difficult to interpret because they may be related to different causes, such as magma cooling and degassing [2], lateral magma migration in sills [3], depressurization of hydrothermal reservoirs [4], regional extension related to tectonics, and gravity instability processes [5]. This is particularly difficult when the subsidence is associated with seismicity. As a result, our understanding of the processes responsible for the subsidence of resurgent calderas represents a primary target to decipher their dynamics. Well-known resurgent calderas include the Campi Flegrei caldera, Italy [6], Yellowstone, USA [7], and Santorini, Greece [8]. All these calderas are characterized by uplift episodes followed by short to long subsidence (deflation) periods with seismicity generally associated with the uplift phases. The identification of the mechanisms responsible for deflation may give us information on the dynamics of the underlying magmatic and hydrothermal systems and on the role of tectonics and gravity processes in modulating the deformation of volcanoes. Aiming to recognize the different causes of the recorded subsidence of Ischia's (Italy) resurgent caldera (Figure 1), we analyze deformation data from the past twenty years. Ischia Island is located in the northern secto of the Gulf of Naples, Italy, which also includes the Somma-Vesuvius volcano and Camp Flegrei caldera (Figure 1). Nowadays a geodetic GPS survey style network of 22 vertice: operates on the island to monitor its deformation behavior [9,10]. The network is managed by INGV—Osservatorio Vesuviano and includes six permanent GPS vertices belonging to INGV—OV and one vertex belonging to the Regione Campania (Figure 2). For the firs time, the deformation pattern of the island is investigated by using a GPS dataset from 1997 to 2018. Our analysis also includes leveling data reported by Trasatti et al. (2019) [11] Results reveal the spatial and temporal displacements; subsequent GPS and levelling data inversion allow us to identify the different causes of measured displacements, revealing the role of passive degassing processes in controlling the deformation of resurgent block: at calderas. Finally, we discuss the relationships between earthquakes and deformatior pattern.

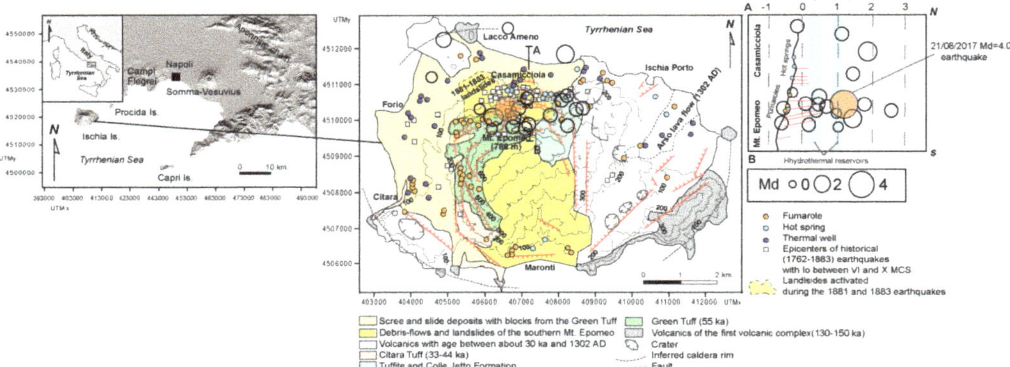

**Figure 1.** Location and geological map of Ischia Island from [12] with epicentral and hypocentral (N-S cross section) distributions of the 1999–2017 earthquakes from [24]. Historical earthquakes are from [18]. Hydrothermal manifestations are from [17,20]. The depth of the hydrothermal reservoirs in the N-S cross section is from [20,21].

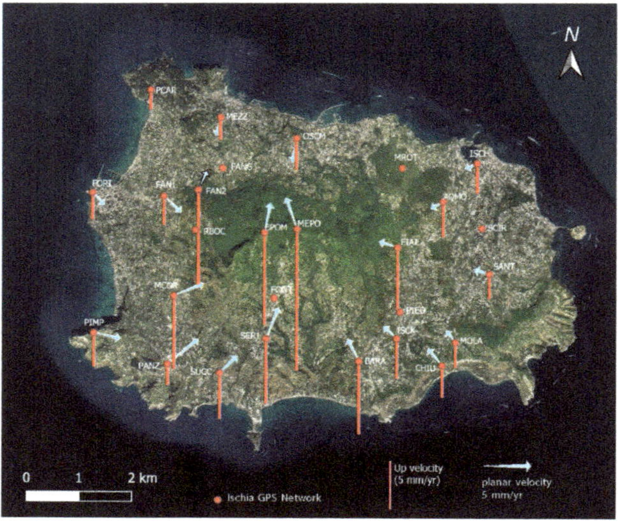

**Figure 2.** Ischia GPS network and vertical and planar velocity field in mm/yr calculated in the time span 1997–2018.

## 2. Geological Setting

The volcanism on Ischia occurred between 130–150 ka and 1302 A.D (Figure 1) [12]. The most recent activity (<10 ka) is concentrated in the eastern sector of the island. The resurgent block of Mt. Epomeo (786 m a.s.l.), located in the western sector, uplifted between 55 ka and 19 ka and has been followed by a still active subsidence [13–15]. In the period 1993–2003, measured subsidence rate varies between 1 and 5 mm/year [16,17], and it has been related to fracture closure due to depressurization of the hydrothermal system [17]. The northern flank of Mt. Epomeo is affected by WSW-ESE to E-W striking, north-dipping normal faults (Figure 1), some of which reactivated during the 1883 earthquake (MCS intensity = X) [18,19]. Ischia is also affected by significant hydrothermal activity [20]. A close spatial relationship occurs between the Casamicciola Terme fumaroles, hot springs, and the above-mentioned E-W striking faults (Figure 1). The highest $CO_2$ (9 t d-1) and steam (135 t d-1) release are measured along the faults that delimit Mt. Epomeo's resurgent structure. Some of these faults are sealed by secondary minerals like gypsum and anhydrite, which are related to the deposition of supersaturated hydrothermal solutions and leaching of the tuffs and lava flows [20,21]. Moreover, a hydrothermal system with heat flow and temperatures up to 40 MW and 250 °C [21], respectively, characterizes the island [22]. The hydrothermal system consists of two sub-horizontal water-vapor reservoirs located at about 400 and 900 m depth with temperatures between 250 and 350 °C and fluid pressures of 4 to 9 MPa, respectively (Figure 1) [20]. These reservoirs are laterally confined by the Mt. Epomeo faults, which, in the Casamicciola Terme area and west of the resurgent block, act as pathways for the rise of hot fluids and gas release (Figure 1). On the basis of resistivity data, the magmatic source responsible for the hydrothermal system heating is supposed to be a laccolith located at depth $\geq$ 3 km, below Mt. Epomeo [23]. Above the laccolith, subvertical vapor zones cross and feed two hydrothermal reservoirs (Figure 1) [21]. Seismic activity is rather low and generally with low (Md < 2) magnitude events. These events mainly occur along the E-W and NNW-SSE faults delimiting the northern and western sectors of the island where gas emissions and hot water springs also concentrate, and occur within the first 3 km of the crust (Figure 1) [24]. However, the historical record shows that Ischia was struck by some larger magnitude earthquakes, a few of them causing severe damages and fatalities. Remarkable events occurred in 1881 and 1883 in the Casamicciola area and on 21 August 2017, (Md = 4; Mw 3.9) [25]. This latter event occurred on the E-W fault system bounding the northern sector of Mt. Epomeo, damaging a limited portion of the Casamicciola area (Iomax = 8) and causing two fatalities. The 'shallow' source of the Io = X 1881, 1883, and Md = 4 2017 (depth 1.7 km) [26] damaging earthquakes, the 'explosion' in 1995 of a geothermal well drilled between 1940 and 1950, and the occurrence of boiling phreatic aquifers are interpreted to be related to the dynamics of the hydrothermal system [20,25,27], and, in particular, to its pressurization phases. The relevant CLVD component of the focal mechanism of the 21 August, 2017 Md = 4 event, its low frequency character, the low S/P spectral ratio, and the characteristics of the seismic noise also provide evidence of the involvement of pressurized fluids in the source of the Ischia earthquakes [25,26]. The temporal trends of selected geochemical parameters (groundwater discharge temperatures, groundwater Mg/Cl ratios and $CO_2$ partial pressure) do not show significant variations from 1984 to 2007, suggesting, according to [21], a stable system with a nearly constant thermal and mass transport from depth. As concerns the deformation field, results from SAR data [16] and levelling lines collected in the 1987–2010 [11] and 1990–2003 [17] periods, evidence a subsidence trend, with values up to 1.2 cm/yr centered on the Mt. Epomeo resurgent block, while little or no subsidence occurs on the coastal areas. The observed pattern has been interpreted as due to degassing processes along the faults bounding Mt. Epomeo from a shallow, depressurizing magma reservoir [11,16] or along two main E-W striking, closing cracks corresponding to the Casamicciola fault and the faults bounding the top of Mt. Epomeo [17]. The role of gravity instability processes in controlling the deformation pattern at a local scale has been also recognized (Manzo et al., 2006) [16]. Main processes are the slide of the northwestern sector, activated during the

1881 and 1883 earthquakes, and the rock-fall zones affecting the steep eastern and northern slopes and the summit of Mt. Epomeo (Figure 1).

### 3. GPS Network and Dataset Processing

Since October 1996, a survey style network of 27 GPS vertices with an inter-distance of 1 to 3 km has been established on Ischia to define the deformation field of the island (Figure 2) [9]. From 1999 to 2013, five of them (AQMO, FORI, OSCM, SANT and SER1) were upgraded into continuous GNSS (CGNSS) stations. The current network configuration also includes the CGNSS station on the top of Mt. Epomeo (MEPO), since March 2017, and one CGNSS station (ISCK) managed by the Campania Region, since 2008 (Figure 2).

The survey style network vertices were realized using 3D-type GPS monuments [9,28]. Siting was performed taking into account the geological and structural features of the island and its urbanization; the benchmarks are located on outcrops or buildings.

During the 1997–2017 time span, all discontinuous stations of the network were repeatedly measured, equipping the GPS stations with TRIMBLE and LEICA receivers and antennas. Each station was occupied for an average observation window ranging from 15 to 24 h for at least 4 survey sessions per station. GPS data were collected at 30 s sampling rate (Table 1).

The complete analyzed dataset, which includes unpublished survey-style data from 2010 to 2017, also considers CGNSS data provided by the CGNSS networks located in the Campania and south Lazio regions. The CGNSS stations belong to different GNSS network providers: the International Global Navigation Satellite Systems GNSS Service IGS [http//igs.org, accessed on 26 October 2021, daily available.], the Integrated National GPS Network, RING [29], the Italian Space Agency (Agenzia Spaziale Italiana; ASI) [30] and Leica Geosystems (Italian Positioning Service [ItalPos] network). The CGNSS data cover the period from 1999 to 2018. They are arranged into several clusters, each sharing common fiducial CGNSS stations used as anchor stations in the subsequent combinations. Each cluster was independently processed and then integrated through a least-squares combination into a single daily solution. The GPS observations were processed using the Bernese 5.0 [31] software, based on the Bernese Processing Engine (BPE) procedure following a standard analysis for regional networks. The daily station coordinates, together with the hourly troposphere parameters, are solved using the a priori Dry Neill troposphere model with the corrections estimated by the Wet Neill mapping function. The ionosphere is neither estimated nor modeled as we used the L3 (ionosphere free) linear combination of L1 and L2. The a priori GPS orbits and the Earth orientation parameters are fixed to the precise IGS products. We applied the ocean-loading finite element solution model FES2004 and used the IGS absolute antenna phase-center corrections. The daily solutions were obtained in a loosely constrained reference frame; i.e., all the a priori station coordinates were left free to 10 m a priori sigma. The time series were obtained by applying minimal inner constraints and a four-parameter Helmert transformation to have the coordinates and errors expressed in the IGS08 reference framework (the IGS realization of the ITRF2008 reference framework). We obtained a velocity field with the estimation of a linear drift (velocity), annual sinusoid, and occasional offsets due to changes in the station equipment in each time series. The formal velocity uncertainties are the white-noise standard deviation multiplied by the variance factors obtained from the time series at each site. We preferred to issue the formal solution without any covariance manipulation, thus focusing on the relative uncertainty between velocities rather than absolute errors.

**Table 1.** Summary of the geodetic network, velocities, time span acquisition, and number of campaigns.

| SITE | East (mm/yr) | sigE (mm/yr) | North (mm/yr) | sigN (mm/yr) | Up (mm/yr) | sigUp (mm/yr) | Tin (initial; yr) | Tfi (final; yr) | # Campaign |
|---|---|---|---|---|---|---|---|---|---|
| AQMO | −1.3185 | 1.0561 | −0.8085 | 1.0672 | −3.7278 | 0.2328 | 2001 | 2018 | CGNSS |
| BARA | −1.2012 | 1.2281 | 2.4108 | 1.4395 | −7.6995 | 5.8372 | 1998 | 2017 | 6 |
| CHIU | −1.5085 | 0.5462 | 1.8581 | 0.6677 | −3.2606 | 2.471 | 1997 | 2017 | 7 |
| EPOM | 0.6406 | 1.138 | 3.1905 | 1.3249 | −9.8474 | 4.9739 | 1997 | 2017 | 7 |
| FAN1 | 1.8555 | 1.5572 | −1.8048 | 1.6763 | −2.9954 | 7.4253 | 1998 | 2017 | 6 |
| FAN2 | 0.9177 | 1.335 | 2.0247 | 1.6369 | −10.1203 | 6.5496 | 1998 | 2017 | 6 |
| FIAI | −1.9689 | 1.3448 | 0.5986 | 1.5824 | −6.6067 | 6.636 | 1999 | 2017 | 5 |
| FORI | 1.2557 | 1.0692 | −1.3669 | 1.0853 | −2.7434 | 0.5228 | 1997 | 2018 | CGNSS |
| ISCH | −1.2798 | 1.03 | −0.621 | 1.2643 | −3.0754 | 4.726 | 1997 | 2017 | 7 |
| ISCK | −1.4056 | 1.057 | 1.3132 | 1.1506 | −4.2302 | 0.7274 | 2008 | 2018 | CGNSS |
| MCOR | 3.1704 | 1.5771 | 1.2881 | 1.6606 | −7.7989 | 7.4588 | 2003 | 2017 | 3 |
| MEPO | −1.1809 | 1.0394 | 3.3943 | 1.1477 | −15.1859 | 5.2276 | 2017 | 2018 | CGNSS |
| MEZZ | −0.512 | 1.199 | −2.0104 | 1.4277 | −2.2951 | 5.6208 | 1997 | 2017 | 7 |
| MOLA | −1.0144 | 0.9449 | 1.3155 | 1.1494 | −2.5843 | 4.4062 | 1998 | 2017 | 7 |
| OSCM | −0.4383 | 1.0813 | −2.5699 | 1.1087 | −3.2291 | 0.3768 | 2010 | 2018 | CGNSS |
| PANZ | 3.2801 | 1.2677 | 2.7378 | 1.5565 | −2.1145 | 6.0143 | 2010 | 2017 | 3 |
| PCAR | −0.289 | 1.2238 | −1.6892 | 1.3691 | −1.9985 | 5.5785 | 1997 | 2017 | 7 |
| PIMP | 2.8468 | 1.3238 | −0.6643 | 1.6352 | −3.5494 | 6.2291 | 1998 | 2017 | 6 |
| SANT | −1.6167 | 1.1062 | 0.4773 | 1.1268 | −2.5306 | 0.4933 | 2013 | 2018 | CGNSS |
| SER1 | 1.3107 | 0.062 | 3.3351 | 0.0748 | −6.8572 | 0.265 | 2001 | 2018 | CGNSS |
| SUCC | 1.8514 | 1.2647 | 1.8193 | 1.6016 | −4.8232 | 6.0664 | 1998 | 2017 | 7 |

## 4. Results

The horizontal velocity field of the complete dataset including southern Lazio region, Campania region with Ponza, Ventotene, and Ischia islands, expressed with respect to a fixed Eurasian plate [32], shows a good coherence with the velocities previously estimated for the Italian peninsula [33,34]. The wide time span of the data set includes the earthquake that struck the island on 21 August 2017. It produced significant offsets only in two continuous GPS stations: MEPO and OSCM (see Figure 3 as an example of time series at OSCM).

The observed displacements were modelled introducing an offset in the respective time series. To better understand the Ischia velocity field, we focused on the island, showing the horizontal velocity field with respect to the island centroid. This latter was obtained by minimizing the horizontal velocities of the GPS stations (Figure 2; Table 1). We also show the vertical velocity field with respect to the reference ellipsoid (Figure 2). The estimated Euler pole and rotation rate for the centroid are at −39.751°N, −165.247°W and 13.5395 ± 9.974 deg/Myr in modulus, respectively. Six vertices (FANG, FONT, MROT, PIED, RBOC and SCIR) have been eliminated due to incomplete time series or because they are installed on unstable areas (landslide and rockfall) or have been destroyed. The velocity vectors, reported in Figure 2, converge towards Mt. Epomeo with values ranging from a maximum of 4.27 mm/yr, in the SW, to a minimum of 1.42 mm/yr, in the northern part of the island. The three stations of FAN2, EPOM, and MEPO show patterns deviating from the dominant one. The above-described planar velocity field is accompanied by a general subsidence, ranging from −15.2 mm/yr at Mt. Epomeo to −1.99 mm/yr in the northeastern side of the island. The largest values of subsidence were been observed in the central and southern sectors of Ischia.

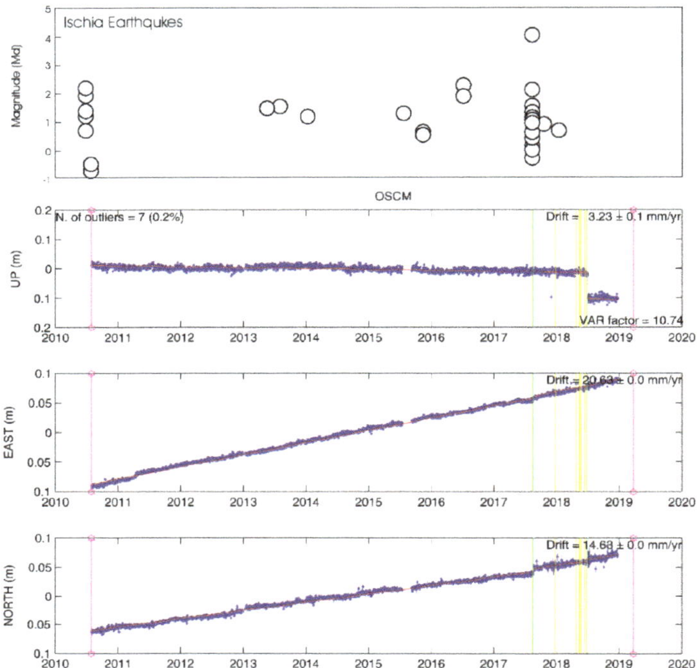

**Figure 3.** Ischia earthquakes (data from [24]) and OSCM (CGNSS) time series. Violet vertical bars are the starting and ending elaboration times. Yellow bars identify the removal of outlier solutions. Green vertical bar recorded in 2017 testifies of the Md 4 earthquake, whereas the green one recorded in 2018 is due to technical causes.

## 5. Displacement Field Modelling

In order to identify and characterize the main ground deformation source of Ischia we performed an analytical modelling by inverting the available ground deformation data. In particular, we used the CGNSS data and the leveling data reported in [11]. Levelling campaigns start from an initial reference point that is assumed to be stable (velocity equal to zero). However, the CGNSS stations close to the leveling reference point measure downwards displacements of about $-3$ mm/yr. Thus, we calibrated the two datasets, adding to the leveling data a rigid offset of $-3.08$ mm/yr. To verify the dataset agreement, we compared the leveling velocities at measure points close to the CGNSS benchmarks with CGNSS velocities (see scatter plot in Figure 4). The comparison highlights a good agreement between the two datasets with a correlation index of 0.83. Because Ischia undergoes localized landslide phenomena mainly concentrated in the central and northern sectors of the island [35], we excluded from the modelling some leveling and CGNSS data as specified in the previous section.

To retrieve the source parameters, we jointly inverted the CGNSS and leveling datasets by using an elastic, isotropic, and homogeneous model [36]. We adopt a two-step approach [37] consisting of a non-linear inversion to define the source parameters (north, east, depth, length, width, dip, strike, and homogeneous closuring along the plane) by means of a non-linear, least-squares inversion algorithm based on the Levemberg–Marquardt approach. Values of 30 GPa and 0.25 are assumed for the shear modulus and Poisson's ratio in the half-space, respectively. Because of the complex geological, hydrothermal, and volcanic structural features of Ischia, the deformation source can be assumed to be a horizontal closuring layer [21,23,25] whose unknown parameters are the depth, the closuring rate, and spatial extent. During the first step, we defined the source depth, position and homogeneous closuring. The parameter uncertainties, best fit and trade-offs (Figure 5) were estimated with 150 restarts of the inversion, adding, each time, a synthetic noise [37].

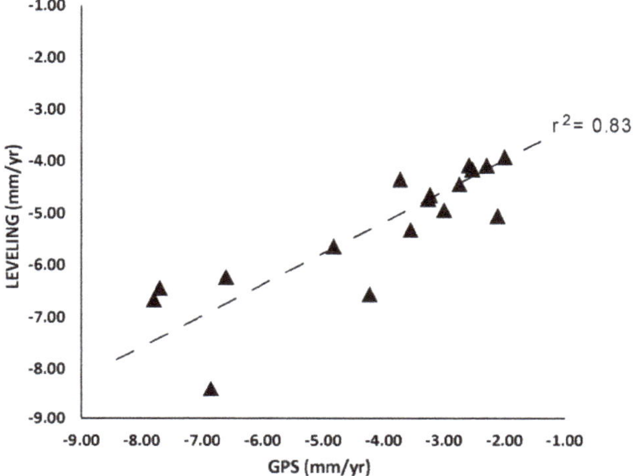


**Figure 4.** Vertical GPS vs. levelling velocities in mm/yr. The linear regression is reported as a dashed line.

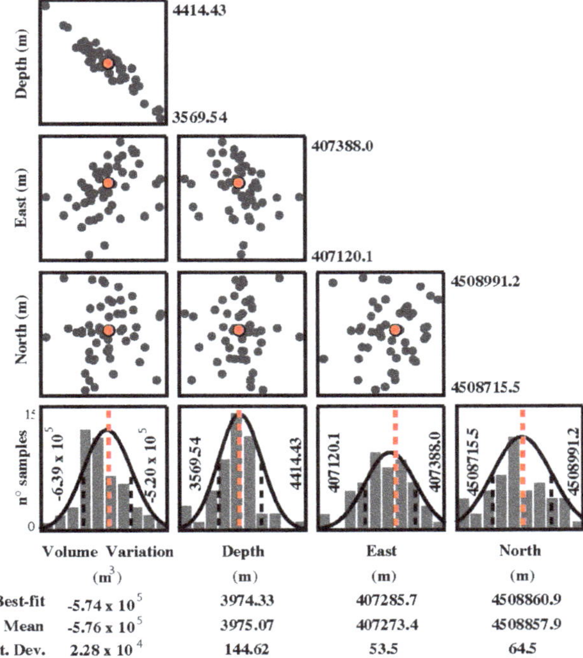

| | Volume Variation (m³) | Depth (m) | East (m) | North (m) |
|---|---|---|---|---|
| Best-fit | -5.74 x 10⁵ | 3974.33 | 407285.7 | 4508860.9 |
| Mean | -5.76 x 10⁵ | 3975.07 | 407273.4 | 4508857.9 |
| St. Dev. | 2.28 x 10⁴ | 144.62 | 53.5 | 64.5 |

**Figure 5.** Source depth, position, and volume variation of a homogeneous closuring (red symbols). The parameter uncertainties, best fit (in red), and trade-offs are shown.

The best-fit model reveals a horizontal closing crack at about 4 km depth located below the southern flank of Mt. Epomeo. The depth value appears different from the 2 km obtained by [11], whose inversion includes only levelling data. Although the non-linear inversion concentrates best solutions around 4 km depth, during the second step, we took into account both 2 km and 4 km depth results, and performed a ground velocity linear inversion exploring both solutions. Source closuring distributions, and observed, modeled and the residual (observed minus modelled) velocities from both linear inversions are reported in Figure 6. Closuring distribution is calculated as closuring rate (mm/yr) on a 0.5 × 0.5 km grid along the horizontal plane. The resulting closuring rate corresponds to $-3.66 \times 10^5$ m$^3$/yr and $-3.19 \times 10^5$ m$^3$/yr for 4 and 2 km depth models, respectively (Table 2). Both models well reproduce the quasi-centripetal horizontal velocity pattern and the concentric vertical one. However, the 2 km depth model underestimates the central subsidence pattern of leveling data (Panel j in Figure 6), the horizontal CGNSS centripetal pattern (Panel a in Figure 6), and overestimates the vertical CGNSS components (Panel b in Figure 6). Therefore, we favor the 4 km depth model, with the source of deformation located in the southern sector of Mt. Epomeo.

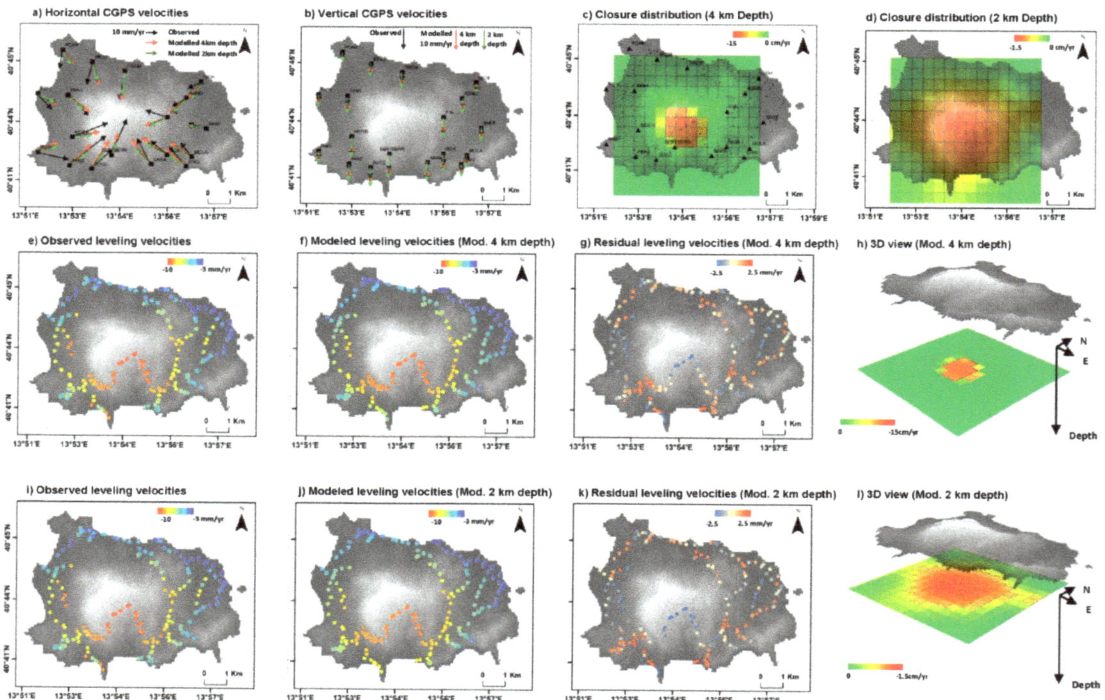



**Figure 6.** (**a**) Horizontal and (**b**) vertical GNSS velocities on Ischia Island. Black arrows represent the observed vectors; red and green ones are the modelled velocities for 4 km and 2 km depth models respectively; in (**c**,**d**), the corresponding closuring distributions for both models. Panels in second and third rows (**e**–**k**), report the observed, modelled, and residual levelling velocities for 4 km and 2 km depth models respectively; in panel (**h**,**l**) the corresponding 3D views of both displacement models are shown.

**Table 2.** Parameters of the retried deformation sources (depth, mesh, and mean volume).

| Model | Mesh (m) | Mean Δ Volume (m$^3$) |
|---|---|---|
| 2 km Depth | 7000 × 7000 | −319,225 |
| 4 km Depth | 7000 × 7000 | −366,325 |

## 6. Discussion

The 1997–2017 GNSS and leveling data show that Ischia is mainly characterized by a deformation field recording a general contraction, a feature observed in other volcanic areas, e.g., at Dallol and Askja [38,39], among others. At Ischia, this is testified by the general subsidence and a centripetal pattern of the horizontal deformation field, with maximum velocities, in both horizontal and vertical components, recorded in the Mt. Epomeo area. Only a few GPS stations diverge from the centripetal pattern, because of gravity instability phenomena [35]. The deformation rate remains nearly constant at each station and neither horizontal and vertical deformations showed significant changes in their rate after the Md = 4, 21 August 2017 earthquake (Figures 1 and 3). According to Calderoni et al. (2019) [25], the Ischia earthquakes are associated with hydrothermal pressurization processes and degassing and not with tectonics or the Mt. Epomeo subsidence. Taking also into account that historical and instrumental earthquakes concentrate in the northern sector of the island, we realistically conclude that the long term deformation field is due neither to earthquake dynamics nor to gravity processes. This latter conclusion is also consistent with the lack of relationships between topography and deformation rate. We also exclude the possibility that the recorded deformation field is due to the closure of two subvertical, E-W striking cracks located in the southern and northern sectors of Mt. Epomeo, as proposed by [17], because the recorded centripetal pattern of the deformation is not consistent with this geometry of sources. Moreover, the lack of significant changes in the chemical compositions of the gas and hot water emissions [21] and of uplift phases exclude the possible emplacement of magma and, possibly, of its lateral migration at depth. Therefore, the recorded contraction could be related to the huge degassing [20] and cooling, possibly reflecting crystallization of a magma reservoir, as also suggested by Trasatti et al. (2019) [11] on the basis of levelling data and coupled geochemical-deformation models. The results of our modeling of the combined GPS and levelling data at Ischia indicate two possible solutions: a sill-like source located at 2 km or 4 km depth. The occurrence of a sill at adepth of 4 km fits the observed pattern of vertical and horizontal deformation better. In addition, available resistivity data [23] indicate a high resistivity body extending from the surface to a depth of at least 3 km below Mt. Epomeo and the Ischia earthquakes are confined within the upper 3 km of the crust [24]. Moreover, the rheological model proposed by Castaldo et al. [40] shows a dome-like elasto-plastic transition in which the main magmatic body lies below a depth of 3 km. Our data indicate that this sill is located below the southern flank of Mt. Epomeo, in an area not affected by earthquakes. In addition, the nearly constant subsidence of the resurgent block does not re-activate its eastern and western bounding faults. Therefore, the deflation dynamics of such a sill does not explain the seismicity of the island. The only seismically active structures on Ischia are represented by E-W striking faults located at the base of the northern flank of the resurgence block, which are frequently re-activated by earthquakes [18,41]. Indeed, most earthquakes occur in areas where fumaroles and hot water emissions concentrate, i.e., in the Casamicciola area (Figure 1). In this framework, the cooling magma at a depth of 4 km releases fluids which mainly expand laterally from the sill, because the rocks overlying the sill include the impermeable 'Green Tuff': hydrothermally altered tuffs and lava flows with sealed fractures [40]. These fluids feed the two hydrothermal reservoirs located at about 1 km and 400 m depth [21] and upraise to the surface along the E-W striking fault of the northern flank of Mt. Epomeo and, to a lesser extent, at the northern tip of the eastern NNW-SSE striking, Mt. Epomeo bounding faults. The upraise and pressurization of these fluids within the hydrothermal reservoirs may trigger seismicity, which, at least for the

August Md 4 event, is very shallow (1.7 km depth) [25]. The above-reported conceptual model is summarized in Figure 7.

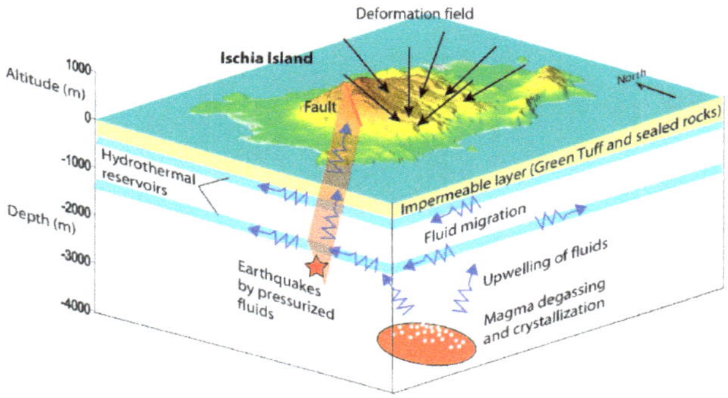

**Figure 7.** Sketch summarizing the conceptual model of the dynamics of Ischia Island and in particular the relationship among magma degassing, seismicity. and deformation.

Our data show that the eastern sector of Ischia, where the volcanic activity <30 ka concentrated, is characterized by the lower values of subsidence and horizontal deformation, this latter consistent with the above described centripetal arrangement (Figure 2). This indicates that this sector of the island is 'inactive', at least from a deformation point of view. Taking into account that only few hydrothermal manifestations and no earthquakes characterizes the eastern sector, we suggest that (a) dynamic processes related to possible magma or hydrothermal activity at depth are virtually lacking in this sector and (b) the most active sector of the island is the western one, where the inferred source of deformation has been identified and modeled, and where hydrothermal manifestations and earthquakes concentrate. In this framework, the possible signs of reactivation of the Ischia magmatic system are expected to include a decrease in the subsidence rate or an uplift, a centrifugal pattern of the horizontal deformation, a deformation pattern of the eastern sector of the island not compatible with a source located below Mt. Epomeo.

## 7. Conclusions

The main conclusions of this study may be summarized in the following points:

(a) Ischia is characterized by a 1997–2017 deformation pattern indicating subsidence and contraction of its western sector.

(b) The observed pattern is consistent with a deflating and contracting sill-like source located at 4 km depth. The sill is affected by cooling and degassing processes.

(c) The recorded deformations are not compatible with the seismicity of the island, which is mainly associated with the dynamics of the hydrothermal system and not with that of the deflating sill.

(d) A change in the rate of the recorded 1997–2017 deformation in the western sector of the island and/or the occurrence of a deformation pattern compatible with a source located in the eastern sector could be signs of the reactivation of the Ischia magmatic system.

Our results provide evidence that the subsidence of resurgent calderas may be associated with the deflation of a residual magma reservoir and that a decoupling between the dynamics of a magma reservoir, in this case, deflation, and of the overlying hydrothermal system may occur. In addition, we highlight that in quiescent calderas, the seismicity may not be related to the deformation pattern induced by magmatic processes.

**Author Contributions:** Conceptualization, A.G., G.P., G.V.; data collection, V.S.; methodology, A.G., G.P.; software, A.G., G.P.; validation, G.V.; writing—original draft preparation, A.G., G.P., G.V.; writing-review and editing, A.G., G.P., G.V.; visualization, A.G., G.P., G.V.; supervision, A.G., G.P., G.V.; project administration, V.S. All authors have read and agreed to the published version of the manuscript.

**Funding:** The present work is supported by INGV.

**Institutional Review Board Statement:** Not applicable.

**Informed Consent Statement:** Not applicable.

**Data Availability Statement:** The data presented in this study are available on request from the corresponding author.

**Acknowledgments:** We are very grateful to Roberto Devoti and Grazia Pietrantonio and Simone Atzori for supporting and data modellingfor discussions and computational support.

**Conflicts of Interest:** The authors declare no conflict of interest. The funders had no role in the design of the study; in the collection, analyses, or interpretation of data; in the writing of the manuscript, or in the decision to publish the results.

## References


1. Lowenstern, J.B.; Smith, R.B.; Hill, D.P. Monitoring super-volcanoes: Geophysical and geochemical signals at Yellowstone and other large caldera systems. *Philos. Trans. Royal Soc. A* **2006**, *364*, 2055–2072. [CrossRef] [PubMed]
2. Caricchi, L.; Biggs, J.; Annen, C.; Ebmeier, S. The influence of cooling, crystallisation and re-melting on the interpretation of geodetic signals in volcanic systems. *Earth Planet. Sci. Lett.* **2014**, *388*, 166–174. [CrossRef]
3. Macedonio, G.; Giudicepietro, F.; D'Auria, L.; Martini, M. Sill intrusion as a source mechanism of unrest at volcanic calderas. *J. Geophys. Res. Solid Earth* **2014**, *119*, 3986–4000. [CrossRef]
4. Kwoun, O.I.; Lu, Z.; Neal, C.; Wicks, C. Quiescent deformation of the Aniakchak Caldera, Alaska, mapped by InSAR. *Geology* **2006**, *34*, 5–8. [CrossRef]
5. Newhall, C.G.; Dzurisin, D. Historical Unrest at large calderas of the world. *US Geol. Survey Bull.* **1988**, *1855*, 1–27.
6. Trasatti, E.; Polcari, M.; Bonafede, M.; Stramondo, S. Geodetic constraints to the source mechanism of the 2011–2013 unrest at Campi Flegrei (Italy) caldera. *Geophys. Res. Lett.* **2015**, *42*, 3847–3854. [CrossRef]
7. Chang, W.L.; Smith, R.B.; Farrell, J.; Puskas, C.M. An extraordinary episode of Yellowstone caldera uplift, 2004–2010, from GPS and InSAR observations. *Geophys. Res. Lett.* **2010**, *37*, L23302. [CrossRef]
8. Foumelis, M.; Trasatti, E.; Papageorgiou, E.; Stramondo, S.; Parcharidis, I. Monitoring Santorini volcano (Greece) breathing from space. *Geophys. J. Inter.* **2013**, *193*, 161–170. [CrossRef]
9. De Martino, P.; Tammaro, U.; Obrizzo, F.; Sepe, V.; Brandi, G.; D'Alessandro, A.; Pingue, F. The GPS network of Ischia Island: Ground deformations in an active volcanic area (1998–2010). *Quad. Geofis.* **2011**, *95*, 95.
10. Tammaro, U.; De Martino, P.; Obrizzo, F.; Brandi, G.; D'Alessandro, A.; Dolce, M.; Malaspina, S.; Serio, C.; Pingue, F. Somma Vesuvius volcano: Ground deformations from CGPS observations (2001–2012). *Ann. Geophys.* **2013**, *56*, S0456. [CrossRef]
11. Trasatti, E.; Acocella, V.; Di Vito, M.A.; Del Gaudio, C.; Weber, G.; Aquino, I. Magma degassing as a source of long-term seismicity at volcanoes: The Ischia island (Italy) case. *Geophys. Res. Lett.* **2019**, *46*, 14421–14429. [CrossRef]
12. Vezzoli, L. Island of Ischia, Quad. *Ric. Sci.* **1988**, *114*, 133.
13. Civetta, L.; Gallo, G.; Orsi, G. Sr- and Nd-isotope and trace-element constraints on the chemical evolution of the magmatic system of Ischia (Italy) in the last 55 ka. *J. Volcanol. Geotherm. Res.* **1991**, *46*, 213–230. [CrossRef]
14. Gillot, P.-Y.; Chiesa, S.; Pasquaré, G.; Vezzoli, L. 33,000-yr K–Ar dating of the volcano–tectonic horst of the Isle of Ischia, Gulf of Naples. *Nature* **1982**, *299*, 242–245. [CrossRef]
15. Sbrana, A.; Toccaceli, R.M.; Biagio, G.; Cubellis, E.; Faccenna, C.; Fedi, M. *Carta Geologica n. 464 Isola di Ischia Scala 1:10.000-Cartografia e Note Illustrative*; Regione Campania: Napoli, Italy, 2011.
16. Manzo, M.; Ricciardi, G.P.; Casu, F.; Ventura, G.; Zeni, G.; Borgström, S. Surface deformation analysis in the Ischia Island (Italy) based on spaceborne radar interferometry. *J. Volcanol. Geotherm. Res.* **2006**, *151*, 399–416. [CrossRef]
17. Sepe, V.; Atzori, S.; Ventura, G. Subsidence due to crack closure and depressurization of hydrothermal systems: A case study from Mt Epomeo (Ischia Island, Italy). *Terra Nova* **2007**, *19*, 127–132. [CrossRef]
18. Alessio, G.; Esposito, E.; Ferranti, L.; Mastrolorenzo, G.; Porfido, S. Correlazione tra sismicità ed elementi strutturali nell'isola d'Ischia. *Il Quat.* **1996**, *9*, 303–308.
19. Tibaldi, A.; Vezzoli, L. A new type of volcano flank failure: The resurgent caldera sector collapse, Ischia, Italy. *Geophys. Res. Lett* **2004**, *31*, L14605. [CrossRef]
20. Chiodini, G.; Avino, R.; Brombach, T.; Caliro, S.; Cardellini, C.; De Vita, S. Fumarolic and diffuse soil degassing west of Mount Epomeo, Ischia, Italy. *J. Volcanol. Geotherm. Res.* **2004**, *133*, 291–309. [CrossRef]

21. Di Napoli, R.; Aiuppa, A.; Bellomo, S.; Brusca, L.; D'Alessandro, W.; Candela, E.G. A model for Ischia hydrothermal system: Evidences from the chemistry of thermal groundwaters. *J. Volcanol. Geotherm. Res.* **2009**, *186*, 133–159. [CrossRef]
22. Panichi, C.; Bolognesi, L.; Ghiara, M.R.; Noto, P.; Stanzione, D. Geothermal assessment of the island of ischia (southern Italy) from isotopic and chemical composition of the delivered fluids. *J. Volcanol. Geotherm. Res.* **1992**, *49*, 329–348. [CrossRef]
23. Di Giuseppe, M.G.; Troiano, A.; Carlino, S. Magnetotelluric imaging of the resurgent caldera on the island of Ischia (southern Italy): Inferences for its structure and activity. *Bull. Volcanol.* **2017**, *79*, 85. [CrossRef]
24. D'Auria, L.; Giudicepietro, F.; Tramelli, A.; Ricciolino, P.; Bascio, D.L.; Orazi, M. The seismicity of Ischia Island. *Seism. Res. Lett.* **2018**, *89*, 1750–1760. [CrossRef]
25. Calderoni, G.; Di Giovambattista, R.; Pezzo, G.; Albano, M.; Atzori, S.; Tolomei, C.; Ventura, G. Seismic and geodetic evidences of a hydrothermal source in the Md 4.0, 2017, Ischia earthquake (Italy). *J. Geophys. Res. Solid Earth* **2019**, *124*, 5014–5029. [CrossRef]
26. Braun, T.; Famiani, D.; Cesca, S. Seismological constraints on the source mechanism of the damaging seismic event of 21 August 2017 on Ischia Island (Southern Italy). *Seism. Res. Lett.* **2018**, *89*, 1741–1749. [CrossRef]
27. Cubellis, E.; Luongo, G. *Il terremoto del 28 luglio 1883 a Casamicciola nell'isola d'Ischia—"Il contesto fisico" Monografia n.1, Presidenza del Consiglio dei Ministri*; Servizio Sismico Nazionale, Istituto Poligrafico e Zecca dello Stato: Roma, Italy, 1998; pp. 49–123.
28. Anzidei, M.; Baldi, P.; Pesci, A.; Del Mese, S.; Esposito, A.; Galvani, A.; Loddo, F.; Massucci, A.; Cristofoletti, P. La rete geodetica dell'appennino centrale CA- GeoNet. *Quad. Geofis.* **2008**, *54*, 1–41.
29. Avallone, A.; Selvaggi, G.; D'Anastasio, E.; D'Agostino, N.; Pietrantonio, G.; Riguzzi, F.; Serpelloni, E.; Anzidei, M.; Casula, G.; Cecere, G.; et al. The RING network: Improvement of a GPS velocity field in the central Mediterranean. *Ann. Geophys.* **2010**, *53*, 39–54. [CrossRef]
30. Vespe, F.; Bianco, G.; Fermi, M.; Ferraro, C.; Nardi, A.; Sciarretta, C. The Italian GPS fiducial network: Services and products. *J. Geodyn.* **2000**, *30*, 327–336. [CrossRef]
31. Beutler, G.; Bock, H.; Dach, R.; Fridez, P.; Gäde, A.; Hugentobler, U.P.; Jäggi, A.; Meindl, M.; Mervart, L.; Prange, L.; et al. *Bernese GPS Software Version 5.0*; Dach, R., Hugentobler, U., Fridez, P., Meindl, M., Eds.; Astronomical Institute, University of Bern: Bern, Switzerland, January 2007.
32. Altamimi, Z.; Métivier, L.; Collilieux, X. ITRF2008 plate motion model. *J. Geophys. Res. Solid Earth* **2012**, *117*, B07402. [CrossRef]
33. Devoti, R.; D'Agostino, N.; Serpelloni, E.; Pietrantonio, G.; Riguzzi, F.; Avallone, A.; Cavaliere, A.; Cheloni, D.; Cecere, G.; D'Ambrosio, C.; et al. A Combined Velocity Field of the Mediterranean Region. *Ann. Geophys.* **2017**, *60*, S0215. [CrossRef]
34. Devoti, R.; Riguzzi, F. The velocity field of the Italian area. *Rend. Fis. Acc. Lincei* **2018**, *29*, 51–58. [CrossRef]
35. Della Seta, M.; Marotta, E.; Orsi, G.; de Vita, S.; Sansivero, F.; Fredi, P. Slope instability induced by volcano-tectonics as an additional source of hazard in active volcanic areas: The case of Ischia island (Italy). *Bull. Volcanol.* **2012**, *74*, 79–106. [CrossRef]
36. Okada, Y. Surface deformation due to shear and tensile faults in a half-space. *Bull. Seismol. Soc. Am.* **1985**, *75*, 1135–1154. [CrossRef]
37. Atzori, S.; Manunta, M.; Fornaro, G.; Ganas, A.; Salvi, S. Postseismic displacement of the 1999 Athens earthquake retrieved by the Differential Interferometry by Synthetic Aperture Radartime series. *J. Geophys. Res. Solid Earth* **2008**, *113*, B09309. [CrossRef]
38. Battaglia, M.; Pagli, C.; Meuti, S. The 2008–2010 Subsidence of Dallol Volcano on the Spreading Erta Ale Ridge: InSAR Observations and Source Models. *Remote Sens.* **2021**, *13*, 1991. [CrossRef]
39. de Zeeuw-van Dalfsen, E.; Pedersen, R.; Hooper, A.; Sigmundsson, F. Subsidence of Askja caldera 2000–2009: Modelling of deformation processes at an extensional plate boundary, constrained by time series InSAR analysis. *J. Volcanol. Geotherm. Res.* **2021**, *213*, 72–82. [CrossRef]
40. Castaldo, R.; Gola, G.; Santilano, A.; De Novellis, V.; Pepe, S.; Manzo, M.; Tizzani, P. The role of thermo-rheological properties of the crust beneath Ischia Island (Southern Italy) in the modulation of the ground deformation pattern. *J. Volcanol. Geotherm. Res.* **2017**, *344*, 154–173. [CrossRef]
41. Sbrana, A.; Marianelli, P.; Pasquini, G. Volcanology of Ischia (Italy). *J. Maps* **2018**, *14*, 494–503. [CrossRef]

*remote sensing*

*Article*

# The 2008–2010 Subsidence of Dallol Volcano on the Spreading Erta Ale Ridge: InSAR Observations and Source Models

Maurizio Battaglia [1,2,*], Carolina Pagli [3] and Stefano Meuti [1]

[1] Department of Earth Sciences, Sapienza University of Rome, 00185 Rome, Italy; stefs_@hotmail.it
[2] U.S. Geological Survey, Volcano Disaster Assistance Program, Moffet Field, Mountain View, CA 94035, USA
[3] Department of Earth Sciences, University of Pisa, 56126 Pisa, Italy; carolina.pagli@unipi.it
* Correspondence: mbattaglia@usgs.gov

**Abstract:** In this work, we study the subsidence of Dallol, an explosive crater and hydrothermal area along the spreading Erta Ale ridge of Afar (Ethiopia). No volcanic products exist at the surface. However, a diking episode in 2004, accompanied by dike-induced faulting, indicates that Dallol is an active volcanic area. The 2004 diking episode was followed by quiescence until subsidence started in 2008. We use InSAR to measure the deformation, and inverse, thermoelastic and poroelastic modelling to understand the possible causes of the subsidence. Analysis of InSAR data from 2004–2010 shows that subsidence, centered at Dallol, initiated in October 2008, and continued at least until February 2010 at an approximately regular rate of up to 10 cm/year. The inversion of InSAR average velocities finds that the source causing the subsidence is shallow (depth between 0.5 and 1.5 km), located under Dallol and with a volume decrease between $-0.63$ and $-0.26 \times 10^6$ km$^3$/year. The most likely explanation for the subsidence of Dallol volcano is a combination of outgassing (depressurization), cooling and contraction of the roof of a shallow crustal magma chamber or of the hydrothermal system.

**Keywords:** volcano deformation; interferometric synthetic aperture radar; ground deformation modelling; volcano geodesy; Dallol volcano

**Citation:** Battaglia, M.; Pagli, C.; Meuti, S. The 2008–2010 Subsidence of Dallol Volcano on the Spreading Erta Ale Ridge: InSAR Observations and Source Models. *Remote Sens.* 2021, 13, 1991. https://doi.org/10.3390/rs13101991

Academic Editor: Mimmo Palano

Received: 9 April 2021
Accepted: 15 May 2021
Published: 19 May 2021

**Publisher's Note:** MDPI stays neutral with regard to jurisdictional claims in published maps and institutional affiliations.

## 1. Introduction

Volcanoes commonly subside during, or after, eruptions as magma flows out of a chamber, or during post-eruptive periods when subsidence can occur because of cooling of magma, outgassing from the magma chamber, or viscoelastic relaxation of a spherical shell surrounding the chamber [1–4]). Cyclic uplift-subsidence periods have been explained as sudden inflows and successive cooling of magma, or they have been attributed to hydrothermal activity [5–8]. However, continuous subsidence which is not triggered by an eruption or an episode of deformation is more difficult to explain. The Askja volcano in Iceland has been subsiding for years without any eruption and the subsidence has been attributed to magma drainage, or cooling and crystallization of magma [9,10]. At Quaternary silicic calderas like Campi Flegrei (Italy) or Yellowstone (USA), the subsidence has been attributed to mixed magmatic–hydrothermal interaction [11,12].

Afar is a late-stage continental rift accommodating the divergence between the Arabian, Nubian, and Somalia plates along three rift arms: the Red Sea, the Gulf of Aden, and the Main Ethiopian rifts. Volcanism in Afar is focused on en echelon magmatic rift segments [13]. Dyke injections along the rift axis occur together with eruptions at central volcanoes. The Dallol magmatic rift segment is in the northern part of the Erta Ale ridge, where the Red Sea ridge moves inland creating the Afar continental rift (Figure 1). Extension rates in northern Erta Ale are relatively low, ~7 mm/year, increasing progressively to 20 mm/year further south in central Afar. The Dallol segment is a plain below sea level, filled by salt deposits. The Dallol explosive crater corresponds to a small topographic high rising ~50 m above the salt plain. The associated hydrothermal system is the lowest elevation subaerial volcano-hydrothermal system on Earth. Hydrothermal features include brightly colored hot springs that are dramatic. Seasonal brine springs are active in the plain around the Dallol crater, depositing salt crusts that have been mined for hundreds of years. No volcanic deposits outcrop at the surface; however, a ~20 m layer of basalt/dolerite was found in a borehole drilled in the salt plain around Dallol at 170 m beneath the surface [14] testifying to the presence of unerupted magma bodies at shallow depth. Indeed, the discovery of an active magma plumbing under Dallol, and its recognition as a nascent volcano with a rift segment was due to an InSAR study that captured a dyke intrusion in 2004 along the rift segment south of Dallol, fed by a magma chamber at 2.5 km depth [15].

In this paper, InSAR measurements were used to probe the deformation at Dallol. We show that since 2004 Dallol has been quiet, until continuous subsidence started in 2008 without any clear triggering. The satellite-based, remote sensing technique interferometric synthetic aperture radar (InSAR) was used to measure ground motions at Dallol. InSAR provides an all-weather, day/night capability to collect measurements, hence it has an edge over other remote sensing techniques such as multispectral imaging. Therefore InSAR has been extensively used to detect ground motions at active volcanoes since the launch of the first European SAR satellites in the early 1990s by ESA (European Space Agency) (i.e., [4,7–11,15]). InSAR missions collect measurements with high spatial resolution (~20-by-20 m pixel) and temporal resolution, with revisit times ranging from ~one month for previous European satellites (ERS1/2 and Envisat) to a few days for the recent missions of Sentinel 1a/b. Time-series analysis also allows accuracy in estimating ground motion of only a few mm/year [16–19].

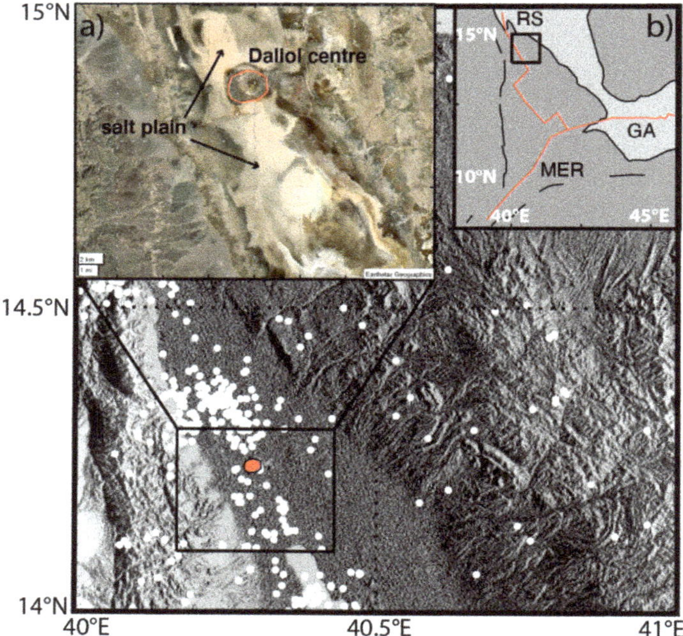


**Figure 1.** The main figure shows the area of Dallol. The red oval is the Dallol explosive crater and hydrothermal area (https://volcano.si.edu/volcano.cfm?vn=221041, accessed on 17 May 2021), and the white filled circles are earthquakes from [20]. (**a**) Panchromatic satellite image of the Dallol area. (**b**) Afar rift—the red line marks the rift axis and black lines are the rift-bounding faults. RS-Red Sea rift, GA-Gulf of Aden rift and MER-Main Ethiopian rift. The black box marks the location of the main figure.

## 2. InSAR Processing and Time-Series

We processed a series of interferograms from SAR images, acquired by the ENVISAT satellite between 2004 and 2010 in ascending and descending orbits, with the JPL/Caltech ROI_PAC software [21]. We used an external 3-arc sec (~90 m resolution) SRTM DEM for topographic correction. We filtered the interferograms using a power spectrum filter, before unwrapping them with a branch-cut algorithm. The final geocoding employed the same 3-arc sec SRTM DEM used for topographic correction [18].

The interferograms show no significant deformation in Dallol from 2004 until October 2008 when a concentric pattern, consistent with subsidence centered at Dallol, started, and continued at least until the end of ENVISAT acquisitions in early 2010. We selected the interferograms spanning the time of the subsidence at Dallol and inverted these to obtain maps of average surface velocities, and incremental time-series of displacements along the satellite Line-of-Sight (LOS), using the pi-rate time-series analysis program [18–20], (Figure 2). In the time-series analysis, any eventual residual unwrapping error was identified by using a phase closure method on Minimum Spanning Trees [18]. We applied orbital filtering to the geocoded interferograms by fitting them with a linear function, using a network approach [18]. We removed topographically correlated atmospheric noise [22] and applied the Atmospheric Phase Screen (APS) filter with a Gaussian temporal high-pass filter with 1σ of 0.5 years, to minimize atmospheric disturbances [23]. Finally, the interferograms were inverted for average velocity, incremental time-series of displacement and RMS error maps, using a linear least-square inversion with Laplacian smoothing. The temporal and spatial correlation between the interferograms were accounted for through the variance–covariance matrix [16,17,23]. We removed the unstable pixels that were not coherent in at least ten independent epochs and noisy pixels with an RMS misfit larger than 4 mm/year.

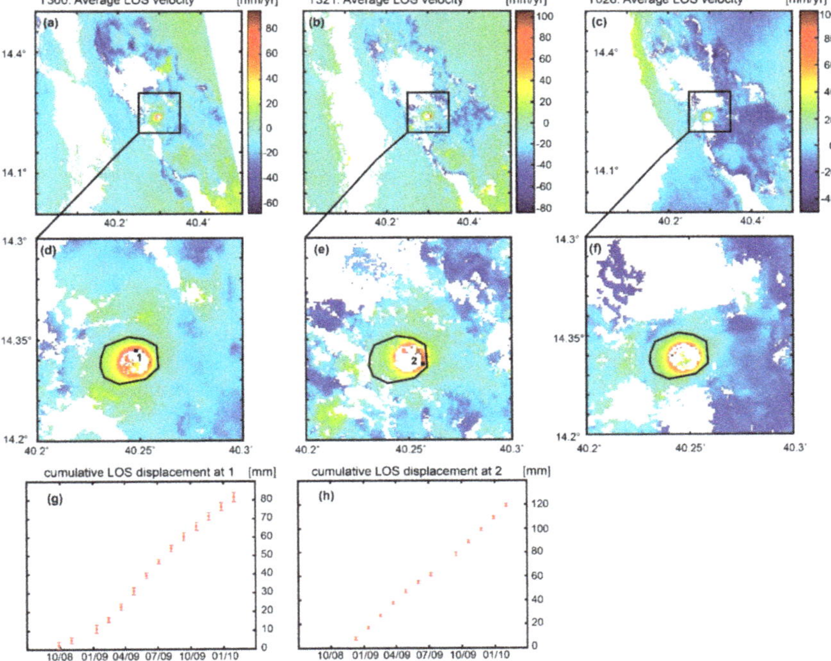

**Figure 2.** InSAR line-of-sight (LOS) average velocity maps of deformation at Dallol crater and surrounding areas from ENVISAT data. In this map, a LOS increase implies a subsidence. T300: ascending orbit; T321: descending orbit; T028: ascending orbit.

(**a–c**) InSAR average LOS velocity maps from 2008 to 2010, showing subsidence (increase in the LOS length) in the topographic depression of Dallol; the box marks the area shown in (**d–f**). (**d–f**) The black line marks the approximate boundary of the topographic depression marking the Dallol crater and hydrothermal area. The volcano formed by the intrusion of basaltic magma into the salt plains of Dallol (Figure 1), a vast area of uplifted thick salt deposits affected by intense fumarolic activity (https://volcano.si.edu/volcano.cfm?vn=221041 accessed on 09 April 2021). (**g,h**) LOS cumulative displacements at the numbered pixels in (**d,e**). No significant deformation is measured before October 2008 and after January 2010.

The average velocity maps from three different tracks exhibit good coherence at Dallol, though some noise occurs in the salt plain. All the maps show a consistent circular subsidence (LOS range increase, Figure 2a–f) centered at the Dallol crater of up to 100 mm/year during 2008–2010. The pattern is focused on a small area with a diameter of 4 km indicating that the source causing the Dallol subsidence is shallow. The pattern of cumulative displacement around Dallol (Figure 2g,h) shows that the rate of subsidence was approximately linear during 2008–2010 without any significant seasonal fluctuation.

The salt plain around the subsidence at Dallol crater is generally incoherent probably because of seasonal salt deposition. However, where coherence is maintained a signal of range decrease (line-of-sight uplift) is observed (Figure 2a–c). This signal is only seen over the salt plain. Therefore, we exclude any correlation to tectonic or volcanic processes. It is likely to be a radar propagation artefact caused by changes in backscattering coefficient in areas of high soil salinity and moisture [24].

## 3. InSAR Modelling

We tested different deformation mechanisms to explain the subsidence at Dallol including reservoir contraction due to magma cooling as well as depressurization of the hydrothermal system.

### 3.1. Models of Reservoir Contraction

The InSAR deformation velocities were inverted using the dMODELS software package ([25]; https://pubs.usgs.gov/tm/13/b1/, accessed on 17 May 2021). dMODELS implements analytical solutions of different types of deformation sources embedded in a homogeneous, elastic half space. The software inverts for the best-fit model parameters from surface deformation data. For the InSAR inversion a broad range of source types commonly used to approximate magma chambers was assumed: a spherical source with correction for the topography [26,27], a spheroidal source [28], and two types of magma chambers and dikes: a horizontal penny-shaped crack [29] and a tensile dislocation [30]. The inversion scheme implements a weighted least-squares algorithm combined with a random search grid to infer the minimum of the penalty function [31]. Measurement errors are coded in the covariance matrix and the penalty function is the chi-square per degrees of freedom, $\chi_v^2$. The local minimum of the penalty function $\chi_v^2$ is determined using a constrained, nonlinear, multivariable interior-point algorithm (function *fmincon*, [32]). The data were decimated using a regular sub-sampling scheme. We employ a sub-sampling step equal to 2, corresponding to an inversion of 50% of the pixels of the InSAR image (Table 1; Figure 3).

The quantitative analysis to determine the best fit model is based on the comparison between the semivariograms of models ($\gamma_m$) and dataset ($\gamma_d$) (Table 1; Figure 4). The semivariogram is an essential tool in any geostatistical analysis [33]. It provides a graphical and numerical measure of the spatial distribution of the InSAR dataset and the results from the different models. By using the semivariogram, the comparison between the fit of different models to the experimental dataset is not based on the value of a single number, the $\chi_v^2$, but on the ability of the model to mimic the spatial distribution of the data (Figure 4).

**Table 1.** Summary of modeling results. Number of random searches 256. Selection radius for data set: 10 km from center of Dallol crater. Regular sub-sampling.

| Data Set | Description | Orbits | # of Pixels | Source | # of PA-RAME-TERS | Orbits | $X_v^2$ | RMSE (1) Semi Variogram | ΔX0 (2)m | ΔY0 (2)m | Depthm b.s.l. | Radiusm | ΔV $10^6$ m$^3$/Year |
|---|---|---|---|---|---|---|---|---|---|---|---|---|---|
| DALLOL 1 | Original data set | T321 &T300 | 78,467 | Sphere | 5 | T321 & T300 | 14.1 | 0.31 | 1647 | 841 | 24,437 | 172 | 5.29 |
| | | | | | | T321 & T028 | 18.2 | 0.28 | 1604 | 726 | 13,827 | 69 | 0.001 |
| | | | | Spheroid | 8 | T321 & T300 | 13.0 | 0.32 | −205 | 215 | 14,814 | 1228 | 45.3 |
| | | | | | | T321 & T028 | 15.4 | 0.30 | 1630 | 821 | 17,443 | 1049 | 46.2 |
| | | T321 &T028 | 77,008 | Penny-shaped crack | 5 | T321 & T300 | 14.0 | 0.87 | 1720 | 912 | 805 | 128 | 0.009 |
| | | | | | | T321 & T028 | 18.0 | 0.27 | 1720 | 912 | 5849 | 169 | 0.45 |
| | | | | Dike (3) | 8 | T321 & T300 | 10.4 | 0.22 | 1502 | 139 | 2684 | - | −0.93 |
| | | | | | | T321 & T028 | 11.6 | 0.20 | 461 | 912 | 5416 | - | −3.10 |
| DALLOL 2 | Reference point defined such that average deformation far away from crater is zero | T321 &T300 | 79,188 | Sphere | 5 | T321 & T300 | 10.4 | 0.12 | −96 | −92 | 1234 | 556 | −0.56 |
| | | | | | | T321 & T028 | 13.6 | 0.15 | −143 | −124 | 1277 | 60 | −0.59 |
| | | | | Spheroid | 8 | T321 & T300 | 10.3 | 0.10 | −112 | −127 | 1274 | 287 | −0.63 |
| | | | | | | *T321 & T028 (4)* | *13.4* | *0.63* | *−53* | *561* | *21,083* | *1103* | *−117* |
| | | T321 &T028 | 75,864 | Penny-shaped crack | 5 | T321 & T300 | 10.3 | 0.13 | −168 | −108 | 1470 | 1560 | −0.62 |
| | | | | | | T321 & T028 | 13.9 | 0.17 | −183 | −168 | 1241 | 1628 | −0.53 |
| | | | | Dike (3) | 8 | T321 & T300 | 11.8 | 0.20 | 63 | −278 | 551 | - | −0.48 |
| | | | | | | *T321 & T028 (4)* | *11.0* | *0.26* | *1526* | *815* | *7023* | *-* | *−2.68* |
| DALLOL 3 | Masked (only data that show subsidence) | T321 &T300 | 6127 | Sphere | 5 | T321 & T300 | 5.1 | 0.09 | −32 | −71 | 915 | 449 | −0.30 |
| | | | | | | T321 & T028 | 4.2 | 0.07 | −79 | −135 | 1046 | 463 | −0.32 |
| | | | | Spheroid | 8 | T321 & T300 | 5.0 | 0.10 | −83 | −77 | 798 | 582 | −0.31 |
| | | | | | | T321 & T028 | 4.2 | 0.08 | −82 | −142 | 1037 | 159 | −0.39 |
| | | T321 &T028 | 17,075 | Penny-shaped crack | 5 | T321 & T300 | 4.8 | 0.12 | −166 | −196 | 566 | 1430 | −0.26 |
| | | | | | | T321 & T028 | 3.8 | 0.12 | −216 | −223 | 516 | 1557 | −0.26 |
| | | | | Dike (3) | 8 | T321 & T300 | 3.9 | 0.06 | 605 | 305 | 707 | - | −0.29 |
| | | | | | | T321 & T028 | 3.6 | 0.09 | 764 | 196 | 724 | - | −0.28 |

step: 2 (inverted of 50% of data). T321: descending orbit. T300, T028: ascending orbits. (1) Normalized root mean square error of the fit between the variograms for models and data. (2) Source offset relative to center of crater: [640,251, 1,574,845] (UTM coordinates, zone 37P). (3) Location of midpoint of the dike. (4) We think this solution is an outlier: a valid mathematical solution with no geological meaning (italics).

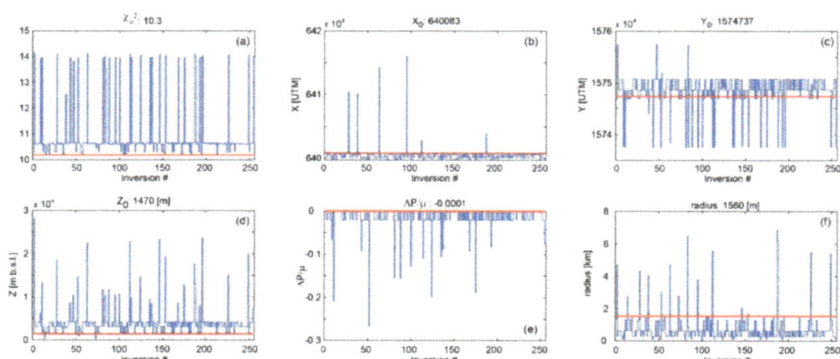

**Figure 3.** Stairs plot of the grid searches for the joint inversion of the InSAR data from orbits T300 and T321 (dataset DALLOL 2, sill source; Table 1). The inversion code implements a weighted least-squares algorithm combined with a random search grid to infer the minimum of the penalty function [31]. The red line points out the best fit solution. (**a**) penalty function $\chi_v^2$; (**b**) source location, $X_0$; (**c**) source location, $Y_0$; (**d**) source depth, $Z_0$; (**e**) dimensionless pressure change, $\Delta P/\mu$; (**f**) source radius.

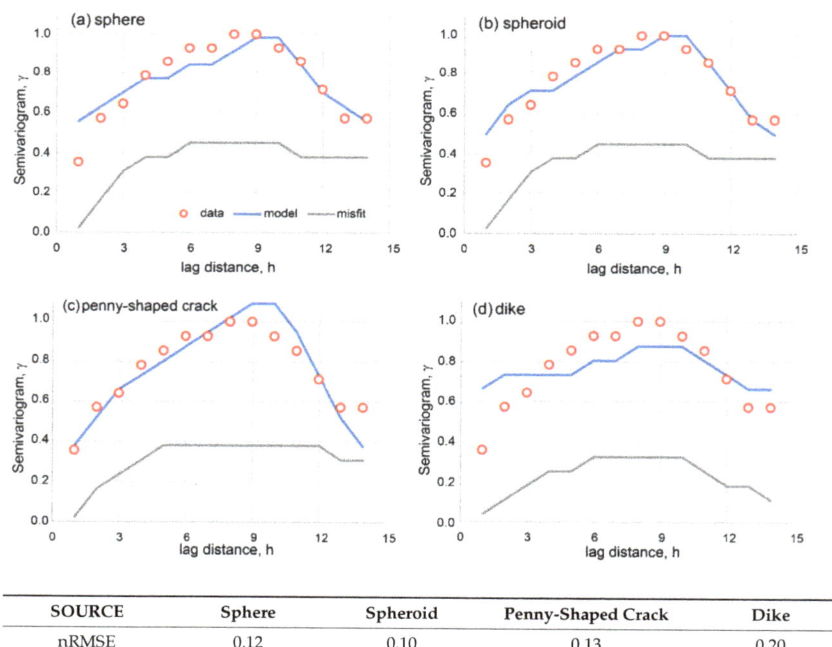

| SOURCE | Sphere | Spheroid | Penny-Shaped Crack | Dike |
|---|---|---|---|---|
| nRMSE | 0.12 | 0.10 | 0.13 | 0.20 |
| slope of misfit | 0.018 | 0.018 | 0.016 | 0.004 |

**Figure 4.** Example of semi-variograms of the deformation data (Figure 2) and models (Table 1) for the inversion of InSAR data from orbits T321 and T300 (dataset DALLOL 2). If the misfit (difference between data and model) is completely random (white noise), its variogram is a flat line (slope ~0). Full results are in Table 2.

**Table 2.** Average source parameters. Parameters are estimated from the solutions for DALLOL2 and DALLOL3 (Table 1) using a weighted average; uncertainties are the standard deviation of the weighted average. Uncertainties σ are one standard deviation.

| Source | X2v (1) | RMSE (1) | ΔX0 (2) | ±σ | ΔY0 (2) | ±σ | Depth | ±σ | Radius | ±σ | ΔV | ±σ | A | ±σ |
|---|---|---|---|---|---|---|---|---|---|---|---|---|---|---|
| | | Variogram | m | | m | | m b.s.l. | | m | | $10^6$ $m^3$/Year | | | |
| Sphere | 9.0 | 0.11 | −110 | 24 | −106 | 17 | 1232 | 63 | 380 | 230 | −0.55 | 0.07 | | |
| Spheroid | 9.0 | 0.10 | −110 | 8 | −128 | 5 | 1255 | 69 | 280 | 39 | −0.61 | 0.06 | 0.34 | 0.15 |
| Penny-shaped crack | 8.9 | 0.13 | −175 | 10 | −132 | 33 | 1357 | 199 | 1582 | 34 | −0.58 | 0.08 | | |
| Dike (3) | 10.0 | 0.17 | 215 | 281 | −164 | 210 | 589 | 71 | - | - | −0.44 | 0.08 | | |

(1) Weight given by number of pixels of the InSAR image. (2) Source offset relative to center of Dallol mountain: [640,251, 1,574,845] (UTM coordinates, 37P). (3) Location of midpoint of the dike.

The fit of the semivariograms was compared using the normalized Root Mean Square Error, *nRMSE*.

$$nRMSE = \frac{1}{\max(\gamma_d) - \min(\gamma_d)} \sqrt{\frac{\sum_{i=1}^{N} \left(\gamma_d^i - \gamma_m^i\right)^2}{N}},\tag{1}$$

where $\gamma_d$ is the semi-variogram of the dataset, $\gamma_m$ the semi-variogram of the model and $N$ the number of pixels in the InSAR image (Table 1, Figure 4). The statistical F-test [34], usually employed to determine if the better fit of a model with more parameters is an actual improvement, is not useful in this case because of the large number of data points.

We jointly invert the InSAR velocity maps from tracks T321 (descending orbit) and T300 (ascending orbit), and tracks T321 (descending orbit) and T028 (ascending orbit) to infer the parameters of a magma body beneath Dallol volcano (Figure 2d,e; Tables 1 and 2), assuming different source geometries: a sphere, a spheroid, a horizontal penny-shaped crack, and a tensile dislocation. For the inversion, we selected the InSAR data over a 20 × 20 km area centered at Dallol volcano. However, the Dallol subsidence decays away from its center and interacts with the salt plain signal, making it difficult to identify the area where the subsidence has completely decayed to zero. To overcome this problem, we tested two different ways to minimize the noise from the salt plain: (a) we selected a reference point outside the Dallol displacement field, setting its value to be equal to the mean of the displacement outside the Dallol subsidence; (b) we masked the interferogram over the salt plain area. We obtained three data sets–the original data set (DALLOL 1), the dataset relative to the reference point outside the displacement field (DALLOL 2), and the dataset with the mask (DALLOL 3), which we inverted to infer the best fit deformation source (Table 1).

The approach described above creates six different data sets that we can use to first calibrate and then verify our models (Table 1). We calibrate our models by inverting the dataset from orbits T321–T300; we validate our models by inverting the dataset from orbits T321–028. The difference between the calibrated and verified model offers an estimate of the uncertainties in the source parameters (Table 2).

The modelling results from inverting the original dataset (DALLOL 1, Table 1) do not return a good match to the data. The solutions mostly indicate very deep sources > 10 km, which is unlikely for the shallow volcanoes of the Erta Ale ridge. These results are probably caused by the inversion fitting the signal around the plain rather than the subsidence centered at Dallol volcano. On the other hand, changing the reference point, or masking the salt plain signal, makes it possible for the inversion algorithm to fit he Dallol volcano subsidence assuming different source types (datasets DALLOL 2 and DALLOL 3). In Table 1 we show the best-fit inversion solutions for the three datasets (DALLOL 1–3) and in Table 2 the statistically significant solutions, estimated by the weighted average of the model parameters for the DALLOL 2–3 datasets. These results clearly show that irrespective of the geometry, the source causing the subsidence is located under the Dallol volcano at shallow depth, between 0.5 and 1.5 km, with a volume decrease between $-0.63$ and $-0.26$ $10^6$ m$^3$/year (Table 1). The fit between model and data for a spherical source is shown in Figures 5–7, for illustrative purposes. Overall, the RMSE of the sphere, spheroid and penny-shaped crack are similar (Table 2), likely because of the limited number of pixels covering the central part of Dallol, which makes it difficult to discriminate between the three different geometries.

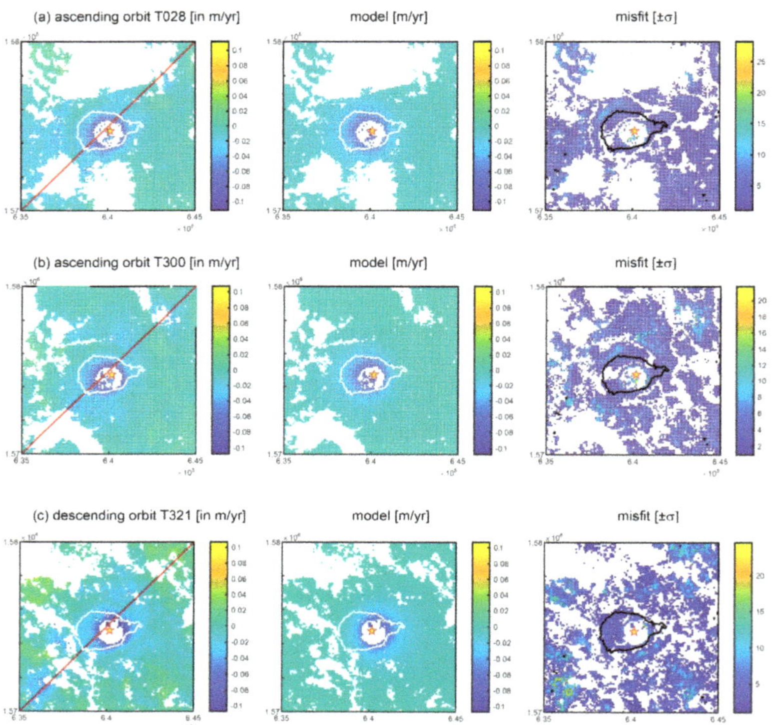



**Figure 5.** Spherical source solution: comparison between InSAR line-of-sight deformation, model (see Table 1, DALLOL 2), and misfit for the three orbits. The image coordinates are in UTM [m]. The scale of the data and model is in m/year. The scale of the misfit is based on measurement errors (5 means the misfit is five times the measurement error); all the pixels where the misfit is smaller than 2 error bars are in white. Dallol volcano is at the center of the images (white/black contour line). The red star with yellow fill is the location of the source. The red line in the data plot identifies the deformation profiles shown in Figure 6.

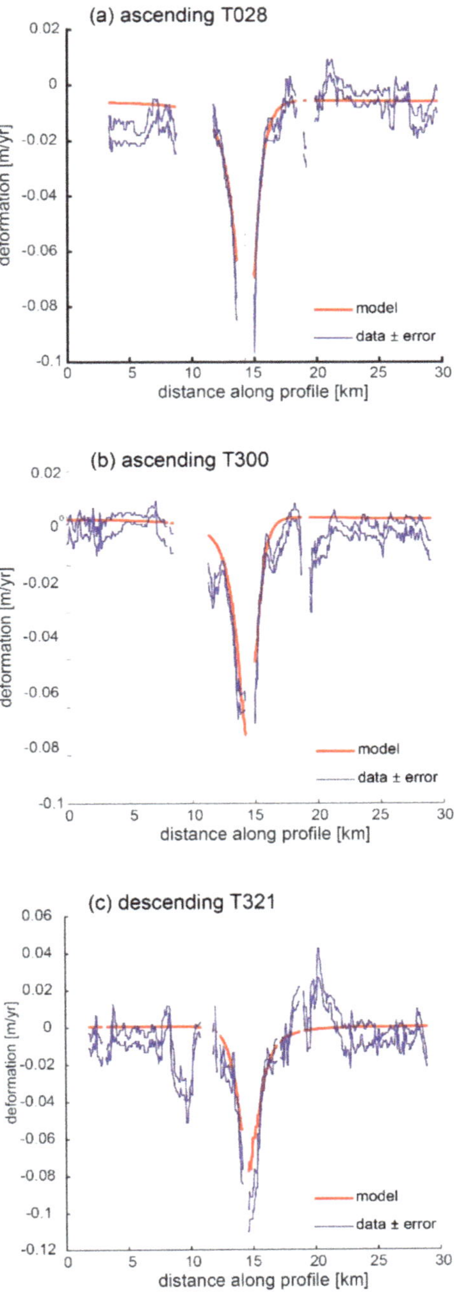



**Figure 6.** Spherical source solution for the three orbits: comparison between InSAR line-of-sight deformation and model (see Table 1; DALLOL 2) along the profile (red line) shown in Figure 5.

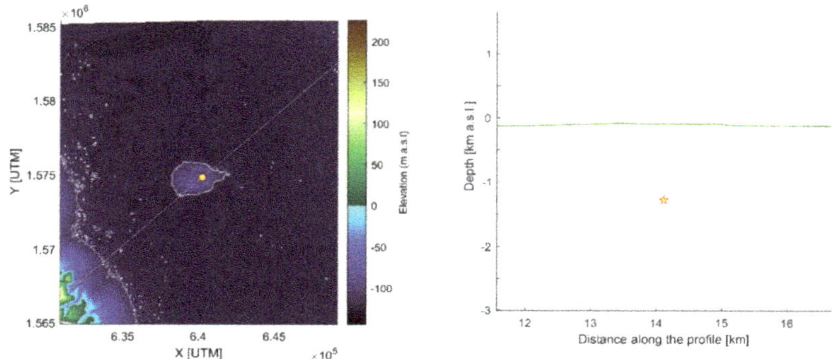

**Figure 7.** Spherical solution (Table 1, DALLOL 2). (**Left**) Source location (red circle, yellow fill); Dallol mountain is at the center of the image (white contour line); DEM from 1 Arc-Second Global SRTM (https://earthexplorer.usgs.gov/ accessed on 09 April 2021). (**Right**) Source location and depth (red star, yellow fill); the green line is the topography along the profile identified by the white diagonal line on the left panel.

### 3.2. Thermomechanical Models

In the absence of eruptions, the observed subsidence could be explained by a decrease in pore fluid pressure in a confined hydrothermal aquifer, or by thermoelastic contraction caused by the cooling of a volume of rock. The biggest challenge in investigating the cause of the subsidence is the scarcity of information on the subsurface and hydrothermal system of Dallol. Hydrothermal activity is fueled by water heated and enriched in gases by a heat source that lies primarily beneath the volcano. Stratigraphic data from the area are lacking, and the definition of the main stratigraphic units is unclear [35,36]. Because of the lack of information about the physical parameters (coefficient of linear thermal expansion, poroelastic expansion coefficient, thickness of the hydrothermal aquifer) of the Dallol geothermal system, we employed the parameters from a natural analog, the Hellisheidi geothermal area in Iceland. Hellisheidi has similarities with Dallol since it is a basaltic volcano with a hydrothermal field in a divergent tectonic setting ([37]; Table 3). The Hellisheidi's geothermal field experienced subsidence from a source that is modelled by a penny-shaped crack at a depth of 1.3 km and with a radius of 1.6 km [37].

In simple models for poroelastic and thermoelastic deformation, the source strength for a deforming cavity is given by [38].

Poroelastic deformation:

$$s = (1 - \nu)\frac{c_f V_f \Delta P_f}{\pi} \tag{2}$$

Thermoelastic deformation:

$$s = (1 - \nu)\frac{\alpha_t V_t \Delta T}{\pi} \tag{3}$$

Deformation from the geodetic volume change:

$$s = (1 - \nu)\frac{\Delta V}{\pi} \tag{4}$$

where $\nu$ is the Poisson's ratio, $\Delta V$ is the geodetic volume change, $c_f$ is Geertsma's uniaxial poro-elastic expansion coefficient, $V_f$ is the volume of the hydrothermal aquifer experiencing the pressure change $\Delta P_f$, $\alpha_t$ is a coefficient of linear thermal expansion, equivalent to one third of the volumetric thermal expansion coefficient, and $V_t$ is the volume of rock experiencing the temperature change $\Delta T$. Equations (2)–(4) are exact for a spherical source but only an approximation for other source geometries. Since we have estimated the source geodetic volume change $\Delta V$, the radius $a$ and their uncertainties $\sigma_{\Delta V}$ and $\sigma_a$ (Table 2), we can use Equations (2)–(4), and error propagation to derive estimates of the changes in the poro-elastic pressure change $\Delta P_f$ and thermoelastic temperature change $\Delta T$ and their uncertainties (Table 3)

$$\Delta P_f = \frac{\Delta V}{c_f V_f} \quad \sigma^2_{\Delta P} = \left(\frac{1}{c_f V_f}\right)^2 \sigma^2_{\Delta V} + \left(\frac{\Delta V}{c_f^2 V_f}\right)^2 \sigma^2_c + \left(\frac{\Delta V}{c_f V_f^2}\right)^2 \sigma^2_V , \tag{5}$$

$$\Delta T = \frac{\Delta V}{\alpha_t V_t} \quad \sigma^2_{\Delta T} = \left(\frac{1}{\alpha_t V_t}\right)^2 \sigma^2_{\Delta V} + \left(\frac{\Delta V}{\alpha_t^2 V_t}\right)^2 \sigma^2_\alpha + \left(\frac{\Delta V}{\alpha_f V_t^2}\right)^2 \sigma^2_V , \tag{6}$$

where $\sigma$ indicates the uncertainties of the parameters. The volume $V_f$ and $V_t$ are approximately

$$V_f \approx V_t \approx \pi a^2 h, \quad \sigma^2_V \approx (2\pi a h \sigma_a)^2 + (\pi a^2 \sigma_h)^2 \quad sill-like \tag{7}$$

where $h$ is the thickness of the penny-shaped crack. The estimates for pressure and temperature decreases are in Table 3.

**Table 3.** Estimate of poro-elastic pressure and thermoelastic temperature changes for a penny-shaped crack, sub-surface structure. Errors are one standard deviation. Parameters for the hydrothermal system are from Hellisheidi volcano, a possible natural analog [37].

| Parameter | Source | Value | Error | Units |
|---|---|---|---|---|
| $a$ | Table 2 | 1582 | 34 | m |
| $h$ | [39] | 750 | 250 | m |
| $V_f, V_t$ | | 5.9 | 2.0 | km$^3$ |
| $\Delta V$ | Table 2 | −0.58 | 0.08 | $10^6$ m$^3$ |

| *Poroelastic medium* | | | | *Thermoelastic medium* | | | |
|---|---|---|---|---|---|---|---|
| Parameter | Value | Error | Units | Parameter | Value | Error | Units |
| $c_f$ | 4.5 | 3.0 | $10^{-10}$ Pa$^{-1}$ | $\alpha_t$ | 1.0 | 0.5 | $10^{-5}$ K$^{\circ-1}$ |
| $\Delta P^f$ | −0.22 | 0.17 | MPa | $\Delta T$ | −10 | 6 | K$^\circ$ |

$\alpha_t$ after [39]; $c_f$, $h$ after [37].

## 4. Discussion

InSAR shows that linear subsidence started at Dallol volcano in October 2008 and continued at least until early 2010, when the ENVISAT acquisitions in the area stopped. More recent Sentinel-1 observations show that the Dallol subsidence was still ongoing at least until March 2015 [40,41]. The Dallol subsidence between October 2008 and February 2010 is best fit by a radially symmetric magma chamber at 0.5–1.5 km depth at 66% confidence interval, experiencing a volume decrease between −0.63 and −0.26 $10^6$ m$^3$/year. Our best-fit depth for the Dallol magma chamber is consistent with a previous InSAR study of the 2004 dyke-induced subsidence which placed the Dallol chamber between 1.5–3.3 km of depth [15]. Although the difference in magma chamber depths between the two InSAR studies is not significant, our model places the deformation source at the roof of the source inferred by [15].

Minor cooling and contraction, or minor depressurization, of the roof in the Dallol chamber is a reasonable explanation for the observed ground subsidence, but the reason why the subsidence suddenly started in October 2008 is less obvious.

An important factor to consider is the role of volatiles. Magma rich in exsolved volatiles is highly compressible to the point that the compression of residing magma may have offset the pressurization caused by magma inflow, resulting in no ground inflation prior to 2008 [42–44]. Outgassing of volatiles from the roof of the chamber [45] agrees with the somewhat shallower source depth inferred for the 2008–2010 Dallol subsidence compared to the 2004 dike injection that tapped the Dallol magma system. The intense Dallol hydrothermal field could be responsible for the observed subsidence since we estimate that only minor drops in pressure and temperature could explain the observed deformation (Table 3). Alternative explanations like (a) magma draining from a shallow crustal chamber into the rift axis, (b) crustal thinning coupled with loading by the volcano and dyke intrusions and weakening of the crust by heating [46], or (c) a deep connection with other active volcanoes in the Erta Ale region, are less likely. The extensional tectonic regime in northern Afar did not change in 2008 when subsidence suddenly started, nor is Dallol volcano likely to exert much loading, as it is only a small 50 m high mountain. The onset of the Dallol subsidence in October 2008 nearly coincides with the Alu-Dalafilla eruption along the Erta Ale magma in November 2008, but it is unlikely that a deep magma or pressure connection exists between volcanoes over 50 km apart and in any case no deformation in the area between the volcanoes was observed. Earthquakes recorded in the Erta Ale region by a local network 2005–2009 showed a clear seismicity increase in November 2008 confined to the eruptive site of Alu-Dalafilla.

## 5. Conclusions

InSAR observations show that a magmatic/hydrothermal system is active beneath the Dallol hydrothermal area. Inverse modeling of InSAR velocity maps demonstrate that the system is shallow, like the other active volcanoes of the Erta Ale ridge. Our new observations together with a previous InSAR study of a dike injection and seismicity along the Dallol rift indicate that Dallol is an active magmatic rift segment with a magma chamber and a hydrothermal field. The most likely explanations for the subsidence of Dallol volcano is a combination of outgassing (depressurization), and cooling and contraction of the roof of a shallow crustal magma chamber and its hydrothermal system (Table 3). Although it may seem unlikely that this would start so quickly in 2008, there are examples of deformation abruptly switching to subsidence in other hydrothermal areas, such as the Norris Geyser Basin area in Yellowstone caldera [47].

**Author Contributions:** Conceptualization, M.B. and C.P.; methodology, M.B. and C.P.; software, M.B.; validation, M.B.; formal analysis, M.B. and C.P.; investigation, S.M.; data curation, C.P.; writing— original draft preparation, M.B. and C.P.; writing—review and editing, M.B. and C.P.; supervision, M.B. and C.P.; funding acquisition, M.B. All authors have read and agreed to the published version of the manuscript.

**Funding:** This research was funded by Sapienza–University of Rome, Piccoli Progetti Universitari 2015 and 2020, USAID via the Volcano Disaster Assistance Program and the U.S. Geological Survey Volcano Hazards Program.

**Data Availability Statement:** Deformation data from InSAR and modeling software are available from the authors. ENVISAT data are freely and openly available from the ESA website https //earth.esa.int/eogateway/missions/envisat/data, (accessed on 9 April 2021).

**Acknowledgments:** Comments by Mike Poland (USGS), Mike Clynne (USGS) and two anonymous reviewers greatly helped to improve the manuscript. The original version of dMODELS is available online at https://pubs.usgs.gov/tm/13/b1/, (accessed on 9 April 2021). Any use of trade, firm, or product names is for descriptive purposes only and does not imply endorsement by the U.S. Government.

**Conflicts of Interest:** The authors declare no conflict of interest.

## References

1. Lu, Z.; Dzurisin, D.; Biggs, J.; Wicks, C., Jr.; McNutt, S. Ground surface deformation patterns, magma supply, and magma storage at Okmok volcano, Alaska, from InSAR analysis: 1. Interruption deformation, 1997–2008. *J. Geophys. Res.* **2010**, *115*. [CrossRef]
2. Segall, P. *Earthquake and Volcano Deformation*; Princeton University Press: Princeton, NJ, USA, 2010.
3. Girona, T.; Costa, F.; Schubert, G. Degassing during quiescence as a trigger of magma ascent and volcanic eruptions. *Sci. Rep.* **2015**, *5*, 18212. [CrossRef] [PubMed]

Hamlyn, J.; Wright, T.; Walters, R.; Pagli, C.; Sansosti, E.; Casu, F.; Pepe, S.; Edmonds, M.; McCormick, B.; Kilbride, G.; et al. What causes subsidence following the 2011 eruption at Nabro (Eritrea)? *Prog. Earth Planet Sci.* **2018**, *5*, 31. [CrossRef]

Lundgren, P.; Usai, S.; Sansosti, E.; Lanari, R.; Tesauro, M.; Fornaro, G.; Berardino, P. Modeling surface deformation observed with synthetic aperture radar interferometry at Campi Flegrei caldera. *J. Geophys. Res.* **2001**, *106*, 19355–19366. [CrossRef]

Gottsmann, J.; Rymer, H.; Berrino, G. Unrest at the Campi Flegrei caldera (Italy): A critical evaluation of source parameters from geodetic data inversion. *J. Volcanol. Geotherm. Res.* **2006**, *150*, 132–145. [CrossRef]

Fournier, T.J.; Pritchard, M.E.; Riddick, S.N. Duration, magnitude, and frequency of subaerial volcano deformation events: New results from Latin America using InSAR and a global synthesis. *Geochem. Geophys. Geosyst.* **2010**, *11*. [CrossRef]

Biggs, J.; Anthony, E.Y.; Ebinger, C.J. Multiple inflation and deflation events at Kenyan volcanoes. East African Rift. *Geology* **2009**, *37*, 979–982. [CrossRef]

Pagli, C.; Sigmundsson, F.; Árnadóttir, T.; Einarsson, P.; Sturkell, E. Deformation of the Askja volcanic system: Constraints on the deformation source from combined inversion of satellite radar interferograms and GPS measurements. *J. Volcanol. Geotherm. Res.* **2006**, *152*, 97–108. [CrossRef]

10. de Zeeuw-van Dalfsen, E.; Pedersen, R.; Hooper, A.; Sigmundsson, F. Subsidence of Askja caldera 2000–2009: Modelling of deformation processes at an extensional plate boundary, constrained by time series InSAR analysis. *J. Volcanol. Geotherm. Res.* **2021**, *213*, 72–82. [CrossRef]

11. Tizzani, P.; Battaglia, M.; Castaldo, R.; Pepe, A.; Zeni, G.; Lanari, R. Magma and fluid migration at Yellowstone Caldera in the last three decades inferred from InSAR, leveling, and gravity measurements. *J. Geophys. Res.* **2015**, *120*, 2627–2647. [CrossRef]

12. Troise, C.; De Natale, G.; Schiavone, R.; Somma, R.; Moretti, R. The Campi Flegrei caldera unrest: Discriminating magma intrusions from hydrothermal effects and implications for possible evolution. *Earth-Sci. Rev.* **2019**, *188*, 108–122. [CrossRef]

13. Ferguson, D.J.; Barnie, T.D.; Pyle, D.M.; Oppenheimer, C.; Yirgu, G.; Lewi, E.; Hamling, I. Recent rift-related volcanism in Afar, Ethiopia. *Earth Planet. Sc. Lett.* **2010**, *292*, 409–418. [CrossRef]

14. Bastow, I.D.; Booth, A.D.; Corti, G.; Keir, D.; Magee, C.; Jackson, C.A.-L. The development of late-stage continental breakup: Seismic reflection and borehole evidence from the Danakil Depression, Ethiopia. *Tectonics* **2018**, *37*, 2848–2862. [CrossRef]

15. Nobile, A.; Pagli, C.; Keir, D.; Wright, T.; Ayele, A.; Ruch, J.; Acocella, V. Dike-fault interaction during the 2004 Dallol intrusion at the northern edge of the Erta Ale Ridge (Afar, Ethiopia). *Geophys. Res. Lett.* **2012**, *39*, L19305. [CrossRef]

16. Pagli, C.; Wang, H.; Wright, T.J.; Calais, E.; Lewi, E. Current plate boundary deformation of the Afar rift from a 3-D velocity field inversion of InSAR and GPS. *J. Geophys. Res.* **2014**, *119*, 8562–8575. [CrossRef]

17. Wang, H.; Wright, T.J. Satellite geodetic imaging reveals internal deformation of western Tibet. *Geophys. Res. Lett.* **2012**, *39*, L07303. [CrossRef]

18. Biggs, J.; Wright, T.; Lu, Z.; Parsons, B. Multi-interferogram method for measuring interseismic deformation: Denali fault, Alaska. *Geophys. J. Int.* **2007**, *170*, 1165–1179. [CrossRef]

19. Ferretti, A.; Prati, C.; Rocca, F. Permanent scatterers in SAR interferometry. *IEEE Trans. Geosci. Remote* **2001**, *39*, 8–20. [CrossRef]

20. Ilsley-Kemp, F.; Keir, D.; Bull, J.M.; Gernon, T.M.; Ebinger, C.; Ayele, A. Seismicity during continental breakup in the Red Sea rift of Northern Afar. *J. Geophys. Res.* **2018**, *123*, 2345–2362. [CrossRef]

21. Rosen, P.A.; Hensley, S.; Peltzer, G.; Simons, M. Updated repeat orbit interferometry package released. *Eos Trans. AGU* **2004**, *85*, 47. [CrossRef]

22. Elliott, J.R.; Biggs, J.; Parsons, B.; Wright, T.J. InSAR slip rate determination on the Altyn Tagh Fault, northern Tibet, in the presence of topographically correlated atmospheric delays. *Geophys. Res. Lett.* **2008**, *35*, L12309. [CrossRef]

23. . Ng, A.H.-M.; Wang, H.; Dai, Y.; Pagli, C.; Chen, W.; Ge, L.; Du, Z.; Zhang, K. InSAR Reveals Land Deformation at Guangzhou and Foshan, China between 2011 and 2017 with COSMO-SkyMed Data. *Remote Sens.* **2018**, *10*, 813. [CrossRef]

24. Aly, Z.; Bonn, F.J.; Magagi, R. Analysis of the Backscattering Coefficient of Salt-Affected Soils Using Modeling and RADARSAT-1 SAR Data. *IEEE Trans. Geosci. Remote* **2007**, *45*, 332–341. [CrossRef]

25. Battaglia, M.; Cervelli, P.; Murray, J. dMODELS: A MATLAB software package for modeling crustal deformation near active faults and volcanic centers. *J. Volcanol. Geotherm. Res.* **2013**, *254*, 1–4. [CrossRef]

26. McTigue, D.F. Elastic stress and deformation near a finite spherical magma body: Resolution of the point source paradox. *J. Geophys. Res.* **1987**, *92*, 12931–12940. [CrossRef]

27. Williams, C.A.; Wadge, G. The effects of topography on magma chamber deformation models: Application to Mt. Etna and radar interferometry. *Geophys. Res. Lett.* **1998**, *25*, 1549–1552. [CrossRef]

28. Yang, X.-M.; Davis, P.M.; Dieterich, J.H. Deformation from inflation of a dipping finite prolate spheroid in an elastic half-space as a model for volcanic stressing. *J. Geophys. Res.* **1988**, *93*, 4249–4257. [CrossRef]

29. Fialko, Y.; Khazan, Y.; Simons, M. Deformation due to a pressurized horizontal circular crack in an elastic half-space, with applications to volcano geodesy. *Geophys. J. Int.* **2001**, *146*, 181–190. [CrossRef]

30. Okada, Y. Surface deformation due to shear and tensile faults in a half-space. *Bull. Seismol. Soc. Am.* **1985**, *75*, 1135–1154.

31. Bergstra, J.; Bengio, Y. Random search for hyper-parameter optimization. *J. Mach. Learn. Res.* **2012**, *13*, 281–305.

32. Mathworks. Optimization Toolbox™ User's Guide (R2020b). 2020. Available online: www.mathworks.com (accessed on 17 May 2021).

33. Deutsch, V.C. Geostatics. In *Encyclopedia of Physical Science and Technology*, 3rd ed.; Meyers, R.A., Ed.; Academic Press: Cambridge, MA, USA, 2003; pp. 697–707. [CrossRef]

34. Gordon, R.G.; Stein, S.; DeMets, C.; Argus, D.F. Statistical tests for closure of plate motion circuits. *Geophys. Res. Lett.* **1987**, *14*, 587–590. [CrossRef]
35. Carniel, R.; Jolis, E.M.; Jones, J. A geophysical multi-parametric analysis of hydrothermal activity at Dallol, Ethiopia. *J. Afr. Earth Sci.* **2010**, *58*, 812–819. [CrossRef]
36. Cavalazzi, B.; Barbieri, R.; Gómez, F.; Capaccioni, B.; Olsson-Francis, K.; Pondrelli, M.; Rossi, A.P.; Hickman-Lewis, K.; Agangi, A.; Gasparotto, G.; et al. The Dallol Geothermal Area, Northern Afar (Ethiopia)—An Exceptional Planetary Field Analog on Earth. *Astrobiology* **2019**, *19*, 553–578. [CrossRef] [PubMed]
37. Juncu, D.; Árnadóttir, T.; Geirsson, H.; Gunnarsson, G. The effect of fluid compressibility and elastic rock properties on deformation of geothermal reservoirs. *Geophys. J. Int.* **2019**, *217*, 122–134. [CrossRef]
38. Masterlark, T.; Lu, Z. Transient volcano deformation sources imaged with interferometric synthetic aperture radar: Application to Seguam Island, Alaska. *J. Geophys. Res.* **2004**, *109*, B01401. [CrossRef]
39. Hutnak, M.; Hurwitz, S.; Ingebritsen, S.E.; Hsieh, P.A. Numerical models of caldera deformation: Effects of multiphase and multicomponent hydrothermal fluid flow. *J. Geophys. Res.* **2009**, *114*, B04411. [CrossRef]
40. Meuti, S.; Pagli, C.; Pepe, S.; Battaglia, M.; Casu, F.; De Luca, C.; Pepe, A. Subsidence at Dallol proto-volcano, Afar (Ethiopia): Cooling of the magma chamber or deep interconnection? *Geophys. Res. Abs* **2017**, *19*, EGU2017.
41. Albino, F.; Biggs, J. Magmatic Processes in the East African Rift System: Insights From a 2015–2020 Sentinel-1 InSAR Survey. *Geochem. Geophys. Geosyst.* **2020**, *22*, e2020GC009488. [CrossRef]
42. Johnson, D.J. Dynamics of magma storage in the summit reservoir of Kilauea Volcano, Hawaii. *J. Geophys. Res.* **1992**, *97*, 1807–1820. [CrossRef]
43. Johnson, D.; Sigmundsson, F.; Delaney, P. Comment on "Volume of magma accumulation or withdrawal estimated from surface uplift or subsidence, with application to the 1960 collapse of Kīlauea volcano" by P. T. Delaney and D. F. McTigue. *Bull. Volcanol.* **2000**, *61*, 491–493. [CrossRef]
44. Rivalta, E.; Segall, P. Magma compressibility and the missing source for some dike intrusions. *Geophys. Res. Lett.* **2008**, *35*, L04306. [CrossRef]
45. Iddon, F.; Edmonds, M. Volatile-rich magmas distributed through the upper crust in the Main Ethiopian Rift. *Geochem. Geophys. Geosyst.* **2020**, *21*, e2019GC008904. [CrossRef]
46. Dzurisin, D.; Poland, M.P.; Bürgmann, R. Steady subsidence of Medicine Lake Volcano, Northern California, revealed by repeated leveling surveys. *J. Geophys. Res.* **2002**, *107*, 2372. [CrossRef]
47. Wicks, C.W.; Dzurisin, D.; Lowenstern, J.B.; Svarc, J. Magma intrusion and volatile ascent beneath Norris Geyser Basin, Yellowstone National Park. *J. Geophys. Res.* **2020**, *125*, e2019JB018208. [CrossRef]

*Article*

# Source Model for Sabancaya Volcano Constrained by DInSAR and GNSS Surface Deformation Observation

Gregorio Boixart [1], Luis F. Cruz [2,3], Rafael Miranda Cruz [2], Pablo A. Euillades [4], Leonardo D. Euillades [4] and Maurizio Battaglia [5,6,*]

[1]  Instituto de Estudios Andinos, Universidad de Buenos Aires-CONICET, Buenos Aires 1428, Argentina; gboixart@gl.fcen.uba.ar
[2]  Escuela Profesional de Ingeniería Geofísica, Universidad Nacional de San Agustín de Arequipa, Arequipa 04001, Peru; lcruzma@unsa.edu.pe (L.F.C.); rmiranda@ingemmet.gob.pe (R.M.C.)
[3]  Observatorio Vulcanológico del INGEMMET, Instituto Geológico Minero y Metalúrgico, Arequipa 04001, Peru
[4]  Facultad de Ingeniería, Instituto CEDIAC & CONICET, Universidad Nacional de Cuyo, Mendoza M5502JMA, Argentina; pablo.euillades@ingenieria.uncuyo.edu.ar (P.A.E.); leonardo.euillades@ingenieria.uncuyo.edu.ar (L.D.E.)
[5]  US Geological Survey, Volcano Disaster Assistance Program, NASA Ames Research Center, Moffett Field, CA 94035, USA
[6]  Department of Earth Sciences, Sapienza-University of Rome, 00185 Rome, Italy
*   Correspondence: mbattaglia@usgs.gov

Received: 23 April 2020; Accepted: 3 June 2020; Published: 8 June 2020

**Abstract:** Sabancaya is the most active volcano of the Ampato-Sabancaya Volcanic Complex (ASVC) in southern Perú and has been erupting since 2016. The analysis of ascending and descending Sentinel-1 orbits (DInSAR) and Global Navigation Satellite System (GNSS) datasets from 2014 to 2019 imaged a radially symmetric inflating area, uplifting at a rate of 35 to 50 mm/yr and centered 5 km north of Sabancaya. The DInSAR and GNSS data were modeled independently. We inverted the DInSAR data to infer the location, depth, and volume change of the deformation source. Then, we verified the DInSAR deformation model against the results from the inversion of the GNSS data. Our modelling results suggest that the imaged inflation pattern can be explained by a source 12 to 15 km deep, with a volume change rate between $26 \times 10^6$ m$^3$/yr and $46 \times 10^6$ m$^3$/yr, located between the Sabancaya and Hualca Hualca volcano. The observed regional inflation pattern, concentration of earthquake epicenters north of the ASVC, and inferred location of the deformation source indicate that the current eruptive activity at Sabancaya is fed by a deep regional reservoir through a lateral magmatic plumbing system.

**Keywords:** volcano deformation; interferometric synthetic aperture radar; ground deformation modelling; GNSS; volcano geodesy

---

## 1. Introduction

Sabancaya is a 5980 m-high stratovolcano located in the Central Volcanic Zone (CVZ) of the Andes, 75 km northwest of the city of Arequipa, Perú. It is the youngest and most recently active of the three volcanoes of the Ampato-Sabancaya Volcanic Complex (ASVC), which includes Hualca Hualca to the north and Ampato to the south (Figure 1). ASVC's volcanoes are mostly composed of andesitic and dacitic lava flows and pyroclastic deposits [1], surrounded by an extensive system of active faults and lineaments (Figure 1). Ampato is a dormant stratovolcano with no historical activity, consisting of three volcanic cones overlying an older eroded volcanic edifice [2]. The Pleistocene Hualca Hualca

*Remote Sens.* **2020**, *12*, 1852; doi:10.3390/rs12111852        www.mdpi.com/journal/remotesensing

volcano is believed to be extinct. There is hydrothermal activity observed near Pinchollo (Hualca Hualca), but it could be of tectonic origin [3].

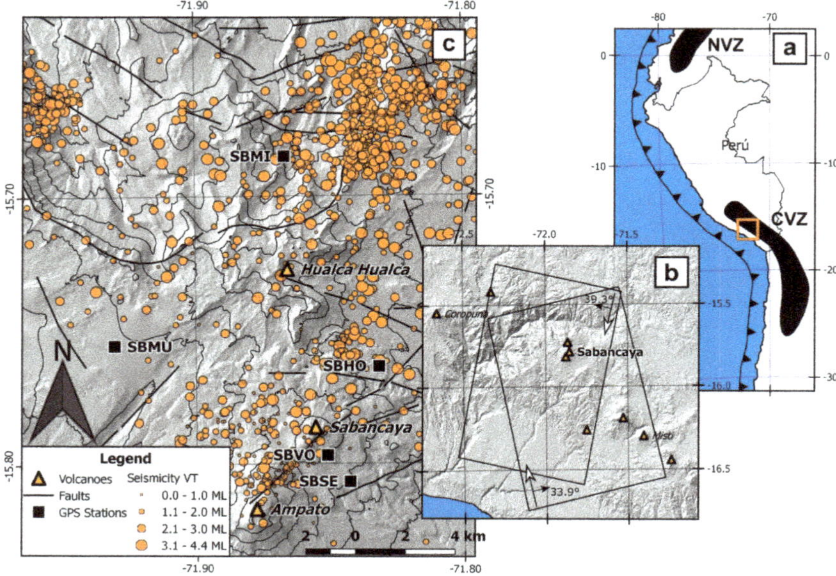

**Figure 1.** (**a**) Overview map with the North Volcanic Zone and the Central Volcanic Zone of the Andes (black). (**b**) Satellite tracks used in this study and the main volcanoes in southern Perú. (**c**) Studied area, showing the main volcanoes of the Ampato-Sabancaya Volcanic Complex (orange triangles), the location of the Global Navigation Satellite System (GNSS) stations (black squares), earthquake epicenters for 2014–2018 (orange circles), and the main faults and lineaments in the zone.

Sabancaya has had five historical eruptive episodes recorded since 1750 CE [4]. Between 1985 and 1997, it produced several Vulcanian eruptions with ash columns 2–3 km high and several small surges and pyroclastic flows [1,5,6]. Crustal deformation related to this eruptive cycle has not been well characterized because of a lack of data. The volcano has not been instrumented yet and very few Synthetic Aperture Radar (SAR) scenes suitable for Differential Interferometry of Synthetic Aperture Radar (DInSAR) studies were acquired by the European Space Agency's ERS Mission over South America. However, Pritchard and Simons [7] were able to create three useful interferograms with scenes acquired between July 1992 and October 1997. These scenes showed an uplift pattern centered near Hualca Hualca, circular in shape and with a radius of ~20 km. The deformation rate reached a mean velocity of 20 mm/yr at the point of maximum displacement. Reference [7] suggests that the inflation rate was constant during the analyzed time span. Unfortunately, the poor temporal resolution of the data prevented linking the observed deformation to a specific eruptive cycle.

In 2013, Sabancaya experienced a significant increase in seismic activity. Reference [8,9] studied the crustal deformation between 2002 and 2013 by applying the DInSAR processing to data from ERS, ENVISAT, and TerraSAR-X missions. They observed several co-seismic deformation episodes and creep along the Solarpampa fault, interpreting them as tectonic in origin. The regional circular pattern previously detected in 1992 to 1997 was absent.

The present eruptive cycle of Sabancaya began in November 2016 and remains ongoing through May 2020, when this paper was accepted. It is characterized by phreatic and Vulcanian explosions associated with constant $SO_2$ and ash emissions, accelerated lava dome growth, and thermal anomalies [10]. Seismicity during this period is concentrated mainly north of Sabancaya, around Hualca

Hualca (Figure 1), similar to previous distributions [9]. AN analysis of the Sentinel-1, TerraSAR-X, and COSMO-SkyMed missions from 2013–2018 found inflation northwest of Sabancaya as well as creep and rupture on multiple faults [11]. Modelling of the inflation source showed that the location (~7 km N of Sabancaya) and depth (~15 km) were consistent with the source inferred by [7].

In this work, we present results relevant to the present volcanic unrest (October 2014 to March 2019). The DInSAR processing of an ascending and descending Sentinel 1 dataset and Global Navigation Satellite System (GNSS) data analysis shows a radially symmetric inflation similar to the one observed for 1992–1997 by [7]. The observed deformation field is compatible with a deep source, 12–15 km below the surface and 5 km north of Sabancaya, that has possibly been feeding the current Sabancaya eruptions.

## 2. Deformation from DInSAR and GNSS

### 2.1. Synthetic Aperture Radar Data

The SAR dataset consists of 84 ascending (Path 47, Frame 1125) and 79 descending orbit (Path 25, Frames 614-646) scenes acquired by the Sentinel-1 Mission (European Space Agency, C-Band) between October 2014 and March 2019. The acquisition mode is Interferometric Wide (IW). Sub-swath 2 of the ascending orbit and sub-swath 1 of the ascending orbit cover the whole area of interest (Figure 1). They were processed using our own implementation of the DInSAR Small Baseline Subsets (SBAS) time-series approach [12], which allows a proper spatial and temporal characterization of the deformation patterns occurring within the studied area.

We treated ascending and descending orbit scenes independently and computed deformation time series and mean deformation velocity maps. We processed seven bursts of the sub-swath 2 for the ascending orbit and seven bursts of the sub-swath 1 for the descending orbit, and analyzed 254 ascending and 232 descending differential interferograms with maximum spatial and temporal baselines of 163 m and 1374 days, respectively. Multilooking factors of 10 and 40 were applied in the azimuth and range directions, respectively, giving a pixel of roughly $150 \times 150$ m. The topographic fringes were removed via a 30 m Digital Elevation Model from NASA Shuttle Radar Topography Mission (SRTM DEM) [13].

We detected a significant contribution of topography-related fringes in most of the interferograms, attributable to a stratified atmospheric phase component. This is particularly noticeable in the Cañon del Colca region, a 1700 m-deep canyon to the North and West of the ASVC. To compensate for the effect, which can cause a sinusoidal signal in the time series [14], we corrected each differential interferogram using the Zenith Time Delay (ZTD) corrections provided by Newcastle University through the Generic Atmospheric Correction Online Service (GACOS) [15] before a phase unwrapping and time-series calculation.

Because of the contribution from the atmosphere, the delay anomalies, the spatial averaging used to down-sample the DInSAR data, and the structure of the noise in the DInSAR data estimate, it is difficult to specify in a rigorous way the uncertainty in displacement measurements made with DInSAR. On the basis of our own experience and consultations with colleagues who specialize in DInSAR, we assigned an uncertainty to the DInSAR displacement measurements reported here of ±10 mm.

### 2.2. GNSS Data

The GNSS volcanic deformation monitoring network, set up in ASVC by the Volcano Observatory of the Instituto Geológico Minero y Metalúrgico (OVI-INGEMMET), has been collecting continuous data since 2014. The network consists of five continuous GNSS stations that have been progressively installed since October 2014. Three stations (SBVO, SBSE, and SBHN/SBHO) are within a 4 km radius from Sabancaya, whereas the others (SBMI and SBMU) are 9 to 12 km away from the volcano. SBHN and SBHO are the same station, but the name was changed in October 2016. Not all the stations were operative during the time span covered by the DInSAR time-series (see Table 1).

**Table 1.** GNSS stations with their corresponding data acquisition time.

| GNSS Station | Start Date | End Date |
|:---:|:---:|:---:|
| SBVO | 25 September 2014 | 25 October 2015 |
| SBSE | 3 October 2015 | 31 October 2018 |
| SBMU | 24 September 2014 | 31 October 2018 |
| SBMI | 5 October 2016 | 31 October 2018 |
| SBHO | 6 October 2016 | 31 October 2018 |
| SBHN | 26 September 2014 | 24 September 2016 |

We used GAMIT/GLOBK v.10.70 [16] for processing the GNSS data. The carrier phase and other observable data from satellites visible above a 10° elevation mask were sampled at 30 s intervals. The daily position solutions were computed using precise ephemerides provided by the International GNSS Service (IGS). The initial velocity solution was computed in the South America (SOAM) reference frame defined by [17]. The position time series and preliminary velocities were adjusted to 14 IGS stations located in the Stable South American (SSA) Plate, like POVE (Porto Velo, Brazil), UNSA (Salta, Argentina), KOUR (Kourou, French Guiana), and MTV2 (Montevideo, Uruguay), among others. The uncertainties of the GNSS velocities were derived by scaling the formal error by the square root of the residual chi-square per the degrees of freedom of the solution (Table 2). We also applied additional corrections, considering the monument instability (random walk), instrumental error (white noise), and flicker noise [18].

To eliminate the tectonic component of the deformation velocities observed at Sabancaya, we employed the Euler pole and angular velocity of the Peruvian block. We first estimated the tectonic deformation velocities from historical data from ten GPS stations active in southern Perú between 2012 and 2014. Then, using these velocities, we calculated a Euler Pole [19] located in northern Perú at 6.81° S and 72.52° W, with an angular velocity of 0.62°/My. Finally, the tectonic deformation velocities for each GPS site of the Sabancaya monitoring network (located 1000 km away the Euler Pole) are 11 ± 9 mm/yr east and 1 ± 1 mm/yr north.

Given the stations' availability (Table 1), the observed time span was split in two periods to improve the outcomes from processing: October 2014 to October 2016, and October 2016 to October 2018. The deformation velocity solutions for the first period include the SBMU, SBVO, SBSE, and SBHN stations, whereas for the second period we have solutions from SBMU, SBSE, SBHO, and SBMI.

### 2.3. Modelling

We inverted the deformation velocities for GNSS and DInSAR using the dMODELS package [20]. It implements analytical solutions for sources embedded in a homogeneous elastic half space. This approach employs pressurized cavities of simple geometry to mimic/approximate the crustal stress field produced by the actual source. None of these geometries reproduced an actual source. We employed the inversion codes for spherical, spheroidal, and sill-like sources. dMODELS implements a weighted least-squares algorithm combined with a random search grid to infer the minimum of the penalty function, the chi-square per degree of freedom $\chi_v^2$.

We inverted the GNSS and DInSAR data independently. This allowed us to verify our modelling results. For DInSAR, we searched for the parameters characterizing each model by jointly inverting the ascending and descending orbits' deformation velocities. This approach allowed us to better constrain the source geometry, given that the deformation was projected in the Line-of-Sight (LOS, positive for uplift in our images) direction. In order to reduce the computational cost, we down-sampled the original data by a factor of 25 using a linear uniform subsampling method. In this case, given the low spatial gradient of the displacement, we felt that it was not necessary to employ more complex algorithms, such as gradient-based ones [21]. The inversion was performed using 17,640 measurements (pixels).

## 3. Results

Figure 2 displays the results obtained by DInSAR processing. The mean deformation velocity in the descending (Figure 2a) and ascending (Figure 2b) orbit reveals a deformation field centered on Hualca Hualca volcano but affecting Ampato and Sabancaya volcanoes as well. Its shape is almost circular, with a diameter of ~45 km, and it appears skewed in the opposite direction within Figure 2 because of the SAR lateral view. The flight and LOS direction are indicated in the panels of Figure 2. The LOS-projected maximum deformation is observed in the ascending orbit and reaches a cumulative displacement of 20 cm.

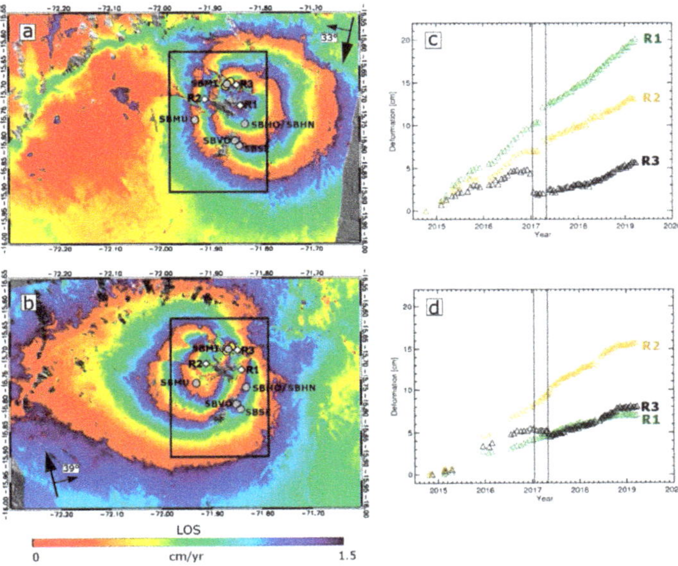

**Figure 2.** Time series extracted at the positions marked R1, R2, and R3 (white squares) for deformation along the line of sight. Ascending and descending Sentinel-1 orbits (DInSAR) processing results for (**a**) descending and (**b**) ascending orbits. SBMU, SBVO, SBSE, SBMI, and SBHO/SBHN are the locations of the GNSS stations. (**c**) and (**d**) are the deformation time series of the descending and ascending orbits, respectively, extracted from R1, R2, and R3 points. The black vertical lines mark the time of the two 2017 earthquakes. The shape of the deformed area is almost circular, with a diameter of ~45 km, and appears skewed in the opposite direction within Figure 2 because of the SAR lateral view.

In the descending orbit (Figure 2a,c), the significant shift observed in the first months of 2017 is related to two earthquakes of magnitude Mw = 3.8 and Mw = 3.4 which occurred on January 15 and 30 April 2017, respectively. The time series at R3, the pixel nearest to the epicenters, shows an offset of almost 2.5 cm of LOS decrease linked to the first earthquake. The other two pixels (R1 and R2), located at the centers of the LOS-projected regional volcanic deformation patterns, show shifts linked to the second earthquake. In the latter case, the fringes suggest a LOS increase around 1 cm.

The series taken from the ascending orbit do not present such a clear interpretation due to the low data availability (no acquisition) between 2014 and mid-2016, resulting in a low temporal resolution. The earthquake signals are not as strong as in the descending orbit, probably because of the relative orientation between the deformation and the sensor flight direction. However, minor shifts coincident with the 30 April 2017 earthquake are observed.

The earthquakes' signals are superimposed on the time series to the linear trend of the regional volcanic deformation. The regional volcanic deformation would be almost linear during the analyzed time span and is perturbed by the two rupture events shown in Figure 2c,d. The deformation mean velocity before and after the earthquakes is very similar, about 35 to 50 mm/yr. The pixels near the epicenters—R3 in Figure 2c—also show a year-long post-seismic relaxation effect followed by a progressive velocity increase.

**Table 2.** Summary of the calculated GNSS velocities per pre-eruptive and eruptive stage (before and after the beginning of the eruption of Sabancaya). The values are in millimeters per year (mm/yr).

|  |  | SBVO | SBMU | SBHN/SBHO | SBSE | SBMI |
|---|---|---|---|---|---|---|
| 10/14–10/16 | E | $5 \pm 2$ | $-24 \pm 2$ | $11 \pm 1$ | $2 \pm 2$ |  |
|  | N | $-21 \pm 2$ | $-7 \pm 1$ | $-10 \pm 1$ | $-16 \pm 1$ |  |
|  | U | $25 \pm 6$ | $33 \pm 4$ | $34 \pm 3$ | $27 \pm 7$ |  |
| **Sabancaya Eruption (6 November 2016)** | | | | | | |
| 10/16–10/18 | E |  | $-28 \pm 1$ | $9 \pm 1$ | $1 \pm 1$ | $10 \pm 2$ |
|  | N |  | $-7 \pm 1$ | $-11 \pm 1$ | $-15 \pm 1$ | $14 \pm 1$ |
|  | U |  | $26 \pm 6$ | $29 \pm 3$ | $17 \pm 5$ | $26 \pm 6$ |

Figure 3 and Table 2 display the mean deformation velocity (red vectors in Figure 3) computed from the GNSS data. The horizontal components, excluding station SBMI, show a radial deformation centered near Hualca Hualca between October 2016 and October 2018, affecting the whole Ampato-Sabancaya volcanic complex. SBMI is located between the epicenters of the two earthquakes in 2017, so this velocity vector is affected by co- and post-seismic displacement. The vertical components are positive, varying between 1.7 and 3.4 cm/year. These results are consistent with the deformation measured by DInSAR (Figure 3c–g).

*Modelling*

Table 3 shows the inversion results using both the GNSS and the DInSAR data. To avoid the local effect of the earthquakes (e.g., post-seismic relaxation), we divided the deformation from the DInSAR time series into two periods. The first period is from October 2014 to December 2016, the second from January 2018 to March 2019. The GNSS data have also been split into two time spans linked to station availability (Table 2).

The results for both the time spans are detailed in Table 3. The independent inversion of the GNSS and DInSAR deformation data infers a deep source close to Hualca Hualca as the most probable result in all the modelled geometries.

The sill-like geometry, more appropriate for local deformation patterns produced by shallow sources, yielded parameters geologically implausible (e.g., extremely large radius or near surface depths), so we discarded it.

Of the remaining geometries, the spherical source model better fits both periods with GNSS and DInSAR data. Since the misfit is very similar, we prefer the simpler spherical source (four parameters) to the spheroidal source (seven parameters). The locations from the four best-fit inversions are within a 2 km radius area beneath Hualca Hualca. The dimensionless pressure change is well within the validity range of the linear elasticity assumption of the analytical model (0–0.1). The best fit models employing a spheroidal geometry infer a deeper source (~16 km) and produce a greater geographic dispersion between the best-fit results of the GNSS and DInSAR data.

The best fit for the independent inversion of GNSS and DInSAR data for the first period has a 12–14 km deep and 1.5 km-radius spherical source located beneath Hualca Hualca. Its volume change rate is between $26 \times 10^6$ m$^3$/yr and $43 \times 10^6$ m$^3$/yr, the dimensionless pressure change between 0.002 and 0.004, and the $\chi^2_v$ between 1.0 and 2.8.

**Table 3.** Parameters of the best-fit sources for each period for both data sets for all geometries. Depth is relative to Sabancaya's vent.

| Period | Geometry | Data | $\chi^2_v$ | Lat $\circ$ | Long $\circ$ | Depth km | Radius km | $\delta P$ | $\delta V$ $\times 10^6$ m$^3$ |
|---|---|---|---|---|---|---|---|---|---|
| 10/2014–12/2016 | Sphere | DInSAR | 1.0 | $-15.7225 \pm 0.0027$ | $-71.8727°$ | $15 \pm 1$ | 1.5 | $0.004 \pm 0.001$ | $43 \pm 5$ |
| 10/2014–12/2016 | Spheroid | DInSAR | 6.6 | $-15.7407$ | $-71.9345°$ | 15.1 | 1.5 | 0.009 | 32 |
| 10/2014–12/2016 | Sill | DInSAR | 6.6 | $-15.7438$ | $-71.8468°$ | 15.6 | 10.9 | 0.0001 | 26 |
| 09/2014–11/2016 | Sphere | GNSS | 2.8 | $-15.7335 \pm 0.0027$ | $-71.8648°$ | $12 \pm 1$ | 1.5 | $0.002 \pm 0.001$ | $26 \pm 4$ |
| 09/2014–11/2016 | Spheroid | GNSS | 1.0 | $-15.7391$ | $-71.8624°$ | 9.3 | 1.5 | 0.481 | 20 |
| 11/2016–11/2018 | Sphere | GNSS | 25.5 | $-15.7340 \pm 0.0027$ | $-71.8618°$ | $11 \pm 1$ | 1.5 | $0.002 \pm 0.001$ | $22 \pm 4$ |
| 11/2016–11/2018 | Sill | GNSS | 17.9 | $-15.7306$ | $-71.8680°$ | 1.2 | 1.5 | 0.052 | 18 |
| 01/2018–4/2019 | Sphere | DInSAR | 1.7 | $-15.7159 \pm 0.0027$ | $-71.8563°$ | $15.3 \pm 0.9$ | 1.5 | $0.005 \pm 0.001$ | $51 \pm 4$ |

The independent inversion for the second GNSS and DInSAR deformation velocity series yields similar results (Table 3). The spherical source is located at nearly the same position and depth (12–15 km) for both data sets. Despite a much higher $\chi^2_v$ for the GNSS inversion, the volume change rate is of the same order of magnitude (see Table 3). We calculated the errors for the best-fit parameters using the Monte Carlo simulation technique [22]. We added white noise to the original data set and then inverted it for the best fit solution. We repeated this process 100 times. The errors presented in Table 3 reflect the distribution of the solutions found.

Figure 3 presents the horizontal velocity vectors produced by the GNSS best-fit spherical sources (blue vectors). Note how the SBMI station shows a noticeable difference between the observed and modelled velocity. This station was affected by the 2017 seismic events, so it recorded the combined effect of the underlying regional pattern and the local earthquakes' deformation field. Using data provided by the SBMI station results in a higher $\chi^2_v$ value (~25) and a smaller fit. The modelled and observed data present a significantly better fit at the other stations.

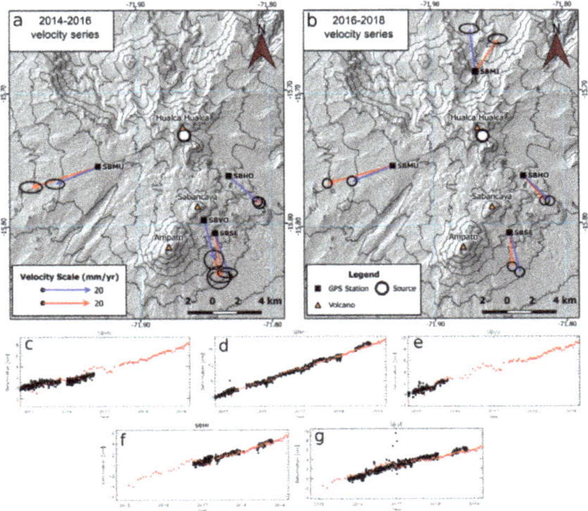

**Figure 3.** Horizontal velocity vectors produced by the best-fit spherical source (blue) and GNSS data (red) for (**a**) the 2014–2016 velocity series and (**b**) the 2016–2018 velocity series. Note how the data and modelled vectors of the SBMI station differ because of a nearby seismic event. (**c**–**g**) GNSS deformation series, projected into the ascending Line-of-Sight (LOS), compared to the ascending DInSAR series.

The pattern and residuals obtained by modelling the DInSAR data are presented in Figure 4. After subtracting the modeled component, the possible seismic effects are observed in the residual from both the analyzed periods and are especially noteworthy in 2014–2017. The main seismic deformation pattern located NE of Sabancaya volcano could be associated with a post-seismic effect related to the Mw = 5.9 earthquake on 17 July 2013 [9].

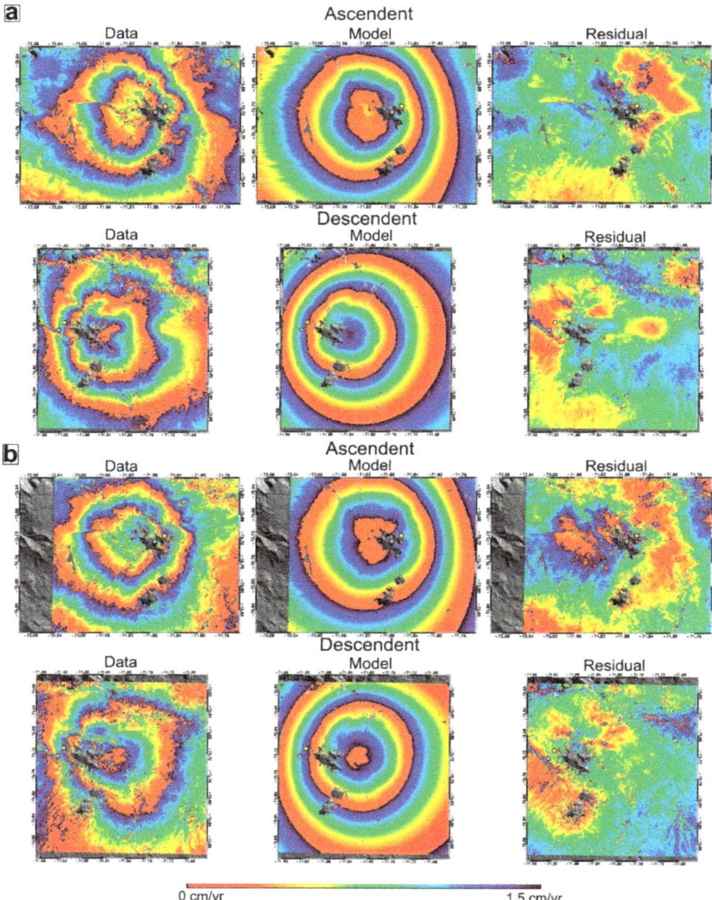


**Figure 4.** Spherical source model results for the DInSAR data (**a**) 2014–2016 ascending and descending series and (**b**) 2018–2019 ascending and descending series. First column is the full resolution DInSAR data, the second column is the inverse modelling result, and the third column is the residual between the data and model. Note how the seismic deformation, still ongoing, is evident in the residuals of both series.

## 4. Discussion

The analysis and interpretation of the DInSAR and GNSS deformation data allow us to better understand the Sabancaya volcanic system. In particular, our analysis shows that ground deformation affects the three volcanic edifices of the Ampato-Sabancaya volcanic complex (Figures 2 and 3). Deformation patterns and best-fit sources (Figure 5) are centered more than 5 km away from the eruptive vent, close to Hualca Hualca. This lateral magmatic deformation zone is a common feature of volcanic unrest. According to a survey by [23], 24% of monitored volcanoes have had similar deformation signals.

A linear deformation trend is observed throughout the entire time span. This trend is caused by magmatic deformation disturbed by local seismicity (Figure 2). To minimize the influence of seismic deformation on our models, we split the DInSAR and GNSS velocity series into two periods—before and after the 2017 seismic events (Figure 2). The offset and long-term trends observed in the time series after the 2017 seismic events (Figure 2) are the effect of a rapid co-seismic deformation followed by a possible slower post-seismic relaxation.

The deformation source inferred by the independent inversion of the DInSAR and GNSS deformation velocities shows a deep source (12–15 km below the surface) beneath Hualca Hualca with a volume change rate of $26$–$43 \times 10^6$ m$^3$/yr, in agreement with the results obtained by [24] for the same period. The residuals in the DInSAR data are related to seismic deformation and provide a clear view of the faults' movement. The model fits very well also with the deformation velocity vectors from the GNSS (Figure 3a,b). The station SMBI is affected by the earthquakes and has a combined signal from the local volcanic source and the regional seismic deformation.

The results obtained by the inversion of the GNSS data validate the DInSAR best-fit model, with both having similar locations, depths, and volume change rates. Furthermore, the similar results for both the deformation velocity series reinforce the linear trend hypothesis for the whole period (Figure 3c–g).

The best-fit source, located between 12 and 15 km beneath Hualca Hualca, reproduces the stress field and volume change of a magma reservoir that might have fed Sabancaya's eruptions since the mid-1990s.

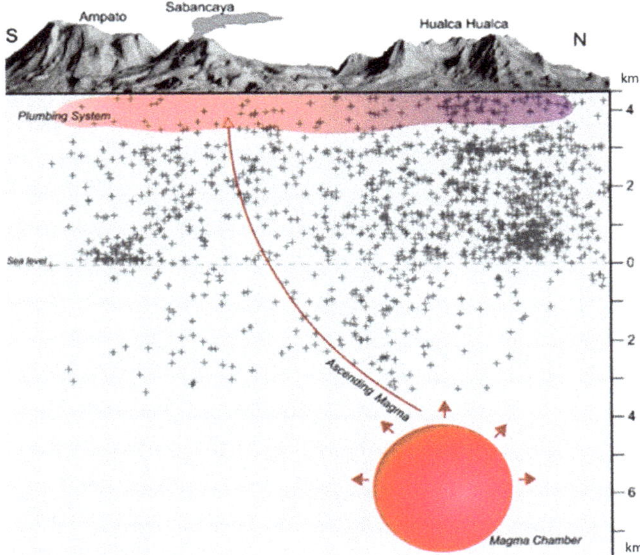


**Figure 5.** Conceptual model of the Sabancaya magmatic system. The 1.5 km radius sphere reproduces the stress field and volume change of a magma chamber at 12–15 km depth beneath Hualca Hualca. Red lines are possible ascending paths of magmatic fluids to the Sabancaya conduit and the extensive plumbing system. Black crosses represent earthquakes occurred during the analyzed time span.

The deep intrusion of magma into an existing reservoir can cause the migration of magmatic fluids into the ASVC plumbing system and a change in the local heat flow [10]. The volume/pressure change started a period of volcanic unrest with inflation accompanied by seismicity. The epicenters of the earthquakes produced by fluid movements along permeable pathways and fault structures concentrate north of Hualca Hualca. Magma and hydrothermal fluids, driven by convection, can reach Sabancaya's conduit (Figure 5).

Our results confirm the existence of a previous inflation episode in the 1990's [7], with ground deformation centered on Hualca Hualca. In our opinion, Sabancaya is fed by a laterally extensive magmatic plumbing system (Figure 5) which allows fluids to move along a highly fractured (Figure 1) and pressured crust from the "offset" deep magma chamber to their emission point.

## 5. Conclusions

Continuous GNSS measurements and DInSAR observation from the ascending and descending Sentinel-1 tracks show a large inflation pattern (~45 km diameter) affecting the ASVC. The beginning date of the inflation is uncertain. The analyzed deformation velocity series have an almost linear trend for the whole period, interrupted by local seismic events. To minimize the local effect of the earthquakes on modelling, we divide both the DInSAR and GNSS data into two series: one before the 2017 earthquakes (October 2014 to December 2016) and one after the post-seismic relaxation (January 2018 to March 2019). For both datasets, the best-fit source is a deep intrusion (12 to 15 km below the surface) beneath Hualca Hualca. This is the same source inferred by [7] for the unrest experienced by Sabancaya in the mid-1990s. The regional inflation pattern centered on Hualca Hualca, the "offset" magma chamber, and the concentration of seismic activity north of ASVC suggest that a laterally extensive magmatic plumbing system may be responsible for this deformation scenario.

**Author Contributions:** Data processing, modelling, and writing—original draft preparation, G.B. and L.F.C.; resources and project supervision, R.M.C.; data curation and advising, P.A.E. and L.D.E.; software, conceptualization, writing—review and editing, M.B. All authors have read and agreed to the published version of the manuscript.

**Funding:** L. Cruz was funded by the Universidad Nacional de San Agustín de Arequipa and the UNSA-INVESTIGA program under contract No. TP-013-2018-UNSA. Funding for this work came from USAID via the Volcano Disaster Assistance Program and from the U.S. Geological Survey (USGS) Volcano Hazards Program.

**Acknowledgments:** We thank Chris Harpel (USGS), Chuck Wicks (USGS), and four anonymous reviewers for improving this work with their constructive comments. Any use of trade, firm, or product names is for descriptive purposes only and does not imply endorsement by the U.S. Government.

**Conflicts of Interest:** The authors declare no conflict of interest.

## References and Note

1. Rivera, M.; Mariño, J.; Samaniego, P.; Delgado, R.; Manrique, N. Geología y evaluación de peligros del complejo volcánico Ampato-Sabancaya (Arequipa). *INGEMMET Boletín Serie C Geodinámica e Ingeniería Geológica* **2015**, *61*, 122.

2. Samaniego, P.; Rivera, M.; Mariño, J.; Guillou, H.; Liorzou, C.; Zerathe, S.; Delgado, R.; Valderrama, P.; Scao, V. The eruptive chronology of the Ampato–Sabancaya volcanic complex (Southern Peru). *J. Volcanol. Geotherm. Res.* **2016**, *323*, 110–128. [CrossRef]

3. Ciesielczuk, J.; Żaba, J.; Bzowska, G.; Gaidzik, K.; Głogowska, M. Sulphate efflorescences at the geyser near Pinchollo, southern Peru. *J. South Am. Earth Sci.* **2013**, *42*, 186–193. [CrossRef]

4. Siebert, L.; Simkin, T.; Kimberly, P. *Volcanoes of the World*, 3rd ed.; University of California Press: London, UK, 2010.

5. Thouret, J.; Guillande, R.; Huamán, D.; Gourgaud, A.; Salas, G.; Chorowicz, J. L'activité actuelle du Nevado Sabancaya (Sud Pérou): Reconnaissance géologique et satellitaire, évaluation et cartographie des menaces volcaniques. *Bulletin de la Société Géologique de France* **1994**, *165*, 49–63.

6. Huamán, D. Métodos y Aplicaciones de las Imágenes de Satélite en la Cartografía Geológica: El caso del Seguimiento y Evolución de la Amenaza Volcánica del Sabancaya (región del Colca, Arequipa, Perú). Tesis de Ingeniero Geólogo, Universidad Nacional de San Agustín, Arequipa, Peru, 1995.

7. Pritchard, M.E.; Simons, M. An InSAR-based survey of volcanic deformation in the central Andes. *Geochem. Geophys. Geosystems* **2004**, *5*, 2. [CrossRef]

8.  Ramos, D.; Masías, P.; Apaza, F.; Lazarte, I.; Taipe, E.; Miranda, R.; Ortega, M.; Anccasi, R.; Ccallata, B.; Calderón, J.; et al. Los inicios de la actividad eruptiva 2016 del volcán Sabancaya. *INGEMMET Informe Técnico n° A6735*, **2016**.

9.  Jay, J.; Delgado, F.; Torres, J.; Pritchard, M.; Macedo, O.; Aguilar, V. Deformation and seismicity near Sabancaya volcano, southern Peru, from 2002 to 2015. *Geophys. Res. Lett.* **2015**, *42*, 2780–2788. [CrossRef]

10. Reath, K.; Pritchard, M.E.; Moruzzi, S.; Alcott, A.; Coppola, D.; Pieri, D. The AVTOD (ASTER Volcanic Thermal Output Database) Latin America archive. *J. Volcanol. Geotherm. Res.* **2019**, *376*, 62–74. [CrossRef]

11. MacQueen, P.; Delgado, F.; Reath, K.; Pritchard, M.; Lundgren, P.; Milillo, P.; Macedo, O.; Aguilar, V.; Zerpa, I.; Machacca, R.; et al. Volcano-tectonic interactions at Sabancaya volcano, Peru (2013–2018): Eruptions, magmatic inflation, moderate earthquakes, and aseismic slip. *Am. Geophys. Union Fall Meet V23C-03* **2018**.

12. Berardino, P.; Fornaro, G.; Lanari, R.; Sansosti, E. A new algorithm for surface deformation monitoring based on small baseline differential SAR interferograms. *IEEE TGARS* **2002**, *40*, 2375–2383. [CrossRef]

13. Farr, T.G.; Rosen, P.A.; Caro, E.; Crippen, R.; Duren, R.; Hensley, S.; Kobrick, M.; Paller, M.; Rodriguez, E.; Roth, L.; et al. The Shuttle Radar Topography Mission. *Rev. Geophys.* **2007**, *45*, RG2004. [CrossRef]

14. Samsonov, S.V.; Trishchenko, A.P.; Tiampo, K.; González, P.J.; Zhang, Y.; Fernández, J. Removal of systematic seasonal atmospheric signal from interferometric synthetic aperture radar ground deformation time series. *Geophys. Res. Lett.* **2014**, *41*, 6123–6130. [CrossRef]

15. Yu, C.; Li, Z.; Penna, N.T.; Crippa, P. Generic Atmospheric Correction Model for Interferometric Synthetic Aperture Radar Observations. *J. Geophys. Res. Solid Earth* **2018**, *123*, 9202–9222. [CrossRef]

16. Herring, T.; King, R.W.; McCluskey, S.M. Introduction to GAMIT/GLOBK Release 10.7. In *Massachusetts Institute of Technology Technical Report*; Massachusetts Institute of Technology: Cambridge, MA, USA, 2018. Available online: http://geoweb.mit.edu/gg/ (accessed on 5 May 2020).

17. Altamimi, Z.; Métivier, L.; Collilieux, X. ITRF2008 plate motion model. *J. Geophys. Res. Solid Earth* **2012**, *117*. [CrossRef]

18. Williams, S.D.; Bock, Y.; Fang, P.; Jamason, P.; Nikolaidis, R.M.; Prawirodirdjo, L.; Miller, M.; Johnson, D.J. Error analysis of continuous GPS position time series. *J. Geophys. Res. Solid Earth* **2004**, *109*. [CrossRef]

19. Forsythe, R.D.; Davidson, J.; Mpodozis, C.; Jesinkey, C. Lower Paleozoic relative motion of the Arequipa Block and Gondwana; paleomagnetic evidence from Sierra de Almeida of northern Chile. *Tectonics* **1993**, *12*, 219–236. [CrossRef]

20. Battaglia, M.; Cervelli, P.F.; Murray, J.R. dMODELS: A MATLAB software package for modelling crustal deformation near active faults and volcanic centers. *J. Volcanol. Geotherm. Res.* **2013**, *254*, 1–4. [CrossRef]

21. Decriem, J.; Árnadóttir, T.; Hooper, A.; Geirsson, H.; Sigmundsson, F.; Keiding, M. The 2008 May 29 earthquake doublet in SW Iceland. *Geophys. J. Int.* **2010**, *181*, 1128–1146. [CrossRef]

22. Wright, T.; Fielding, E.; Parsons, B. Triggered slip: Observations of the 17 August 1999 Izmit (Turkey) earthquake using radar interferometry. *Geophys. Res. Lett.* **2001**, *28*, 1079–1082. [CrossRef]

23. Ebmeier, S.K.; Andrews, B.J.; Araya, M.C.; Arnold, D.W.D.; Biggs, J.; Cooper, C.; Cottrell, E.; Furtney, M.; Hickey, J.; Jay, J.; et al. Synthesis of global satellite observations of magmatic and volcanic deformation: Implications for volcano monitoring & the lateral extent of magmatic domains. *J. Appl. Volcanol.* **2018**, *7*, 2. [CrossRef]

24. MacQueen, P.; Delgado, F.; Reath, K.; Pritchard, M.E.; Bagnardi, M.; Milillo, P.; Lundgren, P.; Macedo, O.; Aguilar, V.; Ortega, M.; et al. Volcano-tectonic interactions at Sabancaya volcano, Peru: Eruptions, magmatic inflation, moderate earthquakes, and fault creep. *J. Geophys. Res. Solid Earth* **2020**. [CrossRef]

 *remote sensing*

Article

# Detection of Terrain Deformations Using InSAR Techniques in Relation to Results on Terrain Subsidence (Ciudad de Zaruma, Ecuador)

**Marcelo Cando Jácome, A. M. Martinez-Graña and V. Valdés ***

Department of Geology, Faculty of Sciences, University of Salamanca, 37008 Plaza de la Caidos s / n., Spain; id00709713@usal.es (M.C.J.); amgranna@usal.es (A.M.M.-G.)
* Correspondence: vvaldes@usal.es; Tel.: +34-923-294-400; Fax: +34-923-294-514

Received: 14 April 2020; Accepted: 15 May 2020; Published: 17 May 2020

**Abstract:** In Zaruma city, located in the El Oro province, Ecuador, gold mines have been exploited since before the colonial period. According to the chroniclers of that time, 2700 tons of gold were sent to Spain. This exploitation continued in the colonial, republican, and current periods. The legalized mining operation, with foreign companies such as South Development Company (SADCO) and national companies such as the Associated Industrial Mining Company (CIMA), exploited the mines legally until they dissolved and gave rise to small associations, artisanal mining, and, with them, illegal mining. Illegal underground mining is generated without order and technical direction, and cuts mineralized veins in andesitic rocks, volcanic breccia, tuffs and dacitic porphyry that have been intensely weatherized from surface to more than 80 meters depth. These rocks have become totally altered soils and saprolites, which have caused the destabilization of the mining galleries and the superficial collapse of the topographic relief. The illegal miners, called "Sableros", after a period of exploitation at one site, when the gold grade decreased, abandon these illegal mines to begin other mining work at other sites near mineralized veins or near legalized mining galleries in operation. Due to this anthropic activity of illegal exploitation through the mining galleries and "piques" that remain under the colonial center of the city, sinkings have occurred in various sectors detected and reported in various technical reports since 1995. The Ecuadorian Government has been unable to control these illegal mining activities. The indicators of initial subsidence of the terrain are small movements that accumulate over a time and that can be detected with InSAR technology in large areas, improving the traditional detection performed with geodetic instrumentation such as total stations and geodetic marks. Recent subsidence at Fe y Alegría-La Immaculada School, the city's hospital and Gonzalo Pizarro Street, indicates that there is active subsidence in these and other sectors of the city. The dynamic triggers that have possibly accelerated the rate of subsidence and landslides on the slopes are earthquakes (5 to 6 Mw) and heavy rains in deforested areas. Although several sinks and active subsidence caused by underground mining were detected in these sectors and in other sectors in previous decades, which were detailed in various reports of geological hazards prepared by specialized institutions, underground mining has continued under the colonial city center. In view of the existing risk, this article presents a forecasting methodology for the constant monitoring of long-term soil subsidence, especially in the center of the colonial city, which is a national cultural heritage and candidate for the cultural heritage of humanity. This is a proposal for the use of synthetic aperture radar interferometry (InSAR) for the subsidence analysis of topographic relief in the colonial area of the city of Zaruma by illegal mining galleries.

**Keywords:** terrain deformation; mining galleries; subsidence; InSAR

## 1. Introduction

The city of Zaruma is in the foothills of the Vizcaya Mountain Range of the Northern Andes, in the south of Ecuador; see Figure 1.

**Figure 1.** Location of the study area in the South of the country, in the Andes Mountains.

Its relief is very irregular, with high areas dissected by deep valleys and canyons, and several rivers, such as Salvias, Ortega, Amarillo and El Salado river. Mining exploitation has a history cataloged since before the conquest of America. The city located within the Puyango river basin has provided abundant gold reserves. According to data from the Mining Regulation and Control Agency (ARCOM), 288,000 tons of quartz is extracted annually with an average of 4 grams of gold per ton, reaching a production of 1152 kilos of gold per year. [1].

One of the hazards for the city caused by underground mining is related to the old abandoned galleries and pikes carried out by illegal miners, which is causing the sinking of the land and the collapse of the buildings in the urban area of the city.

Recently, in the years 2016 and 2019, considerable subsidence has occurred in some parts of the city, such as the city hospital, Gonzalo Pizarro Street, Fe y Alegría-La Inmaculada, subsidence that was detected in several studies conducted in previous decades [2,3].

The abandoned galleries and without technical closure, as well as the illegal pickets possibly caused the sinks dating from 2001 in several places such as in the sports coliseum, the municipal swimming pool, the Immaculate School, Gonzalo Pizarro Street, Zambrano Street, Colon, the city hospital among other sectors.

The sinking ground is a very common phenomenon around mining operations of underground exploitation. This subsidence is causing difficulties in the socio-economic development of Zaruma city and has directly endangered homes, buildings of the municipal basic services, educational centers, roads and uninterrupted environmental systems, including the possible loss of human lives.

Here, the importance of monitoring the subsidence of the relief in the colonial area of the city is born; monitoring that will strengthen the appropriate municipal ordinances for the protection of the population, the city and the environment.

The measurements of the subsidence of the relief are generally carried out punctually, with traditional methods, mainly by leveling the relief in static and dynamic GPS networks with the difficulty of obtaining regional and continuous sinking trends. A regional survey by leveling the relief with these traditional methods would imply onerous economic expenses. On the other hand, the surface of the soil induced by the excavations of mining galleries without pillars that support the weight of the soil load and the eroded rock, Figure 2(A), has exposed the buildings of the city to deformations and collapse of its structure, as several studies show, carried out since 2001 [2].

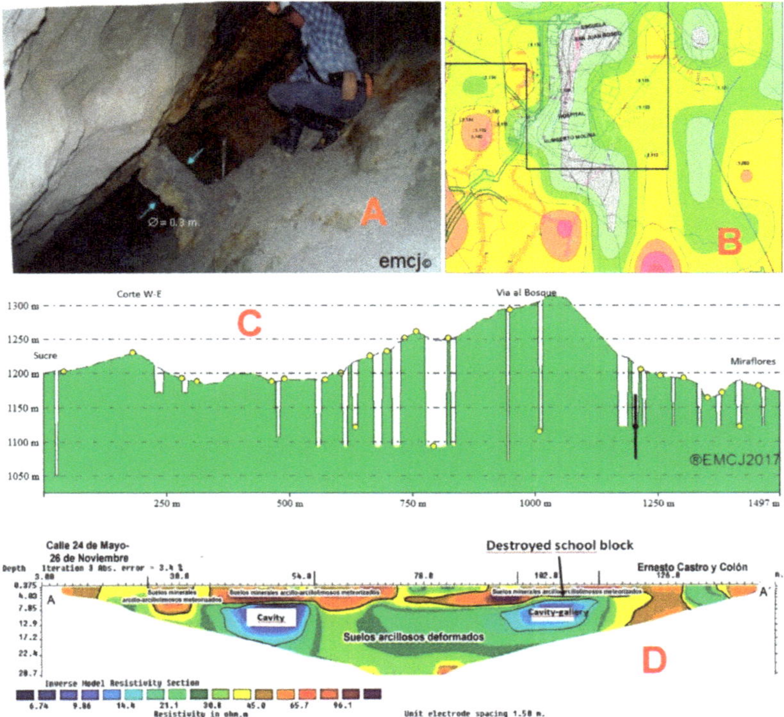


**Figure 2.** (**A**) Pillars exploited and destroyed, without resistance to the load of the upper rock and soil that in the sector reaches more than 50 meters thick. (**B**) Sinking dating back to 2001 using the flow pattern technique and morphological rupture lines. (**C**) W–E Topographic profile and subsidence caused by mining galleries in underground projection to the top of the galleries. (**D**) Electrical tomography of the subsoil in the streets May 24, November 26, Ernesto Castro, Colon and Gonzalo Pizarro. The collapse of the La Immaculada-Fé y Alegría school block was possibly due to the instability caused by the underground gallery that exists below (blue spot). This sinking was reactivated in 2019.

In particular, the stress efforts caused by the bending and cutting component of these deformations have damaged the building materials making them almost fragile. These ground deformations and subsidence were verified in 2001 by inspecting the underground mining galleries, in addition to the geological characterization, without using the DInSAR technique, which did not exist at that time. The technique used and developed was the analysis of flow patterns and morphological rupture lines, developed by the authors of this article, and topographic–geological information of each gallery investigated by ex-National Directorate of Geology, National Directorate of Geology [2–4], among researchers. An example of the state of deformation in 2001 can be seen in Figure 2B in the hospital sector of the city Humberto Molina and the San Juan Bosco school, where there is a anthropogenic subsidence zone (area with black slanted lines), caused by the mining galleries represented by green and cyan lines.

A topographic profile W–E from Sucre Street to Miraflores Street, is presented, Figure 2C, from the surface to the top of the inventoried mining galleries at that time. This topographic profile shows the intensity of subsidence caused by the underground galleries. The last study carried out by the Secretariat of Risk Management by means of electrical tomography of the subsoil in 2016, verified that the areas susceptible to subsidence detected in the previous decade were activated and caused the collapse of the La Inmaculada-Fé y Alegría School. Subsequently in 2019, subsidence occurred on

24 May and 26 November on Ernesto Castro, Colon and Gonzalo Pizarro Streets. Subsidence that were previously detected, Figure 2D.

In December 2016, the Immaculate School-Fe y Alegría, completely collapsed possibly due to being in an old gallery and in illegal exploitation, which sank the ground, as determined by several recent technical reports [3]. The classroom block collapsed as its foundations sank, and consequently the structures fractured and collapsed.

Figure 3 shows the collapse of the school in October 2016 (A), the technical filling done in 2017 (B), the new collapse in the same place (C) and the new collapse in the Gonzalo Pizarro Street sector (D), which occurred in 2019.

**Figure 3.** (**A**) Collapse of the Fe y Alegría School, in October 2916. (**B**) Technical filling done at the site in 2017. (**C**) Reactivation of the collapse in the place where the school was in 2017. (**D**) New collapse near Gonzalo Pizarro Street in 2019. Images taken from El Expreso Newspaper, Ecuador en vivo, Pichincha Universal and Arden los Socavones Facebook page.

The Zaruma canton is developed on the External Slopes of the Western Cordillera, it presents diversified reliefs on ancient volcanic materials, with partial pyroclastic coverage. The city is located on an andesitic–porphyritic volcano-sedimentary series mineralized with metallic sulfides deformed by granodioritic and tonalitic intrusive [5].

The Morphology of the Zaruma Canton is characterized by presenting steep slopes, rounded ridges and numerous hills that result from the dendritic pattern of the secondary drains. The heights are between 1150 and 2800 masl. The main drainages of the area are the Río Luís that merge with the Río Salati and Río Ambocas and which end at the Río Pindo. The Calera River and the Yellow River are also located within the area, which joins a few kilometers before its confluence with the Pindo River that forms the Puyango River downstream. There are numerous relatively large streams. The predominant geoform of the area corresponds to heterogeneous slopes, with medium to strong slopes (>25–40%), relative differences between 200 and 300 m, the drainage density is little dissected, with long lengths greater than 500 m.

Structural lineaments preferential NW–SE directions are fundamentally found by controlling the drainage pattern in the study area; however, important structures especially in the E–W and N–S direction are controlling the mineralization in the sector and are responsible for the external and internal geodynamic processes, which has caused the deformation of the main existing geoforms in the study area. Local geological faults, fracture systems and joints were considered in the concentration pattern model of this type of guidelines, to determine relief deformations in [4] and which were verified in the current methodology.

Locally, the study area is located within the Portovelo Unit, exposed in the work of research of the Zaruma geological sheet, scale 1: 100000, prepared by the National Research Geological, Mining, Metallurgical Institute. The mineralization at Zaruma is housed in intermediate to siliceous volcanics of the Portovelo Unit, which is faulty against the metamorphic rocks of the south, along the Piñas-Portovelo geological fault system and discordantly overlaps the El Oro Matamorphic Complex. This Unit, for the most part, is made up of massive porphyry andesitic lavas to andesitic basalts and gaps with intermediate tuffs. There are also rhyolitic to dacitic "Ash Flow" type tuffs with intercalations of sedimentary rocks (slates, cherts). The andesitic volcanoes present generalized propylitic alteration to epidote, calcite and chlorite. The main structural geological feature of the region is the Piñas-Portovelo fault / thrust system, E–SW direction, which has a great descent in the North block and separates the Saraguro group from the El Oro Metamorphic Complex. The metamorphic rocks of the basement, along this fault system, have been cataclysmically deformed and brecciated [6].

The city has a mining exclusion area, declared EMERGENCY RESOLUTION No. SGR-029-2015, which encloses the colonial center of the city and is the internal limit where it is prohibited to carry out mining work that has not been respected. When the underground mining works and galleries enter the exclusion zone, exploiting and chasing the veins laden with gold, after a while, they abandon those "pikes" without any technical closure, leaving gaps below the surface that have collapsed after several years of settlement. This has caused the collapse of the soil and as in the case of the Fe y Alegría-la Immaculada School, causing a sinkhole that was initially 10 m deep and currently can reach 150 m. Underground galleries in that sector can reach more than 250 meters depth.

The landscape in Zaruma is changing intensely due to the intensification of illegal mining. When galleries are abandoned due to lower ore grade, illegal miners leave the galleries without technical closure, causing the upper layers to collapse. This sinking process continues to the surface and eventually causes the ground to sink, fracturing the houses above them.

The consequence of mining exploitation around the veins that have been exploited indiscriminately for decades, has caused new local surface and underground hydrological conditions, which influenced in equilibrium the species of plants, animals and other organisms that occupy those areas that have had to adapt to the new conditions imposed by the limits of urban mining growth, which has put them at the limit of their extinction [7].

To reduce the threat of subsidence caused by underground mining in this study, in this article, Differential Interferometry techniques used to detect small relief deformations with high precision, using data from the synthetic microwave aperture radar-Synthetic Aperture Radar (SAR) and enhancing the trace or traces of underground galleries that are not found in the mining cadaster of the sector and that are possibly the cause of the collapses in various sectors of the city.

In recent years, DInSAR Differential Interferometry has been used to measure relief deformations very effectively based on large stacks of SAR images, unlike the two classic images used in standard InSAR configurations [8].

The state of the art of DInSAR techniques that make use of the data acquired by spatial SAR sensors exploit the information contained in the radar phase of complex SAR images acquired at different times over the same area. Several studies have contributed to improve the spatial location of subsidence by excavation of tunnels that cause the desiccation of aquifers by underground mineral extraction [9].

The DInSAR technique and its applications have been documented in several articles in high-level scientific journals such as Nature and Science, where it has contributed in different fields of Geosciences such as seismology, with important scientific achievements, including coseismic and post-seismic analysis before the occurrence of an earthquake. Volcanology is another important field of application, with several studies of volcanic deformation (deflation and elevation). These volcanic deformations have allowed determining the possible origins of secondary lahars as can be seen in [10].

The DInSAR analysis on landslides that despite being an important application for the location and reduction of these events, according to several researchers, the DInSAR analysis has less efficient results, mainly due to the loss of coherence due to its heterogeneous character in its composition. Despite this, with the Persistent Scatterers-PS dispersion technique, the geometry of some types of landslides can be determined with good results. The most relevant results are described in [10]. In the case of subsidence and ground lift, the DInSAR analysis has obtained efficient results in several parts of the world that have been described in specialized journals. It has been demonstrated as tunnels and underground galleries that have unbalanced the water systems of the surface runoff as well as the groundwater flow systems, causing the destruction of the buildings on the surface. Other subsidence processes worldwide have been investigated for other factors such as fluid pumping, construction work, geothermal activity, etc. [11]. In most of the published results that refer to urban areas, the DInSAR data is consistent even in long periods, due to the existence of static artificial reflectors such as buildings and other reinforced concrete structures that remain stable. With the advent of Persistent Scatterers techniques with high and pseudo coherence, more efficient results have been obtained to monitor deformations of the relief outside urban, suburban and industrial areas [12–24].

In Ecuador, there are few studies using SAR images, InSAR methodologies and their variants. In 1977 the Ecuadorian Government created the Center for Integrated Surveys of Natural Resources by Remote Sensors (CLIRSEN), a technological and scientific entity aimed at integrating the most advanced technologies related to Geodesy, Natural Resources, Environment, Cadaster, Geographic Information Systems (SIG), Software. Clirsen, used radar images of lateral vision and of real opening (SLAR), acquired by the oil companies in the Amazon region and the first semi-controlled mosaic of Radar Aero transported synthetic opening (SAR), images taken for the exploration of oil fields in 1982.

Subsequently, the Ecuadorian Space Institute replaces Clirsen Center and in 2013 the first InSAR analysis project with Cosmo-SkyMed images is carried out within the Space and Geophysics Technology project in risk management external geodynamics for prevention and mitigation of floods and torrential floods [25]. This was the first project that the regent institution in remote sensing executed in the country. In 2014, one of the first projects of soil deformation analysis was carried out using the DInSAR technique in a mountainous region of the Andean region. This work was done with Roi-Pac software with very good results [26]. Since 2016, after the destructive earthquake of 7.8 Mw on the Ecuadorian coast that caused more than 600 deaths and economic losses of more than 3 miles million dollars, the Risk Management Secretariat, implemented the deformation zone analysis project in the country, using the InSAR methodology.

The authors of this article have developed some projects using the InSAR methodology with a pair of Sentinel images and with piles images to determine deformations of the ground in the Fuego volcano in Guatemala, subsidence in Zaruma, neighborhood in Quito city and combined InSAR analysis with geomorphological, geophysical and geotechnical methods in Manta and Portoviejo cities as can be seen in the articles [27].

The following is a summary of obtaining the Interferometric Phase: The Radar System of Synthetic Opening-SAR, allow the obtaining of images of the reflectivity of the ground, which are subsequently processed by InSAR (Interferometry SAR) techniques for the generation of precise maps of soil deformation. The technology (InSAR) or SAR Interferometry is the technique that is based on the study of the phase interference pattern of the waves of two SAR images to generate maps of displacements of the topographic relief, and digital models of terrain elevation [28]. With this technique, it is possible to determine, displacements of the topographic relief, in mm/year of large extensions of the topographic relief, without the need for field measurements and at low cost. [29]. This technique can also provide more data to understand the geomorphological evolution of the site. An antenna of synthetic or virtual opening consists of a vector of successive and consistent signals of radar that are transmitted and received by a small antenna that moves along a certain flight or orbit path. The signal processing uses the magnitudes and phases of the signal received on successive pulses to create an image [30]. From interferometry, the data obtained are the distances between the satellite and the ground area, calculated by the measurement of times and lags. The main platforms that use Radar Synthetic Aperture are shown in Table 1 [31].

**Table 1.** Resolution of the most common radar images in meters [31].

| SATELLITE | BAND | Acquisition Mode | Nominal Pixel Dimensions: Ground range x Azimuth (m) | Repeat Cycle (day) |
|---|---|---|---|---|
| ERS-1 & ERS-2 | C | Standard Beam | $20 \times 4$ | 35 |
| Envisat | C | Standard Beam | $20 \times 4$ | 35 |
| Radarsat-1 | C | Standard Beam | $20 \times 5$ | 24 |
| Radarsat-1 | C | Fine Beam | $10 \times 5$ | 24 |
| Radarsat-2 | C | Standard Beam | $20 \times 5$ | 24 |
| Radarsat-2 | C | Fine Beam | $10 \times 5$ | 24 |
| Radarsat-2 | C | Ultra-Fine Beam | $3 \times 3$ | 24 |
| TerraSAR-X | X | Standard | $3 \times 3$ | 11 |
| TerraSAR-X | X | Spotlight | $1.5 \times 1.5$ | 11 |
| Cosmo-SkyMed | X | Standard | $3 \times 3$ | 16(8) |
| Cosmo-SkyMed | X | Spotlight | $1.5 \times 1.5$ | 16(8) |
| ALLOS PALSAR | L | Fine Beam | $10 \times 5$ | 46 |

If at least two observations of the same site are obtained from space at different times, the "interferometric phase" is directly proportional to any change in the range of topographic relief characteristics.

In reference to Figure 4 for the acquisition of an interferogram, have followed what M. Crosetto et al., describes about the basic principles of Interferometry from the Synthetic Aperture Radar (SAR) [30].

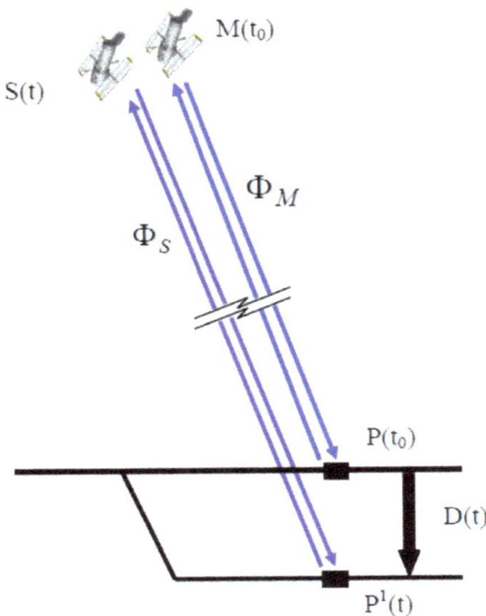


**Figure 4.** Displacement diagram of topographic relief and phase change of the signal [11].

At time $t_0$, the sensor obtains a first SAR image, measuring the phase $\Phi_M$ (Equation (1)), an image called master, M. If a deformation of the ground D occurs, over a time, point P moves to $P^1$. Subsequently, the satellite obtains a second image at time t, and measures the phase $\Phi_S$. This image is called slave, S. The InSAR technique calculates the phase difference between $\Phi_S$ and $\Phi_M$, a difference called interferometric phase $\Delta\Phi_{Int}$.

$$\Delta\Phi_{Int} = \Phi_S - \Phi_M \text{ (eq. 1)} \tag{1}$$

In this study, the DInSAR methodology has been used to determine this type of deformation movement and is an ideal tool to monitor the sinking of the topographic relief in large areas and for prolonged periods, with a high precision of a few millimeters to centimeters per square kilometer. Sentinel images in the C band, were used to determine the deformation and accumulated displacement over a period from 2016 to 2019 [27]. A period in which subsidence occurred as mentioned above.

To study the local subsidence of the soil processed a multitemporal series of interferograms in the C band with one stack of Sentinel 1 images before the collapse of the Fe y Alegría-La Immaculada School and after. The results were verified in those sites that have sunk where abandoned and active underground mines are found.

There are historical studies that report severely damaged buildings, possibly due to being in subsidence areas in the Colonial Center of the Zaruma city. Figure 5 shows the location of the city, the mining exclusion area (ocher color), the underground mining galleries (red lines) in the central part and to the east, the veins and mineralized structures a geological faults (black color lines). In the western part, there are no surveys of mining galleries except for one represented with a red color line (possibly, there is no survey of the mining cadaster in this area).

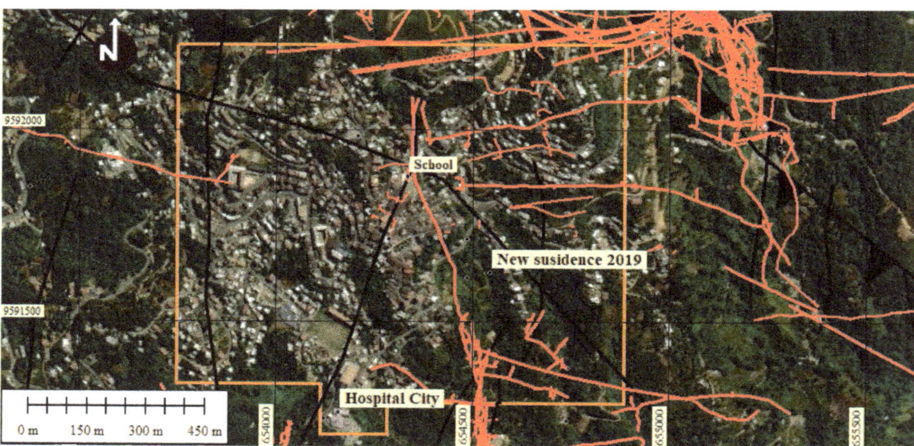


**Figure 5.** Underground locations of mining galleries outside and within the exclusion area of the Zaruma city. In ocher color, the Exclusion Zone polygon, the mining galleries from the center to the east in red lines, the mineralized structures and geological faults in black color lines. Dots black color are bocamines. Data source: Risk Management Secretariat.

## 2. Materials and Methods

Considering that, interferometry can detect minute deformations of the topographic relief with high precision using microwave SAR data. In this article, the DInSAR methodology has been used to determine this type of deformation movement and is an ideal tool to monitor the sinking of the topographic relief in large areas and for prolonged periods of time, with a high precision of a few millimeters to centimeters per square kilometer. Sentinel images in the C band were used to determine the deformation and accumulated displacement over a period from 2016 to 2019 [16–24]. A period in which several subsidence occurred as mentioned above. The calculation of the accumulated displacement may have errors due to atmospheric noise, which in this case was reduced with the filter for Atmospheric Phase Screen reduction (APS) tool, to obtain interferograms with high coherence, which guarantees the accumulated displacement results in mm / year. The residual error is 0.1 mm. The displacement is an average within the period studied.

The images were taken from 25 October 2016, before the sinking, until 22 October 2019, and include images from 2016, 2017, 2018, period within which the beginning of the sinking in the school is located (26 October 2016). This allowed us to analyze how the subsidence evolved and determined possible reactivation of the subsidence in the school (22 October 2019), including the new collapse on Gonzalo Pizarro Street in August 2019 (rate +/- 30 mm/year, lifting/sinking).

One of the factors that causes errors in the capture and interpretation of interferograms with InSAR technology is the influence of atmospheric noise.

To avoid the influence of atmospheric noise, the Atmospheric Phase Screen (APS) filter has been applied in SARPROZ, taking into account that in this central area of the city, the vegetation coverage is minimal and is unlikely to cause atmospheric noise that distorts the quality of acquired images. Other sources of noise have been reduced, applying filters using the SARPROZ program. The objective of the reduction of the atmospheric noise is to obtain a high degree of coherence in the results to guarantee that the accumulated values of displacement and velocity have a high correlation with the real local subsidence caused by underground mining.

In this study, the filter for Atmospheric Phase Screen reduction (APS) was used with the persistent dispersion interferometry methodology PS-InSAR, in a Sentinel 1 B multitemporal imaging stack of 20 available images. The PS-InSAR Persistent Scatter is an InSAR processing technique that uses multiple images taken at regular intervals to achieve better measurement results.

The processing of the image stack allowed determining the topographic relief movement, by varying the time phase for each pixel in a time.

The method focuses on finding stable dispersers over time, not influenced by atmospheric noise and providing a stable response signal, which in the case of the city, correspond to artificial structures such as buildings, roads, obelisks, among others structures, called Persistent Scatters (PS) These Dispersers, provide a history of stable phases during the image acquisition time because they do not suffer from temporal correlation, allowing long-term observation and monitoring of the deformation of the sectors in their surroundings.

To analyze the multitemporal displacement of the Fe y Alegría-La Inmaculada School, the available images of Sentinel 1 B, on the stack [30] were used to obtain results before the beginning of the collapse in the school (26 October 2016), and during the evolution of the deformation of the land, until the reactivation of the sinking in the site and the new sinking in Gonzalo Pizarro Street. The project followed standard interferometric process in the SARPROZ software, according to the flow chart of Figure 6 [32–34].

To analyze the dispersion and coherence of the data and results of the analyzed parameters (velocity of movement mainly) in a time series with 20 images used, the PS technique was applied using the SARPROZ software with the following procedures, which the authors implemented and applied in other sectors and published in scientific articles [28]:

SARPROZ is very powerful and versatile software that implements a wide range of Synthetic Aperture Radar (SAR), Interferometric SAR (InSAR) and Multitemporal InSAR processing techniques [32].

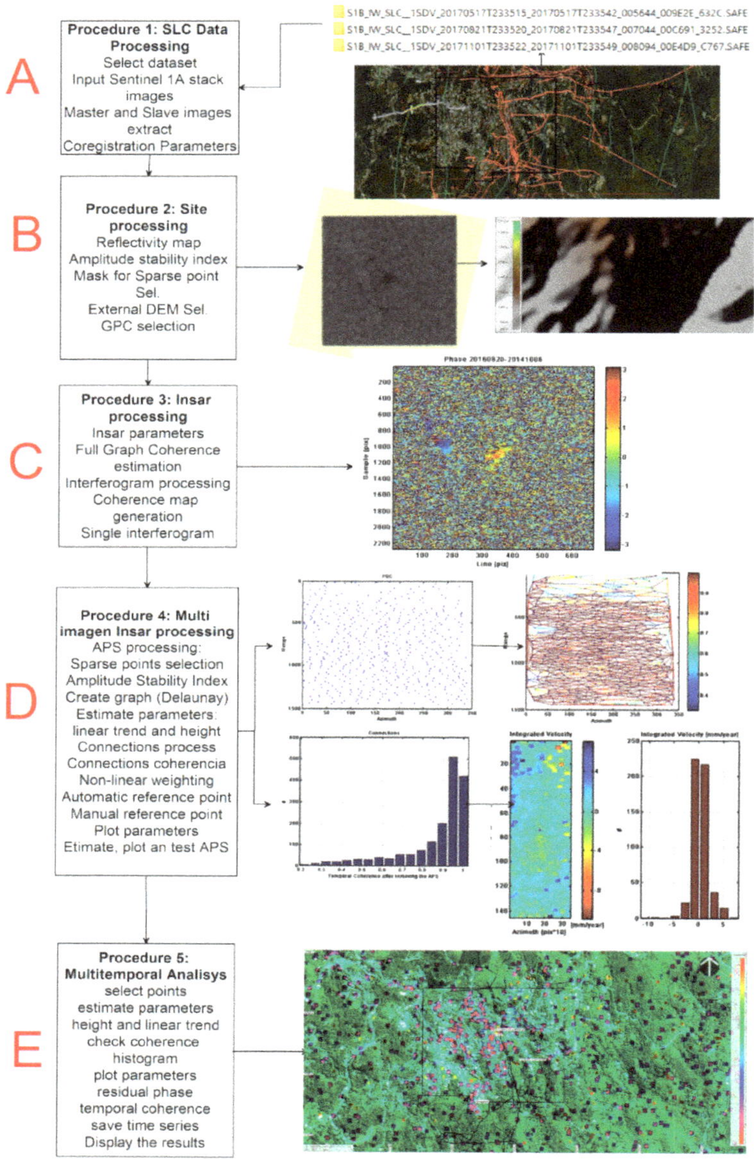



**Figure 6.** The SARPROZ software process used to obtain the relief deformations by DInSAR analysis. Steps (**A–E**) of each portion of the routine. See text for further explanation.

Procedure 1: SLC (A) data processing to import or update extracted data from Sentinel 1 images. Master and slave images are extracted and defined by selecting them manually or automatically. In the star chart in Figure 7A, the master image is close to half the perpendicular and temporal baseline domain between 1B and right, to try to minimize the effects of normal and temporal baselines. Finally, the Co-registration Parameters of the pixels of the corresponding master and slave images are defined for the elaboration of interferograms correctly.

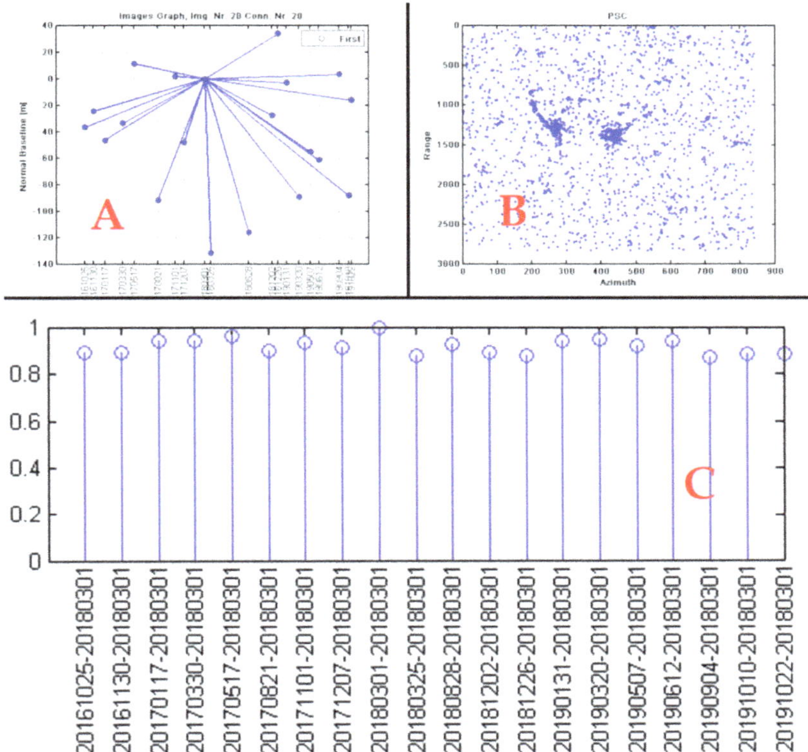

**Figure 7.** (**A**) Representation of the Sentinel 1 B image set, in the normal basal–temporal space, where each point and line represents an image and an interferogram, respectively. (**B**) Scattered points for PS-InSAR analysis. (**C**) Consistency greater than or equal to 0.85 of the Sentinel 1 image stack.

Procedure 2: is called Site Processing (B), in which preliminary parameters of the images such as the reflectivity map are extracted, from the average time reflectivity of all the images in the stack. The amplitude stability index is calculated which is a unique number that provides a statistical property of the amplitude series with ranges between 0 and 1. A mask is generated for the selection of the scattered points based on a threshold value of the index of amplitude stability, whereby values above the threshold enter the interferometric process and values below this threshold will be masked and will not enter the process, Figure 7B. Select an external Digital Elevation Model (DEM), to eliminate the topographic phase and geocode the images with the option Ground Control Point Selection-GPC, for orbital correction, flattening of interferograms and definition of persistent dispersers. The basic characteristic of a DEM is that it has positioned reference points; in this case, it is in relation to the Shuttle Radar Topography Mission -SRTM- that has a high degree of confidence in its position. Other external DEMs of various spatial resolutions can be used, provided they have stable static reference points (stable reflectors) that have not changed over time. Finally, the complete graphic coherence is calculated, to estimate the coherence of all the possible connections (interferograms) in the image scene.

Procedure 3: called InSAR (C) processing in which the InSAR parameters for interferometric processing are defined by estimating the complete graphic coherence according to the loaded image stack and the average coherence of the generated interferograms. From this calculated coherence, an interferogram is generated between the master image and each of the slave images. In this procedure, a single interferogram can also be generated, freely chosen from the master image with any of the slave images.

Procedure 4: called InSAR Multi Image Processing (D) in which the atmospheric noise APS is eliminated and the candidate points are selected to generate the persistent dispersion-PS, those that have a stable position in terms of deformation of the topographic relief. Based on these points, the program calculates the displacement height, relief velocity and the residuals of those parameters that serve to recover the delay of the atmospheric phase. The candidate points of persistent dispersion-PS in the urban area of Zaruma correspond to the constructions of buildings that remain stable over time both in radiometry and in the interferometric phase [34]. The stack of 20 images used is the most important factor for estimating the coherence of the pixels, since it allowed identifying suitable PS for the displacement analysis of the relief. Insufficient use of the images will produce an overestimate of the coherence in the entire scene, a poor estimation of the PS, therefore, in false displacements. PS location is considered reliable when 20 or more images are used. In this phase, the APS ambient noise is also estimated, which can affect the process of generating the interferograms due to different atmospheric conditions at the time of image acquisition. Eliminating atmospheric noise, APS is important as it improves the coherence and phase response signal of the images to obtain more accurate ground displacement data.

Procedure 5: called Multitemporal Analysis (E) is a procedure that, based on the previous atmospheric noise reduction procedure, is used for the analysis of MT-InSAR multitemporal interferometry based on persistent scatter (PS). With this procedure, dispersers were identified, whose signal is dominant within the total dispersion of the analyzed pixels and with which a deformation map of the topographic relief was obtained, in which the deformation rate is represented from the time series obtained. These maps are made up of thousands of PS (persistent dispersions) and each PS is associated with an annual linear velocity value (mm / year), and with the accumulated displacement on each date of image acquisition.

A requirement before performing the analysis with PS-InSAR is that the image signals throughout time series must remain consistent for the extraction of PS points and to analyze their dispersion. In this case, to measure the relative displacement and the accumulated displacement as a function of a reference point, a stable point was selected (an artificial construction anchored to the ground, whose peak value in the histogram has a residual height with value 0 (point in ground) Points that are not on the ground may be unstable. The methodology used calculates the movement of the nearby points relative to this reference point, so it must be very stable.

In this case, the program chose PSC (Persistent Scatter Candidates) points based on their location in a connecting network and with a coherence threshold value greater than> 0.8. As mentioned, these points are parts of artificial civil structures to analyze the dispersion of amplitude stability around those points, drawing a coherent point connection network (Delaunay spatial connection graph). This procedure estimated a high coherence in the connections of each point of the network to obtain the height and relief displacement velocity with high confidence. After estimating the above parameters with high consistency, atmospheric noise was removed. The choice of the number of images in the stack showed coherence greater than or equal to 0.8, as presented in Figure 7C and the estimation of APS atmospheric noise with high coherence for the entire set of points was processed with a nonlinear spatial distribution, ensuring that the final coherence is satisfactory.

The cumulative and integrated velocity and displacement of the relief were obtained considering that the accumulated displacement = velocity × time, so that at the reference point, the peak of the accumulated velocity histogram and displacement tends to zero, which means that most points have relative velocity and zero displacement, when they are approached and compared to the reference point.

The histograms verified that the connection speed and the connection residual height are consistent in the distribution of the residual value of travel velocity (mm/year) / height (m), as presented in Figure 8A–C, where a histogram is seen, without jumps and high coherence (connecting lines with a red tendency), Figure 8D.

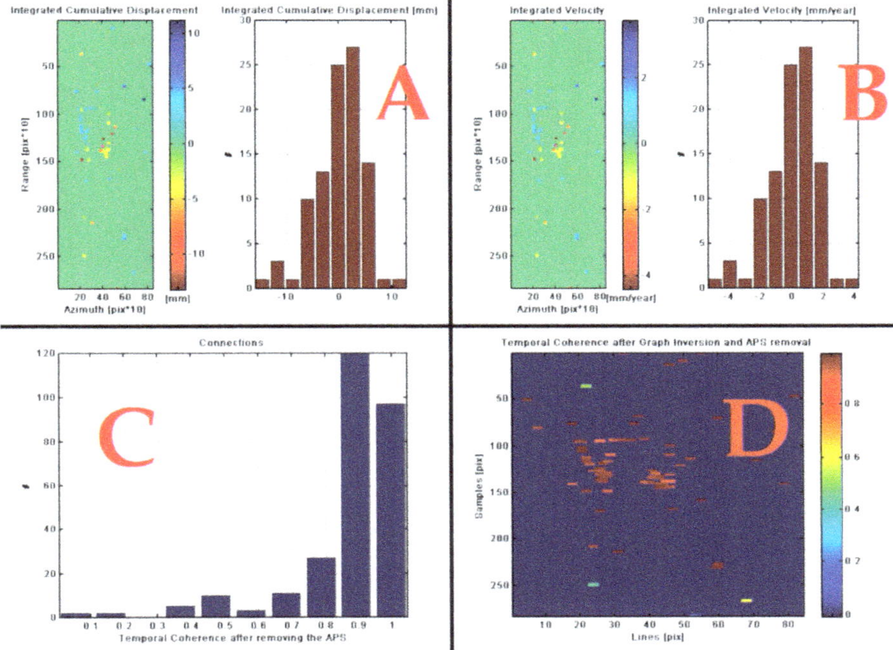

**Figure 8.** (**A**) Histogram of integrated cumulative displacement connections, (**B**) Integrated velocity, (**C**) Histogram and Temporal Coherence after removal of APS noise (**D**).

## 3. Results

The dispersion and coherence analysis of the displacement height relief velocity was obtained over a time, with the availability of a stack of 20 images, applying the PS-InSAR technique, with the procedures mentioned above.

The first result of the interferograms within the period from 25 October 2016 to 22 October 2019, before the sinking in the Fe y Alegría-La Immaculada School, in 26 October 2016 is shown in Figure 9. In (A) the current image of Google Earth with the old school site and the filler created to stabilize the site. In (B) the vertical displacement until October 2016. The elevation and subsidence relief values were within a range between −50 mm of subsidence (blue–cyan color) and 20 mm of the ground lifting (color towards red). The area completely collapsed in October and December 2016 (red polygon). The weak blue color inside the polygon indicates that on 25 October 25, the sinking of the area was in process. In the nearby areas, can see the deep cyan sinks, guided by the deep blue traces possibly demonstrate the presence of underground mining galleries. The white color represents the existing buildings in the school at that time. The year with the highest subsidence was 2016, possibly associated with low–moderate intraplate earthquakes and geological faults in the Gulf of Guayaquil [35,36].

The results of the interferograms obtained from the stack between 25 October 2016 to 22 October 2019, show that the subsidence processes were in progress producing a new subsidence on Gonzalo Pizarro Street (Figure 10).

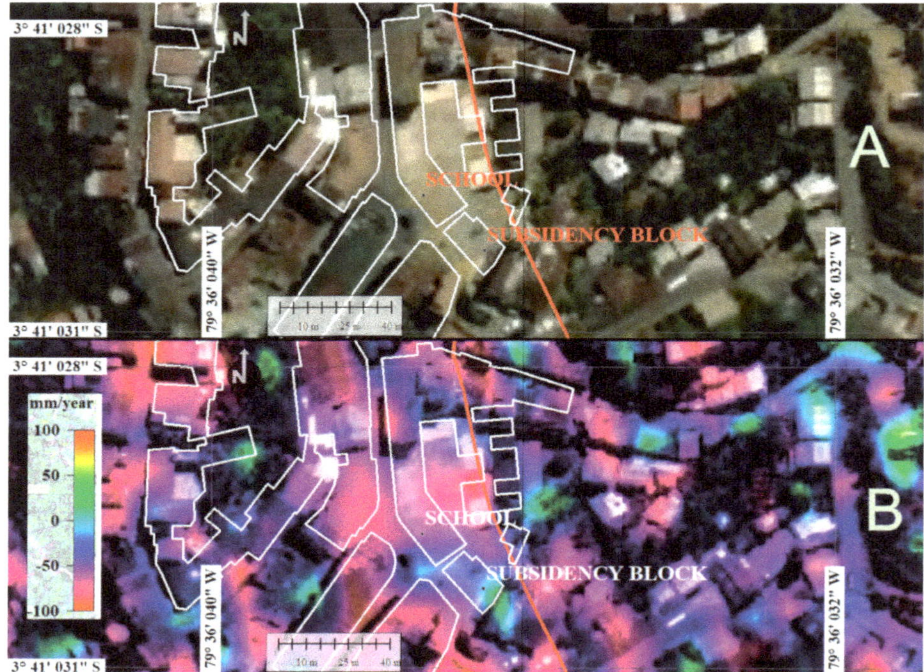


**Figure 9.** (**A**) Location of Fe y Alegría-La Immaculada School in the study area before the collapse of the block of school classrooms. (**B**) Block collapse of school classrooms on October 25 of a day before the collapse that began on October 16, 2016, and completely collapsed in December 2016. The faint blue color indicates that before October 25, 2016 the sinking of the area was in process. Gallery is in red color and blocks houses are in white color.

The reclassified images of subsidences currently occurring in the city obtained by the InSAR method have a more detailed value than those presented in images 9 to 13 in a range of 20 mm/year (lifting relief) to −30 mm/year (subsidence relief) with an error of +/−5 mm, therefore InSAR performs a quantitative analysis since it starts from points that it identifies on the ground as stable and are geo-referenced in a coordinate system (in this case UTM; WGS84; 17 Sur Zone), to determine the elevation difference of the same point taken in two different periods of time and prepare the interferogram. This is presented in the following image. The images obtained demonstrate the ranges of subsidence that currently occur in the sector in a period of time 2016 to 2019, which is in values similar to the ranges of displacement obtained by the INIGEMM that are 15 mm in two months (very short time to have more real data) (Figure 10).

These results were compared with data from subsidence investigations conducted by National Metallurgical Mining Geological Institute-INIGEMM in March 21017 one year after the sinking of 2016 and presented in [37], as part of the studies for the stabilization of the collapse of the school in October 2016.

The technicians of this Institute, made five perforations inside the school and in the southern zone near the sinking area. The perforations reached a depth of 70 m, which in general determined a first layer with an average of 30 cm thick that corresponds to concrete in the foundations. Then there is a layer of about 20 m in average thickness that corresponds to clay, which, according to its lithological composition, corresponds to the saprolite (eroded rock).

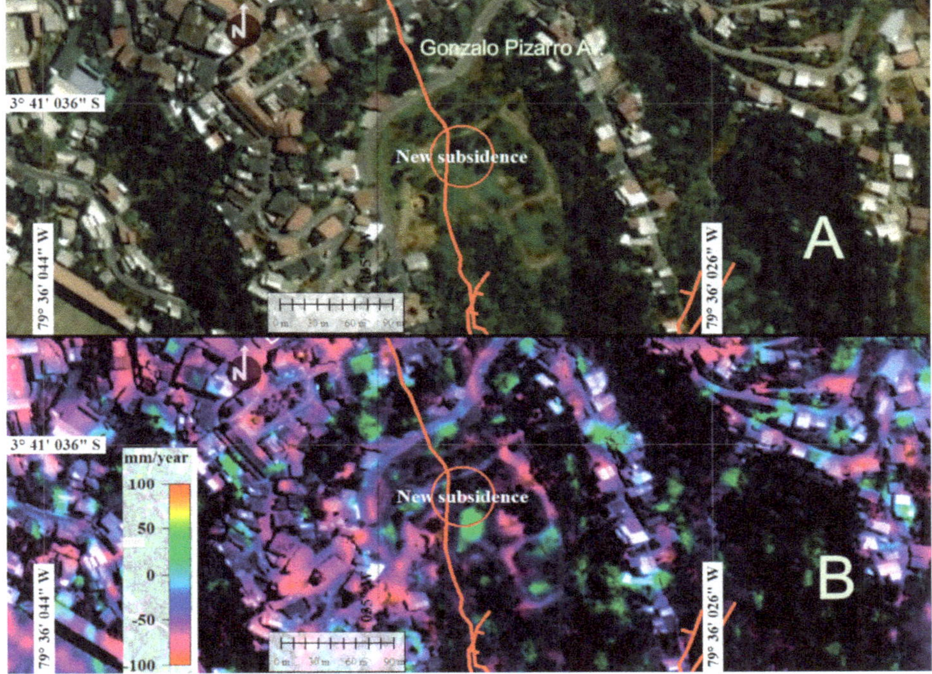



**Figure 10.** (**A**) Sites of new subsidence in Gonzalo Pizarro Street in August–September 2019. (**B**) The intense blue represents a range of high subsidence.

Then there is a layer about 8 m thick on average of andesite strongly weathered and finally, a layer of volcanic breccia with 50 m thickness on average.

The Rock Mass Rating -RMR- values obtained vary in a range from 14 for the clay (saprolite) with the lowest value to a value of 67 for the weathered andesite and the volcanic breccia.

The simple compression strength of these layers proved that the clay (saprolite) has a resistance classified as very soft, while andesite and the volcanic breccia have a resistance classified as hard.

The study of subsidence of the ground that the National Metallurgical Mining Geological Institute-INIGEMM carried out [37] from a monitoring network of 24 points between the months of March and April 2017, using a Total TOPCON GTS-750 station, determined that the average displacements in the school. They have a range between 1 to 15 mm in two months. This range of subsidence is within what the InSAR method.

The subsidence range with InSAR has greater definition and goes up to 50 mm/year in accumulated velocity from 2016 to 2019, at all times. This means that the InSAR methodology is more efficient, free and can be applied in large or regional areas.

Figure 11 shows the velocity of elevation and subsidence of the terrain within a range of −20 to 50 mm, to classify the most accurate rank levels at the site. (A) The new collapse in the Fe y Alegría school, (B) the new subsidence in Gonzalo Pizarro Street and in (C) the subsidence that are occurring in the city hospital, among other places that were analyzed by the authors in 2001 [2] and served to verify the results of this article.

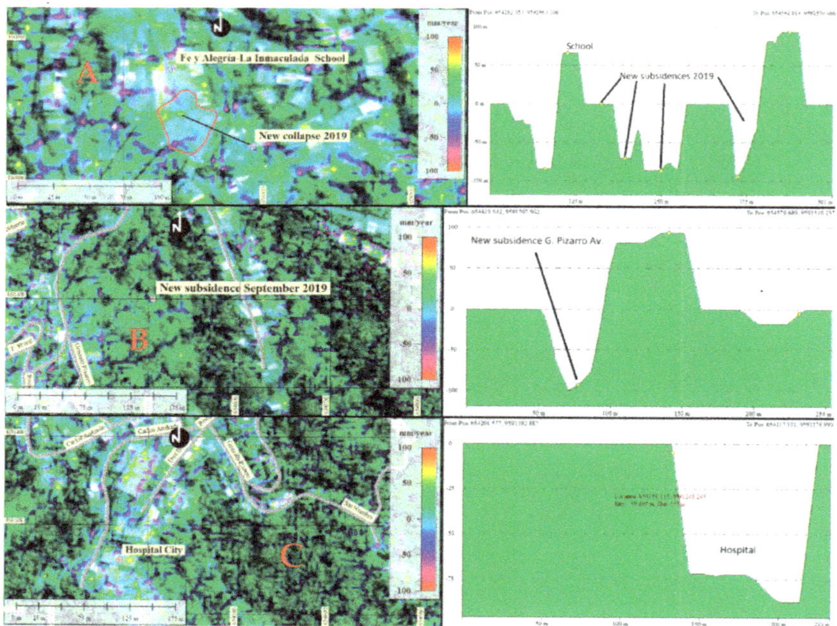

**Figure 11.** (A) Lifting velocity and subsidence of the ground at Fe y Alegría School (**A**), Gonzalo Pizarro Street (**B**), the city hospital(**C**), obtained from the interferometric analysis. These results corroborated previous studies and verified the analysis performed in this article.

The correlation of the subsidence areas that occurred in 2001, 2016 and 2019 through the interferometric analysis performed with the DInSAR methodology and correlated with the geotechnical and subsidence investigations carried out by INIGEMM and the thesis author [36] (Figure 12), allowed the determination of progressive surface deformations that can cause current subsidence mainly in several sectors and its neighborhood.

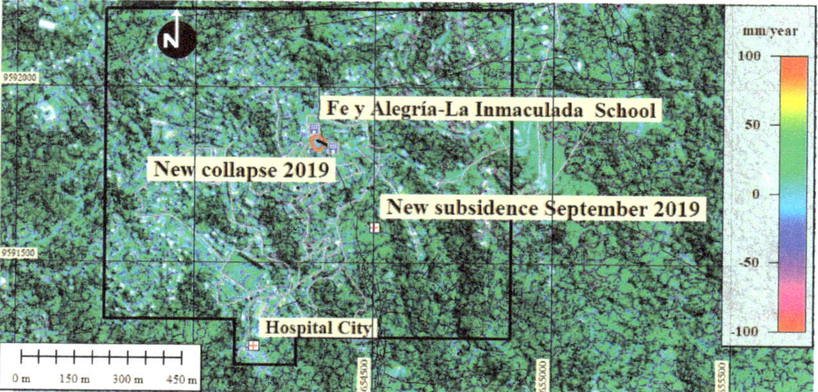


**Figure 12.** Determination of the "traces" (blue stripes) of connections of unmapped mining galleries. These galleries, generally illegal and abandoned, are connected to mineralized veins and legal mapped galleries, breaking the balance of the rock mass, intensifying the subsidence of the topographic surface of the sector.

Note the correlation between the sinking areas CF5 of Figure 13A and the new collapse occurred at the same site in 2019 of Figure 13B.

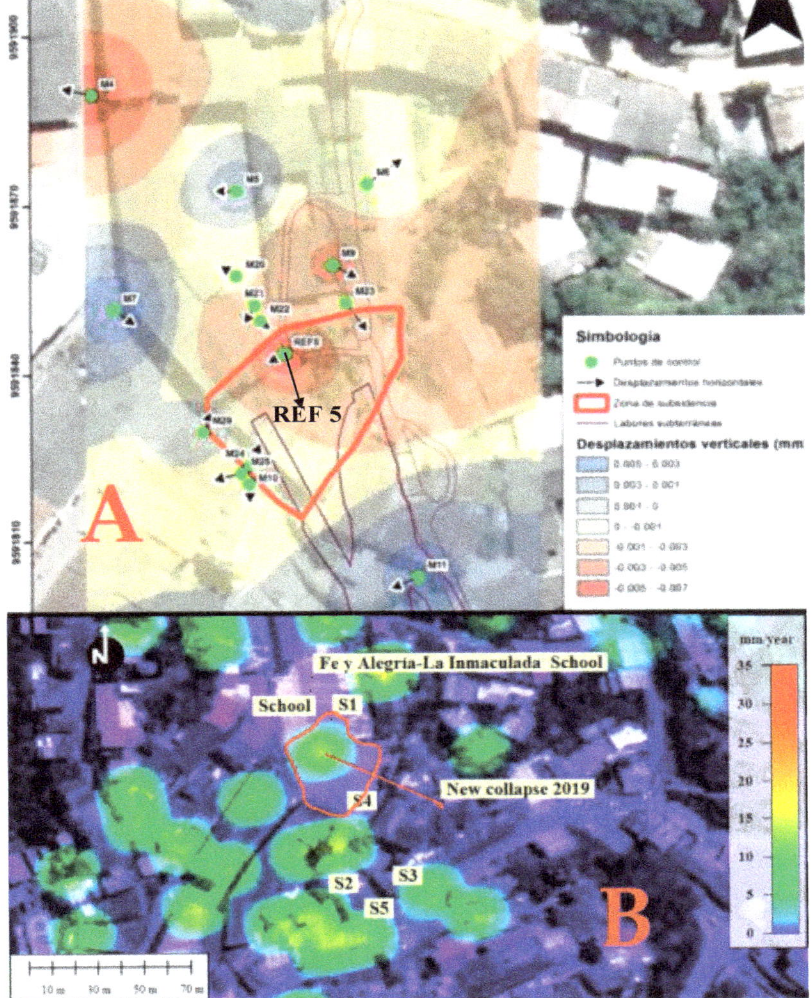


**Figure 13.** Correlation between the results obtained from areas of subsidence by InSAR Interferometry with the geotechnical investigations of subsidence carried out by INIGEMM and the author of the thesis [37]. Image 13A has low resolution since its origin. Note the correlation between the sinking areas CF5 of (**A**) and the new collapse occurred at the same site in 2019 of (**B**). S1 to S2 surveys conducted by INIGEMM.

It is necessary to highlight the correlation that exists between the area of maximum subsidence REF 5 in Fe y Alegría-La Inmaculada School that seen in Figure 13A of the INIGEMM and the new collapse occurred in the same site in the 2019 of Figure 13B elaborated with the InSAR methodology. There is a correlation between subsidence in other nearby sectors. INIGEMM considered geotechnical parameters such as cohesion, angle of friction, UCS classification, and other parameters

## 4. Discussion

From the process and interpretation of the images of the Sentinel 1 C band, unstable areas related to mining activities were obtained in the underground galleries that are within the mining exclusion area, with superficial movements of the relief in some susceptible sites that they require deeper research in geology, geotechnics and geophysics.

The 20 interferograms analyzed in the C band show high spatial coherence due to the high temporal correlation (Figure 8). The reduction of atmospheric noise through the SARPROZ APS process helped to achieve a strong connection between dispersed points with high coherence so that displacement values and travel velocity are reliable and correlated with the actual events that occurred in the years 2016 and 2019.

This high coherence may also be because the study area has no intense vegetation that can cause noise and there is. Before 25 October 2016, interferograms show progressive deformations of the soil that accumulated until the sinking of 26 October 2016 and December 2016 and August–September 2019, dates on which the subsidence occurred.

Most interferograms reveal continuous surface movements in the central part of the exclusion zone of the city of Zaruma, at sites where several underground mining galleries were located (Figure 13).

The sectors that presently present subsidence according to the SAR interferometry study, can be seen in Figure 14 and are the following: the municipal market, the Fe y Alegría-La Inmaculada School (1), sports coliseum (2), municipal pool and jail (3 ), Gonzalo Pizarro Street (4), Church (5), Central Park and Municipality (6), City Stadium (7), El Oro Avenue, (8), City Hospital, San Juan Bosco School (9), Gil Gilbert Street (10), Roground Hotel (11), El Sexmo (12), November 26 School(13), Reinaldo Espinosa Street (14), among other sectors.

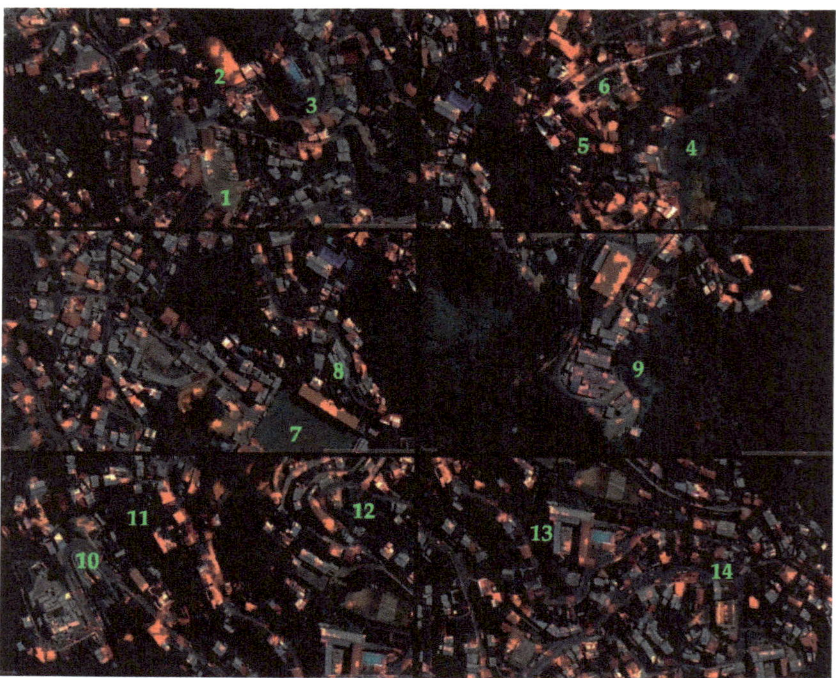


**Figure 14.** DInSAR analysis, determined that there are many constructions with potential sinkholes in some sectors of the city (red spots on the roofs).

## 5. Conclusions

The existence of a layer of clay 25 m thick on average, with low geotechnical characteristics and easily deformable, is the answer for the occurrence of surface subsidence of the ground obtained with the InSAR method.

This means that the response of this clay layer to underground mining excavations in the underlying rock layers (eroded andesite and volcanic gap) is directly related to the sinking of this soft layer therefore from the soil and buildings and other constructions, in several sectors of the city.

The analysis of surface displacement by the DInSAR technique, related to historic, legal and illegal underground galleries corroborated the instability state of the colonial center of the city of Zaruma, where more than 40% of the area is unstable.

20 Sentinel 1 images chosen for the elaboration of the interferograms and the multitemporal analysis of the displacement and its velocity, allowed to obtain reliable values of these parameters coinciding with the real events of subsidence that occurred since 2014 and which were described in studies of various institutions since 2001.

This multitemporal technique for data in the C band also served to extract information on the traces or surface traces of subsidence that have left the tunnels of the underground galleries in the western sector of the exclusion area of the city. Sector where there is no underground mining cadastral information of the mining galleries.

The results of accumulated displacement (mm / year) and deformations of the topographic relief, have determined that the study area is unstable with subsidence of the relief in localized sites, related to underground mining galleries that require further investigation for their stabilization.

SARPROZ efficiently allowed applying the DInSAR Multitemporal Interferometry processing technique as the current study with optimal results.

This technique must be taken by the municipal authorities as a forecasting tool to obtain the displacement and its velocity in intensity ranges over time and propose geotechnical programs for its reduction. This is a home base to strengthen the Territorial Planning.

**Author Contributions:** Conceptualization, M.C.J. and A.M.M.-G.; methodology, M.C.J. and V.V.; software, M.C.J.; validation, M.C.J. and A.M.M.-G.; formal analysis, M.C.J.; investigation, M.C.J. and A.M.M.-G.; resources, M.C.J.; data curation, M.C.J.; writing—original draft preparation, A.M.M.-G.; writing—review and editing, A.M.M.-G. and V.V.; visualization, M.C.J.; supervision, M.C.J.; project administration, A.M.M.-G. All authors have read and agreed to the published version of the manuscript.

**Acknowledgments:** This research was funded by projects Junta Castilla y León SA044G18 and the GEAPAGE research group has participated.

**Conflicts of Interest:** The authors declare no conflict of interest.

## References

1. Decentralized Autonomous Government of Zaruma 2014. Development and Management Territorial Plan Update. *Zaruma, El Oro Province, Ecuador.* Available online: http://app.sni.gob.ec/sni-link/sni/PORTAL_SNI/data_sigad_plus/sigadplusdocumentofinal/0760001150001_PROPUESTA%20ACTUALIZACION%20PDyOT%20del%20CANT%C3%93N%20ZARUMA_19-04-2015_23-06-10.pdf (accessed on 5 May 2020).

2. Vidal Montes, R.; Martinez-Graña, A.M.; Martínez Catalán, J.R.; Ayarza Arribas, P.; Sánchez San Román, F.J. Vulnerability to groundwater contamination, SW Salamanca, Spain. *J. Maps* **2016**, *12*, 147–155. [CrossRef]

3. Martinez-Graña, A.M.; Goy, J.L.; Cimarra, C. 2D to 3D geologic mapping transformation using virtual globes and flight simulators and their applications in the analysis of geodiversity in natural areas. *Environ. Earth Sci.* **2015**, *73*, 8023–8034.

4. Cando-Jácome, M.; Martínez-Graña, A. Numerical modeling of flow patterns applied to the analysis of the susceptibility to movements of the ground. *Geosciences* **2018**, *8*, 340. [CrossRef]

5. Vikentyev, I.; Banda, R.; Tsepin, A.; Prokofiev, V.; Vikentyeva, O. Mineralogy and Formation Conditions of Portovelo-Zaruma Gold-Sulphide Vein Deposit, Ecuador. *Geochem. Mineral. Petrol.* **2005**, *43*, 148–154.

6.  Spencer, R.M.; Montenegro, J.L.; Gaibor, A.; Perez EPViera, F.; Spencer, C.E. The Portovelo-Zaruma mining camp, SW Ecuador: Porphyry and epithermal environments. *SEG Newsl.* **2002**, *49*, 8–14. Available online: https://www.researchgate.net/publication/285751271_The_Portovelo-Zaruma_mining_camp_SW_Ecuador_porphyry_and_epithermal_environments (accessed on 5 May 2020).

7.  Lopez Bravo, M.; Santos Luna, J.; Quezada Abad, C.; Segura Osorio, M.; Perez Rodriguez, J. Actividad Minera y su Impacto en la Salud Humana. Revista Ciencia UNEMI Volumen 9—Número 17, Enero—Abril 2016, 92–100. Available online: https://dialnet.unirioja.es/descarga/articulo/5556797.pdf (accessed on 15 April 2020).

8.  Fárová, K.; Jelének, J.; Kopačková-Strnadová, V.; Kycl, P. Comparing DInSAR and PSI Techniques Employed to Sentinel-1 Data to Monitor Highway Stability: A Case Study of a Massive Dobkovičky Landslide, Czech Republic. *Remote Sens.* **2019**, *11*, 2670. Available online: https://www.mdpi.com/2072-4292/11/22/2670 (accessed on 8 May 2020). [CrossRef]

9.  Crosetto, M.; Crippa, B.; Biescas, E.; Monserrat, O.; Agudo, M. State-of-the-Art of Land Deformation Monitoring Using Differential SAR Interferometry. 2005. Available online: https://www.researchgate.net/publication/228351503_State-of-the-Art_of_Land_Deformation_Monitoring_Using_Differential_SAR_Interferometry (accessed on 8 May 2020).

10. Cando-Jácome, M.; Martínez-Graña, A. Determinación de las rutas de flujo lahar primario y secundario del volcán de Fuego (Guatemala) utilizando parámetros morfométricos. *Remote Sens.* **2019**, *11*, 727. Available online: https://www.mdpi.com/2072-4292/11/6/727 (accessed on 9 May 2020). [CrossRef]

11. Young, N. Applications of Interferometric Synthetic Aperture Radar (InSAR): A Small Research Investigation. 2018. Available online: https://www.researchgate.net/publication/328773243_Applications_of_Interferometric_Synthetic_Aperture_Radar_InSAR_a_small_research_investigation (accessed on 5 May 2020). [CrossRef]

12. Tofani, V.; Raspini, F.; Catani, F.Y.; Casagli, N. Técnica de interferometría de dispersión persistente (PSI) para caracterización y monitoreo de deslizamientos de tierra. *Teledetección* **2013**, *5*, 1045–1065. Available online: https://www.semanticscholar.org/paper/Persistent-Scatterer-Interferometry-(PSI)-Technique-Tofani-Raspini/0f0ed27b2ccaad44e66fdc25c415e91f81b8f1ca (accessed on 5 May 2020).

13. Carlà, T.; Intrieri, E.; Raspini, F.; Bardi, F.; Farina, P.; Ferretti, A.; Colombo, D.; Novali, F.; Casagli, N. Perspectives on the prediction of catastrophic slope failures from satellite InSAR. *Sci. Rep.* **2019**, *9*, 14137. [CrossRef]

14. Zhou, L.; Zhang, D.; Wang, J.; Huang, Z.; Pan, D. Mapping land subsidence related to underground coal fires in the wuda coalfield (north- ern china) using a small stack of alos palsar differential interferograms. *Remote Sens.* **2013**, *5*, 1152–1176. Available online: https://www.semanticscholar.org/paper/Mapping-Land-Subsidence-Related-to-Underground-Coal-Zhou-Zhang/e47448a372084bdb0f1dfd0dd8ad7cf889d41e77 (accessed on 10 April 2020). [CrossRef]

15. Govil, H.; Chatterjee, R.S.; Malik, K.; Diwan, P.; Tripathi, M.K.; Guha, S. Identification and measurement of deformation using sentinel data and psinsar technique in coalmines of korba. *ISPRS Int. Arch. Photogramm. Remote Sens. Spat. Inf. Sci.* **2018**, *XLII-5*, 427–431. Available online: https://www.researchgate.net/publication/329042788_identification_and_measurement_of_deformation_using_sentinel_data_and_psinsar_technique_in_coalmines_of_korba/citation/download (accessed on 14 April 2020).

16. Aobpaet, A.; Caro, C.M.; Hooper, A.; Trisirisatayawong, I. Land Subsidence Evaluation Using Insar Time Series Analysis in Bangkok Metropolitan Area. 2009. Available online: https://www.researchgate.net/publication/228820037_land_subsidence_evaluation_using_insar_time_series_analysis_in_bangkok_metropolitan_area/citation/download (accessed on 10 April 2020).

17. Abdikan, S.; Arıkan, M.; Sanli, F.B.; Cakir, Z. Monitoring of coal mining subsidence in peri-urban area of Zonguldak city (NW Turkey) with persistent scatterer interferometry using ALOS-PALSAR. *Environ. Earth Sci.* **2014**. [CrossRef]

18. Martínez-Graña, A.M.; Goy, J.L.; Zazo, C. Geomorphological applications for susceptibility mapping of landslides in natural parks. *Environ. Eng. Manag. J.* **2016**, *15*, 327–338.

19. Sousa, J.J.; Ruiz, A.M.; Hanssen, R.F.; Perski, Z.; Bastos, L.; Gil, A.J.; Galindo-Zaldívar, J. PS-INSAR measurement of ground subsidence in Granada area (Betic Cordillera, Spain). 2008. Available online: https://www.researchgate.net/publication/228765038_PS-INSAR_measurement_of_ground_subsidence_in_Granada_area_Betic_Cordillera_Spain (accessed on 15 April 2020).

20. Roccheggiani, M.; Piacentini, D.; Tirincanti, E.; Perissin, D.; Menichetti, M. Detección y monitoreo de movimientos de tierra inducidos en túneles usando la interferometría SAR Sentinel-1. *Sens Remote* **2019**, *11*, 639. Available online: https://www.mdpi.com/2072-4292/11/6/639 (accessed on 15 April 2020). [CrossRef]

21. Qin, Y.; Daniele, P. Monitoring underground mining subsidence in South Indiana with C- and L-band InSAR technique. In Proceedings of the 2015 IEEE International Geoscience and Remote Sensing Symposium (IGARSS); 2015; pp. 294–297. Available online: https://www.semanticscholar.org/paper/Monitoring-underground-mining-subsidence-in-South-C-Qin-Perissin/08576df99bb5059e904e808f4502ef6a75e894ca (accessed on 15 April 2020).

22. Colombo, D.; Farina, P.; Moretti, S.; Nico, G.; Prati, C. Land subsidence in the Firenze-Prato-Pistoia basin measured by means of spaceborne SAR interferometry. IGARSS 2003. IEEE International Geoscience and Remote Sensing Symposium. In Proceedings of the Proceedings (IEEE Cat. No.03CH37477), Toulouse, France; 2003; Volume 4, pp. 2927–2929. Available online: http://ieeexplore.ieee.org/stamp/stamp.jsp?tp=&arnumber=1294634&isnumber=28604 (accessed on 15 April 2020). [CrossRef]

23. Thapa, S.; Chatterjee, R.; Singh, K.; Kumar, D. Land subsidence monitoring using ps-InSAR technique for l-band sar data. *ISPRS Int. Arch. Photogramm. Remote Sens. Spat. Inf. Sci.* 2016, XLI-B7, 995–997. Available online: https://www.researchgate.net/publication/309341322_land_subsidence_monitoring_using_ps-insar_technique_for_l-band_sar_data (accessed on 15 April 2020).

24. Zhang, A.; Lu, J.; Kim, J.-W. Detectan la deformación del suelo inducida por la minería y los peligros asociados utilizando técnicas InSAR espaciales, Geomática. *Peligros Riesgos Nat.* **2018**, *9*, 211–223.

25. Muñoz, E.; Flor, C.; Caizaluisa, A.M.R.; Carlos, E. Escenarios Climáticos en presencia del Fenómeno El Niño (FEN) en las Micro-cuencas Cristal, Potosí, Pechiche y Balsas del Ecuador. *Rev. Climatol.* **2015**, *15*, 7–25. Available online: https://www.researchgate.net/publication/279303370_Escenarios_Climaticos_en_presencia_del_Fenomeno_El_Nino_FEN_en_las_Micro-cuencas_Cristal_Potosi_Pechiche_y_Balsas_del_Ecuador (accessed on 1 May 2020).

26. Torres, T.M.; Platzeck, G. Aplicación de interferometría diferencial de radar de apertura sintética (DInSAR) como una herramienta para detectar deslizamientos en una región de los Andes en Ecuador. Pyroclastic Flow. *J. Geol.* **2014**, *4*, 22–52. Available online: http://pyflow.net/joomla30/index.php/archivo/9-all-issues/26-ms042014 (accessed on 10 May 2020).

27. Cando-Jácome, M.; Martínez-Graña, A. Differential interferometry, structural lineaments and terrain deformation analysis applied in Zero Zone 2016 Earthquake (Manta, Ecuador). *Environ. Earth Sci.* **2019**, *78*, 499. Available online: https://link.springer.com/article/10.1007%2Fs12665-019-8517-4 (accessed on 20 April 2020). [CrossRef]

28. Interferometric Synthetic Aperture Radar. An Introduction for Users of InSAR Data." 2010. Available online: https://www.semanticscholar.org/paper/INTERFEROMETRIC-SYNTHETIC-APERTURE-RADAR-An-for-of/0a4e021fb3f247aee3c32c5a89dc4c61774029f2 (accessed on 10 May 2020).

29. Gens, R.; Logan, T. Alaska Satellite Facility software tools Manual. Published by Geophysical Institute, 2003. University of Alaska Fairbanks P.O. Box 7320 Fairbanks, AK-99775 USA. 2003. Available online: https://media.asf.alaska.edu/uploads/Get%20Started/asf_software_tools.pdf (accessed on 10 May 2020).

30. Sillerico, E.; Marchamalo, M.; Rejas, J.G.; Martínez, R. La Técnica DInSAR: Bases y Aplicación a la Medición de Subsidencias del Terreno en la Construcción. *Inf. Constr.* **2010**, *62*. Available online: https://www.researchgate.net/publication/46179345_La_tecnica_DInSAR_bases_y_aplicacion_a_la_medicion_de_subsidencias_del_terreno_en_la_construccion (accessed on 22 April 2020). [CrossRef]

31. Zebker, H.A.; Rosen, P.A.; Goldstein, R.M.; Gabriel, A.; Werner, C.L. On the derivation of coseismic displacement fields using differential radar interferometry: The Landers Earthquake. *J. Geophys. Res.* **1994**, *99*, 19617–19634. Available online: https://ieeexplore.ieee.org/document/399105 (accessed on 20 April 2020). [CrossRef]

32. Perissin, D. SARPROZ Software Manual, 569p. 2009. Available online: http://ihome.cuhk.edu.hk/~{}b122066/manual/index.html (accessed on 26 April 2020).

33. Wegmuller, U.; Werner, C.; Strozzi, T.; Wiesmann, A. Monitoring Mining Induced Surface Deformation. In Proceedings of the Geoscience and Remote Sensing Sym- Posium, 2004. IGARSS'04. Proceedings. IEEE International. IEEE. 2004, Volume 3, pp. 1933–1935. Available online: https://www.researchgate.net/publication/282017903_Monitoring_Underground_Mining_Subsidence_In_South_Indiana_With_C-_And_L-Band_Insar_Technique (accessed on 5 May 2020).

34. Milan, L. Insar Used for Subsidence Monitoring of mining Area Okr, Czech Republic. Advances in the Science and Applications of SAR Interferometry, Fringe. 2009. Available online: https://www.semanticscholar.org/paper/insar-used-for-subsidence-monitoring-of-mining-area-lazecky/b18a7fa05756a494491d20694abb53637a1466c3 (accessed on 10 May 2020).
35. IGEPN. Instituto Geofísico Escuela Politécnica Nacional. *Quito*. Available online: http://www.igepn.edu.ec/solicitud-de-datos (accessed on 10 March 2020).
36. Chunga, K.; Ochoa-Cornejo, F.; Mulas, M.; Toulkeridis, T.; Menéndez, E. Characterization 452 of seismogenic crustal faults in the gulf of Guayaquil, Ecuador. *Andean Geol.* **2019**, *46*, 66–81. [CrossRef]
37. Campoverde, C. Subsidence Analysis for Mining Activity in the Zaruma Urban Helmet. Case of Study la Inmaculada School. Public Thesis Prior to Obtaining the Title of Mining Enginer. Faculty of Engineering in Earth Sciences Faculty. Littoral Polytech High School. 2017. Available online: http://www.fict.espol.edu.ec/sites/fict.espol.edu.ec/files/CAMPOVERDE.pdf (accessed on 10 May 2020).

 *remote sensing*

Article

# Towards a PS-InSAR Based Prediction Model for Building Collapse: Spatiotemporal Patterns of Vertical Surface Motion in Collapsed Building Areas—Case Study of Alexandria, Egypt

**Bahaa Mohamadi [1], Timo Balz [1,*] and Ali Younes [2]**

[1]   State Key Laboratory of Information Engineering in Surveying, Mapping, and Remote Sensing, Wuhan University, Wuhan 430076, China; bh.mo@whu.edu.cn

[2]   Geography and GIS Department, Faculty of Arts, Kafrelsheikh University, Kafrelsheikh 33516, Egypt; ali_ali2011@art.kfs.edu.eg

*   Correspondence: balz@whu.edu.cn; Tel.: +86-27-6877-9986

Received: 27 August 2020; Accepted: 3 October 2020; Published: 12 October 2020

**Abstract:** Buildings are vulnerable to collapse incidents. We adopt a workflow to detect unusual vertical surface motions before building collapses based on PS-InSAR time series analysis and spatiotemporal data mining techniques. Sentinel-1 ascending and descending data are integrated to decompose vertical deformation in the city of Alexandria, Egypt. Collapsed building data were collected from official sources, and overlayed on PS-InSAR vertical deformation results. Time series deformation residuals are used to create a space–time cube in the ArcGIS software environment and analyzed by emerging hot spot analysis to extract spatiotemporal patterns for vertical deformation around collapsed buildings. Our results show two spatiotemporal patterns of new cold spot or new hot spot before the incidents in 66 out of 68 collapsed buildings between May 2015 and December 2018. The method was validated in detail on four collapsed buildings between January and May 2019, proving the applicability of this workflow to create a temporal vulnerability map for building collapse monitoring. This study is a step forward to create a PS-InSAR based model for building collapse prediction in the city.

**Keywords:** building collapse; land subsidence; permanent scatterers interferometry (PSI); Sentinel-1; spatiotemporal data mining; surface deformation

---

## 1. Introduction

Urban areas are often expanding into vulnerable areas and buildings become under risk of collapse due to many reasons in different parts of the world. Therefore, it is essential to monitor surface deformation in cities for abnormal motion. This helps in evacuation before the collapse, to save lives and important properties. Traditionally, building movements are observed by installing equipment such as settlement extensometers into the building structures. Recently, Global Navigation Satellite Systems (GNSS) are another choice to measure building deformation. But, both measurements are limited to a few selected buildings and relatively expensive [1,2].

Now, the wide coverage of remote sensing gives scientists the capability to monitor changes in urban areas rapidly and effectively. Interferometric Synthetic Aperture Radar (SAR) techniques were successfully utilized to detect surface deformation in a millimeter-scale with wide coverage [3], with the capability to create a time series for deformation using permanent scatterers interferometric SAR (PS-InSAR) technique [4]. PS-InSAR is efficient in monitoring slow motion related geohazards. It succeeded in estimating surface motion in regional scales up to country-wide dimensions [5]; in addition, locally the technique can reach down to building scale [6]. Results of PS-InSAR were validated for building deformation monitoring and found to have similar millimeter-level accuracy as traditional leveling technique [7].

In this study, we use PS-InSAR time series results within known collapsed buildings in the city of Alexandria, Egypt, to detect spatiotemporal patterns of vertical deformation right before collapse incidents. Our hypothesis is that the land surface surrounding a collapsed building shows a spatiotemporal pattern of deformation before the collapse. Such patterns would allow for a collapse prediction and therefore a warning system. The patterns detected in this study are tested for the possibility of using as one of the inputs for a building collapse prediction model for collapsed buildings in known vertical deformation regions.

Early on, Ferretti et al. [8] used PS-InSAR to study surface motion within a building before its collapse incident. This was followed by studies interested in building deformation monitoring using InSAR techniques in many countries, including Italy [2,6], France [8,9], Poland and Russia [10], China [7,11,12], and South Korea [1]. However, very few have been applied to already collapsed buildings as the earliest paper, to test the connection of collapse incidents to surface instability [10]. Most of those studies focused on a limited number of buildings and very small areas of interest, except for [1,2,6].

The high resolution and relative short revisit time of X-band, for example, four days for COSMO-SkyMed (CSK), makes it very useful to use in building motion monitoring [1,7,9,12]. C-band has also taken its share in this application [2,6,8,10,11], especially, with the availability of Sentinel-1 data offering global coverage. The literature reveals the higher applicability of SAR X-band data in building damage monitoring studies because of its higher resolution and shorter wavelength, which allows detection of more PS points per building [9]. In this study, the data availability was the reason to use C-band SAR data, as Sentinel-1 open-data coverage over the city of Alexandria is relatively good.

Previous studies depended only on InSAR measurements to evaluate building stability without any additional spatial analysis or modeling approach. Only Ezquerro et al. [2] took the PS-InSAR results to a second level by applying statistical analysis of fragility curve to map damage probability assessment for a part of Pistoia, Italy. Here, we took the analysis to a higher level by applying spatiotemporal analysis for time series measurements surrounding collapsed buildings before the incident. Spatiotemporal analysis discovers interesting and previously unknown patterns from large spatial and spatiotemporal databases, which are potentially useful to understand phenomena [13]. Spatiotemporal pattern analysis was successfully applied in different applications, including climate science, neuroscience, environmental science, precision agriculture, epidemiology and health care, social media, traffic dynamics, heliophysics, criminology, and location-based services [14–16]. To the best of our knowledge, however, it was not used in surface deformation analysis, especially, with PS-InSAR in any previous research. This might be due to the applicability of the PS-InSAR time series results in most applied research.

In a PS-InSAR building collapse application, time series results are sufficient when vertical and horizontal LOS (line-of-sight) deformations are integrated into a building collapse causality model, similar to landslide areas where cumulative displacement is the only significant reason for building cracks and collapse incident [17]. However, when horizontal deformation does not exist, or is very small in an area of interest like Alexandria [18], vertical displacement in a large spatial distribution cannot be taken as a sole reason for the collapse, because the whole building is subjected to subsidence at the same rate. Therefore, the detection of specific spatiotemporal patterns surrounding collapsed buildings before the incident in the city might be helpful in future building collapse prediction.

Alexandria, Egypt, was selected as a case study due to the recently increasing number of building collapse incidents in the city. Alexandria is constructed on a carbonate coastal ridge of Pleistocene age covered by a thin layer of Holocene sediments at the western margin of the Nile delta. Historically, it was susceptible to high rates of subsidence due to short-term catastrophic events of powerful earthquakes and tsunamis, the continuous subsidence of the Quaternary water-saturated substrate, destructive winter wave surges, and anthropogenic influences, such as the load of city constructions [19].

Official records attribute collapse incidents in the city to violations of building specifications, illegal extensions, lax oversight, aging, poor maintenance, and very few incidents to land subsidence [20]. However, this study does not focus on the reason behind the collapse incident, because whatever the reason for the collapse, the effect of all causes are mostly the same [12].

## 2. Materials and Methods

We tested vertical deformation surrounding collapsed buildings of Alexandria, Egypt, between May 2015 and December 2018 to extract common spatiotemporal patterns before the incident. Figure 1 shows the workflow of this study. The workflow consists of three parts: spatiotemporal analysis process, correlation tests, and validation process. The main process in the workflow shown in the middle, correlation, and validation of the analysis of the spatiotemporal patterns are on the left and right of the main process, respectively. Data collection, PS-InSAR processing, spatiotemporal data preparation, and pattern extraction are presented in Sections 2.1–2.3; Sections 2.4 and 2.5 discuss correlation test and validation processing.

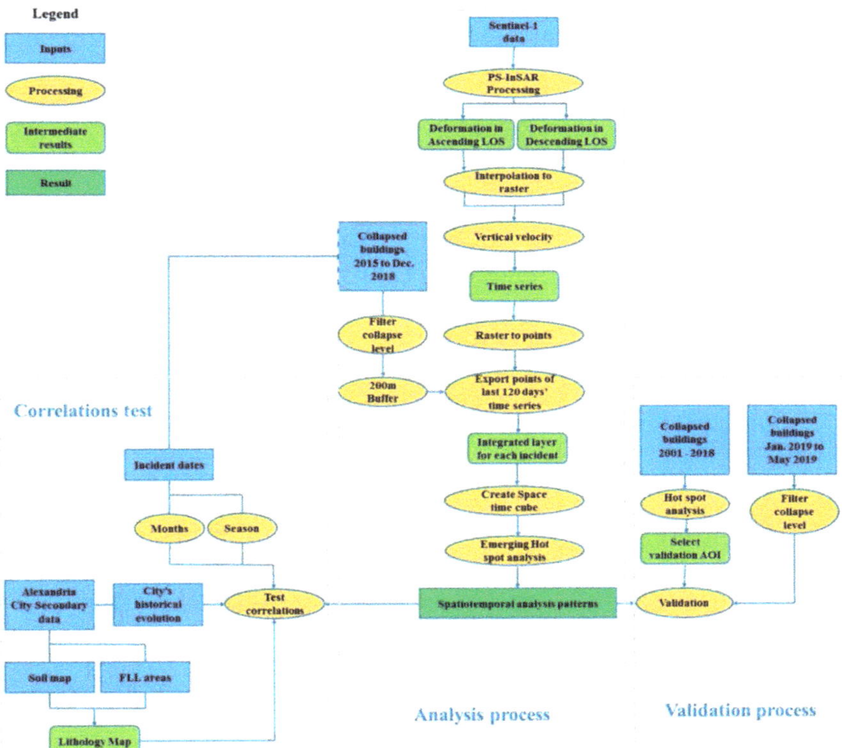


**Figure 1.** Study workflow. LOS refers to the line-of-sight, FLL is the former lakes and lagoons in the study area, and AOI means the area of interest.

### 2.1. Collected Data

The local governorate of Alexandria started to record building collapse data in 2001, as a result of the increasing number of collapsed buildings over time. We collected all collapsed buildings recorded between the January 2001 and May 2019 from local district governmental offices. A total of 255 buildings collapsed during this period, as shown in Figure 2. Recorded data have only a descriptive location based on the street names and building numbers without available x and y coordinates provided in the governmental records. Hence, we used a hand-held Garmin GPSMAP 78 with three meters precision to collect the location of each collapsed building based on their recorded address. We have to address some difficulties that may affect the precision of our GPS points of collapsed buildings, such as narrow streets with high buildings, inaccessible areas for safety, and security reasons that forced us to sometimes read locations a short distance from the incident place.

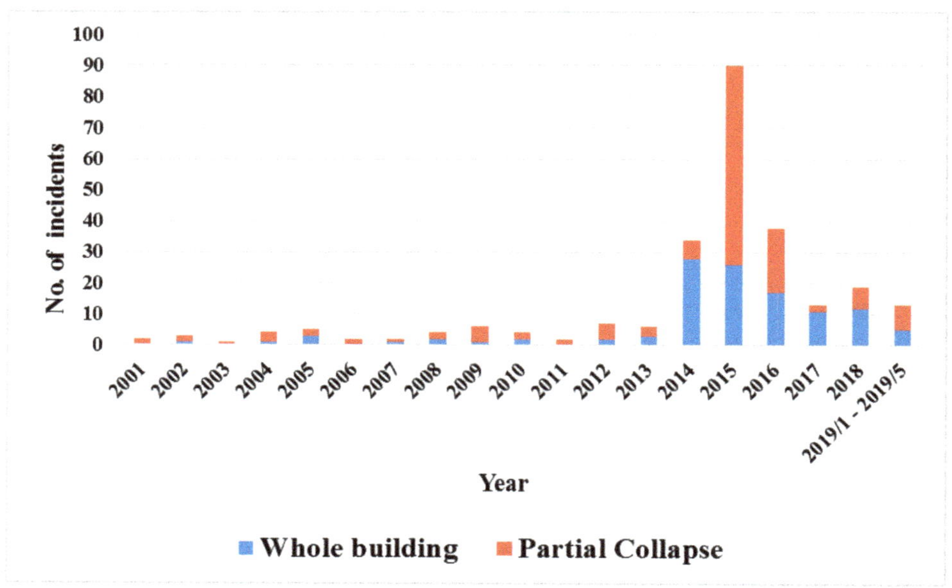


**Figure 2.** Annual distribution of collapsed buildings in Alexandria between January 2001 and May 2019.

A total of 115 buildings completely collapsed, while 140 buildings partially collapsed, varying from a whole side collapse to a wall or balcony collapse. The dramatic increase in collapse incidents conditions, starting from 2014, was interpreted from the recorded data to weather conditions, old buildings' rooftop additions, and several fire and explosion incidents. We excluded ten collapsed buildings from our study processing, because they collapsed due to fire, explosion, and car accidents. Then, incidents were classified based on the collapse degree. Whole building collapse was classified as the highest collapse level of 4. Buildings that have the main vertical collapsed part of it as a whole side of a building were given the second severe level of 3. Level 2 was given for a building with several collapsed parts or inner collapse, such as the stairs of the building. The lowest level of building collapse of 1 was for small collapse parts of the building that are within a room ceiling, balcony, or a wall. Buildings with collapse level 1 were excluded from our analysis, and the remaining collapsed buildings for the analysis were 68, in the period between May 2015 and December 2018.

Sentinel-1 is a C-band two-satellite constellation radar sensor; Sentinel-1A and 1B were launched in 2014 and 2016, respectively. Sentinel-1 data are provided as open-access data with global coverage. It has a geometric resolution of approximately 5 m in range and 20 m in azimuth [21]. In comparison to high-resolution X-band sensor data, the longer wavelength and the moderate geometric resolution result in fewer PS points on each building. However, the density of PS points we found is enough for our study to reach a conclusion on the spatiotemporal patterns related to building collapse incidents in the city. We used the same dataset of Sentinel-1 (S1) Single Look Complex (SLC) of our previous study [18] for the PS-InSAR processing in Alexandria. This dataset consists of 124 ascending images and 129 descending images with an average temporal baseline of 12 days, acquired in the period between January 2015 and May 2019. Additionally, secondary data of a soil map, areas of former lakes and lagoon (FLL), and the historical evolution of the city extent were collected for correlation test purposes.

### 2.2. PS-InSAR Processing and Vertical Velocity Decomposition

With a deformation estimation of millimeter precision, PS-InSAR allows for detecting miniature surface deformation surrounding collapsed buildings before the incident. The PS-InSAR technique relies on the identification of objects smaller than the resolution cell that remained coherent and highly scattered in all the stack's images. The processing of PS-InSAR overcomes temporal decorrelation and phase unwrapping problems by forming a network to separate signals of the atmospheric phase screen (APS), topographic errors, and deformation by spatiotemporal filtering [4].

In this study, we used the SAR PROcessing tool by periZ (SARPROZ) [22]. The process started by selecting an appropriate master image. In this step, we avoid any extreme weather conditions and select a master image with the smallest possible normal baseline. After extracting slave images based on the master one, the coregistration process started. We used a standard PS-InSAR approach, with a threshold of 0.2 for the amplitude stability index, but with a flowered tree graph connection for APS estimation. The estimation of the linear trend was limited between ±160 mm and the estimation of the residual topographic error between ±220 m. We selected the reference point in a stable area based on our background knowledge of the study area, and based on estimated parameter graphs during APS extraction. We also tried to select close reference points for LOS deformation estimation for an effective vertical velocity decomposition process. Finally, to define stable PS points in our study, we selected those points with a temporal coherence of 0.7 or higher.

We used the residual LOS velocity of each image in ascending and descending orbits to decompose a vertical deformation time series. Decomposing vertical velocity by integrating two cross-heading orbits avoids the limitation of the LOS measurements and gives a chance to distinguish vertical and horizontal motion effectively [23]. Each image of an orbit was integrated into its closest image in the other orbit. The time difference was four days in the case of the same S1 sensor (S-1A or S-1B) and two days when using both sensors. The LOS residual velocity of image points were used to interpolate the residual LOS velocity in the city area on that day based on a grid with a pixel size of five meters, which produced a detailed surface of deformation for the city, then integrated to the closest interpolated LOS residual velocity of the other orbit to decompose vertical velocity by using the matrix presented in Equations (1) and (2) [24].

$$\begin{bmatrix} d_{asc} \\ d_{desc} \end{bmatrix} = A \begin{bmatrix} d_{vert} \\ d_{horiz} \end{bmatrix}, \tag{1}$$

where

$$A = \begin{bmatrix} \cos\theta asc. & \sin\theta asc / \cos\theta\Delta\alpha \\ \cos\theta desc. & \sin\theta desc \end{bmatrix}, \tag{2}$$

$d$ is the deformation along LOS for ascending $d_{asc}$ and descending $d_{desc}$, $d_{vert}$ is the vertical velocity, $d_{horiz}$ is the projection of horizontal deformation in descending azimuth look direction, $\theta$ is the incident angle, and $\Delta\alpha$ is the satellite heading difference between ascending and descending mode.

### 2.3. Spatiotemporal Analysis of Collapsed Buildings

With the increasing number of sensors and geospatial data availability worldwide, spatiotemporal data mining becomes an important method to extract new interesting and useful patterns from geographical data [15,16]. However, extracting meaningful spatiotemporal patterns is more difficult than extracting regular patterns from traditional numeric and categorical data. The complexity of spatiotemporal data types and relationships results in challenges in data storage, management, analysis, and knowledge discovery [13,16]. Data storage in a space–time cube is one of the most effective ways to manage and analyze spatiotemporal patterns. Points are aggregated into so-called bins, based on their location and time, where each bin has a specific location. All bins within the same time create a time step (or slice), and time data of each bin creates a bin time series. Each bin has input points based on its location within the bin size and the time step interval. The oldest time step is at the bottom of the cube while the upper time step represents the most recent time data [25].

In this study, we converted the vertical velocity maps of Alexandria for the last 120 days before the collapse incident to points with their corresponding dates. A buffer of 200 meters from the collapsed building was created for processed points to reduce the processing time of each incident. A netCDF data structure cube was then created with a bin size of ten meters and a time step interval of twelve days. The bin size was selected as twice the size of the previously processed vertical velocity maps for Alexandria (5 m). Furthermore, the time step interval value was selected based on the average of Sentinel-1 data availability average over the city during the study period. In case of no data availability for a specific time step, we decided to spatially and temporally interpolate bin values of the slice based on the mean of space–time neighbors.

A Mann–Kendall trend test [26,27] was calculated during the space–time cube creation to investigate the time series trend of each bin. It simply compares the bin value to the previous one. If it is smaller, its result is −1, if it is larger, its result is +1, and if the two bins are equal, then the result is 0. The results of all pairs in the space–time cube are then summed for each bin. If the sum is zero, then it means no trend over time was detected. The other bins are compared to the no-trend ones based on the variance for the values in the bin time series, the number of ties, and the number of time steps to measure its statistical significance based on z-score and $p$-value with confidence levels. The $p$-value of a bin indicates whether it is statistically significant or not, while the z-score shows if the trend is positive or negative based on the increase or decrease of bin values, respectively [28,29].

We used the netCDF space–time cube to extract spatiotemporal patterns of vertical velocity surrounding collapsed buildings by using emerging hot spot analysis. It is a regular hot spot analysis with the addition of time as a third dimension [28]. Emerging hot spot analysis [30,31] calculates the Getis–Ord $Gi^*$ statistic based on the neighborhood in space (bins in the same time step), and time (bins in different time steps) [32]. A simple form of the $Gi^*$ statistic as defined by Getis and Ord [31]:

$$G_i^* \frac{\sum_{j=1}^{n} w_{ij} x_j}{\sum_{j=1}^{n} x_j} \tag{3}$$

where $Gi^*$ is a statistic that describes the spatial dependency of the incident $i$ over all $n$ events, $x_j$ is the magnitude of variable $x$ at incident location $j$ over all $n$ ($j$ may equal $i$), and $w_{ij}$ is a weight value between event $i$ and $j$ that represents their spatial interrelationship.

In this study, we kept the default value of neighborhood distance, which was about 30 meters, and selected one neighborhood time step for the emerging hot spot analysis. The emerging hot spot analysis adds a hot spot classification to all bins in the cube. Trends of hot and cold spots over time are estimated using the Mann–Kendall trend test. The spatiotemporal patterns are then evaluated based on z-score and $p$-value that previously estimated during the space–time cube processing, and the hot spot z-score and $p$-value that processed in the emerging hot spot analysis. Patterns with their statistical significance interpretation are presented in Table 1.

Spatiotemporal patterns of vertical motion within a distance of 200 m from collapsed buildings were defined. We focused on similar patterns within 50 m of collapsed buildings for our final results. The knowledge of similar spatiotemporal patterns of vertical motion in collapsed buildings areas of 2015 and 2016 was then used to define temporal building collapse vulnerability ranks between January 2017 and May 2019. However, due to the large area of the city, we decided to decrease the vulnerability mapping area of interest based on historical building collapse incidents in the city between 2001 and 2016.

**Table 1.** Patterns resulted from emerging hot spot analysis [31].

| Pattern | Statistical Description |
| --- | --- |
| New Hot Spot | A location that is a statistically significant hot spot for the final time step and has never been a statistically significant hot spot before. |
| Consecutive Hot Spot | A location with a single uninterrupted run of statistically significant hot spot bins in the final time-step intervals. The location has never been a statistically significant hot spot prior to the final hot spot run and less than ninety percent of all bins are statistically significant hot spots. |
| Intensifying Hot Spot | A location that has been a statistically significant hot spot for ninety percent of the time-step intervals, including the final time step. In addition, the intensity of clustering of high counts in each time step is increasing overall and that increase is statistically significant. |
| Persistent Hot Spot | A location that has been a statistically significant hot spot for ninety percent of the time-step intervals with no discernible trend indicating an increase or decrease in the intensity of clustering over time. |
| Diminishing Hot Spot | A location that has been a statistically significant hot spot for ninety percent of the time-step intervals, including the final time step. In addition, the intensity of clustering in each time step is decreasing overall and that decrease is statistically significant. |
| Sporadic Hot Spot | A location that is an on-again then off-again hot spot. Less than ninety percent of the time-step intervals have been statistically significant hot spots and none of the time-step intervals have been statistically significant cold spots. |
| Oscillating Hot Spot | A statistically significant hot spot for the final time-step interval that has a history of also being a statistically significant cold spot during a prior time step. Less than ninety percent of the time-step intervals have been statistically significant hot spots. |
| Historical Hot Spot | The most recent time period is not hot, but at least ninety percent of the time-step intervals have been statistically significant hot spots. |
| No Pattern Detected | Has no statistical significance during the study period |
| New Cold Spot | A location that is a statistically significant cold spot for the final time step and has never been a statistically significant cold spot before. |
| Consecutive Cold Spot | A location with a single uninterrupted run of statistically significant cold spot bins in the final time-step intervals. The location has never been a statistically significant cold spot prior to the final cold spot run and less than ninety percent of all bins are statistically significant cold spots. |
| Intensifying Cold Spot | A location that has been a statistically significant cold spot for ninety percent of the time-step intervals, including the final time step. In addition, the intensity of clustering of low counts in each time step is increasing overall and that increase is statistically significant. |
| Persistent Cold Spot | A location that has been a statistically significant cold spot for ninety percent of the time-step intervals with no discernible trend, indicating an increase or decrease in the intensity of clustering of counts over time. |
| Diminishing Cold Spot | A location that has been a statistically significant cold spot for ninety percent of the time-step intervals, including the final time step. In addition, the intensity of clustering of low counts in each time step is decreasing overall and that decrease is statistically significant. |
| Sporadic Cold Spot | A location that is an on-again then off-again cold spot. Less than ninety percent of the time-step intervals have been statistically significant cold spots and none of the time-step intervals have been statistically significant hot spots. |
| Oscillating Cold Spot | A statistically significant cold spot for the final time-step interval that has a history of also being a statistically significant hot spot during a prior time step. Less than ninety percent of the time-step intervals have been statistically significant cold spots. |
| Historical Cold Spot | The most recent time period is not cold, but at least ninety percent of the time-step intervals have been statistically significant cold spots. |

### 2.4. Correlation Tests

Secondary data sources were used to test our results' correlation to some spatial features in the city: the lithology map and the historical extent of the city since its establishment. The lithology map was produced by integrating the city soil map and the formal lakes and lagoon map. For more information on secondary data sources, please refer to Mohamadi et al. [18]. Figure 3 presents the integration process of soil map and FLL map to present the lithology map of Alexandria City.

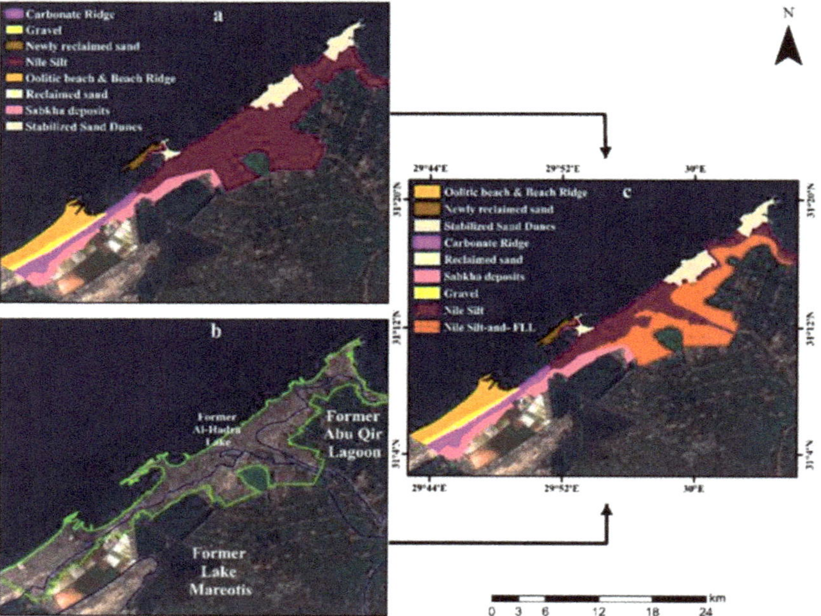


**Figure 3.** Lithology map integration for Alexandria City: (**a**) soil map, (**b**) former lakes and lagoon distribution, and (**c**) lithology map.

The soil map, FLL map, and resulting lithology map of Alexandria are illustrated in Figure 3a–c. The main change in the lithology map from the soil map can be found in the Nile Silt soil. It was divided into two parts: Nile silt soil with and without former lakes and lagoon. Additionally, we classified building collapse incidents based on dates to months and seasons, and test the correlation between these two time-data types and our spatiotemporal pattern results.

### 2.5. Validation

A one square kilometer grid was created for the city, used for the collected building collapse data between 2001 and 2018 to count incidents in each square kilometer, and produced a hot spot map of building collapse in Alexandria during this period. We then selected a four square kilometer area-of-interest (AOI) based on the hot spot analysis for the validation process. Distribution of collapsed buildings and hot spot analysis results are shown in Figure 4a,b.

We depended on collapsed buildings in the period between January and May 2019 for the study validation. In addition to the previous step of selecting an AOI for validation, we also filtered incidents within the AOI based on their collapse level. Only level 4 of collapsed buildings that refer to a whole collapsed building was selected for the validation. This filtering was used to limit the number of incidents for better and more detailed discussion of the study results. After filtering, the four remaining collapsed buildings were used for validation of our study results.

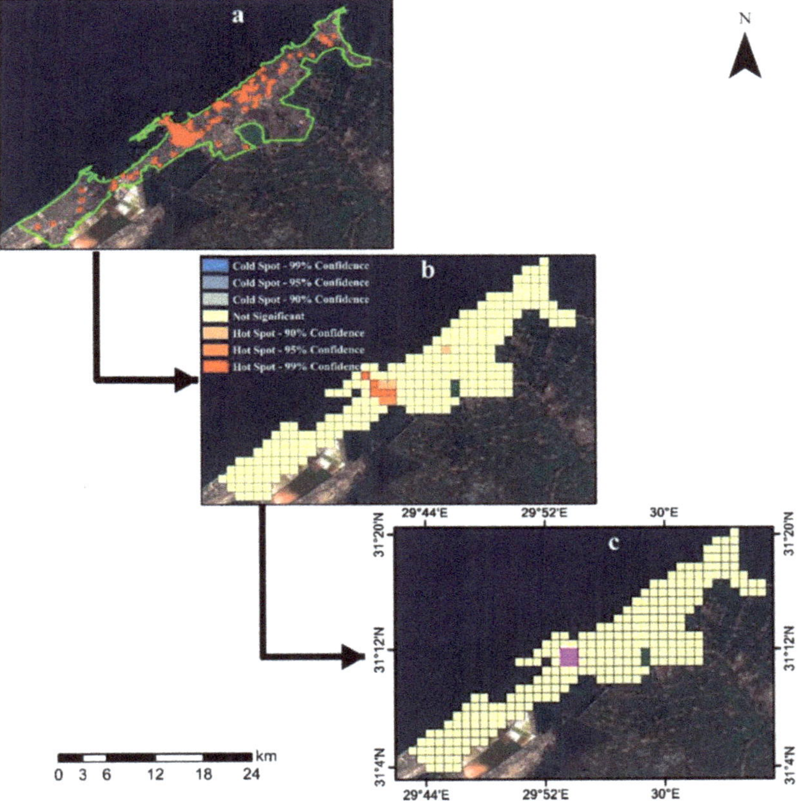


**Figure 4.** Selection procedure validating the area of interest (AOI): (**a**) distribution of historical building collapse incidents from January 2001 to December 2018, (**b**) the hot spot analysis for collapsed building incidents between 2001 and 2018, and (**c**) validation AOI.

## 3. Results

Our working hypothesis is that spatiotemporal patterns close to collapsed areas show sudden subsidence that led the building to collapse. By applying this hypothesis to our results, the expected pattern to find a collapsed building is the "new cold spot" pattern (NCS) in the proximity of collapsed buildings. We created multibuffers of 30, 60, and 90 m around GPS points of the collapsed building, and searched for this pattern from an in-to-out direction. In general, results revealed this pattern in 32 buildings out of the 68 buildings in this study, while interpretation resulted in the existence of a "new hot spot" pattern (NHS), surrounding 34 other buildings in the study. None of these two patterns existed in only the two collapsed buildings of the study.

### 3.1. New Cold Spot Pattern

Table 2 presents the results of all 32 collapsed buildings with the temporal pattern of the new cold spot. Among the 68 studied collapsed buildings, about 47.1% of them have the NCS spatiotemporal pattern within the surrounding area. Fifteen of them show previous statistically hot spot bins while five have no significant hot or cold spot bins during the ten time steps tested. Other NCS temporal patterns were also detected during the 3D temporal interpretation; some collapsed buildings had NCS in the ninth time step, as seen in collapsed buildings with serial numbers between 21 and 25 in the table. Some bins continued in a significant cold spot in the last bin and some others showed a nonsignificant last bin. Two other temporal patterns were found during the interpretation: first was an NCS within the last five or six bins only, and second, an earlier NCS found in the seventh time step and continued as a significant cold spot until the tenth time step.

**Table 2.** Results of collapsed buildings with a new cold spot pattern.

| Serial No. | Collapse Date | Building ID | Collapse Level | Distance of NCS | 1 | 2 | 3 | 4 | 5 | 6 | 7 | 8 | 9 | 10 |
|---|---|---|---|---|---|---|---|---|---|---|---|---|---|---|
| 1 | 2015/6/13 | | 4 | Within 30 m | – | – | H | H | H | H | H | H | H | C |
| 2 | 2016/2/19 | | 4 | Within 30 m | – | – | H | – | – | – | – | – | – | C |
| 3 | 2016/9/22 | | 4 | Within 30 m | – | – | – | – | – | – | H | – | – | C |
| 4 | 2017/1/17 | | 4 | Within 30 m | H | H | H | H | – | – | – | – | – | C |
| 5 | 2018/2/15 | | 4 | Within 30 m | H | H | H | H | – | – | – | H | – | C |
| 6 | 2015/11/5 | 71 | 3 | Within 30 m | H | H | H | H | H | – | – | – | – | C |
| 7 | 2015/11/6 | 54 | 3 | Within 30 m | H | H | H | H | H | – | – | – | – | C |
| 8 | 2015/12/26 | | 3 | Within 30 m | H | H | H | H | H | H | H | – | – | C |
| 9 | 2015/12/27 | | 3 | Within 30 m | H | H | H | H | H | H | H | H | – | C |
| 10 | 2015/12/13 | | 2 | 40 m | H | H | H | H | H | – | H | H | – | C |
| 11 | 2017/2/22 | | 4 | 50 m | H | H | H | H | H | – | – | – | – | C |
| 12 | 2015/5/19 | | 2 | 60 m | – | H | H | – | – | – | – | – | – | C |
| 13 | 2016/1/11 | | 2 | 70 m | – | – | – | H | H | H | – | – | – | C |
| 14 | 2015/11/7 | 114 | 2 | 80 m | H | H | H | H | – | – | H | H | H | C |
| 15 | 2015/11/15 | | 3 | 90 m | – | – | – | – | H | H | H | H | H | C |
| 16 | 2016/6/8 | 192 | 4 | Within 30 m | – | – | – | – | – | – | – | – | – | C |
| 17 | 2016/6/8 | 193 | 4 | Within 30 m | – | – | – | – | – | – | – | – | – | C |
| 18 | 2016/5/21 | | 3 | Within 30 m | – | – | – | – | – | – | – | – | – | C |
| 19 | 2018/8/8 | | 4 | 50 m | – | – | – | – | – | – | – | – | – | C |
| 20 | 2016/9/3 | | 2 | 70 m | – | – | – | – | – | – | – | – | – | C |
| 21 | 2018/11/11 | | 2 | Within 30 m | – | – | – | H | H | – | – | – | C | C |
| 22 | 2015/11/7 | 52 | 2 | 60 m | – | – | H | H | H | H | – | – | C | C |
| 23 | 2017/3/5 | | 4 | 40 m | – | – | H | H | – | – | – | – | C | – |
| 24 | 2018/7/4 | | 4 | Within 30 m | – | – | – | – | – | – | – | – | C | – |
| 25 | 2018/4/19 | | 4 | Within 30 m | – | – | – | – | – | H | – | C | C | – |
| 26 | 2015/12/11 | 241 | 4 | Within 30 m | C | C | C | C | – | – | H | H | – | C |
| 27 | 2018/3/8 | | 4 | 50 m | – | – | – | C | C | H | H | – | – | C |
| 28 | 2018/12/6 | | 4 | Within 30 m | – | – | – | – | C | – | – | – | – | C |
| 29 | 2016/8/11 | | 3 | Within 30 m | – | C | – | – | C | – | – | – | – | C |
| 30 | 2015/7/18 | | 2 | Within 30 m | – | C | C | C | C | – | – | – | – | C |
| 31 | 2015/12/9 | | 4 | Within 30 m | H | H | H | H | – | – | – | C | C | C |
| 32 | 2015/12/11 | 167 | 4 | Within 30 m | – | – | – | – | H | – | – | C | C | C |

Buildings were sorted based on the temporal pattern of the closest NCS to the building in case of multiple NCSs. Red cells with an H letter refer to significant hot spot bins, blue cells with a C letter refer to bins with significant cold spot, and yellow cells with "—" symbol refer to bins that were not significant. Building ID is provided only in the case of more than one building that collapsed on the same day.

In most collapsed buildings, the NCS was found within the closest buffer of 30 m while seven collapsed buildings had the closest NCS within the second buffer of 60 m, and only four collapsed buildings had NCS in the third buffer of 90m. Additionally, the collapse level presents the severity of NCS distance to the collapsed building as nineteen of collapsed buildings in 4 and 3 collapse levels are located within 30 m of the collapsed buildings. Whereas, six of eight buildings with a collapse level 2 are out of the closest buffer of 30 m. Figure 5 presents some samples of collapsed buildings that have NCS spatiotemporal pattern before their collapse.

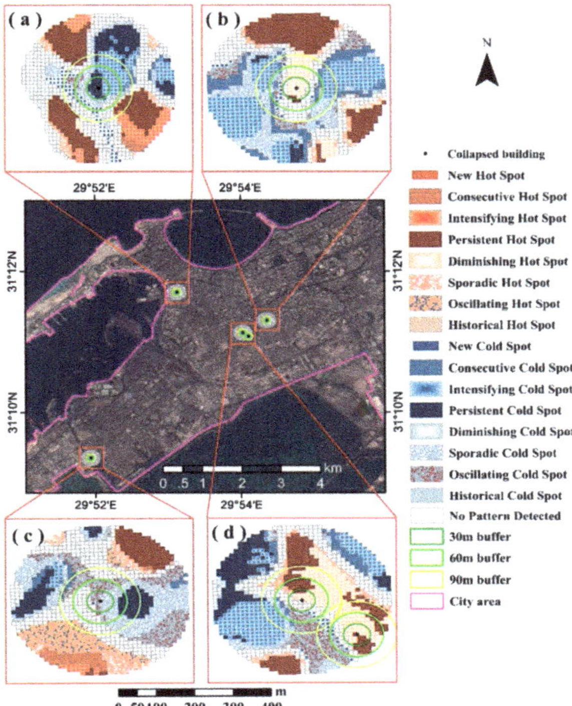



**Figure 5.** Samples of collapsed buildings that have an NCS spatiotemporal pattern before their collapse. The main map shows the location of the five collapsed buildings on a true color Sentinel-2 image acquired on 24 October 2018: (**a**) shows the spatiotemporal analysis for a 200 m buffer of a whole building collapse on 8 August 2018, (**b**) is the result of a partially collapsed building on 6 November 2015, (**c**) illustrates the analysis result of a whole building collapse on 19 February 2016, and (**d**) shows results of two partial collapsed buildings on 26 December 2015 for the middle one, and 27 December 2015 for the southeastern point in the figure.

A whole building collapse on 8 August 2018 is presented in Figure 5a. Spatiotemporal analysis of the area surrounding this building shows many NCS in a linear shape two days before the incident. Figure 5b shows the analysis result for a partial building collapse on 6 November 2015. NCS are distributed on the western side of the GPS point, which matches exactly the geographic description of the collapsed part of the building in the recorded data for this incident. A single NCS that rarely exists in our results was found close to a whole building collapse on 19 February 2016; this is presented in Figure 5c. Figure 5d illustrates the results of the spatiotemporal analysis for vertical deformation surround two partial collapsed buildings; the middle one collapsed on 26 December 2015 and followed by another building collapse during the next day. The geographic description of the collapsed part of the building matches with the location of NCS, as the southern part of the building is the collapsed part. The two collapsed buildings of 6 November 2015 and 26 December 2015 are among very few collapse incidents that have a geographic description for the collapsed part. These two buildings with their geographic description of collapsed parts support the accuracy of our findings for the other buildings in this study. The other building in Figure 5d had a partial collapse in the front part of the building, which, in our opinion, is in the western part of the building, based on the geographical linear distribution of the NCS in the spatiotemporal analysis result.

## 3.2. New Hot Spot Pattern

Results revealed at least one NHS cell close to the collapsed building in 34 incidents before the collapse, as shown in Table 3. Cold spot bins were revealed in the temporal analysis of fourteen incidents before turning to a hot spot in the last bin prior to the collapse; while three incidents had no previous significant hot or cold spots before the final significant hot spot prior to the incident. In some cases, we found a significant hot spot bin for the first time in the ninth bin with a nonsignificant last bin. Additionally, twelve incidents of the 34 NHS had previous hot spot bins during the temporal analysis presented from incident 23 to incident 34, shown in Table 3. Some of these had earlier hot spot bins turn to significant cold spots, and ended with a significant NHS in the last five, six, and seven time steps. Others had mixed significant hot spots and nonsignificant bins, while ending with significant hot spots.

**Table 3.** Results of collapsed buildings with a new hot spot pattern.

| Serial No. | Collapse Date | Building ID | Collapse Level | Distance of NCS | 1 | 2 | 3 | 4 | 5 | 6 | 7 | 8 | 9 | 10 |
|---|---|---|---|---|---|---|---|---|---|---|---|---|---|---|
| 1 | 2015/12/5 | | 4 | Within 30 m | C | C | C | C | C | C | C | C | C | H |
| 2 | 2015/12/16 | | 4 | Within 30 m | – | – | – | – | – | – | C | C | – | H |
| 3 | 2016/8/15 | | 4 | Within 30 m | – | – | – | C | C | C | C | C | – | H |
| 4 | 2018/9/2 | | 4 | Within 30 m | – | – | C | C | – | – | C | C | C | H |
| 5 | 2018/2/3 | | 4 | Within 30 m | – | – | – | – | – | C | – | – | – | H |
| 6 | 2015/11/7 | 10 | 3 | Within 30 m | – | – | – | – | C | C | C | C | – | H |
| 7 | 2015/11/9 | 9 | 3 | Within 30 m | C | C | C | C | C | C | C | C | C | H |
| 8 | 2016/7/14 | | 3 | Within 30 m | C | C | C | C | C | C | C | – | – | H |
| 9 | 2015/11/7 | 50 | 2 | Within 30 m | – | – | – | – | – | C | – | C | C | H |
| 10 | 2015/11/7 | 51 | 2 | Within 30 m | C | – | – | C | C | C | C | C | C | H |
| 11 | 2016/5/1 | | 4 | 40 m | – | – | – | C | – | – | – | – | – | H |
| 12 | 2015/10/13 | 106 | 3 | 40 m | C | C | C | C | C | C | – | – | – | H |
| 13 | 2015/12/11 | 145 | 4 | 60 m | – | – | C | C | C | C | – | – | – | H |
| 14 | 2017/5/15 | | 4 | 60 m | C | C | C | – | – | C | C | C | C | H |
| 15 | 2017/5/3 | | 4 | 40 m | – | – | – | – | – | – | – | – | – | H |
| 16 | 2017/1/22 | | 4 | 60 m | – | – | – | – | – | – | – | – | – | H |
| 17 | 2018/6/26 | | 4 | 80 m | – | – | – | – | – | – | – | – | – | H |
| 18 | 2016/8/24 | | 4 | Within 30 m | – | C | C | – | – | – | – | – | H | – |
| 19 | 2017/5/16 | | 4 | Within 30 m | – | C | C | C | C | C | C | – | H | – |
| 20 | 2016/10/19 | | 3 | Within 30 m | – | – | – | C | C | C | – | – | H | – |
| 21 | 2018/10/15 | | 4 | Within 30 m | – | – | C | – | – | – | – | H | H | – |
| 22 | 2015/11/17 | | 2 | Within 30 m | – | – | – | – | – | – | – | – | H | – |
| 23 | 2018/3/21 | | 4 | Within 30 m | H | H | – | C | – | – | – | – | – | H |
| 24 | 2017/3/1 | | 4 | Within 30 m | H | H | H | H | C | C | C | C | – | H |
| 25 | 2017/3/2 | | 4 | Within 30 m | – | H | H | H | – | C | C | C | – | H |
| 26 | 2016/1/27 | | 4 | Within 30 m | – | H | H | H | H | C | C | – | – | H |
| 27 | 2018/2/18 | | 4 | Within 30 m | H | H | H | – | C | – | – | – | – | H |
| 28 | 2015/11/4 | 73 | 3 | 40 m | – | C | C | C | – | H | – | – | – | H |
| 29 | 2017/7/18 | | 2 | Within 30 m | C | – | C | C | – | C | – | H | H | H |
| 30 | 2016/5/12 | | 4 | Within 30 m | H | H | – | – | – | – | – | – | – | H |
| 31 | 2017/1/2 | | 4 | Within 30 m | H | H | H | – | – | – | – | – | H | H |
| 32 | 2017/5/2 | | 4 | Within 30 m | H | H | – | H | H | – | – | – | – | H |
| 33 | 2016/3/9 | | 4 | 40 m | – | – | – | H | H | – | – | – | – | H |
| 34 | 2015/11/24 | | 3 | 90 m | H | H | H | H | H | – | – | – | – | H |

Buildings were sorted based on the temporal pattern of the closest NCS to the building in case of multiple NCSs. Red cells with and H letter refer to significant hot spot bins, blue cells with a C letter refer to bins with significant cold spot, and yellow cells with "—" symbol refer to bins that were not significant. Building ID is provided only in the case of more than one building that collapsed on the same day.

Table 3 shows a higher number of incidents of whole building collapse (level 4) that had new hot spots before the collapse in comparison to those that have new cold spots, whereas the number of buildings with collapse level 2 is half of those, which have new cold spots. This suggests a higher severity of the new hot spot pattern before the building collapse in comparison to NCS. Figure 6 illustrates some examples of new hot spot spatiotemporal patterns surround collapsed buildings before the incident.

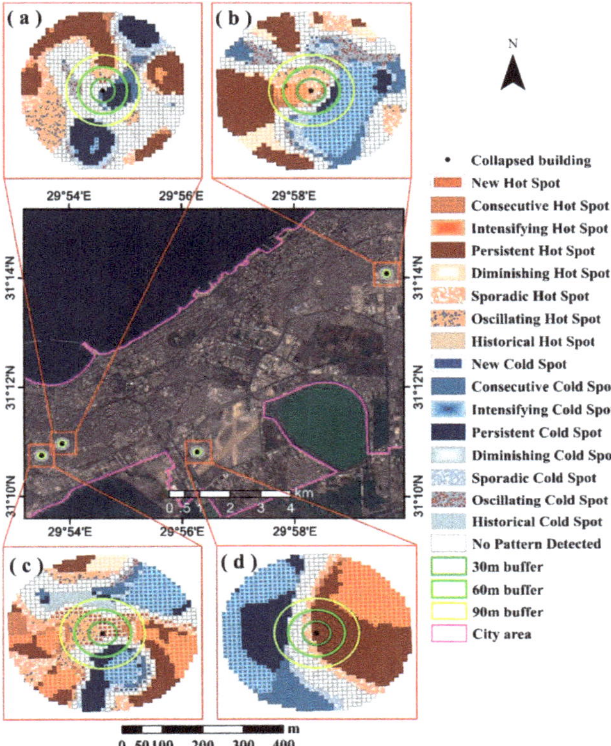



**Figure 6.** Samples of collapsed buildings that have an NHS spatiotemporal pattern before their collapse. The main map shows the location of the four collapsed buildings on a true color Sentinel-2 image acquired on 24 October 2018: (**a**) shows the spatiotemporal analysis for a 200 m buffer of a whole building collapse on 15 August 2016, (**b**) is the result of a whole collapsed building on 1 May 2016, (c) illustrates the analysis result of a partial building collapse on 14 July 2016, and (**d**) shows the result of a whole building collapse on 22 January 2017.

Three of the buildings in Figure 6 were completely collapsed, as shown in Figure 6a for a building collapse on 15 August 2016, Figure 6b for an incident on 1 May 2016, and Figure 6d for a collapsed building on 22 January 2017, and one building with a collapsed front part on 14 July 2016 (Figure 6c). Collapsed buildings were subjected to aging except the building presented in Figure 6d, which was subjected to sudden subsidence resulting in building tilt and collapse. This building was one of five other collapsed buildings have official recorded reason of subsidence, and found to have a new hotspot pattern before the incident in our spatio-temporal pattern analysis. As the subsidence was obvious in those collapsed buildings sites, that suggesting NHS may represent severe surface motion in the study area in comparison to NCS that may represent relatively gentle vertical surface motion.

### 3.3. Correlation Tests

We tested our results for statistical correlation to lithology, historical evolution extent of the city, season, and month of incidents by the chi-square method. None of all tested correlations were statistically significant with the two spatiotemporal patterns of our results, as shown in Table 4. However, the only potential correlation was obtained in the lithology test.

**Table 4.** Correlation results.

| Correlation to | Pearson Chi-Square |
| --- | --- |
| Lithology | 0.143 |
| City's Historical Evolution | 0.338 |
| Season | 0.492 |
| Month | 0.787 |

All incidents that occurred within the stabilized sand dune soil had new hot spots in the vertical motion before the building collapse, while five of six collapsed buildings in the reclaimed sand area in the old city also had new hot spots of vertical motion before the incident. The sandy nature of those two soil types may illustrate a correlation that could be helpful in predicting building collapse. Nevertheless, this potential correlation needs further investigation for better validation.

### 3.4. Validation of Study Results

We validated our results by generating a map for points of new cold spots and hot spots in the tenth and ninth time steps for four whole collapsed buildings in 2019, as illustrated in Figure 7. These lie within an area of high collapse incidence rate in the city. The GPS points for the four buildings used in the validation process were found to be close to at least one NCS or NHS point, with distances that varied from 38 m for a collapsed building on 18 February 2019 (Figure 7a) to 75 m for a collapsed building on 22 March 2019 (Figure 7b).

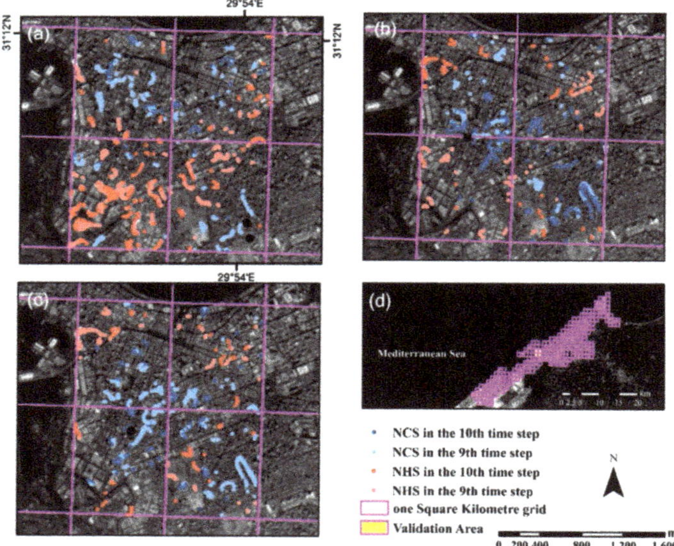


**Figure 7.** Validation of the study results. Maps are presented in a stretch Sentinel-2 image acquired on 24 October 2018: (**a**) presents the validation test of two whole building collapses on 18 February 2019, (**b**) the validation result of a whole collapsed building on 22 March 2019, (**c**) illustrates the validation result of a whole building collapse on 1 April 2019, and (**d**) shows the one square kilometer grid of Alexandria, highlighting the validation area.

The two collapsed buildings of 18 February 2019 were close to new cold spots in the ninth time step, whereas the other two incidents of 22 March 2019 and 1 April 2019, presented in Figure 7c, were close to NCSs of the tenth time step. This result validated the study results since all collapsed buildings' GPS points were within the same distance of our three buffers: 38 and 65 m for the two

18 February 2019 incidents, 41 m for the 1 April 2019 building collapse, and 75 m for the collapsed building on 22 March 2019.

## 4. Discussion

Building damage is one of the frequently occurring geohazards in subsidence zones [1,11]. This study attempts to understand surface vertical motion surrounding collapsed buildings before the incident. The PS-InSAR technique is used to measure the deformation time series, and spatiotemporal data analysis is used to extract surface motion patterns before the collapse. Results show that new cold spot and new hot spot are the two common spatiotemporal patterns shown before 66 buildings collapsed between May 2015 and December 2018, out of a total of 68 buildings that were studied.

Besides the high percentage of collapse incidents having at least one of these two spatiotemporal patterns close in proximity, which is 97% of the studied buildings, the rare distribution of the two spatiotemporal patterns gives confidence to the correlation obtained in this study between building collapse incidents in the city and the NCS and NHS patterns. Table 5 shows the percentage of NCS and NHS in the ninth and tenth time steps among the seventeen spatiotemporal patterns in the 4 km$^2$ area used for the validation process presented in Section 3.4.

**Table 5.** Distribution percentage of NCS and NHS patterns in the validation area.

| Spatiotemporal Pattern | Time Step | Incident Date | | |
| --- | --- | --- | --- | --- |
| | | 18 February 2019 | 22 March 2019 | 1 April 2019 |
| NCS | 10th | 0.3% | 1% | 0.5% |
| | 9th | 0.7% | 0.5% | 1.1% |
| NHS | 10th | 1% | 0.3% | 0.3% |
| | 9th | 1.5% | 0.3% | 0.4% |
| Total percentage | | 3.5% | 2.1% | 2.3% |

The distribution average of NCS and NHS, including the ninth and tenth time step before collapse incident dates, is 2.6% of all spatiotemporal patterns in the four square kilometer of the validation area. The highest distribution of the two patterns was before the 18 February 2019 collapse (3.5% of the validation area), while the lowest distribution obtained was before the collapse incident on 22 March 2019 (2.1%).

The result of the existing new cold spot close to the collapsed building was expected, as it represents sudden subsidence in a very limited space in comparison to the overall velocity of the surrounding area. Although it is also representing a sudden deformation in the surrounding area of the collapsed building, results of the new hot spot pattern was unexpected. Many new hot spots related to collapsed buildings were observed in the winter of 2015, which raised the idea of surface uplift due to the accumulation of underground water and seawater surges during the rainy season of that year. This could be one reason for a new hot spot pattern surrounding collapsed buildings during that period. However, the existence of a new hot spot pattern in the dry season of summer suggested another technical explanation of related collapses. This might be a sudden relative failure on the surface that was higher than the applicability of PS-InSAR slow surface motion detection, which resulted in unwrapping error in this area, and presents the deformation at that time as uplift. However, a sudden failure in the surface cannot be detected due to its high velocity. Regardless, building collapse due to surface alternation between drought–flood seasons might be a reason for a high building collapse ratio [11] in a city like Alexandria.

Ferretti et al. analyzed the deformation time series of PS points close to collapsed buildings in Camaiore, Italy, and found clear subsidence in eight points before the collapse, which returned to random velocity after the incident. The deformation was affected by clay soil erosion and subsequent formation of cavities. In comparison, Perski et al. [10] utilized relatively more images to detect deformation before two collapse incidents in Katowice, Poland, and Moscow, Russia. However, results found no significant deformation in the time series before the collapse. Time series in [8] might include vertical and horizontal motion, which resulted in clear deformation before the collapse. While the result of deformation surrounds the two buildings in [10], it was not clear, which supports the need for further spatial analysis like what has been done in our study.

The existence of subsidence and instability of buildings on clay deposit soils is well known in similar areas [11]. However, our results revealed no correlation between extracted spatiotemporal patterns and lithology. The interaction of lithology and spatiotemporal pattern result needs more investigation in future studies for better preparation of maps showing building collapse hazard.

Although leveling measurement is costly and has limited time and space measurements, those measurements are still the main method to monitor building deformation [7]. The applicability of InSAR measurements to monitor building stability was confirmed in many previous studies. Here, with the help of spatiotemporal analysis, our study shows two common vertical deformations based on spatiotemporal patterns that could be used as inputs for modeling building collapse hazard. To confirm the applicability of those two patterns to initially predict collapse incidents, we applied a buffer of 90 m for new cold/hot spot points, presented in Figure 7, to check how this input layer would look. Figure 8 shows the temporal vulnerability maps of the four validation buildings that could be used as input layers for a temporal building collapse hazard model.

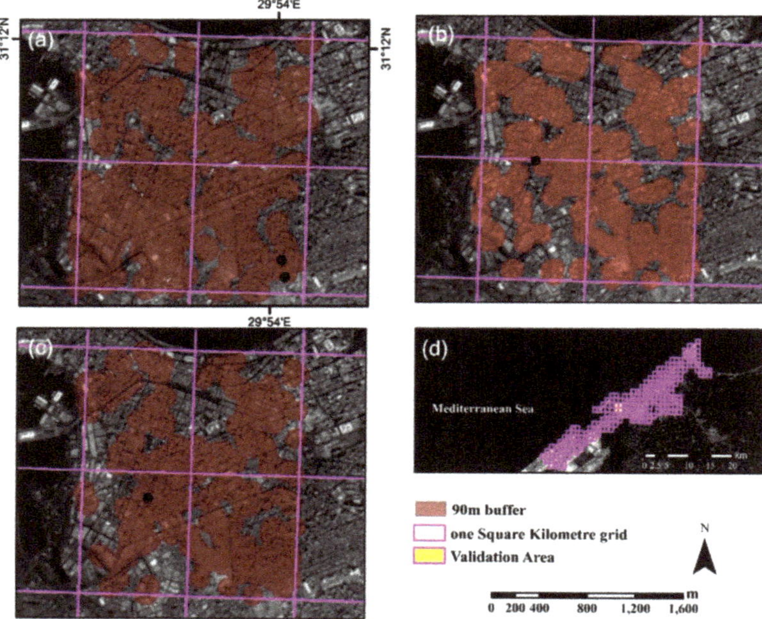


**Figure 8.** Potential input layer for a building collapse hazard model. Maps are presented in a stretch Sentinel-2 image acquired on 24 October 2018: (**a**) building collapse vulnerability map for 18 February 2019 incidents, (**b**) building collapse vulnerability map for the 22 March 2019 incident, (**c**) illustrates building collapse vulnerability map for the 1 April 2019 incident, and (**d**) shows the one square kilometer grid of Alexandria with the validation area highlighted.

The four collapsed buildings were located within the temporal vulnerability map of the area, as shown in Figure 8. However, the three vulnerability map results cover more than 50% of the study area in all cases, which shows a need to further improve the prediction accuracy. Hence, more investigations and correlation tests are needed to reduce the area of vulnerability maps to their minimum distribution. Additionally, land classification would be useful to eliminate unimportant land use and land cover types from our interest, such as the large cemetery area in the mid-south of the AOI of this test. Generally, we believe this study is a step towards a comprehensive building collapse risk assessment model based on PS-InSAR deformation measurement.

## 5. Conclusions

This study was designed to understand the spatiotemporal patterns of vertical deformation surrounding a building before its collapse. We used the PS-InSAR technique to measure surface deformation in the LOS. Then, we decomposed vertical deformation to analyze spatiotemporal patterns around buildings before the collapse incidents. From the results two patterns could be seen before collapse—new cold spot and new hot spot patterns. The finite distribution of these two spatiotemporal patterns in only 2.6% of the area, on average, and their existence in 97% of the studied collapsed buildings, build a confidence in the relationship between building collapse and these two spatiotemporal patterns of surface vertical deformation in the city of Alexandria. Although the distribution of these two patterns is limited in number, the diffusion of the NCS and NHS in space is relatively higher. The workflow in this study succeeded in relating detection between the two patterns and building collapse, but failed to define the reason why those buildings collapsed while other buildings that have the same spatiotemporal pattern at the same time were not. We think the reason may be the situation and structure of the building itself, such as aging, violations of building specifications, and illegal extensions. However, we cannot scientifically confirm this, due to the lack of appropriate data to do that kind of analysis. To build a powerful building collapse prediction model, however, these kinds of data are necessary. Thus, this will soon be our aim to collect this type of data, analyze land use and land cover of the study area, and test more correlations to produce a model for the city of Alexandria, and test its applicability in other cities. We suggest application of our workflow in other similar areas that have high rates of building collapse incidents in subsiding areas. To find out if NCS and NHS patterns are distributed in a global, regional, or local scale, it is important to confirm the existence of these two patterns in areas surrounding buildings before their collapse.

**Author Contributions:** Conceptualization, B.M.; methodology, B.M.; validation, B.M. and A.Y.; formal analysis, B.M.; investigation, A.Y.; resources, T.B.; data curation, B.M.; writing—original draft preparation, B.M.; writing—review and editing, T.B.; visualization, B.M.; supervision, T.B.; project administration, T.B.; funding acquisition, T.B. All authors have read and agreed to the published version of the manuscript.

**Funding:** This work was supported by LIESMARS Special Research Funding.

**Conflicts of Interest:** The authors declare no conflict of interest.

## References

1. Kim, S.-W.; Choi, J.-H.; Hong, S.-H.; Lee, J.-H.; Cho, J.; Lee, M.-J. Monitoring the risk of large building collapse using persistent scatterer interferometry and GIS. *Terr. Atmos. Ocean. Sci.* **2018**, *29*, 535–545. [CrossRef]
2. Ezquerro, P.; Del Soldato, M.; Solari, L.; Tomás, R.; Raspini, F.; Ceccatelli, M.; Fernández-Merodo, J.A.; Casagli, N.; Herrera, G. Vulnerability assessment of buildings due to land subsidence using InSAR data in the ancient historical city of Pistoia (Italy). *Sensors* **2020**, *20*, 2749. [CrossRef]
3. Nolesini, T.; Frodella, W.; Bianchini, S.; Casagli, N. Detecting slope and urban potential unstable areas by means of multi-platform remote sensing techniques: The Volterra (Italy) case study. *Remote Sens.* **2016**, *8*, 746. [CrossRef]
4. Ferretti, A.; Prati, C.; Rocca, F. Permanent scatterers in SAR interferometry. *IEEE Trans. Geosci. Remote. Sens.* **2001**, *39*, 8–20. [CrossRef]
5. Novellino, A.; Cigna, F.; Brahmi, M.; Marsh, S.; Bateson, L.; Marsh, S. Assessing the feasibility of a national InSAR ground deformation map of great Britain with Sentinel-1. *Geoscience* **2017**, *7*, 19. [CrossRef]
6. Comerci, V.; Vittori, E.; Cipolloni, C.; Di Manna, P.; Guerrieri, L.; Nisio, S.; Succhiarelli, C.; Ciuffreda, M.; Bertoletti, E. Geohazards monitoring in Roma from InSAR and in situ data: Outcomes of the PanGeo project. *Pure Appl. Geophys.* **2015**, *172*, 2997–3028. [CrossRef]
7. Yang, K.; Yan, L.; Huang, G.; Chen, C.; Wu, Z. Monitoring building deformation with InSAR: Experiments and validation. *Sensors* **2016**, *16*, 2182. [CrossRef]

8. Ferretti, A.; Ferrucci, F.; Prati, C.; Rocca, F. SAR analysis of building collapse by means of the permanent scatterers technique. In Proceedings of the IGARSS 2000. IEEE 2000 International Geoscience and Remote Sensing Symposium. Taking the Pulse of the Planet: The Role of Remote Sensing in Managing the Environment. Proceedings (Cat. No.00CH37120), Honolulu, HI, USA, 24–28 July 2002; Volume 7, pp. 3219–3221.

9. Weissgerber, F.; Koeniguer, E.C.; Nicolas, J.-M.; Trouvé, N. 3D monitoring of buildings using TerraSAR-X InSAR, DInSAR and PolSAR capacities. *Remote Sens.* **2017**, *9*, 1010. [CrossRef]

10. Perski, Z.; van Leijen, F.; Hanssen, R. Applicability of PS-InSAR for building hazard identification. Study of the 29 January 2006 Katowice exhibition hall collapse and the 24 February 2006 Moscow basmanny market collapse. In Proceedings of the ESA ENVISAT Symposium, Montreux, Switzerland, 23–27 April 2006.

11. Chen, F.; Lin, H.; Zhang, Y.; Lu, Z. Ground subsidence geo-hazards induced by rapid urbanization: Implications from InSAR observation and geological analysis. *Nat. Hazards Earth Syst. Sci.* **2012**, *12*, 935–942. [CrossRef]

12. Wu, W.; Cui, H.; Hu, J.; Yao, L. Detection and 3D visualization of deformations for high-rise buildings in Shenzhen, China from high-resolution TerraSAR-X datasets. *Appl. Sci.* **2019**, *9*, 3818. [CrossRef]

13. Shekhar, S.; Jiang, Z.; Ali, R.Y.; Eftelioglu, E.; Tang, X.; Gunturi, V.M.V.; Zhou, X. Spatiotemporal data mining: A computational perspective. *ISPRS Int. J. Geo-Inf.* **2015**, *4*, 2306–2338. [CrossRef]

14. Atluri, G.; Karpatne, A.; Kumar, V. Spatio-temporal data mining: A survey of problems and methods. *ACM Comput. Surv.* **2018**, *51*, 1–41. [CrossRef]

15. He, Z.; Deng, M.; Cai, J.; Xie, Z.; Guan, Q.; Yang, C. Mining spatiotemporal association patterns from complex geographic phenomena. *Int. J. Geogr. Inf. Sci.* **2019**, *34*, 1162–1187. [CrossRef]

16. Ansari, M.Y.; Ahmad, A.; Khan, S.S.; Bhushan, G. Mainuddin Spatiotemporal clustering: A review. *Artif. Intell. Rev.* **2019**, *53*, 2381–2423. [CrossRef]

17. Beladam, O.; Balz, T.; Mohamadi, B.; Abdalhak, M. Using PS-InSAR with Sentinel-1 images for deformation monitoring in Northeast Algeria. *Geoscience* **2019**, *9*, 315. [CrossRef]

18. Mohamadi, B.; Balz, T.; Younes, A. A model for complex subsidence causality interpretation based on PS-InSAR cross-heading orbits analysis. *Remote Sens.* **2019**, *11*, 2014. [CrossRef]

19. Stanley, J.D.; Toscano, M.A. Ancient archaeological sites buried and submerged along Egypt's Nile delta coast: Gauges of Holocene delta margin subsidence. *J. Coast. Res.* **2009**, *25*, 158–170. [CrossRef]

20. Ahram Online. Building Collapse in Alexandria. Available online: http://english.ahram.org.eg (accessed on 8 August 2018).

21. Torres, R.; Snoeij, P.; Geudtner, D.; Bibby, D.; Davidson, M.; Attema, E.; Potin, P.; Rommen, B.; Floury, N.; Brown, M.; et al. GMES Sentinel-1 mission. *Remote Sens. Environ.* **2012**, *120*, 9–24. [CrossRef]

22. Perissin, D.; Wang, Z.; Wang, T. The SARPROZ InSAR tool for urban subsidence/manmade structure stability monitoring in China. In Proceedings of the ISRSE, Sidney, Australia, 10–15 April 2011; Volume 1015.

23. Crosetto, M.; Monserrat, O.; Cuevas-González, M.; Devanthéry, N.; Crippa, B. Persistent Scatterer Interferometry: A review. *ISPRS J. Photogramm. Remote Sens.* **2016**, *115*, 78–89. [CrossRef]

24. Samieie-Esfahany, S.; Hanssen, R.; van Thienen-Visser, K.; Muntendam-Bos, A. On the effect of horizontal deformation on InSAR subsidence estimates. In Proceedings of the Fringe 2009 Workshop, Frascati, Italy, 30 November–4 December 2009; Volume 30.

25. ESRI. Create Space Time Cube by Aggregating Points. Available online: https://desktop.arcgis.com/en/arcmap/latest/tools/space-time-pattern-mining-toolbox/create-space-time-cube.htm (accessed on 21 April 2020).

26. Kendall, M.; Gibbons, J.D. *Rank Correlation Methods*, 5th ed.; Griffin: London, UK, 1990.

27. Mann, H. BNonparametric tests against trend. Econometrica. *J. Econom. Soc.* **1945**, *13*, 245–259. [CrossRef]

28. Harris, N.L.; Goldman, E.; Gabris, C.; Nordling, J.; Minnemeyer, S.; Ansari, S.; Lippmann, M.; Bennett, L.; Raad, M.; Hansen, M.; et al. Using spatial statistics to identify emerging hot spots of forest loss. *Environ. Res. Lett.* **2017**, *12*, 024012. [CrossRef]

29. ESRI. How Create Space Time Cube by Aggregating Points works. Available online: https://desktop.arcgis.com/en/arcmap/latest/tools/space-time-pattern-mining-toolbox/learnmorecreatecube.htm#ESRI_SECTION1_F1EA94A3BA8940E0B56AB08A302D1C08 (accessed on 22 April 2020).

30. Ord, J.K.; Getis, A. Local spatial autocorrelation statistics: Distributional issues and an application. *Geogr. Anal.* **1995**, *27*, 286–306. [CrossRef]

31. Getis, A.; Ord, J.K. The analysis of spatial association by use of distance statistics. In *Perspectives on Spatial Data Analysis*; Springer: Berlin/Heidelberg, Germany, 2010; pp. 127–145.

32.  ESRI. How Emerging Hot Spot Analysis Works. Available online: https://desktop.arcgis.com/en/arcmap/latest/tools/space-time-pattern-mining-toolbox/learnmoreemerging.htm (accessed on 22 April 2020).

MDPI

St. Alban-Anlage 66

4052 Basel

Switzerland

Tel. +41 61 683 77 34

Fax +41 61 302 89 18

www.mdpi.com

*Remote Sensing* Editorial Office

E-mail: remotesensing@mdpi.com

www.mdpi.com/journal/remotesensing